I0072260

Spectroscopy: A Methodological Approach

Spectroscopy: A Methodological Approach

Edited by
Conrad Sinclair

www.willfordpress.com

Published by Willford Press,
118-35 Queens Blvd., Suite 400,
Forest Hills, NY 11375, USA

ISBN: 978-1-68285-623-9

Cataloging-in-Publication Data

Spectroscopy : a methodological approach / edited by Conrad Sinclair.
p. cm.
Includes bibliographical references and index.
ISBN 978-1-68285-623-9
1. Spectrum analysis. 2. Spectroscope. I. Sinclair, Conrad.
QD95 .S64 2019
543.5--dc23

For information on all Willford Press publications
visit our website at www.willfordpress.com

Contents

Preface

Spectroscopy is the scientific study of interactions between matter and electromagnetic radiation. Spectroscopy is classified according to the type of radiative energy involved like electromagnetic radiation, radiated pressure waves, particles that can be a source of radiative energy, like electrons and neutrons. The interaction between matter and energy can be put under the bracket of absorption, emission, elastic scattering, inelastic scattering, etc. Various other forms of spectroscopy include auger spectroscopy, correlation spectroscopy, Fourier transform spectroscopy and Hyperspectral imaging which are distinguished by their specific applications. The topics included in this book on spectroscopy are of the utmost significance and bound to provide incredible insights to readers. It aims to shed light on some of the unexplored aspects of this field and the recent researches in this field. This book strives to assist students and researchers with a goal of delving into the concepts and applications of spectroscopy.

This book unites the global concepts and researches in an organized manner for a comprehensive understanding of the subject. It is a ripe text for all researchers, students, scientists or anyone else who is interested in acquiring a better knowledge of this dynamic field.

I extend my sincere thanks to the contributors for such eloquent research chapters. Finally, I thank my family for being a source of support and help.

Editor

1

Photoreflectance and Raman Study of Surface Electric States on AlGaAs/GaAs Heterostructures

Luis Zamora-Peredo,[1] Leandro García-González,[1] Julián Hernández-Torres,[1] Irving E. Cortes-Mestizo,[2] Víctor H. Méndez-García,[2] and Máximo López-López[3]

[1] *Centro de Investigación en Micro y Nanotecnología, Universidad Veracruzana, Calzada Adolfo Ruiz Cortines 455, Fracc. Costa Verde, 94292 Boca del Río, VER, Mexico*
[2] *Universidad Autónoma de San Luis Potosí, Center for the Innovation and Application of Science and Technology, Sierra Leona 550, Lomas 4a Secc., 78210 San Luis Potosí, SLP, Mexico*
[3] *Centro de Investigación y Estudios Avanzados del IPN, Apartado Postal 14-740, 07360 Ciudad de México, Mexico*

Correspondence should be addressed to Luis Zamora-Peredo; luiszamora@uv.mx

Academic Editor: Carlos Andres Palacio

Photoreflectance (PR) and Raman are two very useful spectroscopy techniques that usually are used to know the surface electronic states in GaAs-based semiconductor devices. However, although they are exceptional tools there are few reports where both techniques were used in these kinds of devices. In this work, the surface electronic states on AlGaAs/GaAs heterostructures were studied in order to identify the effect of factors like laser penetration depth, cap layer thickness, and surface passivation over PR and Raman spectra. PR measurements were performed alternately with two lasers (532 nm and 375 nm wavelength) as the modulation sources in order to identify internal and surface features. The surface electric field calculated by PR analysis decreased whereas the GaAs cap layer thickness increased, in good agreement with a similar behavior observed in Raman measurements (I_{L-}/I_{LO} ratio). When the heterostructures were treated by Si-flux, these techniques showed contrary behaviors. PR analysis revealed a diminution in the surface electric field due to a passivation process whereas the I_{L-}/I_{LO} ratio did not present the same behavior because it was dominated by the depletion layers width (cap layer thickness) and the laser penetration depth.

1. Introduction

GaAs, with five times higher electron mobility compared to silicon, has a potential to achieve ultrafast electronic and optoelectronic devices. However, this semiconductor suffers from pronounced effects associated with its surface or interfaces: in particular, the large densities of GaAs surface electronic states that pin the Fermi level at midgap and result in large surface recombination velocity (10^6 cm/s). In order to eliminate chemical instability that may cause undesired effects, it is well established that the surfaces of GaAs-based devices have to be treated suitably. Frequently, the surfaces of semiconductor devices are passivated in order to stabilize their chemical nature and to eliminate reactivity. Ammonium polysulfide $(NH_4)_2S_x$ has been frequently used to passivate the surface of GaAs with covalently bonded sulfur atoms [1–9]. However, sulfur passivation only provides short-term surface stability. In order to get more stability, a silicon interface control layers (Si ICL) approach has been proposed by Hasegawa and Akazawa [10–12] as a complement to high-k dielectric oxide surface layers such as HfO_2 [13–16], AlO_3 [16, 17], and SiO [18]. Other studies have been made with self-assembled monolayers (SAMs) of octadecanethiol (ODT) and dodecanethiol (DDT) [19, 20].

Specifically, with AlGaAs/GaAs heterostructures there are reports about the effect of the surface on the electronic properties. In particular, the relationship between the surface electric field and electron mobility in the two-dimensional electron gas (2DEG) achieved in the AlGaAs/GaAs interface has been studied [21–31]. Other studies with heterojunction bipolar transistors [32], high electron mobility transistors

[33–35], quantum Hall effect [36], and nanostructure-based devices [37–41] have been reported.

Photoreflectance and Raman spectroscopy are very useful approaches that have been widely used for the study of GaAs-based devices [4, 23–31]. The PR technique has been used to study the AlGaAs/GaAs system in numerous reports, which were focused on determining the origin of Franz-Keldysh oscillations (FKO) that usually are observed at the PR spectra [24–31]. Today we know that wide-period FKO observed between 1.42 and 1.7 eV are associated with the surface electric field and short-period FKO just above 1.42 eV originate from internal AlGaAs/GaAs interfaces [31]. In the Raman spectroscopy case, there are many reports where it is employed as a useful tool to explore the surface passivation of GaAs films and AlGaAs/GaAs heterostructures. It is well accepted that the relative intensity between longitudinal optical (LO) and coupled plasmon-phonon (L−) modes can reveal useful information about the surface passivation of GaAs [5–7, 20, 42–45].

In this work, we studied a set of AlGaAs/GaAs heterostructures by photoreflectance and Raman spectroscopy in order to make a comparative study with both techniques. Samples with different thickness of the GaAs cap layer were grown in order to change the surface electric field. PR measurements with two different laser wavelengths were used in order to identify internal and surfaces features. In addition, we studied the effect generated by the in situ deposition of silicon monolayers (Si ML) over the cap layer of the heterostructure.

2. Experimental Details

A set of AlGaAs/GaAs heterostructures was grown on semi-insulating GaAs (100) substrate by molecular beam epitaxy. All samples have a 1 μm thick GaAs buffer layer (BL), a first spacer layer (1SL) of 7 nm undoped $Al_xGa_{1-x}As$, followed by an 80 nm thick Si-doped AlGaAs barrier (doping with 1.4×10^{18} atoms/cm^3). Next, there is a second spacer layer (2SL) of 7 nm undoped $Al_xGa_{1-x}As$. Nominal Al concentration of 32% was used for the AlGaAs layers. Finally, the structure was capped with an undoped GaAs layer (see Figure 1). The thickness of the top layer was 25, 60, and 80 nm for samples M1, M2, and M3, respectively. Additionally, two more samples were grown with a 25 nm-GaAs cap layer and an in situ surface passivation treatment with Si-flux in order to get a nominal thickness of 1 and 2 ML, labeled M4 and M5, respectively. When the silicon was deposited the substrate temperature was fixed at 600°C and the arsenic flux was maintained in order to get a good quality surface.

Photoreflectance (PR) measurements were carried out alternately with two solid-state lasers as modulation source (543 nm and 375 nm wavelengths, with a maxima output power of 12 and 10 mW, resp.) and chopped with a frequency of 200 Hz. A Sciencetech monochromator of 0.5 m focal distance was used. The experimental setup used was similar to those that are described elsewhere [23]. Using the 543 nm laser, it is possible to get a PR signal from the GaAs buffer layer because the penetration depth is close to 120 nm; however with a 375 nm line laser, the penetration depth is reduced significantly (<50 nm) [46]. All measurements were carried out at room temperature. Raman spectra were acquired with a Thermo Scientific confocal microscopy system arranged in a 180° backscattering configuration and equipped with a 532 nm solid-state laser with output power of 10 mW, 100x objective, and a charged-coupled device (CCD). 100 exposures and 10 s as collection time were used to collect Raman scattering. All spectra were normalized using the LO-GaAs mode intensity.

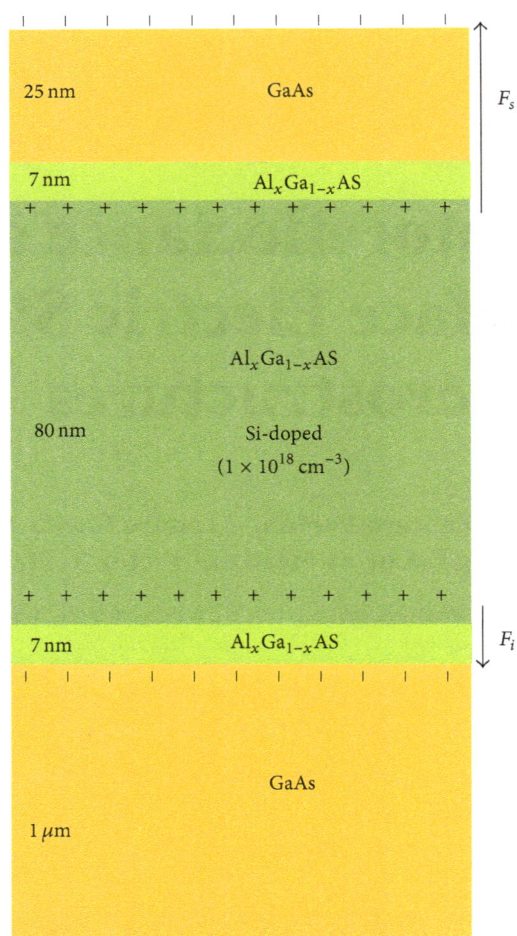

Figure 1: Schematic diagram of semiconductor layers in the heterostructures. F_s and F_i label the surface and internal electric field region, respectively.

3. Results and Discussion

3.1. Effect of the Capping Layer Thickness. Figure 2 shows room temperature PR spectra of M1 obtained with two different lasers as modulation source: with 532 nm (black line) and with 375 nm (red line). In the PR spectrum obtained with 532 nm, it is possible to see three features: a short-period Franz-Keldysh oscillation (s-FKO) at 1.42 eV associated with the GaAs energy band gap, a wide-period Franz-Keldysh oscillation (w-FKO) between 1.42 and 1.6 eV, and a wide oscillation associated with the AlGaAs energy band gap. PR

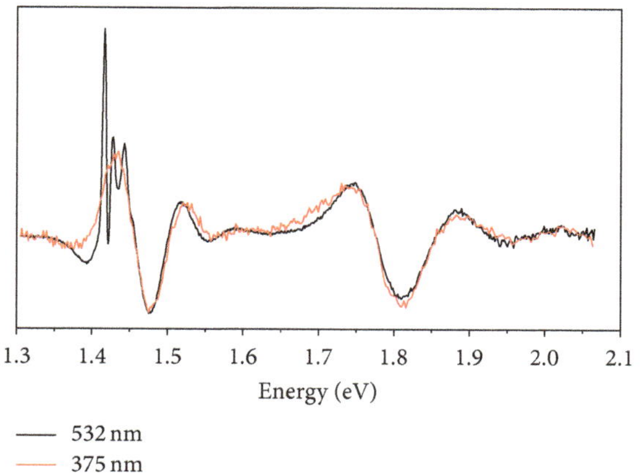

FIGURE 2: PR spectra of M1 heterostructure obtained with 532 nm (black line) and 375 nm (red line) lasers as modulation source.

obtained with a 375 nm laser does not have the first feature, which indicates that it originates from an internal AlGaAs/GaAs-BL interface considering that the 375 nm laser has a penetration depth smaller than 50 nm. Bessolov et al. estimated a penetration depth of 108 nm with a 514.5 nm line laser and 50.8 nm for a 457.9 nm line [6]. The w-FKO is observable with both lasers, which indicates that it originates from the surface GaAs layer. In previous reports this feature has been associated with the surface electric field [31]. Finally, at 1.8 eV we can see the feature originates from the AlGaAs layers, which offers information about energy band gap and consequently of the Al concentration. Considering that PR spectra originate mainly from interfaces, it is possible to establish that this oscillation originates from the GaAs/AlGaAs interface nearest to the surface because it is observed with both lasers.

In order to diminish the surface electric field (F_s) originating from the electron migration from the AlGaAs:Si layer to the surface (see Figure 1), the M2 and M3 samples were grown with a GaAs cap layer of 60 and 80 nm thickness, respectively. Figure 3 shows PR spectra obtained with the 375 nm laser. The s-FKO feature vanishes for all samples. The w-FKO shows a reduction on its periodicity originating from the diminution of the surface electric field. The feature associated with the AlGaAs layer is only observed in M1 and almost disappears in the M2 and M3 spectra, due to the increase in the distance between the surface and the AlGaAs layer (2SL).

To determine the electric field magnitude associated with the FKO, we considered the asymptotic modulation expression for the electroreflectance proposed by Aspnes and Studna [47]:

$$\frac{\Delta R}{R} \propto \frac{1}{E^2\left(E-E_g\right)} \exp\left[\frac{-2\Gamma\sqrt{E-E_g}}{(\hbar\theta)^{3/2}}\right] \cdot \cos\left[\frac{4}{3}\left(\frac{E-E_g}{\hbar\theta}\right)^{3/2}+\chi\right], \tag{1}$$

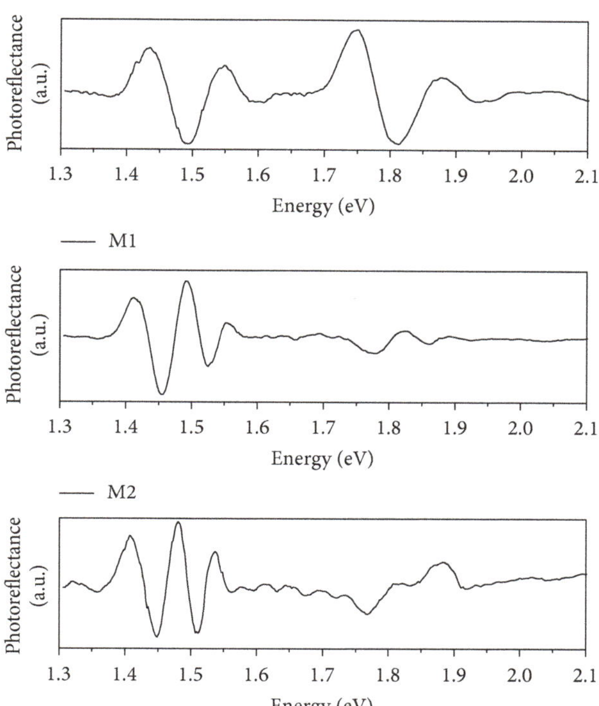

FIGURE 3: PR spectra of M1, M2, and M3 heterostructures, capped with 25, 60, and 80 nm GaAs layer, respectively. The 375 nm laser was used as modulation source.

where $\hbar\theta$ is the electrooptic energy, Γ is the linewidth, E_g is the band gap energy, and χ is an arbitrary phase factor. The electric field F is related to $\hbar\theta$ by the expression

$$\hbar\theta = \left(\frac{e^2\hbar^2F^2}{2\mu}\right)^{1/3}, \tag{2}$$

where μ is the electron-hole reduced mass and e is the electron charge.

In this model, the position of an nth extreme in the FKO is given by

$$n\pi = \frac{4}{3}\left(\frac{E_n-E_g}{\hbar\theta}\right)^{3/2}+\chi, \tag{3}$$

where n is the index and E_n is the corresponding energy.

Equation (3) can be rearranged as

$$\frac{4}{3\pi}\left(E-E_g\right)^{3/2} = (\hbar\theta)^{3/2}n - \frac{(\hbar\theta)\chi}{\pi}. \tag{4}$$

As we can see, (4) corresponds to a linear function with slope $(\hbar\theta)^{3/2}$, which can be determined using experimental data by a linear fitting of the plot of $(4/3\pi)(E-E_g)^{3/2}$ versus the index number n. Next, F can be determined using (2).

Figure 4 shows the linear fit obtained with experimental data from the w-FKO extreme in PR spectra of M1, M2, and M3 where it is evidence of the reduction of the surface electric

TABLE 1: Cap layer thickness, surface electric field (F_s) obtained from PR measurements, and I_{L-}/I_{LO} ratio calculated from Raman measurements.

Sample	Cap (nm)	Si (ML)	F_s (10^7 V/m)	I_{L-}/I_{LO}
M1	25	0	5.99	0.1699
M2	60	0	3.57	0.1407
M3	80	0	3.40	0.1345
M4	25	1	5.91	0.1896
M5	25	2	5.08	0.1772

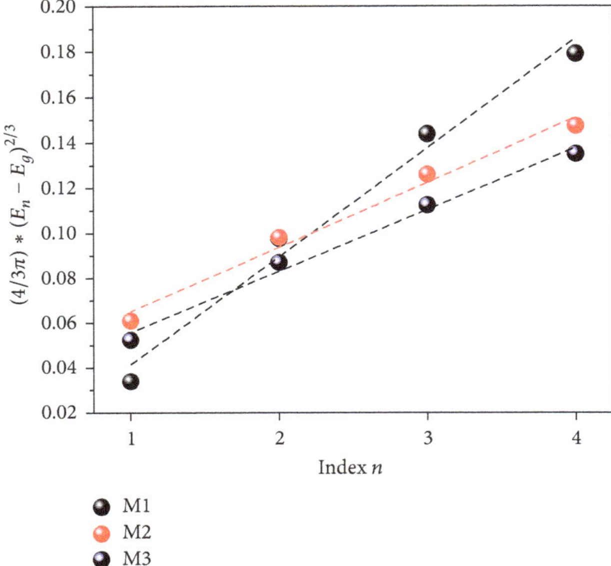

FIGURE 4: Linear fit in w-FKO analysis from PR spectra of M1, M2, and M3.

field (F_s) as the cap layer thickness increases. As we can see in Table 1, the magnitude of F_s decreases from 5.99 to 3.57 and 3.40 × 10^5 V/cm (reduction of 43.2%) for M1, M2, and M3, respectively.

Figure 5 shows Raman spectra of heterostructures M1, M2, and M3 (cap layer thickness of 25, 60, and 80 nm, resp.) where there are observed four vibration modes: the coupled plasmon-phonon (L−) and longitudinal optical (LO) from the GaAs cap layer localized at 268 and 291 cm^{-1}, respectively and two modes originating from the AlGaAs layers (LO GaAs-like and LO AlAs-like) located at 281 and 377 cm^{-1}, respectively. All Raman spectra were normalized at LO GaAs mode intensity. L− mode intensity is lower than the LO GaAs mode because the GaAs layers (cap and BL) are undoped. The effect of increasing the cap layer thickness is palpable with the LO GaAs-like and LO AlAs-like modes intensity because both decrease as the surface GaAs thickness increases.

In n-doped GaAs films, the LO peak (291 cm^{-1}) is attributed to the surface depletion layer whereas L− mode originates from the bulk where free carriers exist [3–7, 20, 42–44]. In this case, the cap layer thickness is the same as the wide depletion layer as we can see in Figure 1. Then the LO intensity (I_{LO}) will be increased as the cap layer thickness increases and

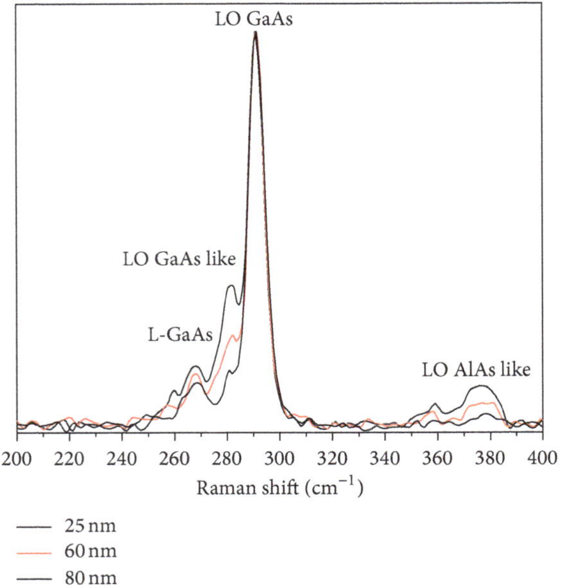

FIGURE 5: Raman scattering of M1, M2, and M3 heterostructures, capped with GaAs layer of 25, 60, and 80 nm, respectively.

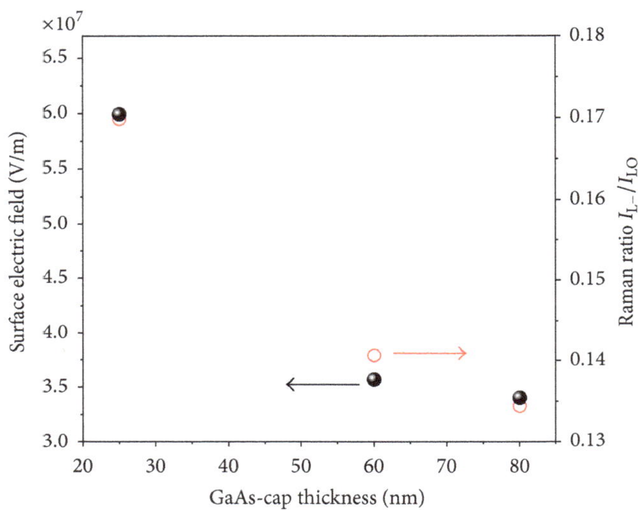

FIGURE 6: Behavior comparison of the surface electric field obtained by PR (black point) and Raman ratio I_{L-}/I_{LO} (red circle) as a function of cap layer thickness.

therefore we can use the I_{L-}/I_{LO} ratio to study the surface states. A summary of the measured values for the I_{L-}/I_{LO} ratio acquired from Raman spectra in Figure 5 is presented in Table 1. Figure 6 plots the surface electric field obtained by PR (black point) and Raman ratio I_{L-}/I_{LO} (red circle) as a function of cap layer thickness. As we can see, both techniques show a similar behavior of the surface electric field in these heterostructures. Similar comportment of the surface electric field has been found by Kudrawiec et al. in GaN Van Hoof structures studied by contactless electroreflectance [48].

3.2. Effect of Si Passivation. An analogous study was made with heterostructures passivated with Si-ML. In this case,

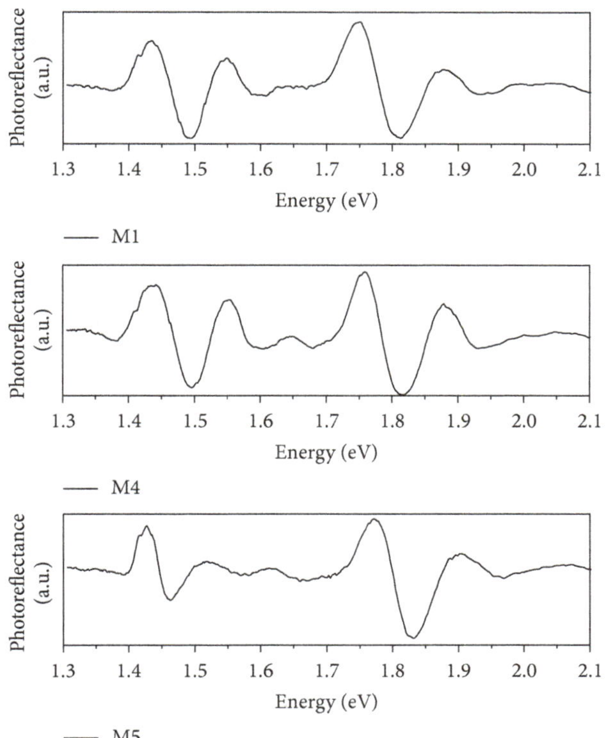

FIGURE 7: PR spectra of M1, M4, and M5 heterostructures passivated with 0, 1, and 2 silicon monolayers, respectively. The 375 nm laser was used as modulation source.

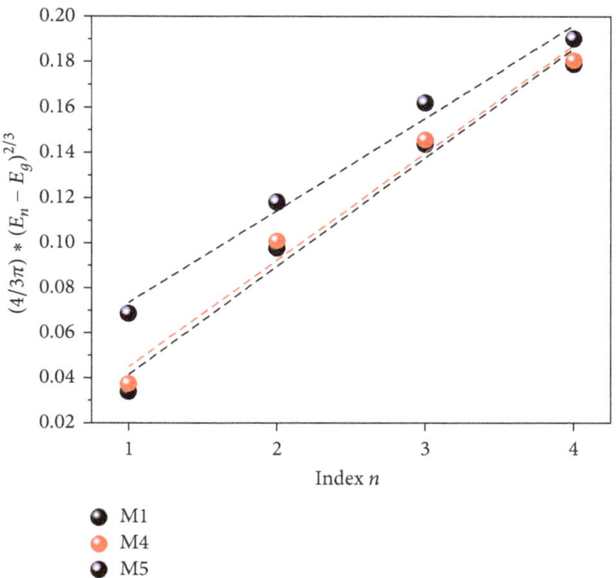

FIGURE 8: Linear fit in w-FKO analysis from PR spectra of M1, M4, and M5.

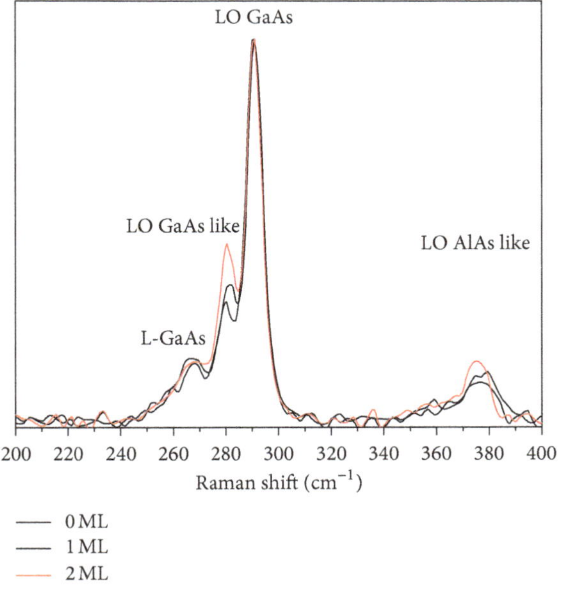

FIGURE 9: Raman scattering of M1, M4, and M5 heterostructures capped with a 25 nm GaAs layer and passivated with 0, 1, and 2 Si-ML, respectively. A 532 nm laser was used as excitation source.

the cap layer thickness remains constant and therefore a reduction of the surface states density is expected due to the passivation process originating from the formation of a SiO monolayer when the samples were exposed to the atmosphere. Figure 7 shows PR spectra of M1, M4, and M5 in order to compare samples without (M1) and with Si-ML (M4 and M5). The oscillation at $E > 1.7$ eV associated with the AlGaAs layer did not disappear because the cap layer thickness was 25 nm for the three samples and the laser penetration depth is bigger. The w-FKO period has a similar behavior for M1 and M4; however it presents a notable change in M5. Linear fit with experimental data from w-FKO extreme is shown in Figure 8. As we can see in Table 1, the surface electric field magnitude changes from 5.99 to 5.91 and 5.08 $\times 10^5$ V/cm for M1, M4, and M5, respectively. These results suggest that the surface passivation process is occurring but could be insufficient to eliminate the surface state because the Si deposition was made at high temperature (~600°C).

Figure 9 shows Raman spectra of heterostructures M1, M4, and M5 (treated with 0, 1, and 2 Si-ML, resp.) where the same four vibration modes that were observed in Figure 5 are observed. All Raman spectra were normalized at LO GaAs mode. When the heterostructure is treated with Si monolayers it is possible to see some changes. In this case the LO AlAs-like peak originating from the AlGaAs layer does not decrease because the cap layer is of a thicknesses of 25 nm for the three samples. The I_{L-}/I_{LO} ratio changes from 0.170 to 0.189 and 0.177 for M1, M4, and M5, respectively (Table 1).

Figure 10 plots the surface electric field obtained by PR (black point) and Raman intensity ratio I_{L-}/I_{LO} (red circle) as a function of the Si monolayers. In this case, the PR measurement analysis gives a slight decrease of the surface electric field for samples with 1 and 2 Si-ML but Raman scattering insinuates a contrary situation. Considering that the surface depletion layer remains constant, we would expect an unchanged ratio I_{L-}/I_{LO}. This disagreement between PR and Raman spectroscopy could be associated with the laser penetration depth because the laser used in Raman system is

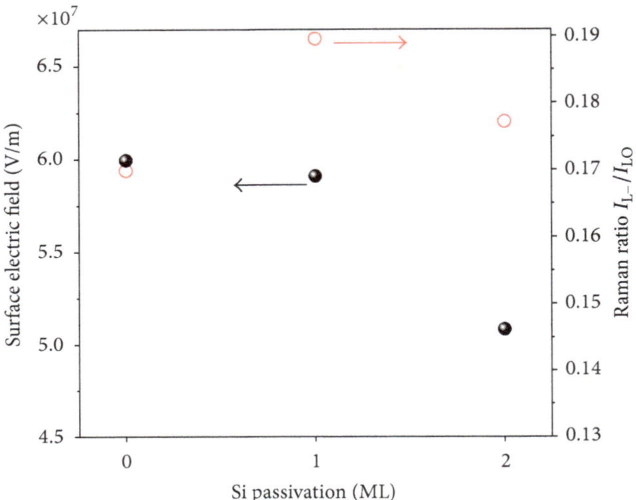

FIGURE 10: Behavior comparison of the surface electric field obtained by PR (black point) and Raman ratio I_{L-}/I_{LO} (red circle) as a function of Si-ML.

similar to PR measurements (Figure 2) so that it excites the GaAs buffer layer. Consequently, both cap and deeper GaAs layers influence Raman spectra. In previous studies with these kinds of heterostructures, we found that if the surface electric field decreases (as is suggested by PR measurements) a wider depletion zone is originated in the AlGaAs/GaAs-BL interface [49], which changes the Raman intensity ratio I_{L-}/I_{LO}. That means that PR measurements are the best approach to study the surface electric field because Raman spectra are influenced by deeper GaAs interface.

The PR analysis in this work is in good agreement with the engineering of electric field distribution in AlGaN/GaN heterostructures that widely has been studied by other authors where electroreflectance spectroscopy has shown similar behavior when the cap layer is modified [48, 50, 51]. The above suggests that PR spectroscopy is an excellent tool that could be used to explore more complex heterostructures like GaN/graphene/Si [52] or GaAs/Graphene/Si [53] where the internal electric field can give information about crystal quality.

4. Conclusions

The surface electronic states on AlGaAs/GaAs heterostructures were studied by photoreflectance and Raman spectroscopy techniques. The surface electric field calculated by PR analysis decreased whereas the GaAs cap layer thickness increases in good agreement with a similar behavior observed in Raman measurements (I_{L-}/I_{LO} ratio). When the heterostructures were treated with a Si-flux, these techniques showed contrary behaviors originating from the penetration depth of the laser used in the PR and Raman measurements. PR analysis found a slow diminishment in the surface electric field whereas the I_{L-}/I_{LO} ratio observed in Raman analysis showed higher values than the sample without a Si-ML. This work illustrates how it is possible to use PR and Raman spectroscopy to study the surface electronic states in AlGaAs/GaAs heterostructures and the passivation process in this kind of semiconductor device.

Competing Interests

The authors declare that they have no competing interests.

Acknowledgments

The authors want to thank DGDAIE-UV, DGI-UV, and SENER-CONACYT for their support of this work.

References

[1] Y. Nannichi, J.-F. Fan, H. Oigawa, and A. Koma, "A model to explain the effective passivation of the GaAs surface by $(NH_4)_2S_x$ treatment," *Japanese Journal of Applied Physics*, vol. 27, no. 12, pp. L2367–L2369, 1988.

[2] H. Oigawa, J.-F. Fan, Y. Nannichi, H. Sugahara, and M. Oshima, "Universal passivation effect of $(NH_4)_2S_x$ treatment on the surface of III-V compound semiconductors," *Japanese Journal of Applied Physics*, vol. 30, no. 3, pp. L322–L325, 1991.

[3] R. V. Ghita, C. C. Negrila, C. Cotirlan, and C. Logofatu, "On the passivation of GaAs surface by sulfide compounds," *Digest Journal of Nanomaterials and Biostructures*, vol. 8, no. 3, pp. 1335–1344, 2013.

[4] P. Jin, S. H. Pan, Y. G. Li, C. Z. Zhang, and Z. G. Wang, "Electronic properties of sulfur passivated undoped-n$^+$ type GaAs surface studied by photoreflectance," *Applied Surface Science*, vol. 218, no. 1–4, pp. 210–214, 2003.

[5] X. Chen, X. Si, and V. Malhotra, "Measurement of reduced surface barrier height in sulfur passivated InP and GaAs using Raman spectroscopy," *Journal of the Electrochemical Society*, vol. 140, no. 7, pp. 2085–2088, 1993.

[6] V. N. Bessolov, M. V. Lebedev, and D. R. T. Zahn, "Raman scattering study of surface barriers in GaAs passivated in alcoholic sulfide solutions," *Journal of Applied Physics*, vol. 82, no. 5, pp. 2640–2642, 1997.

[7] T. Błachowicz, G. Salvan, D. R. T. Zahn, and J. Szuber, "Micro-Raman spectroscopy of disordered and ordered sulfur phases on a passivated GaAs surface," *Applied Surface Science*, vol. 252, no. 21, pp. 7642–7646, 2006.

[8] M. V. Lebedev, "Passivation at semiconductor/electrolyte interface: role of adsorbate solvation and reactivity in surface atomic and electronic structure modification of III-V semiconductor," *Applied Surface Science*, vol. 254, no. 24, pp. 8016–8022, 2008.

[9] J. S. Ha, S.-J. Park, S.-B. Kim, and E.-H. Lee, "Correlation of surface morphology with chemical structures of sulfur–passivated GaAs(100) investigated by scanning tunneling microscopy and x–ray photoelectron spectroscopy," *Journal of Vacuum Science and Technology A: Vacuum, Surfaces and Films*, vol. 13, no. 3, pp. 646–651, 1995.

[10] H. Hasegawa, "MBE growth and applications of silicon interface control layers," *Thin Solid Films*, vol. 367, no. 1-2, pp. 58–67, 2000.

[11] M. Akazawa and H. Hasegawa, "Formation of ultrathin SiN_x/Si interface control double layer on (001) and (111) GaAs surfaces for ex situ deposition of high-k dielectrics," *Journal of Vacuum Science and Technology B: Microelectronics and Nanometer Structures*, vol. 25, no. 4, pp. 1481–1490, 2007.

[12] H. Hasegawa and M. Akazawa, "Surface passivation technology for III-V semiconductor nanoelectronics," *Applied Surface Science*, vol. 255, no. 3, pp. 628–632, 2008.

[13] D. J. Webb, J. Fompeyrine, S. Nakagawa et al., "In-situ MBE Si as passivating interlayer on GaAs for HfO_2 MOSCAP's: effect of GaAs surface reconstruction," *Microelectronic Engineering*, vol. 84, no. 9-10, pp. 2142–2145, 2007.

[14] C. Marchiori, D. J. Webb, C. Rossel et al., "H plasma cleaning and a-Si passivation of GaAs for surface channel device applications," *Journal of Applied Physics*, vol. 106, no. 11, Article ID 114112, 2009.

[15] M. El Kazzi, D. J. Webb, L. Czornomaz et al., "1.2 nm capacitance equivalent thickness gate stacks on Si-passivated GaAs," *Microelectronic Engineering*, vol. 88, no. 7, pp. 1066–1069, 2011.

[16] Y. Xuan, H.-C. Lin, and P. D. Ye, "Simplified surface preparation for GaAs passivation using atomic layer-deposited high-κ dielectrics," *IEEE Transactions on Electron Devices*, vol. 54, no. 8, pp. 1811–1817, 2007.

[17] C. L. Hinkle, A. M. Sonnet, E. M. Vogel et al., "Frequency dispersion reduction and bond conversion on n -type GaAs by in situ surface oxide removal and passivation," *Applied Physics Letters*, vol. 91, no. 16, Article ID 163512, 2007.

[18] D. L. Winn, M. J. Hale, T. J. Grassman, A. C. Kummel, R. Droopad, and M. Passlack, "Direct and indirect causes of Fermi level pinning at the SiO/GaAs interface," *Journal of Chemical Physics*, vol. 126, no. 8, Article ID 084703, 2007.

[19] H. A. Budz, M. C. Biesinger, and R. R. LaPierre, "Passivation of GaAs by octadecanethiol self-assembled monolayers deposited from liquid and vapor phases," *Journal of Vacuum Science and Technology B: Microelectronics and Nanometer Structures*, vol. 27, no. 2, pp. 637–648, 2009.

[20] C. L. Mcguiness, A. Shaporenko, M. Zharnikov, A. V. Walker, and D. L. Allara, "Molecular self-assembly at bare semiconductor surfaces: Investigation of the chemical and electronic properties of the alkanethiolate-GaAs(001) interface," *Journal of Physical Chemistry C*, vol. 111, no. 11, pp. 4226–4234, 2007.

[21] F. Castro, B. Nabet, and X. Chen, "Closed-form electric-field profile model for AlGaAs/GaAs heterostructures," *Journal of Applied Physics*, vol. 92, no. 1, pp. 218–222, 2002.

[22] R. A. Khabibullin, I. S. Vasil'evskii, G. B. Galiev et al., "Effect of the built-in electric field on optical and electrical properties of AlGaAs/InGaAs/GaAs P-HEMT nanoheterostructures," *Semiconductors*, vol. 45, no. 5, pp. 657–662, 2011.

[23] J. Misiewicz, P. Sitarek, G. Sęk, and R. Kudrawiec, "Semiconductor heterostructures and device structures investigated by photoreflectance spectroscopy," *Materials Science*, vol. 21, no. 3, pp. 263–320, 2003.

[24] N. Pan, X. L. Zheng, H. Hendriks, and J. Carter, "Photoreflectance characterization of AlGaAs/GaAs modulation-doped heterostructures," *Journal of Applied Physics*, vol. 68, no. 5, pp. 2355–2360, 1990.

[25] R. A. Novellino, C. Vazquez-López, A. A. Bernussi et al., "On the origin of Franz-Keldysh oscillations in AlGaAs/GaAs modulation-doped heterojunctions," *Journal of Applied Physics*, vol. 70, no. 10, pp. 5577–5581, 1991.

[26] J. A. N. T. Soares, D. Beliaev, R. Enderlein, L. M. R. Scolfaro, M. Saito, and J. R. Leite, "Photoreflectance investigations of semiconductor device structures," *Materials Science and Engineering B*, vol. 35, no. 1-3, pp. 267–272, 1995.

[27] I. Hwang, J.-E. Kim, H. Y. Park, and S. K. Noh, "Photoreflectance study of etching and annealing effect on AlGaAsGaAs heterostructure," *Solid State Communications*, vol. 103, no. 1, pp. 1–3, 1997.

[28] J. A. N. T. Soares, R. Enderlein, D. Beliaev, J. R. Leite, and M. Saito, "Photoreflectance spectra from GaAs HEMT structures reinvestigated: solution of an old controversy," *Semiconductor Science and Technology*, vol. 13, no. 12, pp. 1418–1425, 1998.

[29] E. Estacio, M. Bailon, A. Somintac, R. Sarmiento, and A. Salvador, "Observation of high junction electric fields in modulation-doped GaAs/AlGaAs heterostructures by room temperature photoreflectance spectroscopy," *Journal of Applied Physics*, vol. 91, no. 6, pp. 3717–3720, 2002.

[30] H. Takeuchi, Y. Kamo, Y. Yamamoto, T. Oku, M. Totsuka, and M. Nakayama, "Photovoltaic effects on Franz-Keldysh oscillations in photoreflectance spectra: application to determination of surface Fermi level and surface recombination velocity in undoped GaAs/n-type GaAs epitaxial layer structures," *Journal of Applied Physics*, vol. 97, no. 6, Article ID 063708, 2005.

[31] L. Zamora-Peredo, I. E. Cortes-Mestizo, L. García-González et al., "Determination of surface electric potential by photoreflectance spectroscopy of HEMT heterostructures," *Journal of Crystal Growth*, vol. 378, pp. 100–104, 2013.

[32] R. T. Yoshioka, L. E. M. de Barros Jr., J. A. Diniz, and J. W. Swart, "Improving performance of microwave AlGaAs/GaAs HBTs using novel SiNx passivation process," in *Proceedings of the SBMO & IEEE MTT-S, APS and LEOS-International Microwave and Optoelectronics Conference (IMOC '99)*, vol. 1, pp. 108–111, Rio de Janeiro, Brazil, August 1999.

[33] J.-H. Oh, W.-S. Sul, H.-J. Han et al., "Effects of silicon-nitride passivation on the electrical behavior of 0.1-μm pseudomorphic high-electron-mobility transistors," *Journal of the Korean Physical Society*, vol. 44, no. 4, pp. 899–903, 2004.

[34] D. J. Carrad, A. M. Burke, P. J. Reece et al., "The effect of $(NH_4)_2S_x$ passivation on the (311)A GaAs surface and its use in AlGaAs/GaAs heterostructure devices," *Journal of Physics: Condensed Matter*, vol. 25, no. 32, Article ID 325304, 2013.

[35] H.-C. Chiu, Y.-C. Huang, C.-W. Chen, and L.-B. Chang, "Electrical characteristics of passivated Pseudomorphic HEMTs with $P_2S_5/(NH_4)_2S_X$ pretreatment," *IEEE Transactions on Electron Devices*, vol. 55, no. 3, pp. 721–726, 2008.

[36] G. Kopnov, V. Y. Umansky, H. Cohen, D. Shahar, and R. Naaman, "Effect of the surface on the electronic properties of a two-dimensional electron gas as measured by the quantum Hall effect," *Physical Review B*, vol. 81, no. 4, Article ID 045316, 2010.

[37] C.-C. Chang, C.-Y. Chi, M. Yao et al., "Electrical and optical characterization of surface passivation in GaAs nanowires," *Nano Letters*, vol. 12, no. 9, pp. 4484–4489, 2012.

[38] H. J. Joyce, P. Parkinson, N. Jiang et al., "Electron mobilities approaching bulk limits in 'surface-free' GaAs nanowires," *Nano Letters*, vol. 14, no. 10, pp. 5989–5994, 2014.

[39] P. A. Alekseev, M. S. Dunaevskiy, V. P. Ulin et al., "Nitride surface passivation of GaAs nanowires: impact on surface state density," *Nano Letters*, vol. 15, no. 1, pp. 63–68, 2015.

[40] S. Mokkapati, D. Saxena, N. Jiang, L. Li, H. H. Tan, and C. Jagadish, "An order of magnitude increase in the quantum efficiency of (Al)GaAs nanowires using hybrid photonic-plasmonic modes," *Nano Letters*, vol. 15, no. 1, pp. 307–312, 2015.

[41] J. L. Boland, S. Conesa-Boj, P. Parkinson et al., "Modulation doping of GaAs/AlGaAs core-shell nanowires with effective

defect passivation and high electron mobility," *Nano Letters*, vol. 15, no. 2, pp. 1336–1342, 2015.

[42] L. A. Farrow, C. J. Sandroff, and M. C. Tamargo, "Raman scattering measurements of decreased barrier heights in GaAs following surface chemical passivation," *Applied Physics Letters*, vol. 51, no. 23, pp. 1931–1933, 1987.

[43] A. Osherov, M. Matmor, N. Froumin, N. Ashkenasy, and Y. Golan, "Surface termination control in chemically deposited PbS films: nucleation and growth on GaAs(111)A and GaAs(111)B," *Journal of Physical Chemistry C*, vol. 115, no. 33, pp. 16501–16508, 2011.

[44] S. L. Peczonczyk, J. Mukherjee, A. I. Carim, and S. Maldonado, "Wet chemical functionalization of III–V semiconductor surfaces: alkylation of gallium arsenide and gallium nitride by a Grignard reaction sequence," *Langmuir*, vol. 28, no. 10, pp. 4672–4682, 2012.

[45] S. Arab, C.-Y. Chi, T. Shi et al., "Effects of surface passivation on twin-free GaAs nanosheets," *ACS Nano*, vol. 9, no. 2, pp. 1336–1340, 2015.

[46] L. Pavesi and M. Guzzi, "Photoluminescence of $Al_xGa_{1-x}As$ alloys," *Journal of Applied Physics*, vol. 75, no. 10, pp. 4779–4842, 1994.

[47] D. E. Aspnes and A. A. Studna, "Schottky-barrier electroreflectance: application to GaAs," *Physical Review B*, vol. 7, no. 10, pp. 4605–4625, 1973.

[48] R. Kudrawiec, M. Gladysiewicz, L. Janicki et al., "Contactless electroreflectance studies of Fermi level position on c-plane GaN surface grown by molecular beam epitaxy and metalorganic vapor phase epitaxy," *Applied Physics Letters*, vol. 100, no. 18, Article ID 181603, 2012.

[49] L. Zamora-Peredo, I. Cortes-Mestizo, L. García-Gonzáez et al., "Optical and electrical study of cap layer effect in QHE devices with double-2DEG," *MRS Proceedings*, vol. 1617, pp. 31–36, 2013.

[50] Ł. Janicki, R. Kudrawiec, K. Pakula, R. Stępniewski, and J. Misiewicz, "Contactless electroreflectance studies of surface potential barrier in AlGaN/n-AlGaN structures with various Al concentrations," *Physica Status Solidi B*, vol. 252, no. 5, pp. 1038–1042, 2015.

[51] M. Gladysiewicz, L. Janicki, J. Misiewicz et al., "Engineering of electric field distribution in GaN(cap)/AlGaN/GaN heterostructures: theoretical and experimental studies," *Journal of Physics D: Applied Physics*, vol. 49, no. 34, Article ID 345106, 2016.

[52] Y. Alaskar, S. Arafin, D. Wickramaratne et al., "Towards van der waals epitaxial growth of GaAs on Si using a graphene buffer layer," *Advanced Functional Materials*, vol. 24, no. 42, pp. 6629–6638, 2014.

[53] J. Kim, C. Bayram, H. Park et al., "Principle of direct van der Waals epitaxy of single-crystalline films on epitaxial graphene," *Nature Communications*, vol. 5, article 4836, 2014.

2

Identification and Quantitation of Melamine in Milk by Near-Infrared Spectroscopy and Chemometrics

Tong Wu,[1] Hui Chen,[2] Zan Lin,[1,3] and Chao Tan[1]

[1]*Key Lab of Process Analysis and Control of Sichuan Universities, Yibin University, Yibin, Sichuan 644000, China*
[2]*Yibin University Hospital, Yibin, Sichuan 644000, China*
[3]*The First Affiliated Hospital, Chongqing Medical University, Chongqing 400016, China*

Correspondence should be addressed to Chao Tan; chaotan1112@163.com

Academic Editor: Feride Severcan

Melamine is a nitrogen-rich substance and has been illegally used to increase the apparent protein content in food products such as milk. Therefore, it is imperative to develop sensitive and reliable analytical methods to determine melamine in human foods. Current analytical methods for melamine are mainly chromatography-based methods, which are time-consuming and expensive and require complex pretreatment and well-trained technicians. The present paper investigated the feasibility of using near-infrared (NIR) spectroscopy and chemometrics for identifying and quantifying melamine in liquor milk. A total of 75 samples were prepared. Uninformative variable elimination-partial least square (UVE-PLS) and partial least squares-discriminant analysis (PLS-DA) were used to construct quantitative and qualitative models, respectively. Based on the ratio of performance to standard deviate (RPD), UVE-PLS model with 3 components resulted in a better solution. The PLS-DA model achieved an accuracy of 100% and outperformed the optimal reference model of soft independent modeling of class analogy (SIMCA). Such a method can serve as a potential tool for rapid screening of melamine in milk products.

1. Introduction

Melamine or 2,4,6-triamino-1,3,5-triazine is a nitrogen-rich compound and has been widely used in industry for producing melamine-formaldehyde resin [1]. Such a substance contains a substantial amount of nitrogen, 66.7% by mass, and therefore forms a driving force of the adulteration of milk products with melamine for obtaining high protein content readings. This is also because the conventional Kjeldahl or Dumas test is a method of analyzing total nitrogen content, without identifying its sources [2, 3]. It gives an appearance of protein content, not true protein content, and therefore cannot distinguish nitrogen of melamine from other proteins.

Although melamine is not inherently a carcinogen, it has low oral acute toxicity [4]. High-dose ingestion of melamine or chronic administration can induce urinary calculi and acute renal failure and even results in death, especially in babies and children [5–7]. Studies on toxicity caused by oral ingestion of melamine in humans are nonexistent. LD50, the lethal dose of a given compound resulting in death of 50% of tested animals, is 3.1 g/kg of body mass for melamine in rats [8]. At present, the US Food and Drug Administration (FDA) states limit of melamine in milk of no more than 2.5 ppm [9]. In 2008, a serious adulteration incident, that is, milk powder adulterated with melamine, was made public in China and many other countries [10]. The adulterated products resulted in kidney illness of various degrees affecting about 294,000 individuals, 6 of whom died [11]. This incidence has incurred global concern about melamine and food safety. Therefore, the development of fast and reliable analytical methods to determine melamine in human food is of utmost importance.

In the wake of these incidents, various methods have been developed in the area of melamine detection, including gas chromatography (GC) [12], high-performance liquid chromatography (HPLC) [13], mass spectrometry (MS) [14], matrix-assisted laser desorption/ionization time-of-flight mass spectrometry, enzyme linked immune sorbent assay (ELISA) [15], and capillary electrophoresis (CE) [16]. However, to eliminate the strong interference components from matrix, sophisticated instrument and complicated sample treatment are generally needed for the above techniques, which seriously hampered the speed in field analysis. Also, these methods maybe require well-trained technicians to operate the instrumentation. Currently, spectrophotometry combined with chemometrics seems to be a kind of attractive methods in analytical chemistry [17–19]. On one hand, spectroscopic instruments are widely available in ordinary analytical laboratory; they are also much faster than the usual techniques, present good accuracy and precision, and are nondestructive, and the procedure of sample preparation is relatively simple. On the other hand, chemometrics can help chemists to resolve the constituents of complex chemometrics by introducing of computer and statistical techniques [20, 21]. In particular, Raman, infrared, and near-infrared (NIR) spectroscopy have been successfully used in adulteration detection for various food products such as milk and honey. Several researches have mainly focused on the detection of melamine in milk powder by these techniques [22–24]. A good review related to melamine prediction in milk by vibration spectroscopy is available [25].

Among spectroscopic methods, NIR spectroscopy has obvious advantages because it is noninvasive, requires minimal sample preparation, and can yield a response in real time. NIR spectroscopy corresponds to the absorption of electromagnetic radiation in 780–2500 nm [26–28]. Quantitative and qualitative analysis with NIR spectroscopy depend greatly on the regression/classification model to relate subtle spectral variations to changes of certain component concentration in the sample matrix or to discriminate sample categories. The accuracy and robustness of a model directly decide its applicability in most situations. Even if many modeling methods have been developed, undoubtedly, the most commonly used multivariate technique is still partial least squares (PLS) [29]. Traditionally, PLS is considered being based on latent variables and without having to perform a previous variable selection. However, more and more evidences have indicated that variable selection is still very important and necessary for a parsimonious and robust model [30]. Comparing with many other methods of variable selection, uninformative variable elimination-PLS (UVE-PLS) is a method for variable selection based on an analysis of regression coefficients of PLS, which has been widely applied in spectral analysis and provided satisfactory results.

In this study, the feasibility of identifying and quantifying melamine in liquor milk by NIR spectroscopy in combination with chemometrics was investigated. Several algorithms, that is, PLS, UVE-PLS, partial least square-discriminant analysis (PLS-DA), and the soft independent modeling of class analogy (SIMCA), were used to construct quantitative and qualitative models and the optimal results were obtained.

2. Materials and Methods

2.1. Materials and Reagents. Melamine (1,3,5-triazine-2,4,6-triamine) (99%) was purchased from Aladdin Reagent Co., Ltd. and used without further purification. Nescafe milk powder from Shuangcheng Nestle Co., Ltd., was purchased from a local supermarket. It was confirmed to be melamine-free. Based on the data provided by the manufacturer, protein, fat, and carbohydrates contents in the milk powder were 23.2%, 28.2%, and 37%, respectively. The milk powder was used for the preparation of liquor milk, which was then adulterated by melamine in order to improve apparent protein concentration. A total of 45 samples of liquor milk were prepared. Each of the first 30 samples was obtained by dissolving 15 grams of milk powder in 100 mL deionized water. These samples have the same protein concentration of 3.48% and were used for simulating pure liquor milk unadulterated with melamine. Each of the other 15 samples was generated by first dissolving 11 grams of milk powder in deionized water (protein % = 2.55%) to obtain 100 mL liquor milk sample and then adding 0.01, 0.03, 0.05, 0.07, 0.09, 0.11, 0.13, 0.15, 0.17, 0.19, 0.21, 0.23, 0.25, 0.27, or 0.29 grams of melamine in it. That is, the range of melamine concentration was set to be from 0.01% to 0.29% (i.e., 0.1–2.9 g/L). Such an upper limit set is based on the consideration of the solubility of melamine in water (3.1 g/L at room temperature). In the market, the popular protein concentration of liquid milk is 3.0–3.5%; the adulteration of melamine is to improve the apparent protein content of liquid milk. Each 0.01 g melamine in 100 mL can only improve 0.04% of apparent protein content, below which it will lose driving force for counterfeiters. Apparent protein concentration of the adulterated samples varies from 2.59% to 3.76%. Each time, melamine was added to a milk sample independently so as to ensure appropriate variations in all samples. For constructing quantitative models of melamine, only later were 45 spectra used. Also, these samples with melamine concentration of 0.03%, 0.07%, 0.13%, 0.15%, 0.19%, 0.23%, and 0.27% constituted the test set. For constructing an identification/classification model, all samples/spectra were divided into the training/calibration set and the test set by alternative sampling.

2.2. Instrument and Measurement. A spectrum was collected for each unadulterated sample while 3 spectra were collected for each of the adulterated milk samples. In order to measure the influence of environmental conditions, the adjacent measurement interval was set as 20 minutes. A total of 75 NIR spectra were acquired on Fourier transform NIR spectrometer (Thermo Fisher, USA) equipped with a special optical fiber probe. The collection process was controlled by the Result software of Thermo Fisher. The wavenumber range was set as 4000–10000 cm^{-1}. Each spectrum was the average result of 32 scans with 1557 points. Only 489 points in the region of 4000–5882 cm^{-1} were used for modeling. Figure 1 shows the NIR spectra of all experimental milk samples. All calculation and modeling were performed in MATLAB 7.0 for Windows using PLS toolbox.

2.3. Partial Least Squares. Partial least squares (PLS) [31], as a classic multivariate calibration, is commonly used in spectroscopy, especially for NIR spectroscopy, to correlate spectral data (**X**) with related reference data (**Y**). It is based on latent variables and can therefore handle so-called collinear problem, but for PLS the decomposition of **X** is guided by the variation in **Y**: the explained covariance between **X** and **Y** is maximized, so that the variation in **X** directly correlating

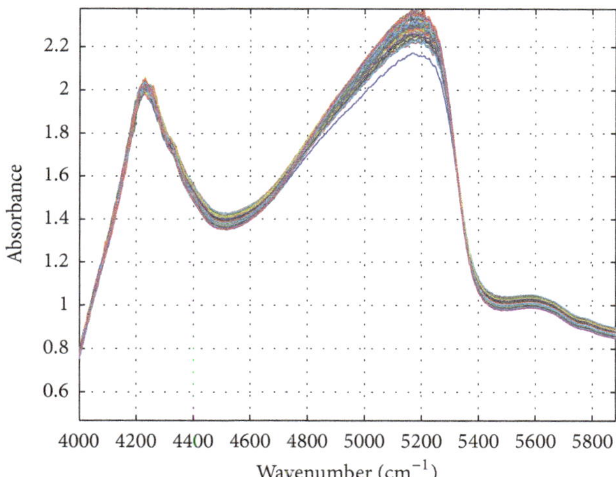

FIGURE 1: The NIR spectra of all experimental milk samples.

with **Y** is extracted. PLS decomposes the matrix of zero-mean variables **X** and the matrix of zero-mean variables **Y** into the form

$$\mathbf{X} = \mathbf{T}\mathbf{P}^{t} + \mathbf{E}$$
$$\mathbf{Y} = \mathbf{U}\mathbf{Q}^{t} + \mathbf{F}, \quad (1)$$

where **X** is the matrix of spectral data; **T** is the **X** score matrix; **P** is the **X** variable loading matrix and **E** is the **X** residual matrix; **Y** is the response matrix; **U** is the **Y** score matrix; **Q** is the **Y** loading matrix; and **F** is the **Y** residual matrix. The **T** and **U** represent information after removing most noise. Based on the correlation between them, the linear regression model can be given by

$$\mathbf{U} = \mathbf{T}\mathbf{b}, \quad (2)$$

where **b** is the regression vector to be determined during calibration. To obtain a good estimation of **b**, it maybe needs to collect a training set that span the variation in **Y** well and in general are representative of the future samples. From the mathematical point of view, a vector **b** means a calibration model even if a few versions of realizing PLS algorithm are available.

2.4. Uninformative Variable Elimination-PLS (UVE-PLS). Nowadays, in spectroscopy, many researchers make use of PLS because it is a full-spectrum based method. However, using full spectral region does not always yield optimal results because it may include variables containing more noise than relevant information to models. So, eliminating the variables which have high variance but small covariance with the dependent variables, so-called uninformative variables, may be helpful for improving the model. Based on this, center proposed an algorithm called uninformative variable elimination-partial least squares (UVE-PLS) algorithm [32]. Compared to other algorithms of variable selection, it is user independent and therefore does not cause any configuration problems. It consists of the following steps:

(1) Using leave-many-out procedure on spectral matrix **X** ($n \times p$) for determining the optimal complexity of the model, with the lowest RMSEP as criterion.

(2) Building up of a matrix (**R**) of artificial variables of the same size as **X**, generated by the rand command. These variables are multiplied by a small constant (10^{-10}) so as to make them obviously smaller than the imprecision of the instrument; such an operation retains the variation of the variables but makes their influence on the final model negligible.

(3) Combining of **R** with **X** to form an extended matrix with twice as many variables as **X**. The resulting matrix can be denoted as **XR** ($n \times 2p$), the p first columns being those of **X** and the p last ones being those of **R**.

(4) Building PLS models for **XR** according to leave-one-out procedure. The complexity, that is, the number of latent variables square discriminan, is set the same as for **X**. This leads to n PLS models, each with a coefficient vector of $2p$ elements. All of them are collected in a matrix **B** ($n \times 2p$).

(5) Calculating the reliability factor (t_{value}) for each variable by dividing the mean value of the coefficient vector by its standard deviation.

(6) Setting the cut-off limit as certain percentage (k) of the largest t_{value} of artificial variables and constructing $\mathbf{X}_{\text{new}}$ consist of the variables with high t_{value} compared to the cut-off limit. This means that all artificial variables and all original variables assumed to contain nothing but noise are eliminated.

(7) Building the final PLS model by leave-one-out cross-validation procedure.

Considering the computational cost, the number of columns of **R** matrix was fixed at 200.

2.5. PLS-DA. The PLS regression is initially developed for the prediction of continuous target variable. But it seems to be useful in the classification problem where one expects to predict the values of discrete attributes. The generic name is the partial least square-discriminant analysis, that is, PLS-DA, which is often used in the literature [33]. As a special form of PLS modeling, PLS-DA aims to find the variables and directions in multivariate space, which discriminates the known classes in a dataset. If there are only two classes to separate, the PLS model uses one dummy variable, which codes for class membership as follows: 0 for members of class A and 1 for members of class B. A discriminant model is developed by regression of the spectral data (**X**) against the assigned dummy variable (**Y**). The model based on experimental data is established so as to assign unknown samples to a previously defined class based on its pattern of measured features. The threshold is set in the model, and a sample is considered to be categorized correctly if the predicted value lies on the same side of the midpoint of the assigned value. A sample is identified as class A if its predicted value is below the threshold and as class B if its predicted value is above the threshold.

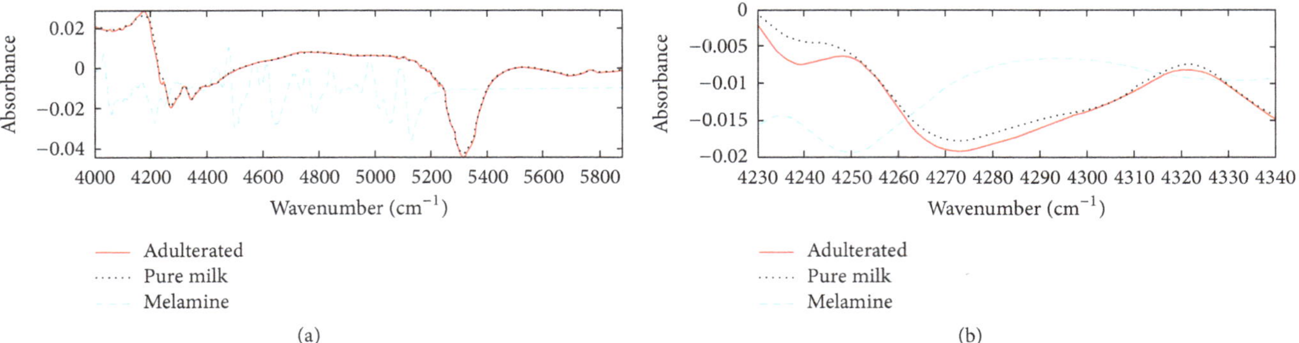

Figure 2: (a) The mean 1st-derivate NIR spectra of adulterated milk, pure milk, and melamine samples in the range of 4000–5880 cm^{-1} and (b) the enlarged version in the range of 4230–4340 cm^{-1}.

2.6. SIMCA. The SIMCA is a popular classification method based on principal component analysis (PCA) [34]. In SIMCA, a separate PCA is performed on each class in the dataset, and a sufficient number of principal components (PCs) are retained to account for most of the variation within each class. The number of PCs retained for each class is usually determined by cross-validation and is maybe different for each class model. Classification in SIMCA is made by comparing the residual variance of a sample with the average residual variance of a class. This comparison provides a direct measure of the goodness of fit of a sample for a particular class model. SIMCA can classify an unknown sample to the class for which it has a high probability.

2.7. Statistical Evaluation. The performance of both PLS and UVE-PLS was assessed using three measures, that is, the standard error of prediction (SEP), the coefficient of determination R^2, and the ratio of performance to standard deviate (RPD). R^2 which can measure the proportion of the variation in the response that may be attributed to the model rather than to random error was also recorded. RPD measures the ratio of the standard deviation of reference values to the root mean squared error (RMSE) of prediction. Generally, the RPD values were classified into four levels of prediction accuracy: RPD < 1.5 indicates very bad model/predictions; RPD between 1.5 and 2.0 indicates poor model/predictions; RPD between 2.0 and 2.5 indicates good model/predictions; and RPD > 2.5 indicates very good/excellent model/predictions [35].

3. Results and Discussions

3.1. Spectral Analyses. Even if the spectra were collected in the range of 4000–10000 cm^{-1}, only the spectral region of 4000–5880 cm^{-1} was selected to construct models since other regions obviously contain little useful information. Figure 2(a) gives the mean 1st-derivate NIR spectra of adulterated milk, pure/unadulterated milk, and melamine samples in the region, where a vertical shift is used to prevent superposition; that is, the spectrum corresponding to melamine is moved down by 0.01 units. As can be seen in Figure 1, there exists spectral difference between adulterated and pure milk, but the difference is too small to find by the naked eye. Further, Figure 2(b) amplifies this difference by only showing a smaller region of 4230–4340 cm^{-1}, which indicates that the difference is still present and can be used for quantitative and qualitative purposes. As the melamine molecule consists of rich NH bonds, the key spectral difference of these samples in the range of 4230–4340 cm^{-1} mainly comes from the combination modes of NH vibrational absorption with other energy transfers. For establishing optimal quantitative/qualitative models, several pretreatment methods including 1st derivate and mean-centering and standard normal variate (SNV) were attempted. As a result, the 1st-derivate spectra were optimal for constructing the quantitative model while the original spectra, that is, without any pretreatment, were optimal for qualitative model.

3.2. Quantitative Model. Based on the original spectra, both PLS and UVE-PLS methods were used to construct calibration models. To obtain a robust model, the calibration/training set must capture the possible variability. It is crucial that the calibration set covers the range of concentrations of the melamine to be predicted. So, as described above, eight groups of samples were used as the calibration set and seven groups of samples constitute the test set. By this way, the test set can be covered by the design space of the calibration set. In the application of PLS-related algorithms, it is generally known that the number of latent variables (Lvs) was decisive. A maximum of 15 Lvs were tested. The selection of the optimal Lvs was based on the minimum root mean squared error of cross-validations (RMSECV). Figure 3 shows the RMSECV plot as a function versus the number of latent variables of PLS and UVE-PLS models. As seen from Figure 3, RMSECV decreases first with initial Lvs and achieves the lowest point and then basically remains with an increase of Lvs. The optimal number of Lvs seems to be 5 and 3 for the PLS and UVE-PLS models, respectively, since the utilization of a more than this number does not improve the results. Compared to the PLS model, the UVE-PLS model used fewer original variables to produce fewer Lvs, meaning a more concise model. In fact, the Lvs can be interpreted as a measure of the complexity of the PLS model.

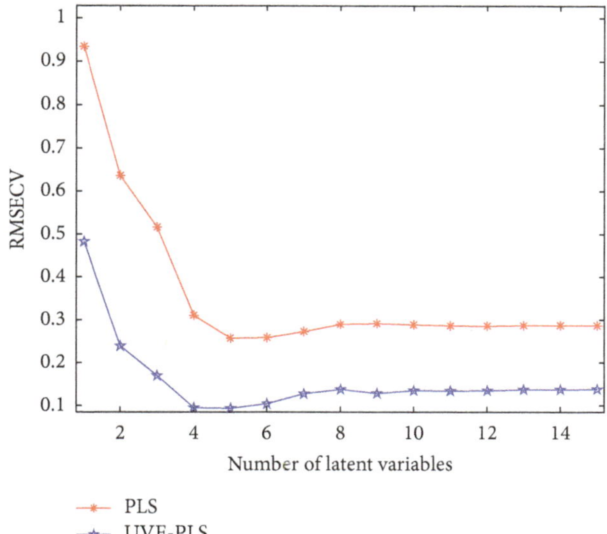

FIGURE 3: The RMSECV plot as a function versus the number of latent variables of PLS and UVE-PLS models.

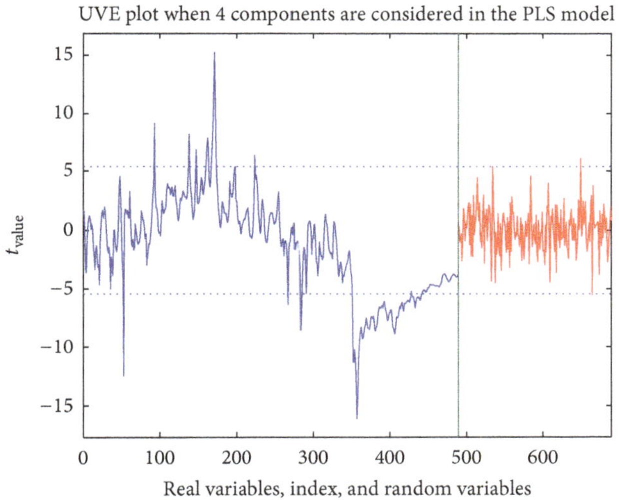

FIGURE 4: Plot of t_{value} for original wavenumber variables and artificial variables. The cut-off limit is indicated by the dotted lines.

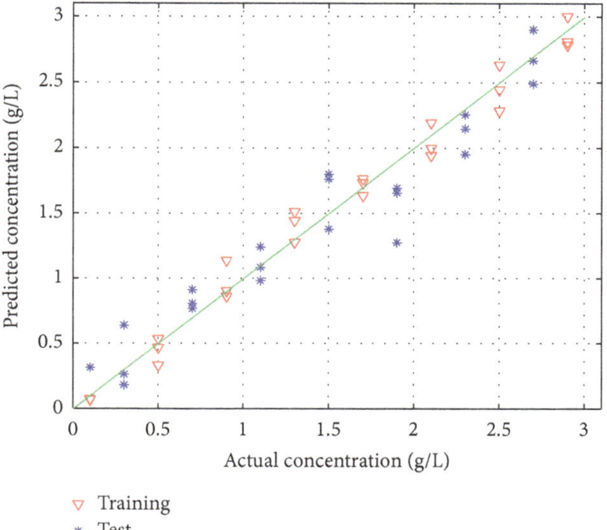

FIGURE 5: Scatter plots of predicted versus actual melamine concentration in milk samples for the PLS model.

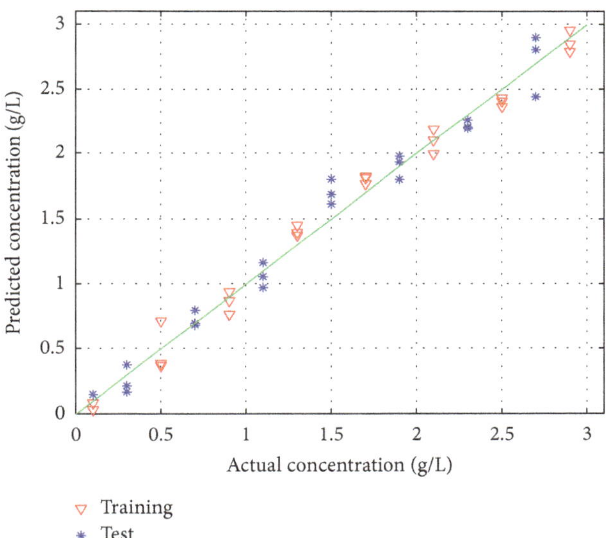

FIGURE 6: Scatter plots of predicted versus actual melamine concentration in milk samples for the UVE-PLS model.

According to the UVE-PLS procedure described above, the stability index (t_{value}) of all variables was calculated. Figure 4 shows the plot of t_{value} for original variables in the region of 4000–5882 cm^{-1} and 200 artificial variables. The dotted lines show the upper and lower boundaries determined with $k = 0.9$. Variables corresponding to stability within the boundaries will be eliminated, and the variables whose stability lies out of the dot lines were used for PLS modeling. Also, the variables selected by UVE method are concentrated on a broad region around 5300 cm^{-1} and three narrow regions in the range of 4000–4700 cm^{-1}. The number of retained variables was investigated since it maybe has an influence on the stability and accuracy of the PLS model. Generally, when the number of retained variables is too small, the robustness and accuracy of the PLS model may be affected due to the loss of informative variables. On the contrary, if the number of retained variables is too large, uninformative variables may be contained in the model and make its performance poor. In this work, different k values were attempted. As a result, $k = 0.9$ was selected and only 89 variables were retained after the UVE procedure.

Figure 5 shows the relationship between the actual melamine concentration and the predicted ones from the PLS model by scatter plot. Similarly, Figure 6 shows the scatter plot for the UVE-PLS model. In these plots, predicted concentration values are plotted against actual ones and the points will fall on the diagonal only if the model predicts

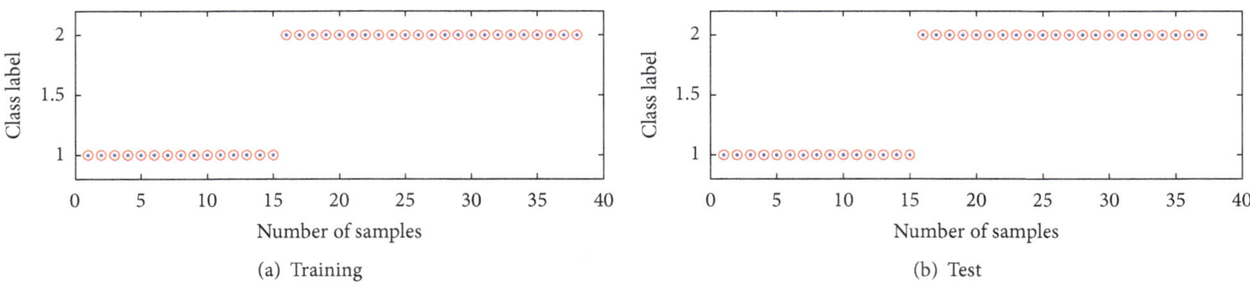

(a) Training (b) Test

FIGURE 7: Identification performance of the PLS-DA classifier.

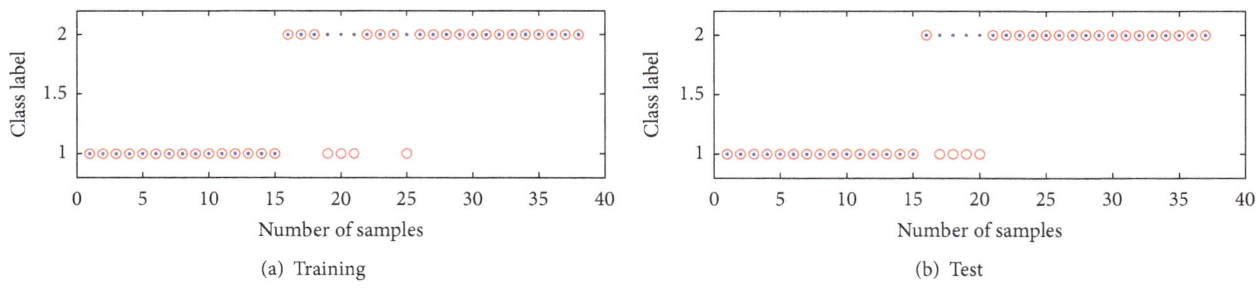

(a) Training (b) Test

FIGURE 8: Identification performance of the SIMCA classifier.

TABLE 1: Performance comparison of the optimal quantitative PLS and UVE-PLS models.

Method	SEP	RPD	R^2	Lvs
PLS	0.023	3.93	0.93	5
UVE-PLS	0.015	6.23	0.97	4

the concentration perfectly. As shown in Figures 5 and 6, for the training set, there is no obvious difference between the PLS and UVE-PLS model since those points corresponding to the training set (expressed by red triangles) have a similar distribution around the diagonal. However, on the test set (expressed by blue stars), the UVE-PLS model outperforms the full-spectrum PLS model. It is also clear in Figures 5 and 6 that the same model can provide different predictions for the same sample due to the spectral diversity. Table 1 summarizes the performance comparison of the optimal quantitative PLS and UVE-PLS models based on three indices. It is clear that UVE-PLS can not only compress the spectral data in a very high ratio but also provide excellent model/predictions corresponding to RPD = 6.23.

3.3. Qualitative Model. To construct a PLS-DA model for identifying/classifying the milk samples, all 75 spectra were divided into a training set with 38 spectra and a test set with 37 spectra. For classification/identification purposes, each sample or spectrum was assigned a class label (1 for unadulterated and 2 for adulterated milk with melamine). The adulterated concentration of melamine was 0.01–0.29% and 0.03%–0.27% for the training set and the test set, respectively. The number of Lvs was optimized in the range of 2–10. It was found that when the number of variables was greater than 4, all samples were classified correctly, as Figure 7 showed, implying the sensitivity of 100% and specificity of 100%. The circles denoted the predicted class labels while the points denoted actual labels in Figure 7. Therefore, the calibrated PLS-DA model can be considered to be reliable and stable, since its performance on future samples is expected to be comparable to those achieved on the training samples. To prove the ability of the PLS-DA, the soft independent modeling of class analogy (SIMCA) was also used and the optimal model contained 4 principal components. As can be seen in Figure 8, there existed four samples with low adulterated concentration misclassified as pure samples for both the training set and the test set. For the optimal SIMCA model, even if the specificity was 100%, the sensitivity was 82.6% and 81.8% for the training set and the test set, respectively. The classification model developed based on PLS-DA and NIR spectroscopy could be used as a potential method for melamine identification in milk products.

4. Conclusions

A fast-screening approach for detecting melamine in liquor milk with NIR spectroscopy was developed. Both quantitative UVE-PLS model and qualitative PLS-DA model were constructed and showed satisfactory performance compared to the optimal reference models, that is, PLS and SIMCA. It can serve as a promising complementary scheme for conventional methods with less interference and lower investment and could be suitably applied to on-site quality control of milk products. Even so, to reduce the detection limit, further research is still needed.

Competing Interests

The authors declare that there is no conflict of interests regarding the publication of this paper.

Acknowledgments

The authors thank the National Natural Science Foundation of China (Grant no. 21375118), the Applied Basic Research Programs of Science and Technology Department of Sichuan Province of China (2013JY0101), Innovative Research and Teaching Team Program of Yibin University (Cx201104), Scientific Research Foundation of Sichuan Provincial Education Department of China (13ZB0300), and Opening Fund of Key Lab of Process Analysis and Control of Sichuan Universities of China (2015006) for supporting this work.

References

[1] Y. Liu, J. Deng, L. An, J. Liang, F. Chen, and H. Wang, "Spectrophotometric determination of melamine in milk by rank annihilation factor analysis based on pH gradual change-UV spectral data," *Food Chemistry*, vol. 126, no. 2, pp. 745–750, 2011.

[2] Z. N. Guo, Z. Y. Cheng, R. Li et al., "One-step detection of melamine in milk by hollow gold chip based on surface-enhanced Raman scattering," *Talanta*, vol. 122, pp. 80–84, 2014.

[3] L. J. Mauer, A. A. Chernyshova, A. Hiatt, A. Deering, and R. Davis, "Melamine detection in infant formula powder using near- and mid-infrared spectroscopy," *Journal of Agricultural and Food Chemistry*, vol. 57, no. 10, pp. 3974–3980, 2009.

[4] Y. Liu, E. E. D. Todd, Q. Zhang, J.-R. Shi, and X.-J. Liu, "Recent developments in the detection of melamine," *Journal of Zhejiang University: Science B*, vol. 13, no. 7, pp. 525–532, 2012.

[5] T.-J. Hsieh, P.-C. Hsieh, Y.-H. Tsai et al., "Melamine induces human renal proximal tubular cell injury via transforming growth factor-β and oxidative stress," *Toxicological Sciences*, vol. 130, no. 1, pp. 17–32, 2012.

[6] Y. Wei and D. Liu, "Review of melamine scandal: still a long way ahead," *Toxicology and Industrial Health*, vol. 28, no. 7, pp. 579–582, 2012.

[7] J. L. Dorne, D. R. Doerge, M. Vandenbroeck et al., "Recent advances in the risk assessment of melamine and cyanuric acid in animal feed," *Toxicology and Applied Pharmacology*, vol. 270, no. 3, pp. 218–229, 2013.

[8] J. Y. W. Chan, C. M. Lau, T. L. Ting et al., "Gestational and lactational transfer of melamine following gavage administration of a single dose to rats," *Food and Chemical Toxicology*, vol. 49, no. 7, pp. 1544–1548, 2011.

[9] C. Lu, B. Xiang, G. Hao, J. Xu, Z. Wang, and C. Chen, "Rapid detection of melamine in milk powder by near infrared spectroscopy," *Journal of Near Infrared Spectroscopy*, vol. 17, no. 2, pp. 59–67, 2009.

[10] N. Yan, L. Zhou, Z. Zhu, and X. Chen, "Determination of melamine in dairy products, fish feed, and fish by capillary zone electrophoresis with diode array detection," *Journal of Agricultural and Food Chemistry*, vol. 57, no. 3, pp. 807–811, 2009.

[11] L. Elvira, J. Rodríguez, and L. C. Lynnworth, "Sound speed and density characterization of milk adulterated with melamine," *The Journal of the Acoustical Society of America*, vol. 125, no. 5, pp. EL178–EL182, 2009.

[12] X.-D. Pan, P.-G. Wu, D.-J. Yang, L.-Y. Wang, X.-H. Shen, and C.-Y. Zhu, "Simultaneous determination of melamine and cyanuric acid in dairy products by mixed-mode solid phase extraction and GC-MS," *Food Control*, vol. 30, no. 2, pp. 545–548, 2013.

[13] A. Filazi, U. T. Sireli, H. Ekici, H. Y. Can, and A. Karagoz, "Determination of melamine in milk and dairy products by high performance liquid chromatography," *Journal of Dairy Science*, vol. 95, no. 2, pp. 602–608, 2012.

[14] A. Singh and V. Panchagnula, "High throughput quantitative analysis of melamine and triazines by MALDI-TOF MS," *Analytical Methods*, vol. 3, no. 10, pp. 2360–2366, 2011.

[15] B. Kim, L. B. Perkins, R. J. Bushway et al., "Determination of melamine in pet food by enzyme immunoassay, high-performance liquid chromatography with diode array detection, and ultra-performance liquid chromatography with tandem mass spectrometry," *Journal of AOAC International*, vol. 91, no. 2, pp. 408–413, 2008.

[16] Z. Chen and X. Yan, "Simultaneous determination of melamine and 5-hydroxymethylfurfural in milk by capillary electrophoresis with diode array detection," *Journal of Agricultural and Food Chemistry*, vol. 57, no. 19, pp. 8742–8747, 2009.

[17] A. Afkhami and L. Khalafi, "Spectrophotometric investigation of the effect of β-cyclodextrin on the intramolecular cyclization reaction of catecholamines using rank annihilation factor analysis," *Analytica Chimica Acta*, vol. 599, no. 2, pp. 241–248, 2007.

[18] S. S. Souza, A. G. Cruz, E. H. M. Walter et al., "Monitoring the authenticity of Brazilian UHT milk: a chemometric approach," *Food Chemistry*, vol. 124, no. 2, pp. 692–695, 2011.

[19] C. Tan, H. Chen, C. Wang, W. Zhu, T. Wu, and Y. Diao, "A multi-model fusion strategy for multivariate calibration using near and mid-infrared spectra of samples from brewing industry," *Spectrochimica Acta Part A: Molecular and Biomolecular Spectroscopy*, vol. 105, pp. 1–7, 2013.

[20] L. An, J. Deng, L. Zhou et al., "Simultaneous spectrophotometric determination of trace amount of malachite green and crystal violet in water after cloud point extraction using partial least squares regression," *Journal of Hazardous Materials*, vol. 175, no. 1–3, pp. 883–888, 2010.

[21] E. Aprea, M. L. Corollaro, E. Betta et al., "Sensory and instrumental profiling of 18 apple cultivars to investigate the relation between perceived quality and odour and flavour," *Food Research International*, vol. 49, no. 2, pp. 677–686, 2012.

[22] Y. Cheng, Y. Dong, J. Wu et al., "Screening melamine adulterant in milk powder with laser Raman spectrometry," *Journal of Food Composition and Analysis*, vol. 23, no. 2, pp. 199–202, 2010.

[23] Y. Huang, K. D. Tian, S. G. Min, Y. M. Xiong, and G. R. Du, "Distribution assessment and quantification of counterfeit melamine in powdered milk by NIR imaging methods," *Food Chemistry*, vol. 177, pp. 174–181, 2015.

[24] S. Jawaid, F. N. Talpur, S. T. H. Sherazi, S. M. Nizamani, and A. A. Khaskheli, "Rapid detection of melamine adulteration in dairy milk by SB-ATR-Fourier transform infrared spectroscopy," *Food Chemistry*, vol. 141, no. 3, pp. 3066–3071, 2013.

[25] E. Domingo, A. A. Tirelli, C. A. Nunes, M. C. Guerreiro, and S. M. Pinto, "Melamine detection in milk using vibrational spectroscopy and chemometrics analysis: a review," *Food Research International*, vol. 60, pp. 131–139, 2014.

[26] T. T. Zou, Y. Dou, H. Mi, J. Y. Zou, and Y. L. Ren, "Support vector regression for determination of component of compound oxytetracycline powder on near-infrared spectroscopy," *Analytical Biochemistry*, vol. 355, no. 1, pp. 1–7, 2006.

[27] M. Sattlecker, R. Baker, N. Stone, and C. Bessant, "Support vector machine ensembles for breast cancer type prediction from mid-FTIR micro-calcification spectra," *Chemometrics and Intelligent Laboratory Systems*, vol. 107, no. 2, pp. 363–370, 2011.

[28] C. Tan, H. Chen, T. Wu, Z. Xu, W. Li, and X. Qin, "Determination of total sugar in Tobacco by near-infrared spectroscopy and wavelet transformation-based calibration," *Analytical Letters*, vol. 46, no. 1, pp. 171–183, 2012.

[29] C. Tan, T. Wu, Z. Xu, W. Li, and K. Zhang, "A simple ensemble strategy of uninformative variable elimination and partial least-squares for near-infrared spectroscopic calibration of pharmaceutical products," *Vibrational Spectroscopy*, vol. 58, pp. 44–49, 2012.

[30] R. Leardi, "Application of genetic algorithm-PLS for feature selection in spectral data sets," *Journal of Chemometrics*, vol. 14, no. 5-6, pp. 643–655, 2000.

[31] M. Sjöström, S. Wold, W. Lindberg, J.-Å. Persson, and H. Martens, "A multivariate calibration problem in analytical chemistry solved by partial least-squares models in latent variables," *Analytica Chimica Acta*, vol. 150, pp. 61–70, 1983.

[32] V. Centner, D.-L. Massart, O. E. de Noord, S. de Jong, B. M. Vandeginste, and C. Sterna, "Elimination of uninformative variables for multivariate calibration," *Analytical Chemistry*, vol. 68, no. 21, pp. 3851–3858, 1996.

[33] M. R. de Almeida, D. N. Correa, W. F. C. Rocha, F. J. O. Scafi, and R. J. Poppi, "Discrimination between authentic and counterfeit banknotes using raman spectroscopy and PLS-DA with uncertainty estimation," *Microchemical Journal*, vol. 109, pp. 170–177, 2013.

[34] K. V. Branden and M. Hubert, "Robust classification in high dimensions based on the SIMCA Method," *Chemometrics and Intelligent Laboratory Systems*, vol. 79, no. 1-2, pp. 10–21, 2005.

[35] R. A. Viscarra Rossel, "Robust modelling of soil diffuse reflectance spectra by 'bagging-partial least squares regression,'" *Journal of Near Infrared Spectroscopy*, vol. 15, no. 1, pp. 39–47, 2007.

3

Feasibility of Terahertz Time-Domain Spectroscopy to Detect Carbendazim Mixtures Wrapped in Paper

Binyi Qin,[1,2] Zhi Li,[1,3] Zhihui Luo,[4] Huo Zhang,[1] and Yun Li[4]

[1]*School of Mechano-Electronic Engineering, Xidian University, Xi'an, Shanxi 710126, China*
[2]*School of Electronics and Communication Engineering, Yulin Normal University, Yulin, Guangxi 537000, China*
[3]*Guangxi Key Laboratory of Automatic Detecting Technology and Instruments, School of Electronic Engineering and Automation, Guilin University of Electronic Technology, Guilin, Guangxi 541004, China*
[4]*Guangxi Key Laboratory for Agricultural Resources Chemistry and Efficient Utilization (Cultivation Base), Colleges and Universities Key Laboratory for Efficient Use of Agricultural Resources in the Southeast of Guangxi, College of Chemistry and Food Science, Yulin Normal University, Yulin, Guangxi 537000, China*

Correspondence should be addressed to Zhi Li; zhili_workemail@163.com

Academic Editor: Khalique Ahmed

The purpose of this work was to detect carbendazim mixtures wrapped in paper. Unlike previous reports of THz-TDS for detecting pesticide residue, this work focused on detecting pesticide residue in packaged foods. Different weight ratios of carbendazim in polyethylene and in rice powder were detected qualitatively and quantitatively. Results show that pure carbendazim, polyethylene, and rice powder can be easily distinguished from each other. However, when the weight ratio was low, the absorbance of the mixture was similar with that of pure polyethylene and rice powder. With the help of SVM, carbendazim could be qualitatively detected in low weight ratio mixture successfully. Moreover, PLS and SVR were selected to quantitatively detect carbendazim mixtures. SVR has higher R and lower RMSECV, RMSEC, and RMSEP than PLS model. Lasty, the results also indicate that THz-TDS is a potential tool to detect pesticide residue in packaged foods qualitatively and quantitatively.

1. Introduction

Pesticide has been widely used in agriculture due to its broad spectrum activity and low production costs. However, overusing pesticide has cased residue of this drug in food product, which is detrimental to the health of consumers. Thus, detecting pesticide residue has become more and more important in the modern world. Traditional methods of pesticide residue detection include gas chromatography, liquid chromatography, and high-performance liquid chromatography. Nevertheless, these methods are time consuming and require tedious sample preparation procedures and cannot offer fast pesticide detection [1–3]. Other rapid detection methods based on immunology assays, such as immunochemistry and biosensors are time saving, sensitive, and highly selective, but they are difficult to establish and usually yield false-positive results [4].

In recent years, spectroscopic methods, such as ultraviolet spectroscopy, near-infrared spectroscopy, and infrared spectroscopy, are applied to detect pesticide residue because of their nondestructive and simple sample pretreatment. But there are limitations to these methods. Ultraviolet spectroscopy is harmful to the active ingredients in food and the human body [5]. Near-infrared spectroscopy is only suitable for the quantitative analysis rather than the qualitative analysis, owing to its complex spectra for intramolecular overtone and combination bands of chemicals [6]. In contrast, infrared spectroscopy is used for a qualitative tool rather than as a quantitative tool as a result of its sensitivity to the scattering effect in the sample and less stable radiation

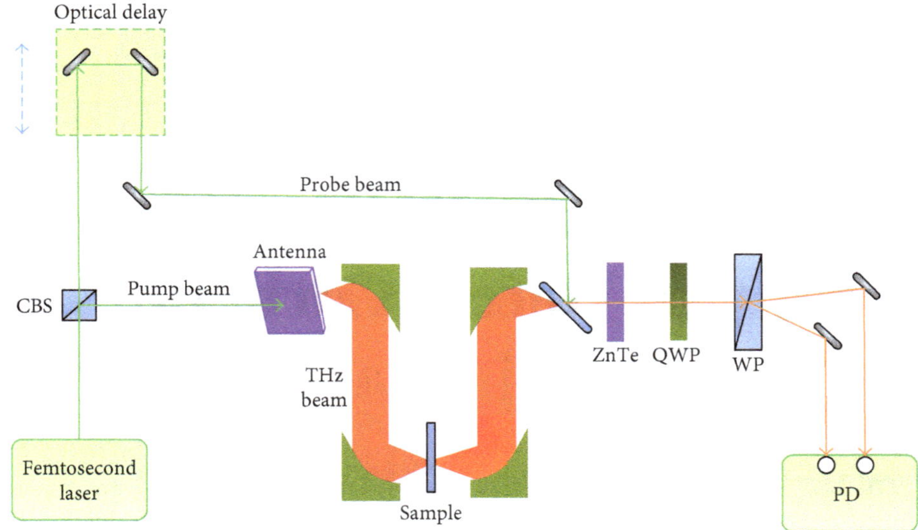

FIGURE 1: Schematic of the experimental apparatus.

source [7]. Therefore, there is an increasing demand to exploit new techniques for rapid, sensitive, and nondestructive detection of pesticide residue in routine assays.

Terahertz (THz) refers to the electromagnetic wave in the frequency range of 0.1 to 10 THz between microwave and infrared. Comparing with other spectroscopic methods, THz has little damage to the target material because of its low photon energy and is less sensitive to noise that comes from thermal background radiation [8]. Moreover, it can give the amplitude and phase information of the sample simultaneously. So, THz has been qualitatively and quantitatively applied in many fields, for example chemistry, biology, medical science, and homeland security [9–13].

In the field of pesticide detection, Bing-Hua et al. analyzed dimethoate using density functional theory and experimental investigations [14]. Hua et al. quantitatively studied cyfluthrin n-hexane solutions based on partial least squares (PLS) and principal component regression (PCR) methods [15]. Qiang et al. used partial least squares (PLS), interval PLS (iPLS), moving window PLS (mwPLS), and backward interval PLS (biPLS) methods to quantify nitrofen [16]. However, few papers are concerned for detecting pesticide residue in packaged foods. It will be a meaningful progress.

In this work, we mixed carbendazim with polyethylene and rice powder, separately, and then employed terahertz time-domain spectroscopy (THz-TDS) to analyze these mixtures wrapped in paper. Firstly, we observed the absorbency of carbendazim in THz range. Secondly, we analyzed the absorbency of mixtures wrapped in paper. And then, the absorbency of the THz spectrum was used as input of support vector machine (SVM), partial least squares regression (PLSR), and support vector regression (SVR) for qualitative and quantitative analysis. The results indicate that pesticide can be qualitatively and quantitatively detected in mixtures and give evidence of a potential of applying THz-TDS to detect the presence of pesticide in packaged foods.

2. Materials and Methods

2.1. Experimental System. The experimental apparatus of the terahertz time-domain spectroscopy (THz-TDS) system, shown in Figure 1, is composed of two parts: a terahertz time-domain spectrometer Z-3 (Zomega Terahertz Corp., USA) and an ultrafast fiber laser FemtoFiber pro NIR (TOPTICA Photonics Inc., Germany). The ultrafast fiber laser generates a laser beam, with a pulse width of about 100 fs, the wavelength centered around 780 nm, the repetition rate of about 80 MHz, and the average power of nearly 100 mW. The laser beam is split into a pump beam and a probe beam by a cubic beam splitter (CBS) for THz generation and detection, respectively. The pump beam elicits the terahertz beam at the emitter, which is composed of a photoconductive antenna. Then, the generated THz beam is focused onto the sample, carries sample characteristics, and meets the probe beam at the ZnTe detector, where the probe beam is modulated by THz radiation through the electro-optic effect. After transmitting through a quater-wave-plate (QWP) and a Wollaston prism (WP), the modulated probe beam is then detected by a set of balanced photodiodes (PD).

In the experiment, we utilized the THz-TDS system in transmission mode. In order to reduce absorption of THz spectra by atmospheric water vapor, the THz beam path was enclosed in a box and dry air was injected into the box. During experiment, the relative humidity within the box was less than 4%, and the temperature was kept at room temperature (about 295 K).

2.2. Sample Preparation. Carbendazim powder with a purity of 98% was purchased from Adamas Company. High-density polyethylene powder was supplied by the Sigma-Aldrich company. Rice powder was obtained from a local market. To remove the water, all the powder-form samples were dried in a vacuum-drying oven for one hour at 323 K.

A series of mixtures with weight ratios ranging from 0% to 100% were prepared by mixing appropriate amounts of carbendazim in polyethylene and rice powder. And 25 replicates were prepared for each weight ratio. All the mixtures were mixed and ground by using a pestle and mortar, filtrated by a 200-mesh sieve, and then pressed into a 1–1.4 mm thick disk by a tablet press.

2.3. Data Acquisition. THz waveform was measured directly by THz-TDS system. Both reference and sample THz time-domain spectra were obtained from free path and sample measurement. To avoid echoing influence, the time-domain waveform was truncated firstly. And then, a fast Fourier transform was applied to obtain the spectral distribution in a frequency domain from time-domain THz waveform. After that, the absorbance spectrum could be acquired as follows [17]:

$$\text{Absorbance}(\omega) = -\log_{10}\left|\frac{E_s(\omega)}{E_{\text{ref}}(\omega)}\right|^2, \quad (1)$$

where $E_s(\omega)$ and $E_{\text{ref}}(\omega)$ were the amplitude of the sample and reference signal in a frequency domain, respectively.

In the experiment, we obtained the absorbance spectrum of pure carbendazim firstly. After that, the absorbance spectrums of mixtures, pure carbendazim, pure rice powder, and pure polyethylene, which were wrapped in paper, were gained.

2.4. Qualitative Modeling and Evaluation. With qualitative detection, SVM was employed. It is a supervised learning model with associated learning algorithms that analyze data used for classification. The SVM model maps the instances into a space, in which the instances of the categories are separated by a distinct gap that is as wide as possible. And then, new instances are mapped into the same space and predicted to belong to a category based on which side of the gap they fall [18, 19]. In addition, SVM can efficiently perform a nonlinear classification using the kernel trick.

Accuracy of cross validation (ACCCV) set was applied to optimize the model parameters. Accuracy of training (ACCTR) set was applied to evaluate the qualitative model. Accuracy of testing (ACCTE) set was used to evaluate generalization ability of the qualitative model. Accuracy is calculated using

$$\text{accuracy} = \frac{\text{correctly predicted data}}{\text{total testing data}} \times 100\%. \quad (2)$$

Generally, a good model should have higher ACCCV, ACCTR, and ACCTE.

2.5. Quantitative Modeling and Evaluation. PLS is a linear regression method, able to deal with collinear variables and accepting a large number of variables. It linearly projects both input data and output data into the subspace, to find a feature matrix that can best represent the input data with as much variance as possible while correlating to the output to the maximum extent [20, 21]. Thus, the feature matrix carries the most useful information of input data, which are correlated to output data.

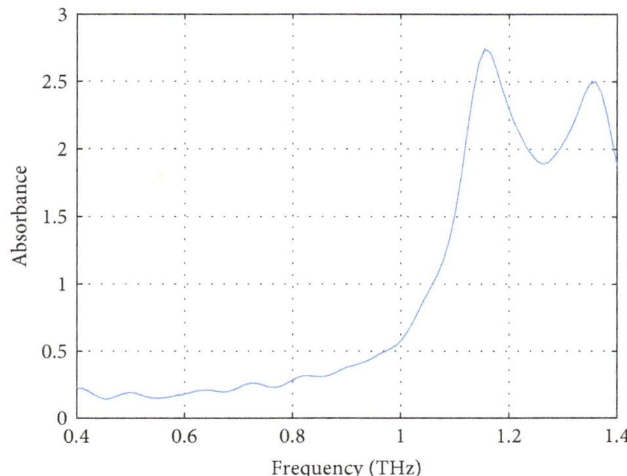

FIGURE 2: Average absorbance spectrum of pure carbendazim.

SVR is a nonlinear regression method, constructed by employing the ε-insensitive loss function, the penalty factor C, and the kernel function of radial basis function (RBF). The ε-insensitive loss function determines a deviation of $\pm\varepsilon$. Instances, those prediction errors that exceed the given deviation of $\pm\varepsilon$, would be penalized by the penalty factor C [22, 23]. In the experiment, ε was equal to 0.001. The kernel function is a nonlinear mapping function, which maps instances to a hyperdimensional feature space where the solution may become linear.

The root mean square error of cross validation (RMSECV) set was applied to optimize the model parameters. The root mean square error of training (RMSEC) set was used to evaluate the quantitative model. The root mean square error of testing (RMSEP) set and the correlation coefficient (R) were used to evaluate generalization ability of the quantitative model. RMSE and R are calculated using

$$\text{RMSE} = \sqrt{\frac{\sum_{i=1}^{N}\left(y_i^{\text{pre}} - y_i\right)^2}{N}},$$
$$R = \frac{\sum_{i=1}^{N}\left(y_i - \bar{y}\right)\left(y_i^{\text{pre}} - \bar{y}^{\text{pre}}\right)}{\sqrt{\sum_{i=1}^{N}\left(y_i - \bar{y}\right)^2 \sum_{i=1}^{N}\left(y_i^{\text{pre}} - \bar{y}^{\text{pre}}\right)^2}}, \quad (3)$$

where N is the sample size, y_i is actual value of the ith sample, y_i^{pre} is the predicted value of the ith sample, $\bar{y}$ is the average of the sample actual value, and $\bar{y}^{\text{pre}}$ is the average of the sample predicted value.

A good model had a great model precision and prediction precision with higher R and lower RMSECV, RMSEC, and RMSEP sets.

3. Results and Discussions

3.1. THz Spectra of Carbendazim Mixtures. Figure 2 shows the average absorbance spectrum of pure carbendazim. Carbendazim exhibited two relatively distinct absorption

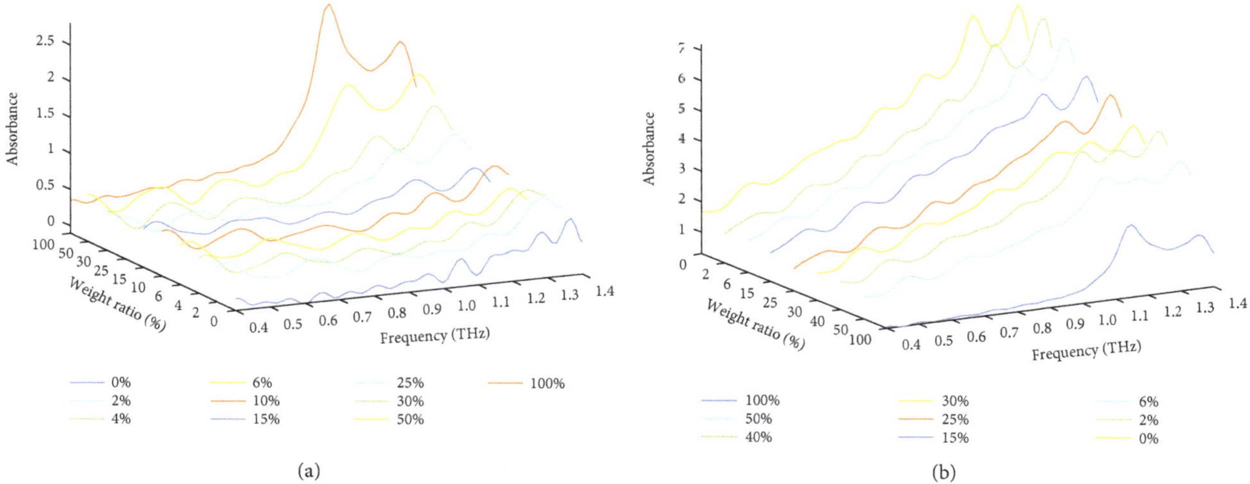

FIGURE 3: Average absorbance spectrum of mixtures wrapped in paper. (a) Carbendazim and polyethylene mixture. (b) Carbendazim and rice powder mixture.

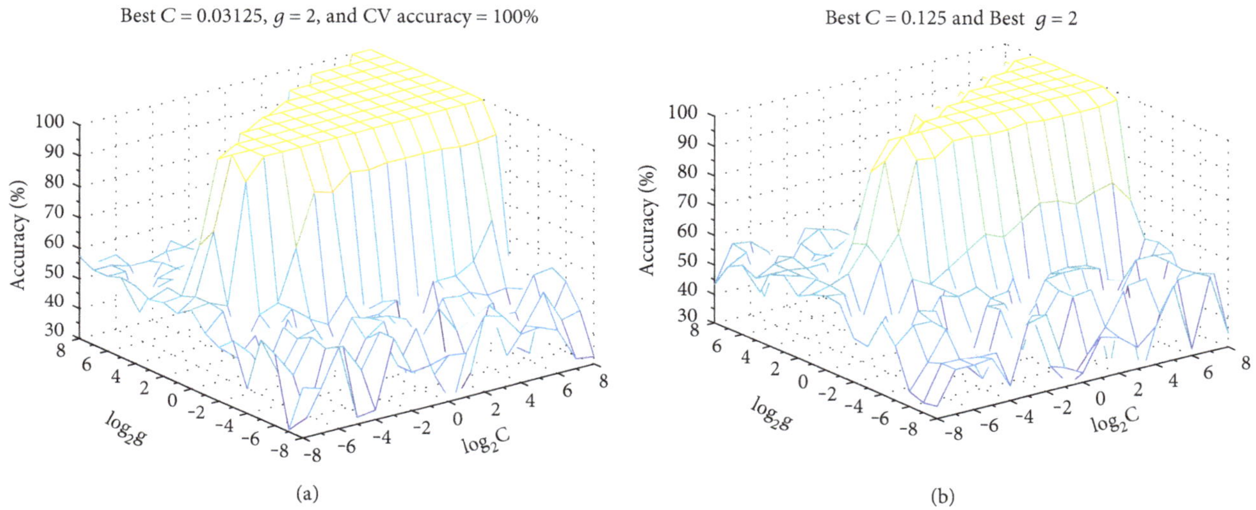

FIGURE 4: Grid search optimization parameter selection results of a 3D chart. (a) Carbendazim and polyethylene mixture. (b) Carbendazim and rice powder mixture.

TABLE 1: Performance of qualitative analyses.

Mixture type	ACCCV (%)	ACCTR (%)	ACCTE (%)
Carbendazim and polyethylene	100	100	100
Carbendazim and rice powder	100	100	100

TABLE 2: Samples in training set and testing set.

Mixture type	Samples in training set	Samples in testing set
Carbendazim and polyethylene	166	84
Carbendazim and rice powder	150	75

peaks at 1.15 and 1.35 THz. These absorption peaks can be regarded as the feature of carbendazim.

To detect carbendazim in mixtures, we mixed carbendazim with polyethylene and, later, rice powder. All the mixtures were wrapped in paper, when they were measured by THz-TDS. Figure 3(a) displays the absorption spectra of mixtures with different weight ratios from 0% to 100% *w*/*w*. The mixture 0% *w*/*w* is pure polyethylene, and the mixture 100% *w*/*w* is pure carbendazim. It shows that these absorbance spectra crossed together, and no obvious correlation between absorbance and weight ratio is observed, except for the region of 1.1–1.2 THz. In the region of 1.1–1.2 THz, the absorbance spectrum increases with the weight ratio increasing.

Figure 3(b) shows the absorption spectra of mixtures with different weight ratios from 0% to 100% *w*/*w*. The mixture 0% *w*/*w* is pure rice powder, and the mixture 100% *w*/*w* is pure carbendazim. The spectral features of mixture with weight ratios ranging from 0% to 25% *w*/*w* are dominated

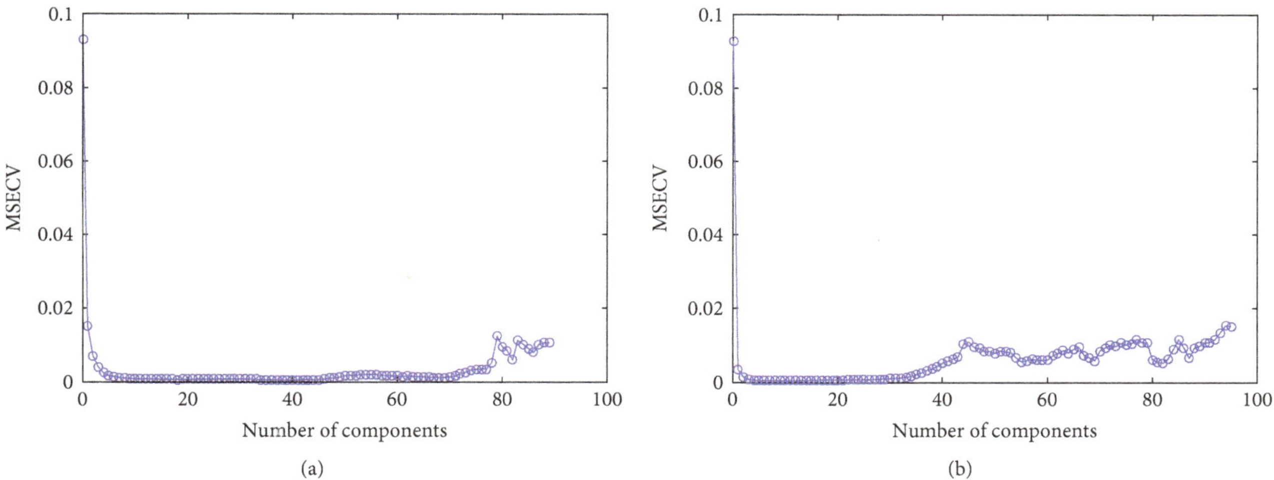

FIGURE 5: The correlation between the number of components and RMSECV for PLS. (a) Carbendazim and polyethylene mixture. (b) Carbendazim and rice powder mixture.

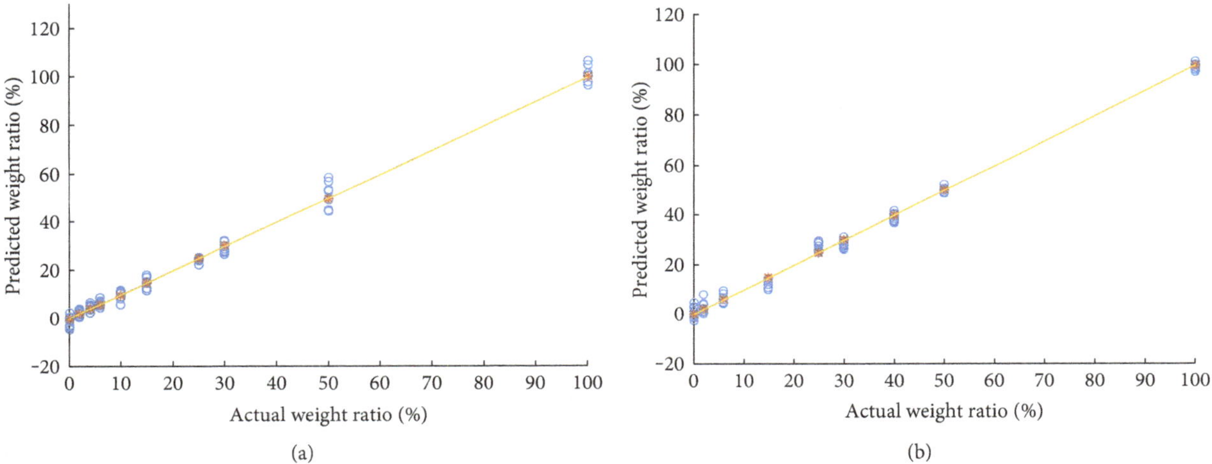

FIGURE 6: Performance of PLS. (a) Carbendazim and polyethylene mixture. (b) Carbendazim and rice powder mixture.

by the strong absorbance of rice powder, and no obvious absorption peak at 1.15 THz is observed. However, as the weight ratio is increasing to 30% *w*/*w*, the peak at 1.15 THz appears. In the region of 1.2–1.4 THz, the absorbance spectrum increases with the weight ratio decreasing. Furthermore, an additional absorption peak at 1.23 THz appears. It is caused by rice powder. Further comparing Figures 2, 3(a), and 3(b), whether wrapped in paper or not, the absorption spectra of pure carbendazim are the same. It indicates that THz could penetrate packaging material.

3.2. Qualitative Detection of Carbendazim Mixtures. In Figures 3(a) and 3(b), it illustrates that carbendazim can be easily detected in high weight ratio mixture by absorbance peaks. While as the weight ratio is low, the absorbance of the mixture is similar with pure polyethylene and rice powder. For example, the absorbance of mixture 2% *w*/*w* resembles polyethylene in Figure 3(a). The situation is similar in Figure 3(b). It is hard to detect carbendazim in mixtures when the weight ratio is low. To solve this problem, SVM with radial basis function was utilized. As mentioned above, the absorbance of the mixture was used as input of SVM. Grid search technique was selected to find optimal model parameter values, and 10-fold crossvalidation was implemented to prevent overfitting.

To qualitatively detect carbendazim in low weight ratio mixture (mix carbendazim with polyethylene), we selected 50 samples (25 pure polyethylene and 25 mixture 2% *w*/*w*). These samples were divided into a training set and a testing set. In the training set, the number of pure polyethylene and mixture 2% *w*/*w* were 18 and 15, respectively. At the same time, in the testing set, the number of pure polyethylene and mixture 2% *w*/*w* were 7 and 10, respectively. Penalty factor C and kernel parameter g are of great effect on SVM after select RBF. The grid search is used to optimize the two parameters C and g based on synchronous optimization idea. The searching interval was $C \in [2^{-8}, 2^{8}]$ and $g \in [2^{-8}, 2^{8}]$. Figure 4(a) displays the process of SVM parameter

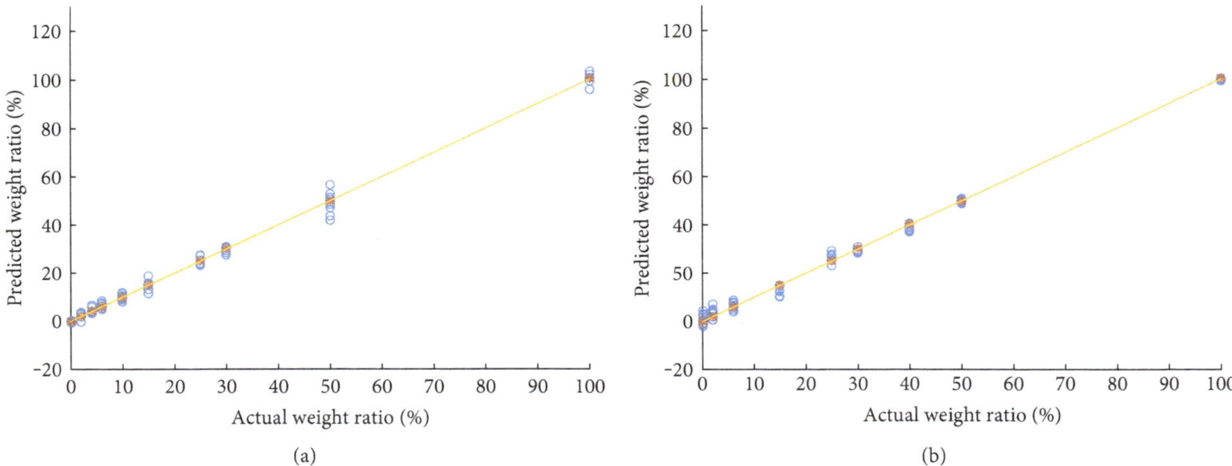

FIGURE 7: Performance of SVR model. (a) Carbendazim and polyethylene mixture. (b) Carbendazim and rice powder mixture.

TABLE 3: Performance of PLS and SVR in carbendazim and polyethylene mixture.

Model	Cross validation set	Training set	Testing set	
	RMSECV	RMSEC	RMSEP	R
PLS	0.0271	0.0172	0.0255	0.9957
SVR	**0.0218**	**0.0087**	**0.02**	**0.9972**

TABLE 4: Performance of PLS and SVR in carbendazim and rice powder mixture.

Model	Cross validation set	Training set	Testing set	
	RMSECV	RMSEC	RMSEP	R
PLS	0.0241	0.0189	0.023	0.9967
SVR	**0.0224**	**0.013**	**0.0188**	**0.9978**

optimization. It shows that accuracy depends on penalty factor C and kernel parameter g. When the C is large, a higher possibility of high accuracy is achieved. However, it should be noticed that if C is too large, it will be likely to be overfitting. So, we make a tradeoff between C and accuracy. At last, C was equal to 0.03125 and g was equal to 2.

Using the same method, we identify mixture 2% *w/w* and rice powder. The searching interval was $C \in [2^{-8}, 2^{8}]$ and $g \in [2^{-8}, 2^{8}]$. The process of SVM parameter optimization is illustrated in Figure 4(b). As the result, C was equal to 0.125 and g was equal to 2. Table 1 shows the performance of detecting carbendazim in low weight ratio mixture. It indicates that carbendazim in low weight ratio mixture can be qualitatively detected successfully with the help of SVM.

3.3. Quantitative Detection of Carbendazim Mixtures. Two kinds of mixtures were, respectively, divided into training set and testing set, and the detail was given in Table 2. PLS and SVR were used for quantificationally detecting carbendazim in mixtures. The absorbance of the mixture was used as input of PLS and SVR. In PLS, 10-fold cross-validation was implemented to prevent overfitting and select the number of components. In SVR, RBF was utilized, 10-fold crossvalidation was applied to prevent overfitting, and grid search technique was selected to find the optimal model parameter values.

Figures 5(a) and 5(b) display the correlation between the number of components and RMSECV for PLS. It turned out that MSECV decreased at the beginning and then became worse with increasing the number of components, which are due to introduced noise. In these two kinds of mixtures, the minimum value of MSECV was 7.4×10^{-4} and 5.87×10^{-4}, and the corresponding components were 41 and 11, accounting for 46.1% and 11.6% of the full spectral data of 89 and 95 wavelengths. As the result, the number of components was selected 41 and 11, respectively.

Figures 6(a) and 6(b) show the scattered plot of the predicted weight ratio in PLS against the actual values. Figures 7(a) and 7(b) show the scattered plot of the predicted weight ratio in SVR against the actual values. The reference line represents zero residuals between the predicted and actual values. The closer the scattered points go to the reference line, the better the performance will be.

Comparing Figures 6 and 7, all the predicted points of SVR are closer to the reference line than PLS. It intuitively depicts that SVR is preferable to PLS in the experiment. Tables 3 and 4 display the performance of PLS and SVR. The result on mixtures constituted carbendazim and rice powder shows better performance than the corresponding result on mixtures constituted carbendazim and polyethylene. Comparing the two results, SVR proves to be better than PLS, with relatively larger R and lower RMSEC, RMSECV, and RMSEP sets.

4. Conclusions

The applicability of THz-TDS to detect carbendazim in the mixture was present in this study. Unlike previous reports of THz-TDS for detecting pesticide, this work focused on detecting pesticide residue in packaged foods.

All the mixtures were wrapped in paper, when they were measured. Carbendazim and rice powder revealed their own distinct absorption peaks, while polyethylene had no distinct absorption peaks. Pure carbendazim, polyethylene, and rice powder could be directly identified by their absorption spectra. But, when the weight ratio was low, the absorbance of the mixture is similar with pure polyethylene and rice powder. With the help of SVM, carbendazim could be qualitatively detected in low weight ratio mixture successfully. Moreover, carbendazim was quantificationally detected with PLS and SVR. The results showed that the performance of SVR was better than that of PLS. The best result of SVR is that R, RMSECV set, RMSEC set, and RMSEP set were 0.9978, 2.24%, 1.3%, and 1.8%, respectively.

Despite the limited sensitivity of the current technique, the current work demonstrates its ability to qualitatively and quantitatively detect pesticide residue in packaged foods. Further study will be undertaken to improve the precision and robustness of the prediction model. At last, we expect that THz-TDS become a rapid detection tool for commercial applications in food quality control.

Acknowledgments

This work was supported by the Yulin Normal University Research Grant (Grant no. 2016YJKY06), National Natural Science Foundation of China (Grant no. 21565028), and Guangxi Colleges and Universities Program of Innovative Research Team and Outstanding Talent.

References

[1] S. H. Baek, J. H. Kang, Y. H. Hwang, K. M. Ok, K. Kwak, and H. S. Chun, "Detection of methomyl, a carbamate insecticide, in food matrices using terahertz time-domain spectroscopy," *Journal of Infrared Millimeter and Terahertz Waves*, vol. 37, pp. 486–497, 2016.

[2] Y. Hua and H. Zhang, "Qualitative and quantitative detection of pesticides with terahertz time-domain spectroscopy," *IEEE Transactions on Microwave Theory and Techniques*, vol. 58, pp. 2064–2070, 2010.

[3] Z. Chen, Z. Zhang, R. Zhu, Y. Xiang, Y. Yang, and P. B. Harrington, "Application of terahertz time-domain spectroscopy combined with chemometrics to quantitative analysis of imidacloprid in rice samples," *Journal of Quantitative Spectroscopy & Radiative Transfer*, vol. 167, pp. 1–9, 2015.

[4] C. Chafer-Pericas, A. Maquieira, and R. Puchades, "Fast screening methods to detect antibiotic residues in food samples," *Tractrends in Analytical Chemistry*, vol. 29, pp. 1038–1049, 2010.

[5] R. Martinez, E. Gonzalo, M. Moran, and J. Mendez, "Sensitive method for the determination of organophosphorus pesticides in fruits and surface waters by high-performance liquid chromatography with ultraviolet detection," *Journal of Chromatography*, vol. 607, pp. 37–45, 1992.

[6] M. Khanmohammadi, S. Armenta, S. Garrigues, and M. de la Guardia, "Mid- and near-infrared determination of metribuzin in agrochemicals," *Vibrational Spectroscopy*, vol. 46, pp. 82–88, 2008.

[7] S. Armenta, S. Garrigues, and M. de la Guardia, "Determination of iprodione in agrochemicals by infrared and Raman spectrometry," *Analytical and Bioanalytical Chemistry*, vol. 387, pp. 2887–2894, 2007.

[8] B. Ferguson and X. Zhang, "Materials for terahertz science and technology," *Nature Materials*, vol. 1, pp. 26–33, 2002.

[9] P. U. Jepsen, D. G. Cooke, and M. Koch, "Terahertz spectroscopy and imaging - modern techniques and applications," *Laser & Photonics Reviews*, vol. 5, pp. 124–166, 2011.

[10] F. Rong, L. Zhe, J. Biao-bing et al., "A study of vibrational spectra of L-, D-, DL-alanine in terahertz domain," *Spectroscopy and Spectral Analysis*, vol. 30, pp. 2023–2026, 2010.

[11] M. D. King, W. D. Buchanan, and T. M. Korter, "Identification and quantification of polymorphism in the pharmaceutical compound diclofenac acid by terahertz spectroscopy and solid-state density functional theory," *Analytical Chemistry*, vol. 83, pp. 3786–3792, 2011.

[12] D. G. Allis and T. M. Korter, "Theoretical analysis of the terahertz spectrum of the high explosive PETN," *Chemphyschem*, vol. 7, pp. 2398–2408, 2006.

[13] M. T. Ruggiero and J. A. Zeitler, "Resolving the origins of crystalline anharmonicity using terahertz time-domain spectroscopy and ab initio simulations," *Journal of Physical Chemistry B*, vol. 120, pp. 11733–11739, 2016.

[14] C. Bing-Hua, M. Guang-Xin, and Z. Ze-Kui, "Terahertz time-domain spectroscopy of dimethoate," *Chinese Journal of Analytical Chemistry*, vol. 36, pp. 623–626, 2008.

[15] Y. Hua, H. Zhang, and H. Zhou, "Quantitative determination of cyfluthrin in N-hexane by terahertz time-domain spectroscopy with chemometrics methods," *IEEE Transactions on Instrumentation and Measurement*, vol. 59, pp. 1414–1423, 2010.

[16] W. Qiang and M. Ye-hao, "Qualitative and quantitative identification of nitrofen in terahertz region," *Chemometrics and Intelligent Laboratory Systems*, vol. 127, pp. 43–48, 2013.

[17] J. B. Sleiman, B. Bousquet, N. Palka, and P. Mounaix, "Quantitative analysis of hexahydro-1,3,5-trinitro-1,3,5, triazine/pentaerythritol tetranitrate (RDX -PETN) mixtures by terahertz time domain spectroscopy," *Applied Spectroscopy*, vol. 69, pp. 1464–1471, 2015.

[18] C. Burges, "A tutorial on support vector machines for pattern recognition," *Data Mining and Knowledge Discovery*, vol. 2, pp. 121–167, 1998.

[19] B. Scholkopf, K. Sung, C. Burges et al., "Comparing support vector machines with Gaussian kernels to radial basis function classifiers," *IEEE Transactions on Signal Processing*, vol. 45, pp. 2758–2765, 1997.

[20] S. Wold, M. Sjostrom, and L. Eriksson, "PLS-regression: a basic tool of chemometrics," *Chemometrics and Intelligent Laboratory Systems*, vol. 58, pp. 109–130, 2001, International Symposium on Partial Least Squares (PLS 99), Jouy En Josas, France, October, 1999.

[21] N. Sorol, E. Arancibia, S. A. Bortolato, and A. C. Olivieri, "Visible/near infrared-partial least-squares analysis of Brix in sugar cane juice: a test field for variable selection methods," *Chemometrics and Intelligent Laboratory Systems*, vol. 102, pp. 100–109, 2010.

[22] H. Drucker, C. Burges, L. Kaufman, A. Smola, and V. Vapnik, "Support vector regression machines," in *Advances in Neural Information Processing Systems 9: Proceedings of the 1996 Conference*, vol. 9 of *Advances in Neural Information Processing Systems, NIPS Fdn, 1997*, pp. 155–161, Denver, CO, December, 1996, 10th Annual Conference on Neural Information Processing Systems (NIPS).

[23] V. Vapnik, S. Golowich, and A. Smola, "Support vector method for function approximation, regression estimation, and signal processing," in *Advances in Neural Information Processing Systems 9: Proceedings of the 1996 Conference*, vol. 9 of *Advances in Neural Information Processing Systems, NIPS Fdn, 1997*, pp. 281–287, Denver, CO, December, 1996, 10th Annual Conference on Neural Information Processing Systems (NIPS).

4

Characterization of an Atmospheric-Pressure Argon Plasma Generated by 915MHz Microwaves Using Optical Emission Spectroscopy

Robert Miotk,[1] Bartosz Hrycak,[1] Mariusz Jasiński,[1] and Jerzy Mizeraczyk[2]

[1] *The Institute of Fluid Flow Machinery, Polish Academy of Sciences, Fiszera 14, 80-231 Gdansk, Poland*
[2] *Department of Marine Electronics, Gdynia Maritime University, Morska 81-87, 81-225 Gdynia, Poland*

Correspondence should be addressed to Robert Miotk; rmiotk@imp.gda.pl

Academic Editor: Nikša Krstulović

The paper presents the investigations of an atmospheric-pressure argon plasma generated at 915 MHz microwaves using the optical emission spectroscopy (OES). The 915 MHz microwave plasma was inducted and sustained in a waveguide-supplied coaxial-line-based nozzleless microwave plasma source. The aim of presented investigations was to estimate parameters of the generated plasma, that is, excitation temperature of electrons T_{exc}, temperature of plasma gas T_g, and concentration of electrons n_e. Assuming that excited levels of argon atoms are in local thermodynamic equilibrium, Boltzmann method allowed in determining the T_{exc} temperature in the range of 8100–11000 K. The temperature of plasma gas T_g was estimated by comparing the simulated spectra of the OH radical to the measured one in LIFBASE program. The obtained T_g temperature ranged in 1200–2800 K. Using a method based on Stark broadening of the H_β line, the concentration of electrons n_e was determined in the range from 1.4×10^{15} to 1.7×10^{15} cm^{-3}, depending on the power absorbed by the microwave plasma.

1. Introduction

The atmospheric-pressure microwave plasma sources (MPSs) found many different physical and technical applications such as decomposition of gaseous pollutants [1–4], deposition thin layers in nanosensors [5, 6], medicine for bacteria inactivation [7], and production of hydrogen via conversion of hydrocarbons or other hydrogen carriers [8–12].

Since in process of the gas treatment by the plasma, the temperature of plasma gas and concentration of electrons play an important role; therefore, the knowledge of these basic parameters is crucial for understanding the chemical kinetics and its optimization.

In this work, an optical emission spectroscopy (OES) [13–17] method has been used for diagnosing the microwave argon plasma. The plasma was induced by microwaves at a frequency of 915 MHz in waveguide-supplied coaxial-line-based nozzleless MPS [1]. The presented device allows the generation of so-called cold plasma which is classified as a partial local thermodynamic equilibrium (PLTE) plasma [13, 14, 18–20].

2. Experiment

In Figure 1(a), a photo of the MPS is shown, whereas a draft of an experimental setup is presented in Figure 1(b). The MPS is supplied via a standard waveguide WR 975 and ended by a movable plunger which allows for an effective transfer of the microwave power from the electric field to the plasma.

Inside the used MPS, a quartz tube is placed. In the tube, two gas flows are formed. The first one is the axial flow (processed gas), and the second one is the swirl flow (cooling gas). In the axial flow, the processed gas is introduced into the quartz tube through the inner electrode. In the swirl flow, the cooling gas is delivered into the MPS by four inlets located tangentially to the quartz tube wall [1]. This resulted in a vortex (swirl) flow inside the quartz tube. In this type of the MPS, plasma in a form of flame occurs inside the quartz

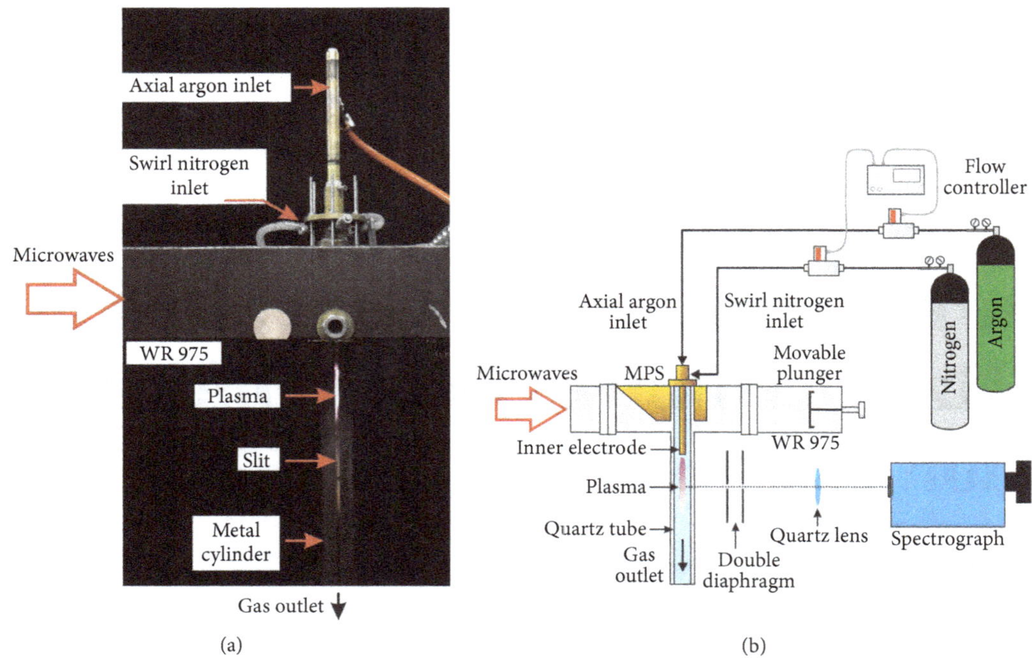

FIGURE 1: (a) Photo of the microwave plasma source and (b) the experimental setup used for the spectroscopic investigations.

tube at the tip of the inner electrode. The additional swirl flow stabilized the discharge in the center of the quartz tube and protects the quartz wall from overheating [1]. Below the MPS waveguide, the quartz tube is surrounded by a metal cylinder with a vertical slit for the observation of the generated discharge.

Symmetrical double convex quartz lens (50 mm in diameter, focal length 75 mm) was used to focus light emitted by the microwave plasma. Additionally to collimate the emitted light, two diaphragms with pinholes of 1 mm in diameter were placed. In these investigations, the spectrum of the microwave plasma was measured by McPherson model 209 spectrometer, equipped with double-pass scanning monochromator. The used spectrograph is equipped with sensitivity-calibrated iCCD camera and diffraction grating 1200 grooves/mm.

Using Hg I lines $\lambda = 365.02$ nm, 435.84 nm, and 546.07 nm emitted from low-pressure calibration Hg-Ne lamp, the instrumental line broadening $\Delta\lambda_I$ has been determined. The obtained values of $\Delta\lambda_I$ was about 0.07 nm.

3. Results

During the experiment, the nitrogen was used as a cooling gas with constant flow rate of 50 NL/min. The investigations were performed with argon as processed gas with flow rate equal to 50 NL/min. The power absorbed by the microwave plasma P_A was calculated as a difference between power incident P_I and power reflected P_R in the MPS [1]. The P_I and P_R were directly measured using a directional coupler. In these investigations, the power P_A was changed from 2 to 4 kW.

The spectra were recorded 15 mm below the tip of the inner electrode. In our measurements, we focused on the range of emission spectra from 300–600 nm. An example of the recorded spectra is shown in Figure 2. To detect emission spectral lines of the H or the OH radicals, a small addition of H_2O vapour (H_2O—0.1 kg/h, H_2O vapour temperature was equal to 400°C) was added to the process gas flow.

In performed investigations, we assumed that microwave plasma at atmospheric pressure is generally in partial local thermodynamic equilibrium [13, 14, 18–20]. This assumption allowed us to use the Boltzmann plot method to determine the T_{exc} [14, 15]. Five transition lines of argon (see Figure 2) were selected to determine the T_{exc}. Selected argon lines with the parameters for the Boltzmann plot method were presented in Table 1. An example of Boltzmann plot is shown in Figure 3. Conformity of a straight line with experimental points indicates balance in the excited states of argon atoms. The obtained T_{exc} was in the range of 8100–11000 K, as shown in Figure 4. The estimated T_{exc} temperature increased with increasing the absorbed microwave power P_A.

It is widely accepted that in the microwave discharges, the rotational temperature of the OH radical T_{rot} corresponds to the translational temperature of heavy particles in the plasma (temperature of plasma gas T_g) [16, 17]. To obtain the molecule rotational temperature, the OH band $A^2\Sigma^+ \rightarrow X^2\Pi$ was used. This band is very sensitive against the changes of the rotational temperature [16]. After measuring the OH spectrum, we simulated this band in the LIFBASE program [21]. This program allows calculating the emitted spectrum of plasma radiation of various gases at individually given rotational and vibrational temperatures. In this program, the simulated OH band was fitted to the experimental one (Figure 5). A good agreement has been found. Spectrum simulations were performed for Gaussian line shapes with a FWHM value equal to

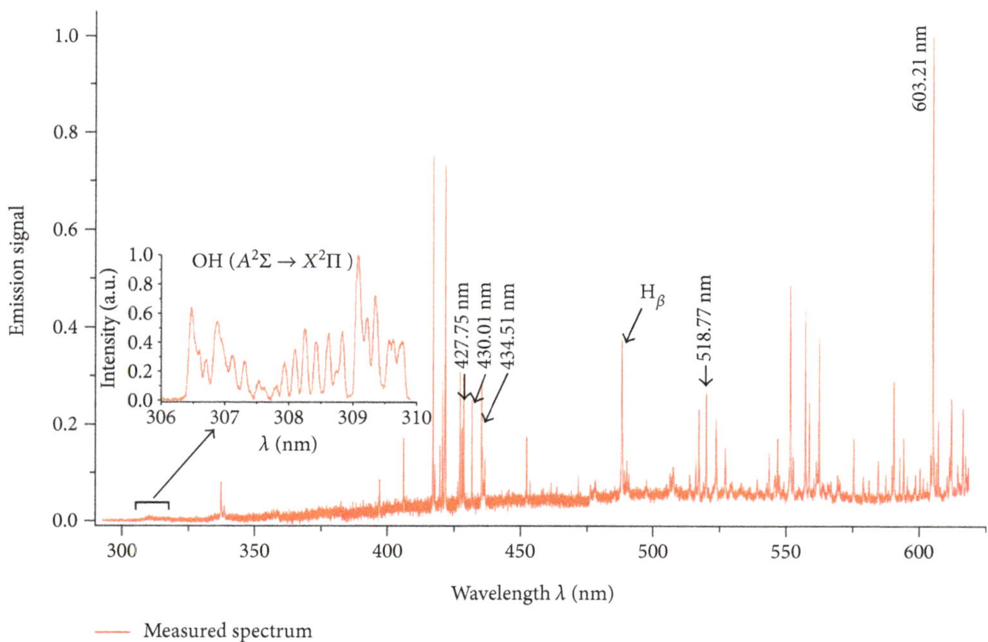

FIGURE 2: Measured emission spectra of argon plasma (with a small amount of H_2O vapour) with selected argon $4p$–$4s$ and $5p$–$4s$ transition lines for Boltzmann plot method. Absorbed microwave power $P_A = 3$ kW and argon flow rate $Q = 50$ l/min.

TABLE 1: Parameters of selected argon emission lines used to determine the excitation temperature of the electrons T_{exc}. n/m: energy levels upper/lower, respectively; λ_{nm}: wavelength of transition $n \rightarrow m$, A_{nm}: Einstein coefficient for transition $n \rightarrow m$, g_n: statistical weight of the upper level n. E_n is the energy of the upper level n.

l.p.	λ_{nm} (nm)	Transition	A_{nm} (10^7 s^{-1})	g_n	E_n (cm^{-1})
1	427.75		0.0797	3	117,151
2	430.01	$5p \rightarrow 4s$	0.0377	5	116,999
3	434.51		0.0297	5	118,407
4	518.77	$5d \rightarrow 4p$	0.1380	5	123,372
5	603.21		0.1400	9	122,036

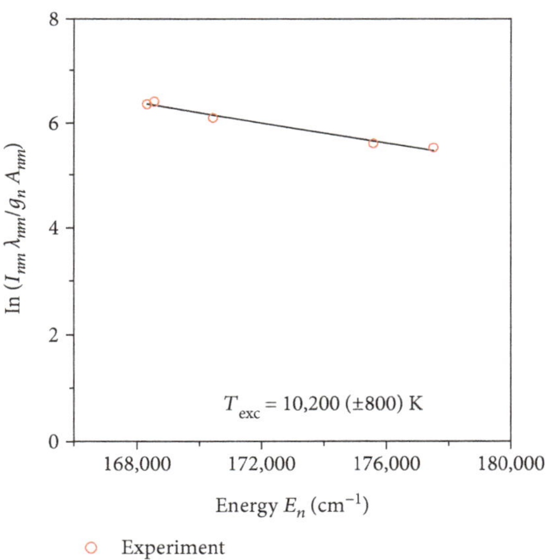

FIGURE 3: Example of Boltzmann plot for determination of the T_{exc}. I_{nm}—intensity of the recorded emission line from transition $n \rightarrow m$, microwave power absorbed $P_A = 3$ kW, and argon flow rate $Q = 50$ l/min.

0.07 nm. The obtained gas temperature T_g ranged from 1200 to 2800 K (Figure 6).

By using the method based on Stark broadening of the hydrogen H_β line, the concentrations of electrons n_e in the plasma were determined [13–15, 17]. The introduction of water vapour caused the emergence of emission lines of hydrogen H_β, H_δ, and H_γ. In our work, we focus only on the H_β (486.13 nm) line. The H_δ and H_γ lines were hardly noticeable or partially overlapped by the argon lines. Therefore, these two lines were not used to determine the concentration of electrons n_e.

The shape of the recorded H_β line is affected by several different mechanisms of broadening (instrumental $\Delta\lambda_I$, Van der Waals $\Delta\lambda_W$, Stark $\Delta\lambda_S$, resonance $\Delta\lambda_R$, Doppler $\Delta\lambda_D$, and natural $\Delta\lambda_N$) which result to a Voigt profile [13–15]. In order to obtain the FWHM of H_β line in investigations to the measured profile, the Voigt function was fitted. The fitting was performed using the Origin software [22].

The Doppler broadening $\Delta\lambda_D$ is a result of atoms' random motions in the plasma. This effect can be calculated from [15, 23]

$$\Delta\lambda_D = 7.17 \times 10^{-7}\lambda_0\sqrt{\frac{T}{M}}, \quad (1)$$

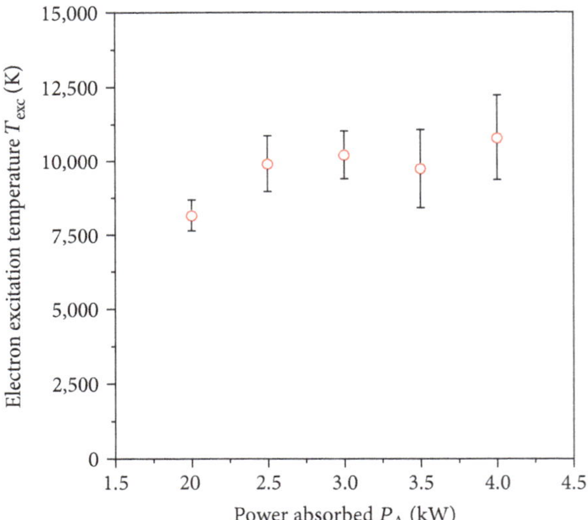

FIGURE 4: Electron excitation temperature as a function of absorbed microwave power P_A.

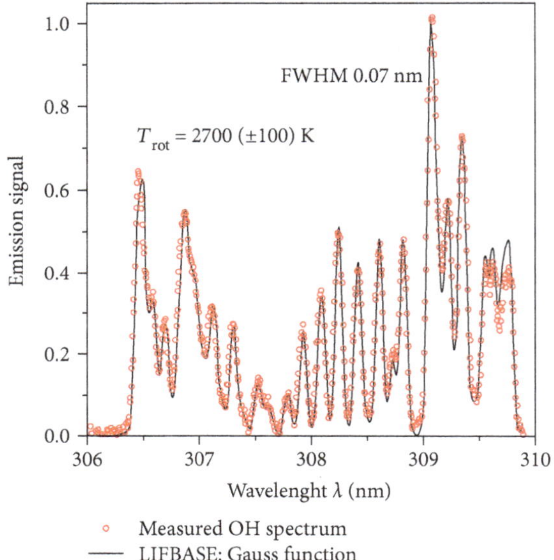

FIGURE 5: Comparison of the measured and simulated in LIFBASE emission spectra of OH band. Absorbed microwave power $P_A = 3$ kW and argon flow rate $Q = 50$ l/min.

where λ_0 is the wavelength, T is the temperature of the emitter in Kelvins, M is the mass of the emitter in a.m.u. In this work, T was assumed equal to the temperature of plasma gas T_g. The Van der Waals broadening $\Delta\lambda_W$ is the effects of dipolar interaction between excited the atoms and the neutral ground state atom [14]. The $\Delta\lambda_W$ broadening can be estimated from [23]

$$\Delta\lambda_W = 6.48 \times 10^{-22} \frac{p}{kT_g^{0.7}}, \quad (2)$$

where p is the pressure and k is the Boltzmann constant. Determination of plasma gas temperature T_g allows to

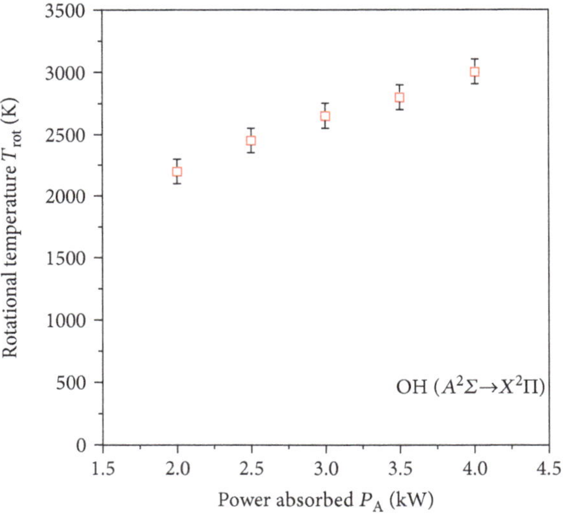

FIGURE 6: Measured rotational temperatures of the OH radical as a function of absorbed microwave power P_A.

estimate values of Van der Waals and Doppler broadening effect. Using the above formulas, the values of $\Delta\lambda_D$ and $\Delta\lambda_W$ broadening of the H_β line were calculated. The obtained value of $\Delta\lambda_D$ was equal to 0.003 nm while $\Delta\lambda_W = 0.02$ nm, respectively. In the tested range of the absorbed microwave power P_A, determined values were constant. In the presented work, resonance and natural broadening have been omitted due to low FWHM values in comparison to the other effects [13–15].

Taking into account the estimated values of $\Delta\lambda_I$, $\Delta\lambda_D$, and $\Delta\lambda_W$ and obtained value FWHM of H_β line, the Stark broadening $\Delta\lambda_S$ was calculated [15, 23]. In the experiment, we observed a linear relationship between the estimated value of Stark broadening $\Delta\lambda_S$ of H_β line and the absorbed microwave power P_A by the plasma. In calculation of the n_e, a Gig-Card theory [24] was used. The measured concentration of electrons n_e ranged from 1.4×10^{15} to 1.7×10^{15} cm^{-3} (Figure 7). The values of the electron concentration indicate that the balance between electrons and heavy particles as a result of collisions cannot be achieved. Thus, the plasma cannot be described by a single temperature.

Adopting a classical ideal gas model and using the measured concentration of electrons, we estimated that the ionization degree in plasma was about ~10^{-4}. This indicates that ionization degree is too low to thermalize the electron energy distribution function. Therefore, there may be a lack of balance between the basic state and the excited states of argon atoms in plasma. This cause that the measured temperatures could be overestimated. In measurements, we record the radiation from the excited states, which are the result of collisions, while the basic states remain neutral.

4. Conclusions

The investigations of an atmospheric-pressure argon plasma generated at 915 MHz microwaves using optical emission

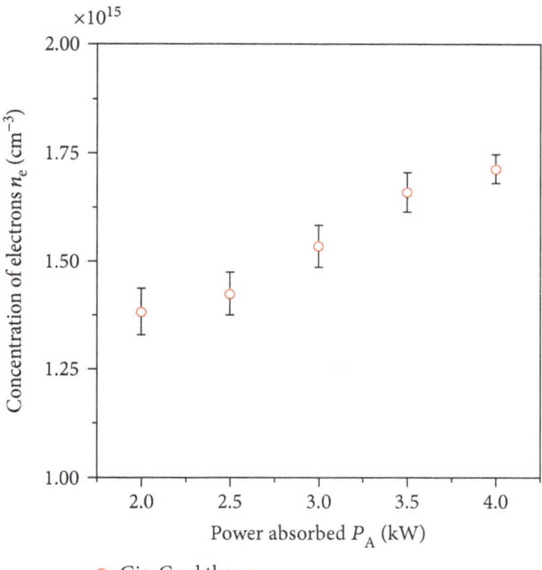

FIGURE 7: The concentration of electrons as a function of absorbed microwave power P_A.

spectroscopy (OES) are presented in this work. These investigations yielded the excitation temperature of electrons T_{exc}, the gas temperature T_g, and the concentration of electrons n_e in the generated argon plasma. In the tested range of the absorbed microwave power P_A by the plasma, we observed an increase in the excitation temperature T_{exc}, the gas temperature T_g, and the concentration of electrons n_e. These results indicate that appropriate selection of the gases and the operating parameters of the MPS (central and the additional flow rate, absorbed microwave power) enables in obtaining the plasma with desired parameters. It should also be mentioned that the investigated MPS works very stable with various processing gases (argon, nitrogen, air, and carbon dioxide) at high flow rates and absorbed microwave power by the plasma can be changed in a wide range. Thus, the above properties make the presented MPS an attractive tool for different gas processing at high flow rates.

Acknowledgments

The authors are grateful to the National Science Centre (Program no. 2013/11/N/ST8/00802) and the Foundation for Polish Science (FNP Program START no. 53.2017) for the financial support of this work.

References

[1] J. Mizeraczyk, M. Jasiński, H. Nowakowska, and M. Dors, "Studies of atmospheric -pressure microwave plasmas used for gas processing," *Nukleonika*, vol. 57, no. 2, pp. 241–247, 2012.

[2] Y. C. Hong, H. S. Uhm, M. J. Kim, H. S. Han, S. C. Ko, and S. K. Park, "Decomposition of phosgene by microwave plasma-torch generated at atmospheric pressure," *IEEE Transactions on Plasma Science*, vol. 33, no. 2, pp. 958–963, 2005.

[3] H. S. Uhm, Y. C. Hong, and D. H. Shin, "A microwave plasma torch and its applications," *Plasma Sources Science and Technology*, vol. 15, pp. 26–34, 2006.

[4] Y. Ko, G. Yang, D. P. Y. Chang, and I. M. Kennedy, "Microwave plasma conversion of volatile organic compounds," *Journal of the Air & Waste Management Association*, vol. 53, no. 5, pp. 580–585, 2003.

[5] X. Landreau, B. Lanfant, T. Merle, C. Dublanche-Tixier, and P. Tristant, "A thorough FT-IR spectroscopy study on micrometric silicon oxide films deposited by atmospheric pressure microwave plasma torch," *The European Physical Journal D*, vol. 66, no. 160, p. 8, 2012.

[6] X. Landreau, B. Lanfant, T. Merle, E. Laborde, C. Dublanche-Tixier, and P. Tristant, "Ordering of $SiO_xH_yC_z$ islands deposited by atmospheric pressure microwave plasma torch on Si(100) substrates patterned by nanoindentation," *The European Physical Journal D*, vol. 65, pp. 421–428, 2011.

[7] J. Mizeraczyk, M. Dors, M. Jasiński, B. Hrycak, and D. Czylkowski, "Atmospheric pressure low-power microwave microplasma source for deactivation of microorganisms," *The European Physical Journal Applied Physics*, vol. 61, article 24309, 2013.

[8] J. Mizeraczyk and M. Jasiński, "Plasma processing methods for hydrogen production," *The European Physical Journal Applied Physics*, vol. 75, article 24702, 2016.

[9] Y. F. Wang, H. Tsai Ch, W. Y. Chang, and Y. M. Kuo, "Methane steam reforming for producing hydrogen in an atmospheric-pressure microwave plasma reactor," *International Journal of Hydrogen Energy*, vol. 35, pp. 135–140, 2010.

[10] M. Jasinski, D. Czylkowski, B. Hrycak, M. Dors, and J. Mizeraczyk, "Atmospheric pressure microwave plasma source for hydrogen production," *International Journal of Hydrogen Energy*, vol. 38, no. 26, pp. 11473–11483, 2013.

[11] R. Rincon, M. Jimenez, J. Munoz, M. Saez, and M. D. Calzada, "Hydrogen production from ethanol decomposition by two microwave atmospheric pressure plasma sources: surfatron and TIAGO torch," *Plasma Chemistry and Plasma Processing*, vol. 34, pp. 145–157, 2014.

[12] D. Czylkowski, B. Hrycak, R. Miotk, M. Jasinski, M. Dors, and J. Mizeraczyk, "Hydrogen production by conversion of ethanol using atmospheric pressure microwave plasmas," *International Journal of Hydrogen Energy*, vol. 40, pp. 14039–14044, 2015.

[13] B. N. Sismanoglu, K. G. Grigorov, R. A. Santos et al., "Spectroscopic diagnostics and electric field measurements in the near-cathode region of an atmospheric pressure microplasma jet," *The European Physical Journal D*, vol. 60, pp. 479–487, 2010.

[14] R. Miotk, B. Hrycak, M. Jasinski, and J. Mizeraczyk, "Spectroscopic study of atmospheric pressure 915 MHz microwave plasma at high argon flow rate," *Journal of Physics: Conference Series*, vol. 406, article 012033, 2012.

[15] B. N. Sismanoglu, K. G. Grigorov, R. Caetano, M. V. O. Rezende, and Y. D. Hoyer, "Spectroscopic measurements and electrical diagnostics of microhollow cathode discharges in argon flow at atmospheric pressure," *The European Physical Journal D*, vol. 60, pp. 505–516, 2010.

[16] C. Izarra, "UV OH spectrum used as a molecular pyrometer," *Journal of Physics D: Applied Physics*, vol. 33, no. 14, pp. 1697–1704, 2000.

[17] B. Hrycak, M. Jasinski, and J. Mizeraczyk, "Spectroscopic investigations of microwave microplasmas in various gases at atmospheric pressure," *The European Physical Journal D*, vol. 60, pp. 609–619, 2010.

[18] M. Capitelli, G. Colonna, and A. D'Angola, *Fundamental Aspects of Plasma Chemical Physics*, vol. 66 of Springer series on Atomic, Optical and Plasma Physics: Thermodynamics, Springer, New York, NY, USA, 2012, chapter 9.

[19] M. Capitelli, R. Celiberto, G. Colonna et al., *Fundamental Aspects of Plasma Chemical Physics*, vol. 85 of Springer series on atomic, Optical and Plasma Physics: Kinetis, Springer, New York, NY, USA, 2016, chapter 5.

[20] H. R. Griem, "Principles of plasma spectroscopy," in *Cambridge Monographs on Plasma Physics, chapter 7*, pp. 187–220, Cambridge University Press, Cambridge, 1997.

[21] J. Luque and D. R. Crosley, *LIFBASE: Database and Spectral Simulation Program (Version 1.5) [Computer Software]*, SRI International, Silicon Valley, CA, USA, 2015.

[22] Origin Lab, *Origin pro 9.1 [Computer Software]*, OriginLab Corporation, Northampton, MA, USA, 2015.

[23] C. Lazzaroni, P. Chabert, A. Rousseau, and N. Sadeghi, "Sheath and electron density dynamics in the normal and self-pulsing regime of a micro hollow cathode discharge in argon gas," *The European Physical Journal D*, vol. 60, pp. 555–563, 2010.

[24] M. A. Gigosos and V. J. Cardenosos, "New plasma diagnosis tables of hydrogen Stark broadening including ion dynamics," *Journal of Physics B: Atomic, Molecular and Optical Physics*, vol. 29, no. 20, pp. 4795–4836, 1996.

5

Measurement of the Euler Angles of Wurtzitic ZnO by Raman Spectroscopy

Wu Liu,[1,2] Qiu Li,[1] Gang Jin,[1] and Wei Qiu[3]

[1] *Tianjin Key Laboratory of High Speed Cutting and Precision Machining, School of Mechanical Engineering, Tianjin University of Technology and Education, Tianjin 300222, China*
[2] *School of Mechanical and Automotive Engineering, Zhejiang University of Water Resources and Electric Power, Hangzhou 310018, China*
[3] *Tianjin Key Laboratory of Modern Engineering Mechanics, School of Mechanical Engineering, Tianjin University, Tianjin 300072, China*

Correspondence should be addressed to Qiu Li; qiuli_tj@163.com and Wei Qiu; daniell_q@hotmail.com

Academic Editor: Jau-Wern Chiou

A Raman spectroscopy-based step-by-step measuring method of Euler angles $\varphi, \theta,$ and ψ was presented for the wurtzitic crystal orientation on a microscopic scale. Based on the polarization selection rule and coordinate transformation theory, a series of analytic expressions for the Euler angle measurement using Raman spectroscopy were derived. Specific experimental measurement processes were presented, and the measurement of Raman tensor elements and Euler angles of the ZnO crystal were implemented. It is deduced that there is a trigonometric functional relationship between the intensity of each Raman bands of wurtzite crystal and Euler angle ψ, the polarization direction of incident light under different polarization configurations, which can be used to measure the Euler angles. The experimental results show that the proposed method can realize the measurement of Euler angles for wurtzite crystal effectively.

1. Introduction

The wide-band gap ZnO semiconductor material with hexagonal wurtzite crystal structure has excellent photoelectric properties. It is expected to become the next generation of the UV photoelectric material with excellent performance. And also, it has found wide extensive application prospects in many areas, such as piezoelectric gas sensor and pressure-sensitive sensor [1–3]. Since crystals often have different performances (e.g., electrical conductivity, thermal expansion coefficient, and mechanical strength) in different crystallographic orientations, fully grasping the crystallographic orientation parameters is necessary in the structural design of ZnO devices and mechanical behavior prediction for material microstructure. Namely, the relationship between crystallographic orientation and the observable characteristic direction of the device should be grasped. In short, the geometric relationship between the crystal coordinate system and specimen coordinate system, that is, Euler angles, needs to be determined.

The existing measurement methods of crystallographic orientations included etch-pit technique [4], neutron diffraction [5], electron backscatter diffraction (EBSD) [6–8], X-ray diffraction [9, 10], and micro-Raman spectroscopy [11]. For the etch-pit technique, it is destructive and restricted by a special erodent. The neutron diffraction method needs a long measurement time, and the neutron source construction and operation are expensive. The X-ray diffraction pattern may show no peaks at all in the whole scanning range when the deviation angle between the crystal plane and the sample surface is relatively large, making it difficult to assess the actual crystal orientation in such a case. The determination by micro-Raman spectroscopy has several advantages: (1) it needs no special preparation for the sample and is nondestructive, (2) the sample can be exposed to the air during the measurement, (3) spatial resolution is high (~1 μm),

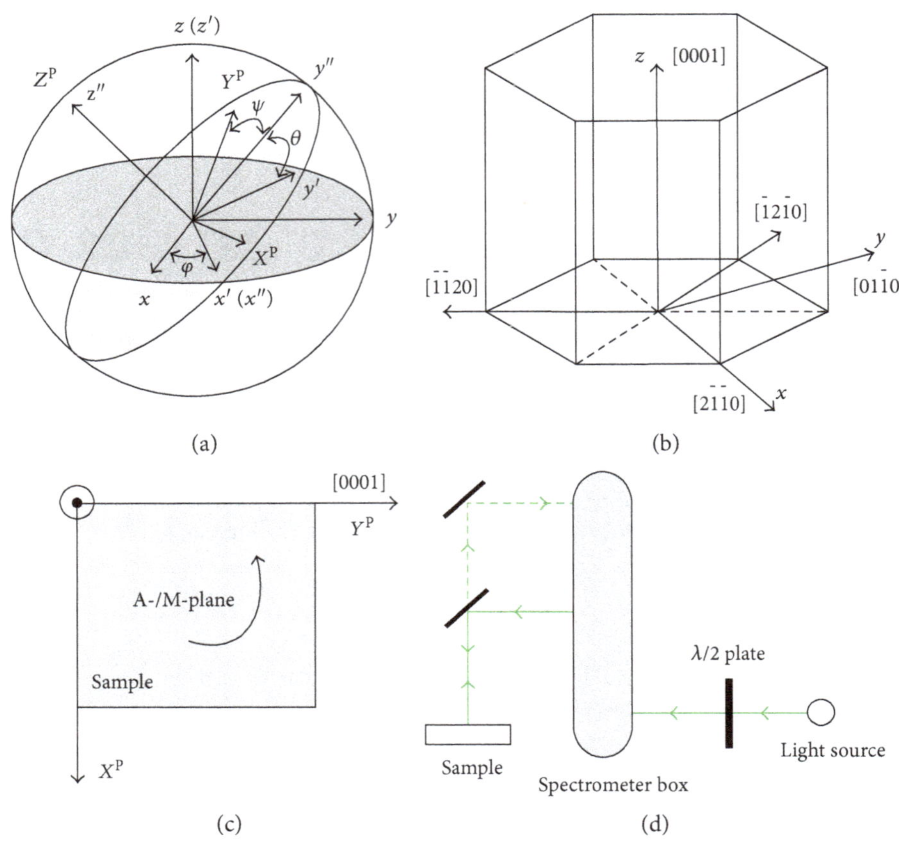

FIGURE 1: (a) The relationship between specimen coordinate system and crystal coordinate system; (b) the relationship between the hexagonal coordinates and Cartesian coordinates for wurtzitic crystal; (c) A-/M-plane ZnO sample initial position; (d) optical path in the Euler angle measurement experiment.

and (4) time required for the determination is relatively short. Hopkins and Farrow have determined the local crystallographic orientations of laser-recrystallized polycrystalline silicon in silicon-on-insulator (SOI) structures using polarized Raman spectroscopy [12]. Munisso et al. also employed this method to determine the crystallographic orientation of polycrystalline alumina [13]. The Euler angles of multicrystalline silicon have been determined by Becker et al. with Raman spectroscopy during the stress measurement, though this method involves complicated formulas and tedious calculations [14]. The validity of the Raman polarization analysis for the crystallographic orientation determination of the diamond and sapphire crystal has been demonstrated. However, the Raman measurement theory and method for the Euler angles of wurtzite crystal are still in the study stage.

In this article, based on the polarization properties of Raman spectra, the step-by-step measuring method for the Euler angle of the wurtzitic crystal is proposed. First, the theoretical formulas for the Euler angle measurement by Raman spectroscopy were deduced for wurtzite crystal. Subsequently, the specific experimental processes were given for the measurement of the Euler angles of a ZnO crystal. Using the sample with a known crystal orientation (A-plane), the Raman tensor elements of ZnO crystal were determined for the first time. Based on the proposed method, the Euler angles of the M-plane ZnO were measured step by step. Then, the experimental results were compared with the theoretical value to verify the validity of the presented method.

2. Theory

Crystal coordinate system is represented by xyz. The relationship between the hexagonal coordinates and Cartesian coordinates of wurtzite crystal is shown in Figure 1(a), where $x \parallel [2\bar{1}\bar{1}0]$, $y \parallel [01\bar{1}0]$, and $z \parallel [0001]$.

Specimen coordinate system is represented by $X^P Y^P Z^P$. The relationship between specimen coordinates and crystal coordinates is shown in Figure 1(b). φ, θ, and ψ are the Euler angles which transform the crystal coordinate system to the specimen coordinate system. The values of φ, θ, and ψ are between 0 and 2π.

The theoretical value of the Euler angles of the crystallographic orientation known sample can be calculated according to the relationship between the Miller indices and the Euler angles. The relationship between the Miller index of the hexagonal system, $(hkil)[uvtw]$, and the Euler angles can be expressed as follows [15]:

$$\begin{bmatrix} \frac{3u}{2d_{uvw}} & \frac{\sqrt{3}h}{d_{hkl}} \\ \frac{\sqrt{3}(u+2v)}{2d_{uvw}} & \frac{h+2k}{d_{hkl}} \\ \frac{w\cdot c}{a\cdot d_{uvw}} & \frac{\sqrt{3}l\cdot a}{c\cdot d_{hkl}} \end{bmatrix} = \begin{bmatrix} \cos\varphi\cdot\cos\psi-\cos\theta\cdot\sin\varphi\cdot\sin\psi & \sin\theta\cdot\sin\psi \\ -\cos\varphi\cdot\cos\psi-\sin\varphi\cdot\cos\theta\cdot\cos\psi & \sin\theta\cdot\cos\psi \\ \sin\varphi\cdot\sin\theta & \cos\theta \end{bmatrix}, \tag{1}$$

where $(hkil)$ represent the crystal orientation index at the X^{P} direction and $[uvtw]$ represent the crystal plane index of the measuring plane; a and c are lattice constants. For ZnO, $d_{uvw} = [9u^2/4 + 3(u+2v)^2/4 + w^2(c/a)^2]^{1/2}$, $d_{hkl} = [3h^2 + 3(h+2k)^2 + 3l^2(a/c)^2]^{1/2}$, and $c/a = 1.602$.

In actual Raman scattering experiments, the geometric configuration of the experiment can be defined according to the Porto formalism [16] as follows: $\mathbf{k}_i(\mathbf{e}_i, \mathbf{e}_s)\mathbf{k}_s$ = propagation direction of incident light (polarization direction of incident light, polarization direction of scattering light) and propagation direction of scattering light. Mathematically, the Raman scattering intensity I depends on the polarization vectors of both the incident ($\mathbf{e}_i$) and scattered ($\mathbf{e}_s$) lights:

$$I = C\left|\mathbf{e}_i \cdot \mathbf{R}_j \cdot \mathbf{e}_s\right|^2, \tag{2}$$

where C is a constant and $\mathbf{R}_j$ is the Raman tensor of the phonon j in crystal coordinate system. Loudon derived the Raman tensors for each of the 32 crystal classes (symmetry point groups). As for wurtzite crystal, there are five Raman tensors that can be expressed in the crystal coordinate system $x\parallel[2\bar{1}\bar{1}0]$, $y\parallel[01\bar{1}0]$, and $z\parallel[0001]$ as follows [17]:

$$\mathbf{R}_{A_1(z)} = \begin{pmatrix} a' & 0 & 0 \\ 0 & a' & 0 \\ 0 & 0 & b' \end{pmatrix}, \quad \mathbf{R}_{E_1(x)} = \begin{pmatrix} 0 & 0 & -c' \\ 0 & 0 & 0 \\ -c' & 0 & 0 \end{pmatrix}, \quad \mathbf{R}_{E_1(y)} = \begin{pmatrix} 0 & 0 & 0 \\ 0 & 0 & c' \\ 0 & c' & 0 \end{pmatrix}, \quad \mathbf{R}_{E_2(x)} = \begin{pmatrix} 0 & d' & 0 \\ d' & 0 & 0 \\ 0 & 0 & 0 \end{pmatrix}, \quad \mathbf{R}_{E_2(y)} = \begin{pmatrix} d' & 0 & 0 \\ 0 & -d' & 0 \\ 0 & 0 & 0 \end{pmatrix}, \tag{3}$$

where a', b', c', and d' are Raman tensor constants and the superscripts $A_1(z)$, $E_1(x)$, $E_1(y)$, $E_2(x)$, and $E_2(y)$ denote Raman vibration modes of wurtzite crystal.

Given that the matrix of Euler angles Φ_{xyz} (and its inverse matrix $\overline{\Phi}_{xyz}$) transforms the Raman tensor in crystal coordinate system into that in specimen coordinate system, the following expression is given:

$$\mathbf{R}_j' = \overline{\Phi}_{xyz}\mathbf{R}_j\Phi_{xyz}, \tag{4}$$

where

$$\Phi_{xyz} = \begin{vmatrix} \cos\varphi\cdot\cos\psi-\cos\theta\cdot\sin\varphi\cdot\sin\psi & -\cos\varphi\cdot\sin\psi-\sin\varphi\cdot\cos\varphi\cdot\cos\psi & \sin\varphi\cdot\sin\theta \\ \cos\varphi\cdot\cos\theta\cdot\sin\psi+\cos\psi\cdot\sin\varphi & \cos\varphi\cdot\cos\theta\cdot\cos\psi-\sin\varphi\cdot\sin\psi & -\cos\varphi\cdot\sin\theta \\ \sin\theta\cdot\sin\psi & \sin\theta\cdot\cos\psi & \cos\theta \end{vmatrix},$$
$$\overline{\Phi}_{xyz} = \begin{vmatrix} \cos\varphi\cdot\cos\psi-\cos\theta\cdot\sin\varphi\cdot\sin\psi & \cos\varphi\cdot\cos\theta\cdot\sin\psi+\cos\psi\cdot\sin\varphi & \sin\theta\cdot\sin\psi \\ -\cos\varphi\cdot\sin\psi-\sin\varphi\cdot\cos\varphi\cdot\cos\psi & \cos\varphi\cdot\cos\theta\cdot\cos\psi-\sin\varphi\cdot\sin\psi & \sin\theta\cdot\cos\psi \\ \sin\varphi\cdot\sin\theta & -\cos\varphi\cdot\sin\theta & \cos\theta \end{vmatrix}. \tag{5}$$

Therefore, the expression, $I = C|\mathbf{e}_i \cdot \overline{\Phi}_{xyz}\mathbf{R}_j\Phi_{xyz} \cdot \mathbf{e}_s|^2$, can be used to calculate the Raman intensity in the specimen coordinate system.

In the parallel polarization $z(yy)\bar{z}$ and cross polarization $z(yx)\bar{z}$ of backscattering geometry, the unit polarization vectors in specimen coordinate system can be expressed as

$$\mathbf{e}_i = (0\ 1\ 0), \quad \mathbf{e}_s^{\parallel} = (0\ 1\ 0)^{\mathrm{T}}, \quad \mathbf{e}_s^{\perp} = (1\ 0\ 0)^{\mathrm{T}}, \tag{6}$$

where the superscript symbols $\parallel$ and $\perp$ refer to parallel and cross polarization configurations, respectively.

Substituting (3), (4), and (6) into (2), the Raman intensities at A_1, E_1, and E_2 bands are obtained as follows:

$$I_A^{\parallel} \propto \left[a' - (a' - b')\sin^2\theta \cdot \cos^2\psi\right]^2, \quad (7)$$

$$I_A^{\perp} \propto \left[-\frac{(a' - b')}{2}\sin 2\psi \cdot \sin^2\theta\right]^2, \quad (8)$$

$$I_{E_1(x)}^{\parallel} \propto \left[2c' \cdot \sin\theta \cdot \cos\psi(\cos\varphi \cdot \sin\psi + \sin\varphi \cdot \cos\theta \cdot \cos\psi)\right]^2, \quad (9)$$

$$I_{E_1(x)}^{\perp} \propto \left[c' \cdot \sin\theta(-\cos\varphi \cdot \cos 2\psi + \sin\varphi \cdot \cos\theta \cdot \sin 2\psi)\right]^2, \quad (10)$$

$$I_{E_1(y)}^{\parallel} \propto \left[2c' \cdot \sin\theta \cdot \cos\psi(\sin\varphi \cdot \sin\psi - \cos\varphi \cdot \cos\theta \cdot \cos\psi)\right]^2, \quad (11)$$

$$I_{E_1(y)}^{\perp} \propto \left[c' \cdot \sin\theta(\sin\varphi \cdot \cos 2\psi + \cos\varphi \cdot \cos\theta \cdot \sin 2\psi)\right]^2, \quad (12)$$

$$I_{E_2(x)}^{\parallel} \propto \left[2d'(\cos\varphi \cdot \sin\psi + \cos\theta \cdot \cos\psi \cdot \sin\varphi) \cdot (\sin\varphi \cdot \sin\psi - \cos\varphi \cdot \cos\theta \cdot \cos\psi)\right]^2, \quad (13)$$

$$I_{E_2(x)}^{\perp} \propto \left[-d'\left(\frac{1}{2}\sin 2\varphi \cdot \sin 2\psi(1 + \cos^2\theta) - \cos\theta \cdot \cos 2\varphi \cdot \cos 2\psi\right)\right]^2, \quad (14)$$

$$I_{E_2(y)}^{\parallel} \propto \left[d'\left((\cos\varphi \cdot \sin\psi + \cos\varphi \cdot \cos\psi \cdot \sin\varphi)^2 - (\sin\varphi \cdot \sin\psi - \cos\varphi \cdot \cos\theta \cdot \cos\psi)^2\right)\right]^2, \quad (15)$$

$$I_{E_2(y)}^{\perp} \propto \left[-d'\left(\frac{1}{2}\cos 2\varphi \cdot \sin 2\psi(1 + \cos^2\theta) + \cos\theta \cdot \sin 2\varphi \cdot \cos 2\psi\right)\right]^2. \quad (16)$$

The proportional coefficients in (7), (8), (9), (10), (11), (12), (13), (14), (15), and (16) are the same. The Raman intensity at E bands (E_1 and E_2) consists of two components, $I_{E(x)}$ and $I_{E(y)}$, and they can be described according to the following equation [18]:

$$I_E^{\perp\parallel} = n \cdot I_{E(x)}^{\perp\parallel} + (1 - n) \cdot I_{E(y)}^{\perp\parallel}, \quad (17)$$

where weight factor $n = 0.5$ [19]. By (17), the Raman intensity at the E bands can be simplified as follows:

$$I_{E_1}^{\parallel} \propto 2c'^2 \cdot \sin^2\theta \cdot \cos^2\psi(\sin^2\psi + \cos^2\theta \cdot \cos^2\psi), \quad (18)$$

$$I_{E_2}^{\parallel} \propto \frac{1}{2}d'^2(\sin^2\theta \cdot \sin^2\psi - \sin^2\theta + 1)^2, \quad (19)$$

$$I_{E_1}^{\perp} \propto \frac{1}{2}c'^2 \cdot \sin^2\theta(\cos^2 2\psi + \cos^2\theta \cdot \sin^2 2\psi), \quad (20)$$

$$I_{E_2}^{\perp} \propto -\frac{1}{2}d'^2(\sin^4\theta \cdot \sin^4\psi - \sin^4\theta \cdot \sin^2\psi - \cos^2\theta). \quad (21)$$

Equations (7), (8), (18), (19), (20), and (21) describe the relationship of Raman intensities with θ and ψ in different polarization configurations. For the sample with known crystallographic orientation, such as A-plane ZnO sample, the Raman intensities of different in-plane rotation angles, ψ, under the same value of θ are measured; the curves of Raman intensity versus ψ were fitted by (7), (8), (18), (19), (20), and (21); the Raman tensor elements (a', b', c', and d') can be obtained. Further, for the sample with unknown crystallographic orientation, the Raman intensities of different in-plane rotation angles, ψ, under the same value of θ are also measured. According to (7), (8), (18), (19), (20), and (21) and combining the Raman tensor elements, the curves of Raman intensity versus ψ were fitted. Thus, the unknown Euler angles θ_c and ψ_c can be estimated.

For the measurement of the Euler angle φ_c, it is necessary to change the polarization direction of incident light for many times. The polarization direction of the incident light can be adjusted by a $\lambda/2$ wave plate. While incident light passes through the $\lambda/2$ wave plate, the mirrors and the notch filter will affect its polarization direction. Suppose the primary optical axis of the $\lambda/2$ wave plate is initially parallel to the polarization direction of incident light. Then, the polarization vector of the incident beam depends on the $\lambda/2$ wave plate's rotation angle (γ) and can be written as [14]

$$\mathbf{e}_s = (\cos\alpha, \quad r \cdot \sin\alpha, \quad 0), \quad (22)$$

where α is twice the $\lambda/2$ wave plate's rotation angle ($\alpha = 2\gamma$) and r is the correction factor of incident light ($0 \le r \le 1$).

The mirrors and the notch filter will also affect the polarization direction of the scattering light. The effective polarization vector $\mathbf{e}_s$ of the scattering beam can be written as [14]

$$\mathbf{e}_s = (m \cdot \cos\beta, \quad \sin\beta, \quad 0)^{\mathrm{T}}, \quad (23)$$

where β is describing the analyzer position and m is the correction factor of the scattered light.

In any coordinate system, the Raman intensity can be expressed as [14]

$$I(\alpha, \beta, \varphi, \theta, \psi) = C\left|\mathbf{e}_i \cdot \bar{\Phi}_{xyz}\mathbf{R}_j\Phi_{xyz} \cdot \mathbf{e}_s\right|^2. \quad (24)$$

Substituting (2), (3), (7), (22), and (23) into (24), a concise matrix equation is finally obtained.

$$I(\alpha,\beta,\varphi,\theta,\psi) = C\cdot\begin{pmatrix}\cos^2\alpha\\ r\cdot\cos\alpha\cdot\sin\alpha\\ r^2\cdot\sin^2\alpha\end{pmatrix}^{\mathrm{T}}\cdot\begin{pmatrix}f_{11} & f_{12} & f_{13}\\ f_{12} & f_{22} & f_{23}\\ f_{13} & f_{23} & f_{33}\end{pmatrix}\cdot\begin{pmatrix}m^2\cdot\cos^2\beta\\ m\cdot\cos\beta\cdot\sin\beta\\ \sin^2\beta\end{pmatrix}, \tag{25}$$

where $f_{ij}=f_{ij}(\varphi,\theta,\psi)$ are the functions of $\varphi,\theta,$ and ψ $(i,j=1,2,3)$. The functions f_{ij} for E_2 band are as follows:

$$\begin{aligned}
f_{11}(\varphi,\theta,\psi) &= 4\sin^2\theta\cdot\sin^2\psi(\cos\psi\cdot\sin\varphi-\cos\varphi\cdot\cos\psi+\cos\varphi\cdot\cos\theta\cdot\sin\psi+\cos\theta\cdot\sin\theta\cdot\sin\psi)^2,\\
f_{12}(\varphi,\theta,\psi) &= 4\sin^2\theta\cdot\sin\psi(\cos\psi\cdot\sin\varphi-\cos\varphi\cdot\cos\psi+\cos\varphi\cdot\cos\theta\cdot\sin\psi+\cos\varphi\cdot\cos\theta\cdot\sin\psi)(\cos\varphi+\sin\varphi)\\
&\quad\cdot(\cos 2\psi+\cos\theta\cdot\sin 2\psi),\\
f_{13}(\varphi,\theta,\psi) &= -\sin^2\theta(-2\cos^4\psi+\cos 2\psi+16\sin\psi\cdot\cos^2\varphi\cdot\cos\theta\cdot\cos^3\psi-8\sin\psi\cdot\cos^2\varphi\cdot\cos\theta\cdot\cos\psi+4\sin 2\varphi\cdot\cos^2\theta\\
&\quad\cdot\cos^4\psi-4\sin 2\varphi\cdot\cos^2\theta\cdot\cos^2\psi+4\sin 2\varphi\cdot\cos^4\psi-4\sin 2\varphi\cdot\cos^2\psi-4\cos^2\theta\cdot\cos^2\psi-8\sin\psi\\
&\quad\cdot\cos\theta\cdot\cos^3\psi+4\sin\psi\cdot\cos\theta\cdot\cos\psi-2\cos^4\psi+4\cos^2\theta\cdot\cos^4\psi+2\cos^2\psi+\sin 2\varphi),\\
f_{22}(\varphi,\theta,\psi) &= 2\sin^2\theta(32\sin\psi\cdot\cos^2\varphi\cdot\cos\theta\cdot\cos^3\psi-16\sin\psi\cdot\cos^2\varphi\cdot\cos\theta\cdot\cos\psi+8\sin 2\varphi\cdot\cos^2\theta\cdot\cos^4\psi-8\sin 2\varphi\cdot\cos^2\theta\\
&\quad\cdot\cos^2\psi+8\sin 2\varphi\cdot\cos^4\psi-8\sin 2\varphi\cdot\cos^2\psi+8\cos^2\theta\cdot\cos^4\psi-8\cos^2\theta\cdot\cos^2\psi-16\sin\psi\cdot\cos\theta\\
&\quad\cdot\cos^3\psi+8\sin\psi\cdot\cos\theta\cdot\cos\psi-8\cos^4\psi+8\cos^2\psi+\sin 2\varphi-1),\\
f_{23}(\varphi,\theta,\psi) &= 2\sin^2\theta\cdot\sin 2\psi\left(2\sin^2\theta\cdot\sin^2\psi-\sin^2\varphi\cdot\sin^2\psi-2\sin^2\theta+\sin^2\varphi\cdot\sin^2\psi+1\right),\\
f_{33}(\varphi,\theta,\psi) &= -4(\cos\varphi\cdot\sin\psi-\sin\varphi\cdot\sin\psi+\cos\varphi\cdot\cos\theta\cdot\cos\psi+\cos\theta\cdot\cos\psi\cdot\sin\varphi)^2\cdot\cos^2\psi\cdot\sin^2\theta_c.
\end{aligned} \tag{26}$$

In order to simplify f_{ij}, the values of θ_c and ψ_c are substituted into the expression f_{ij}. Thus, we can explicitly obtain a set of general expressions that depend on φ.

For the accessible analyzer position ($\beta=0^\circ$), (25) can be simplified as follows:

$$I_x = I(\alpha,0^\circ,\varphi,\theta,\psi) = C(f_{11}\cdot\cos^2\alpha+r\cdot f_{12}\cdot\cos\alpha\sin\alpha+r^2\cdot f_{13}\cdot\sin^2\alpha)\cdot m^2. \tag{27}$$

To fit the experimental data, we use the following fitting functions [14]:

$$I_x = U_1\cdot\cos^2\alpha+U_2\cdot\sin 2\alpha+U_3\cdot\sin^2\alpha, \tag{28}$$

where U_1, U_2, and U_3 are adjustable coefficients.

By comparing the coefficient of (27) and (28), we obtain the equations below:

$$\begin{aligned}
r^2\cdot u_1\cdot f_{13}(\varphi)-f_{11}(\varphi)&=0,\\
r\cdot u_2\cdot f_{12}(\varphi)-f_{11}(\varphi)&=0,
\end{aligned} \tag{29}$$

where $u_1=U_1/U_3$ and $u_2=U_1/U_2$.

Solving the unknown parameters φ and r in (29) using the Gauss-Newton method, the value φ_c can be determined.

3. Experimental

The samples analyzed in this work were the A-plane and M-plane ZnO grown by the hydrothermal method. Both crystals were cut into square-shaped plates with a size of $5\times5\,\mathrm{mm}^2$ and a thickness of 0.5 mm. The measurement surfaces of samples were polished to an optical grade. The crystallographic orientation of the studied crystal faces was preliminarily determined by means of X-ray diffraction analysis.

The macro-Raman measurements were performed at room temperature using a confocal Raman spectrometer (Renishaw InVia Reflex) with the 532 nm emission line of a

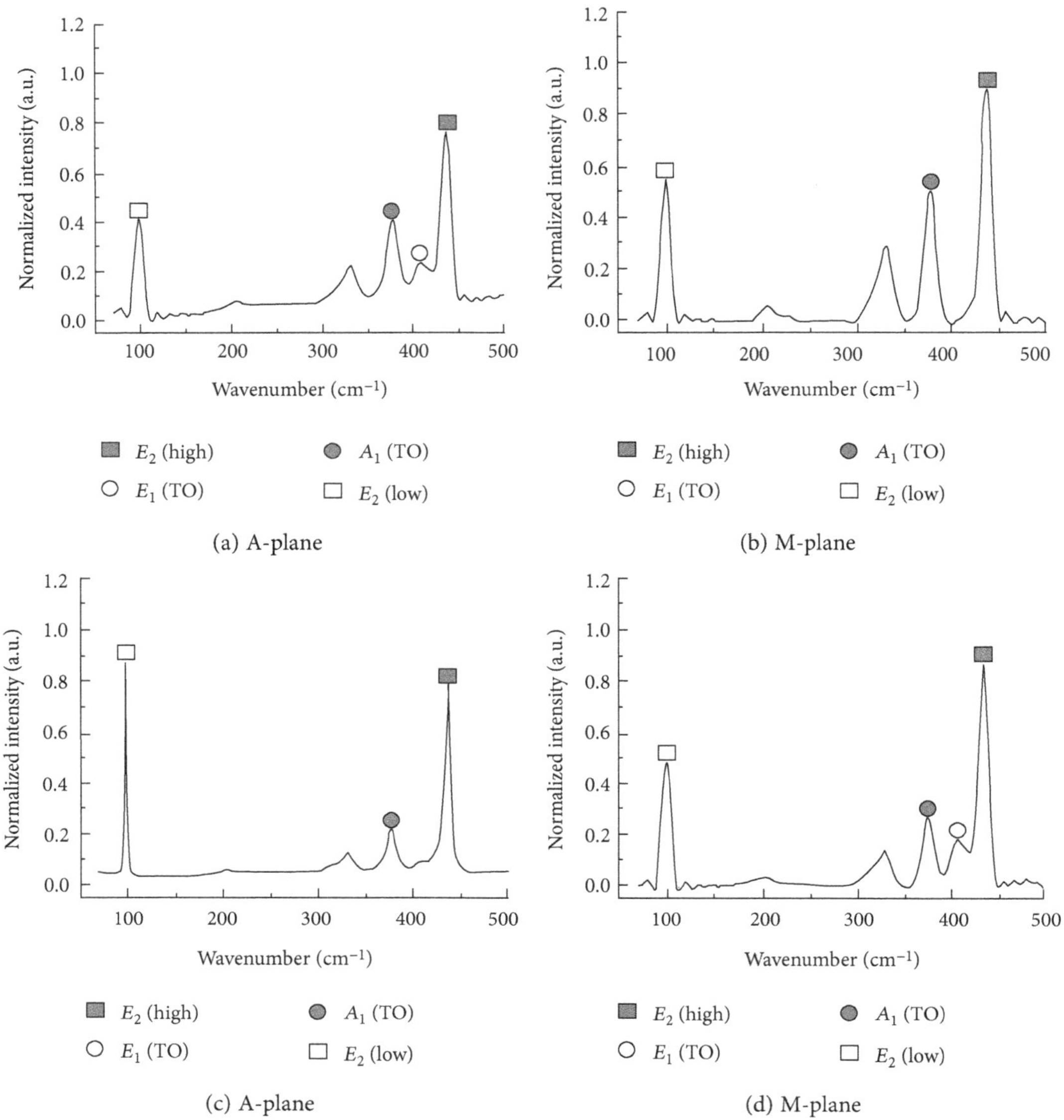

FIGURE 2: Polarized Raman spectra collected on A- and M-plane ZnO samples in parallel (a and b) and cross (c and d) polarization configurations.

semiconductor laser as excitation source. A ×100 objective lens was adopted, and the spot diameter is less than 1 μm.

The Y^P axis (crystal orientation index was [0001]) of the specimen coordinate system of A-plane ZnO was horizontally placed on the experimental platform of Raman spectrometer, as shown in Figure 1(c). The A-plane sample was rotated counter-clockwise by 180 degrees by virtue of parallel polarization and cross polarization configuration, respectively. In the process of sample rotation, the Raman spectral data were collected every 10 degrees. The sample to be measured (M-plane) was placed on the Raman spectrometer in the same way as shown in Figure 1(c). The sample to be measured was rotated counterclockwise by 180 degrees through parallel polarization. In the process of sample rotation, the Raman spectral data were collected every 10 degrees. After that, the sample was rotated counterclockwise by 180 degrees, and 0-degree polarization analyzer ($\beta = 0$ degree) was used for the light path in Raman spectrometer. The $\lambda/2$ wave plate was placed between the light source and the spectrometer box as shown in Figure 1(c). The $\lambda/2$ wave plate was rotated clockwise by 90 degrees along the laser propagation direction. In the rotation process of the $\lambda/2$ wave plate, Raman spectral data were collected every 5 degrees.

Raman spectra collected on the A-plane sample and M-plane sample of the wurtzitic ZnO single crystals with different polarized configurations are shown in Figure 2. We can clearly observe four first-order optical bands (E_2 (low; 99 cm^{-1}), A_1 (TO; 378 cm^{-1}), E_1 (TO; 407 cm^{-1}), and E_2 (high; 437 cm^{-1})) of A-plane ZnO under the parallel polarization; E_2 (low; 99 cm^{-1}), A_1 (TO; 378 cm^{-1}), and E_2 (high; 437 cm^{-1}) bands can be observed under the cross

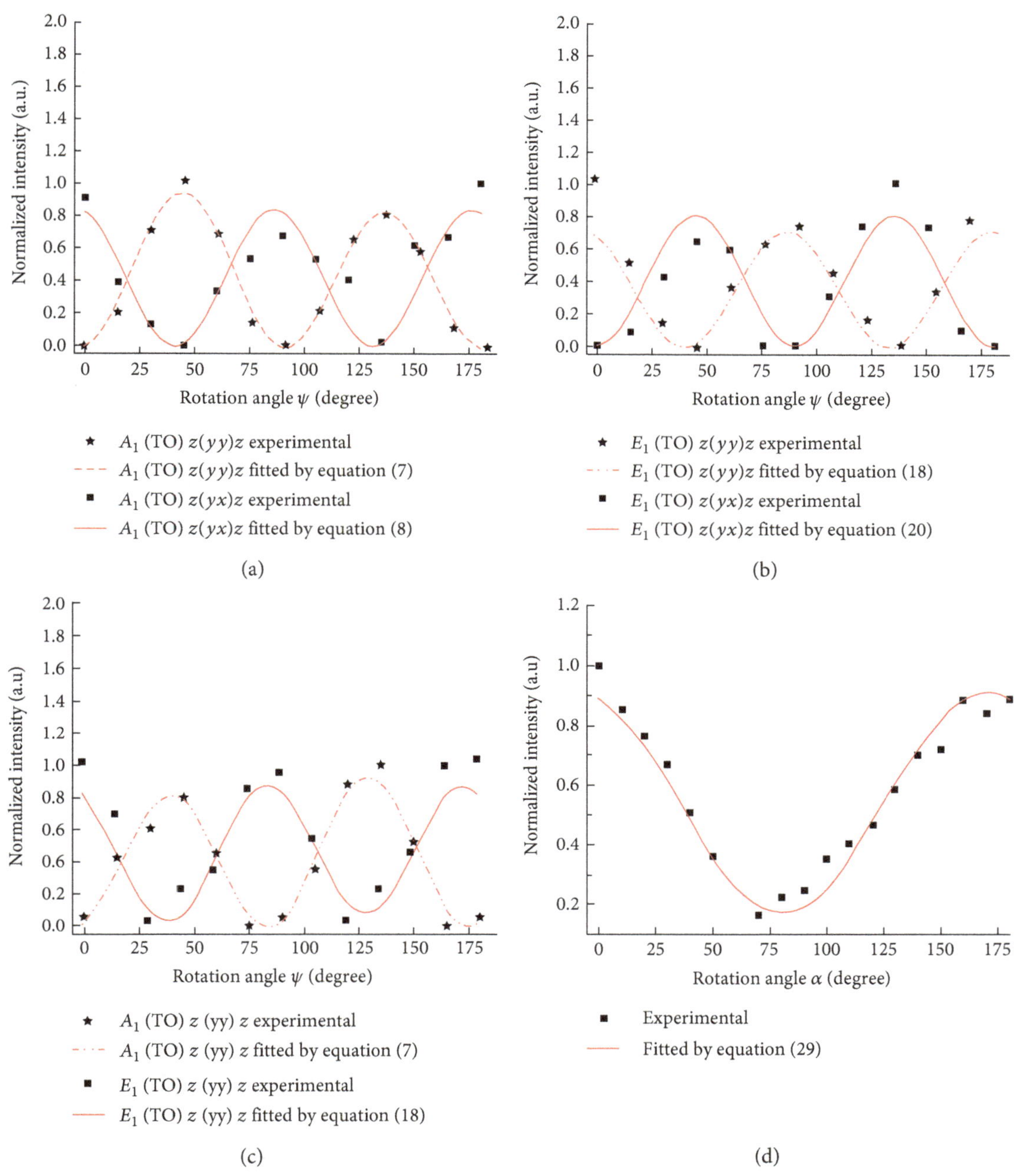

FIGURE 3: Change of Raman intensities at (a) A_1 (TO) and (b) E_1 (TO) bands of A-plane ZnO with the rotation angle ψ; (c) change of intensities at A_1 (TO) and E_1 (TO) bands of M-plane ZnO under parallel polarization with the rotation angle ψ; (d) change of intensities at E_2 (low) band of M-plane ZnO with the polarization direction of incident light.

polarization. As to the M-plane ZnO, we can clearly observe three first-order optical bands, E_2 (low), A_1 (TO), and E_2 (high), under the parallel polarization and four first-order optical bands, E_2 (low), A_1 (TO), E_1 (TO), and E_2 (high), under the cross polarization. The observed vibration bands are in good agreement with the theoretical calculation results based on the polarization selection rules.

Figures 3(a) and 3(b) show the experimentally detected intensity variations at A_1 (TO) and E_1 (TO) bands of the A-plane ZnO single crystal ($\theta = 90^\circ$) as a function of in-plane rotation angle ψ. Raman intensities collected at the A_1 (TO) band under both parallel and cross polarization were fitted by theoretical equations (7) and (8), according to the least-square method. The constants $a' = 0.91$ and $b' = -0.935$ were then obtained. An average value of the constant c', $c' = 1.275$, was extracted from the results of fitting the experimental data in Figure 3(b) by (18) and (20).

Figure 3(c) shows the intensity variations at A_1 (TO) and E_1 (TO) bands of M-plane ZnO single crystal as a function of the in-plane rotation angle ψ. Substituting the values of a', b', and c' into (7) and (18), the angular dependence of the intensity of the Raman bands was fitted according to the least-square method and then we got the values of ψ_c and θ_c ($\psi_c = 128.97°$, $\theta_c = 89.87°$ and $\psi_c = 128.97°$, $\theta_c = 97.0°$). Their average values are $\psi_c = 128.93°$ and $\theta_c = 93.44°$, respectively. Raman intensity at the E_2 (low) band of M-plane ZnO changed with the polarization direction of incident light, α, as shown in Figure 3(d). The values of $U_1 = 0.89$, $U_2 = -0.126$, and $U_3 = 0.194$ were obtained from a least-square fitting of the plot shown in Figure 3(d). Substituting the values of u_1 and u_2 ($u_1 = 4.75$ and $u_2 = -7.06$) into the equation system (29), the value of the Euler angle φ_c can be obtained ($\varphi_c = -3.67°$).

The Miller index of M-plane sample used in the experiment is $(10\bar{1}0)[1\bar{1}00]$. The theoretical values of Euler angles calculated by (1) are $\varphi_c = 0°$, $\theta_c = 90°$, and $\psi_c = 120°$. These values are in agreement with experimental values ($\varphi_c = -3.67°$, $\theta_c = 93.44°$, and $\psi_c = 128.93°$). Therefore, the measurement of Euler angles of wurtzitic ZnO by Raman spectroscopy is feasible.

4. Conclusions

Based on the polarization selection rule and coordinate transformation theory, the step-by-step method for the Euler angle of the crystal of wurtzitic ZnO is proposed. A series of theoretical formulas regarding the Euler angle measurement were deduced. The specific experimental processes were given. The measurement of the Raman tensor elements and Euler angles of a ZnO crystal were achieved. The results show a trigonometric functional relationship between the intensities at each Raman band of wurtzite crystal and the Euler angle, ψ, under different polarization configurations. Fitting the curves of measured Raman intensity versus ψ, the Raman tensor elements and Euler angles θ and ψ can be obtained. There is also a trigonometric functional relationship between the intensities at each Raman band and the polarization direction of incident light, which can be used to measure the Euler angle φ. The experimental results show that the method proposed in this paper can effectively realize the measurement of Euler angles for wurtzite crystal. Our investigation provides the approach for the nondestructive measurement of crystal orientation on the microscale, which can be used to determine the crystallographic orientation of small crystals, local micro regions of crystals, and thin epitaxial films.

Acknowledgments

The authors acknowledge the financial support from the National Natural Science Foundation of China (nos. 11302149 and 11422219), the Tianjin Application Foundation and Frontier Technology Research Program (no. 15JCQNJC05000), the Natural Science Foundation of Hebei Province (no. A2016202195), and the Innovation Team Training Plan of Tianjin Universities and Colleges (no. TD12-5043).

References

[1] P. Zu, Z. K. Tang, G. K. L. Wong et al., "Ultraviolet spontaneous and stimulated emissions from ZnO mic-rocrystallite thin films at room temperature," *Solid State Communications*, vol. 108, no. 8, pp. 459–463, 1997.

[2] M. H. Huang, S. Mao, H. Feick et al., "Room-temperature ultraviolet nanowire nanolasers," *Science*, vol. 292, no. 5523, pp. 1897–1899, 2001.

[3] M. Q. Wang, Y. Q. Lian, and X. G. Wang, "PPV/PVA/ZnO nanocomposite prepared by complex precursor method and its photovoltaic application," *Current Applied Physics*, vol. 9, no. 1, pp. 189–194, 2009.

[4] G. E. Bacon, N. A. Curry, and S. A. Wilson, "A crystallographic study of solid benzene by neutron diffraction," *Proceedings of the Royal Society of London A: Mathematical and Physical Sciences*, vol. 279, no. 1376, pp. 98–110, 1964.

[5] M. Zucali, M. Voltolini, B. Ouladdiaf, L. Mancini, and D. Chateigner, "The 3D quantitative lattice and shape preferred orientation of amylonitised metagranite from Monte Rosa (Western Alps): combining neutron diffraction texture analysis and synchrotron X-ray microtomography," *Journal of Structural Geology*, vol. 63, no. 3, pp. 91–105, 2014.

[6] M. Calcagnotto, D. Ponge, E. Demir, and D. Raabe, "Orientation gradients and geometrically necessary dislocations in ultrafine grained dual-phase steels studied by 2D and 3D EBSD," *Materials Science and Engineering A*, vol. 527, no. 10-11, pp. 2738–2746, 2010.

[7] S. L. Shrestha, A. J. Breen, P. Trimby, G. Proust, S. P. Ringer, and J. M. Cairney, "An automated method of quantifying ferrite microstructures using electron backscatter diffraction (EBSD) data," *Ultramicroscopy*, vol. 137, no. 1, pp. 40–47, 2014.

[8] A. Eghlimi, M. Shamanian, M. Eskandarian, A. Zabolian, M. Nezakat, and J. A. Szpunar, "Evaluation of microstructure and texture across the welded interface of super duplex stainless steel and high strength low alloy steel," *Surface & Coatings Technology*, vol. 264, pp. 150–162, 2015.

[9] K. Gao, S. Li, L. Xu, and H. Fu, "Effect of solidification rate on microstructures and orientations of Al-Cu hypereutectic alloy in thin crucible," *Crystal Research & Technology*, vol. 49, no. 2-3, pp. 164–170, 2014.

[10] F. Li, L. Jin, Z. Xu, and Z. Guo, "Determination of three-dimensional orientations of ferroelectric single crystals by an improved rotating orientation x-ray diffraction method," *Review of Scientific Instruments*, vol. 80, no. 8, pp. 085106-1–085106-5, 2009.

[11] K. Mizoguchi and S. Nakashima, "Determination of crystallographic orientations in silicon films by Raman microprobe polarization measurements," *Journal of Applied Physics*, vol. 65, no. 7, pp. 2583–2590, 1989.

[12] J. B. Hopkins and L. A. Farrow, "Raman microprobe determination of local crystal orientation," *Journal of Applied Physics*, vol. 59, no. 1103, pp. 1103–1110, 1986.

[13] M. C. Munisso, W. L. Zhu, and G. Pezzotti, "Raman tensor analysis of sapphire single crystal and its application to define crystallographic orientation in polycrystalline alumina," *Physica Status Solidi B-Basic Solid State Physcics*, vol. 246, no. 8, pp. 893–1900, 2009.

[14] M. Becker, H. Scheel, and S. Christiansen, "Grain orientation, texture, and internal stress optically evaluated by micro-Raman spectroscopy," *Journal of Applied Physics*, vol. 101, no. 063531, pp. 1–10, 2007.

[15] W. R. Busing and H. A. Levy, "Angle calculations for 3- and 4-circle X-ray and neutron diffractometers," *Acta Crystallographica*, vol. 22, no. 4, pp. 457–464, 1967.

[16] T. D. Tell and S. Porto, "Raman effect in cadmium sulfide," *Physical Review*, vol. 144, no. 2, pp. 771–774, 1966.

[17] R. Loudon, "The Raman effect in crystals," *Advances in Physics*, vol. 13, no. 52, pp. 423–482, 1964.

[18] G. Pezzotti, H. Sueoka, and A. A. Porporati, "Raman tensor elements for wurtzitic GaN and their application to assess crystallographic orientation at film/substrate interfaces," *Journal of Applied Physics*, vol. 110, no. 013527, pp. 1–10, 2009.

[19] K. Nakamoto, *Infrared and Raman Spectra of Inorganic and Coordination Compounds*, pp. 56–98, Wiley, New York, 1986.

6

Spectroscopic Characterization of Omeprazole and Its Salts

Tomislav Vrbanec,[1] Primož Šket,[2] Franci Merzel,[2] Matej Smrkolj,[1] and Jože Grdadolnik[2]

[1]*Krka d.d., Novo Mesto, Šmarješka c. 6, SI-8501 Novo Mesto, Slovenia*
[2]*National Institute of Chemistry, Hajdrihova 19, SI-1000 Ljubljana, Slovenia*

Correspondence should be addressed to Jože Grdadolnik; joze.grdadolnik@ki.si

Academic Editor: Yiannis Sarigiannis

During drug development, it is important to have a suitable crystalline form of the active pharmaceutical ingredient (API). Mostly, the basic options originate in the form of free base, acid, or salt. Substances that are stable only within a certain pH range are a challenge for the formulation. For the prazoles, which are known to be sensitive to degradation in an acid environment, the formulation is stabilized with alkaline additives or with the application of API formulated as basic salts. Therefore, preparation and characterization of basic salts are needed to monitor any possible salinization of free molecules. We synthesized salts of omeprazole from the group of alkali metals (Li, Na, and K) and alkaline earth metals (Mg, Ca). The purpose of the presented work is to demonstrate the applicability of vibrational spectroscopy to discriminate between the OMP and OMP-salt molecules. For this reason, the physicochemical properties of 5 salts were probed using infrared and Raman spectroscopy, NMR, TG, DSC, and theoretical calculation of vibrational frequencies. We found out that vibrational spectroscopy serves as an applicable spectroscopic tool which enables an accurate, quick, and nondestructive way to determine the characteristic of OMP and its salts.

1. Introduction

In drug development, only the most stable crystalline forms of an active substance are suitable for commercial use. In general, active substances can be used in form of an acid, base, or in a form of a salt. Substances, which are stable over a certain pH range, are a challenge for the development of formulations. Prazoles are susceptible to degradation in an acid environment. Therefore, they are stabilized in the formulation with alkaline additives or administered in the form of basic salts [1]. The preparation and complete characterization of basic salt is needed in the case of direct application of the formulation and for monitoring the potential undesired salt formation. For the pharmaceutical industry, the disposal of stable product-specific solubility and bioavailability is crucial. Preparation of the salt of the active substance in the case of omeprazole makes it possible to affect the stability, solubility, and bioavailability. Vibrational spectroscopy is a very useful technique due to its ability to probe a sample in an accurate, fast, and nondestructible way. Moreover, the vibrational spectra possess all information of the structure and dynamics of the probed molecule. Raman and infrared of mapping experiments probe a small area of sample and thus check the homogeneity of sample below micrometer (Raman). However, the limitation of applicability of Raman and infrared spectroscopy lays in the necessity to assign the vibrational bands in the corresponding spectrum.

5-Methoxy-2-[[(4-methoxy-3,5-dimethyl-2-pyridinyl)-methyl] sulfinyl]-^{1}H-benzimidazole (omeprazole, OMP) is used to treat gastric and duodenal ulcers, erosive esophagitis, gastroesophageal reflux disease, Zollinger-Ellison syndrome, and others [2]. It acts as a proton-pump inhibitor which inhibits gastric acid secretion, that is, irreversibly blocks the enzyme system of hydrogen/potassium adenosine triphosphatase (H^+/K^+-ATPase) in gastric parietal cells. The OMP molecule has a stereochemical center and hence it exists as the R and S isomer (Figure 1). Both isomers show activity, but the S isomer is metabolized more slowly, which results in reproducibly extended release of S-omeprazole [3]. The

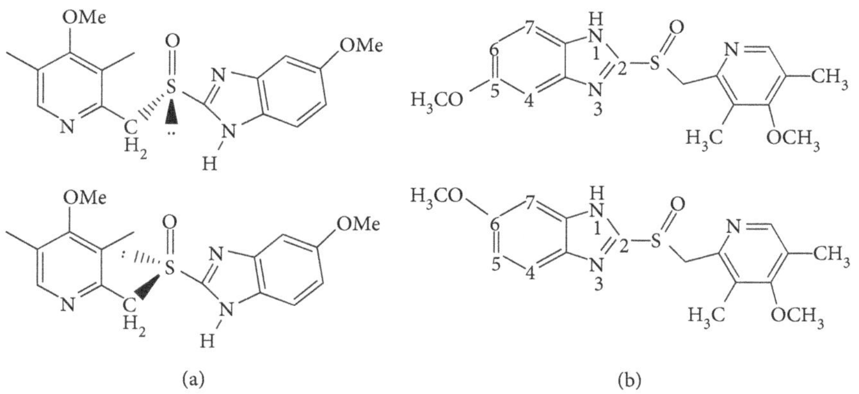

FIGURE 1: Enantiomers (a) and tautomers (b) of OMP. S (upper left molecule) and R (lower left molecule) enantiomers of OMP. 5-Methoxy OMP (upper right) and 6-methoxy (lower right) tautomers of OMP.

reactive side of the OMP molecule is an acid group in the benzimidazole ring. Therefore, in the preparation of the final form (formulation), various stabilizers are used in order to prevent chemical reaction between excipients and OMP. One of the possibilities to reduce the reactivity of the acid group is a reaction with bases to synthesis an OMP in salt form.

Previous vibrational studies [4] analysed only the principal characteristic peaks of OMP. Similar peaks in infrared spectra were also described in various patents [5–7]. Murakami et al. partially extended peak assignment of the solid-state OMP-Na. The analyses revealed that OMP-Na contains one water molecule in its crystal structure. It was suggested that the water molecule in the OMP-Na probably establishes one hydrogen bond with nitrogen of the imidazole ring and the other with a sulfoxide group. Markovic et al. [8] studied both OMP isomers. By using DSC method, they identified that both optical isomers of OMP-Na are thermodynamically more stable (melting of S isomer at 227°C and R isomer at 229°C) than the OMP in neutral form which melts at 159°C.

In some cases, vibration spectroscopy may be preferable with respect to other analytical techniques, considering that the vibrational spectra reflect changes in functional groups that appear after salt formation. This can be very helpful in detection of salts in a mixture with OMP. For example, Raman microspectroscopy (mapping) may be utilized in the determination of salt formation on the edges of a specific layer in an OMP pellet. The amount of salts is in general very low and thus under the detection limit for other techniques [9–11].

The presented study will show the ability to monitor these transformations by the application of various types of spectroscopies (infrared, Raman, and NMR) supported by X-ray powder diffraction, DSC, and thermogravimetric measurements. The formation of OMP salts was confirmed by comparing the recorded infrared and NMR spectra with XRPD diffraction patterns. Since we would like to apply only infrared and/or Raman spectroscopy for identification of OMP in salt form, the assignment of specific vibrational peaks that are characteristic for the formation of salt was analysed in detail. The first candidates are the characteristic modes of the benzimidazole ring (N-H, C=C-N, and S-C=N) and S=O group.

2. Materials and Methods

Omeprazole salts were prepared from the omeprazole purchased from Shouguang Fukang Pharmaceutical. All reactants used were of analytical grade. The alkali salts of omeprazole (Na, K, and Li) were synthesized by applying a similar procedure. Sodium salt was synthesized by reaction of OMP (10 g) with a water solution of NaOH (Aldrich, 1.16 g, 25 mL). After 5 minutes of vigorous stirring, methylene chloride (Merck) was added (50 mL) and stirring was continued for the next 15 minutes, after which two phases were separated and the water phase was evaporated to dry matter using a water bath (35°C). The dry substance was then mixed with 70 mL of ethyl acetate and stirred under reflux (at 77°C) for the next 30 minutes. After an overnight cooling and resting, the residue was dried under vacuum (40°C for 24 h). The final product had a mass of 90.7 g which correlates with a yield of 85%. Potassium and lithium salt of omeprazole were synthesized in a similar procedure by using LiOH (Aldrich, 0.207 g in 40 mL) or KOH (Aldrich, 1.69 g in 60 mL). The yields of OMP-Li was 74% and of OMP-K was 87%.

The alkaline earth salts of OMP (Mg and Ca) were synthesized using OMP-Na as a starting material. Anhydrous calcium (Aldrich, 1.16 g)/magnesium chloride (Aldrich, 2.12 g) was dissolved in distilled water (20 mL), and the solution was added drop by drop to a water solution of OMP-Na followed by vigorous stirring for 60 minutes The remainder in the flask was filtered, washed up with water, and dried on 40°C for 24 h. The yields in case of OMP-Mg and OMP-Ca reactions were 74% and 68%, respectively.

The XRPD patterns and NMR spectra of all salts were recorded and compared with those found in literature [1, 5–7, 12–14] to confirm the formation and purity of prepared salts of omeprazole.

The DSC analysis was performed by Mettler Toledo DSC1. A nitrogen flow of 40 mL min^{-1} was used in all measurements. Samples were analysed in aluminium pans (volume 40 μL) with a punctured cover in a temperature range between 20°C and 200°C and at a heating rate of

10°C min^{-1}. Stare Software 11.00 software package was used for data collecting and processing.

Thermogravimetric experiments were measured on the Mettler Toledo TGA/DSC1 instrument. Samples were analysed in aluminium containers with a perforated aluminium lid with a capacity of 100 μL in the temperature range from room temperature to 180°C with a heating rate of 10°C min^{-1}. Stare Software 11.00 software package was used for data collection and processing.

X-ray diffraction intensities were measured on an X-ray diffractometer PHILIPS X'Pert PRO, PANalytical. The diffractometer is equipped with a detector X'celerator and an X-ray tube with a copper anode k-$\alpha = 1.54$. Data processing was done by applying a software package HighScore Plus 3.0e. Samples were recorded in the range between 3° and 32.5°. The integration time was 100 s, and the step was 0.033°. The patterns were compared with published data using a software package for digitizing named WinDig 2.5.

NMR spectra of solid samples were recorded on Agilent Technologies VNMRS 600 MHz NMR spectrometer equipped with 3.2 mm NB Double Resonance HX MAS Solids Probe. The ^{1}H MAS and ^{13}C CP-MAS NMR spectra were externally referenced using adamantane. Samples were spun at the magic angle with 20 and 16 kHz for ^{1}H MAS and ^{13}C CP-MAS NMR spectra, respectively. The proton spectra were acquired using a composite pulse sequence. Repetition delay in all experiments was 5 s. The number of scans was 16. The pulse sequences used for acquiring the ^{13}C spectra were a standard cross-polarization MAS pulse sequences with high-power proton decoupling during acquisition. The repetition delay was 5 s. The number of scans was between 350 and 760. NMR spectra of the liquid samples were recorded on Agilent Technologies DD2 300 MHz NMR spectrometer, using a 5 mm ID probe equipped with gradients. The spectra were recorded at 25°C. Proton chemical shifts were determined relative to tetramethylsilane (TMS).

Raman spectra were recorded using the Raman spectrometer RAMII attached to Vertex 80 infrared spectrometer (BRUKER). The spectrometer was equipped with lasers emitting at 1064 nm and a LN-Ge diode detector. The spectra were recorded with a laser power of 300 mW, in the spectral region between 3600 and 32 cm^{-1} with a nominal resolution of 4 cm^{-1}. The final spectrum was a result of averaging of 128 scans. The device operates using a software package OPUS 6.5, which is also used for processing the recorded spectra.

Measurements of infrared spectra were performed by FT-IR Vertex 70 spectrometer, manufactured by Bruker. The samples were prepared in the form of KBr pellets. The spectrometer was equipped with a DLaTGS detector. Recorded spectra are the result of averaging 32 scans in the spectral region between 4000–400 cm^{-1} with a nominal resolution of 4 cm^{-1}. The spectra were recorded and processed by the application of the OPUS 6.5 software (Bruker).

The calculation of vibrational spectrum of omeprazole was based on density functional theory as implemented in the Vienna Ab initio simulation package (VASP) [15, 16]. VASP performs an iterative solution of the Kohn–Sham equations in a plane-wave basis; the interaction of the valence electrons with the ionic cores is described within the projector-augmented-wave (PAW) formalism [17]. The cut-off energy value was set to 500 eV. The electronic exchange and correlation were described by the gradient-corrected PBE functional proposed by Perdew et al. [18]. The self-consistency cycle was terminated when the total energies in the next step changed by less than 10^{-6} eV per cell. Brillouin-zone integrations were performed on Monkhorst–Pack grids [19]. The crystal structure of omeprazole was obtained from the Crystallography Open Database (COD id 7101903) [20]. A corresponding triclinic crystal cell (P1) ($a = 9.701$ Å, $b = 10.259$ Å, $c = 10.694$ Å, $\alpha = 91.720°$, $\beta = 112.117°$, and $\gamma = 115.642°$) contains two molecular units. The initial atomic positions (see Supplementary Material available online at https://doi.org/10.1155/2017/6505706) were energy minimized by applying the conjugate gradient algorithm followed by the residual minimization scheme with direct inversion in the iterative subspace (RMM-DIIS) until stopping criterion for forces, Fmax $< 10^{-5}$ eV/Å, was achieved [21]. Vibrational density of states (vDOS) was obtained as a distribution of eigenvalues of the dynamical matrix formed from the force-constant matrix by taking into account periodic boundary conditions. The elements of the force-constant matrix were determined as first derivatives of the forces induced by small ionic displacements, calculated according to the Hellmann–Feynman theorem by using the finite difference method [22]. Assignment of the vibrational spectrum was facilitated by projecting the total vDOS, $g(\omega)$, into partial diatomic vDOS, $g_{\text{bond}}(\omega)$, associated with the stretching vibration of a given chemical bond in the molecule, simply by determining the appropriate relative weight, $w_{\text{bond}}(\omega)$, of the corresponding atomic pair over the entire molecule for a given frequency mode,

$$g_{\text{bond}}(\omega) = w_{\text{bond}}(\omega) g(\omega). \tag{1}$$

As dynamical matrix eigenvectors contain information about individual atomic displacements in a given vibrational mode, we calculate $w_{\text{bond}}(\omega)$ by dividing the sum of the squared displacements along the unit bond direction $\vec{e}_{12} = (\vec{r}_{a1} - \vec{r}_{a2})/|\vec{r}_{a1} - \vec{r}_{a2}|$, between atoms α_1 and α_2 by the square of the eigenvector,

$$w_{\text{bond}}(\omega) = \frac{\left|\vec{e}_{12} \cdot \vec{d}_{a1}(\omega)\right|^2 + \left|\vec{e}_{12} \cdot \vec{d}_{a2}(\omega)\right|^2}{\sum_a \left|\vec{d}_a(\omega)\right|^2}. \tag{2}$$

3. Results and Discussion

3.1. Thermogravimetric Analysis. Thermal measurements started by using TGA and DSC techniques were applied to explore polymorphism of omeprazole and its salts. Each form of prepared omeprazole salts showed the loss of mass which is consistent with the release of crystalline and/or surface-bound water (Figure 2). In OMP-K, the loss in mass was observed in one step from 100°C to 170°C and corresponds to the release of one water molecule. In OMP-Na, the loss in mass was observed in two steps, the first one from 85°C

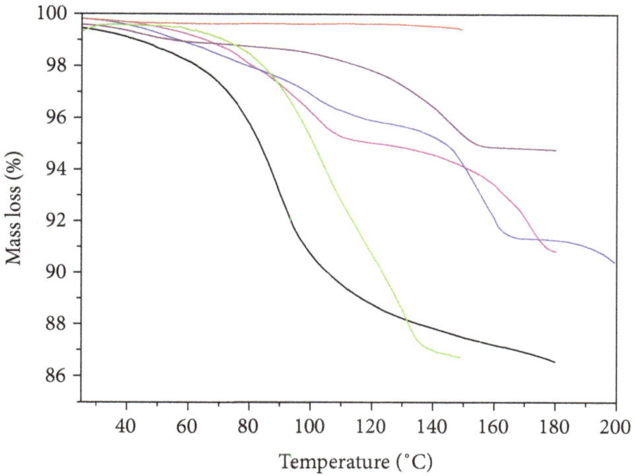

FIGURE 2: Thermo gravimetrical curves of OMP (red) and five OMP salts (OMP-K (violet), OMP-Na (blue), OMP-Mg (pink), OMP-Li (green), and OMP-Ca (black)).

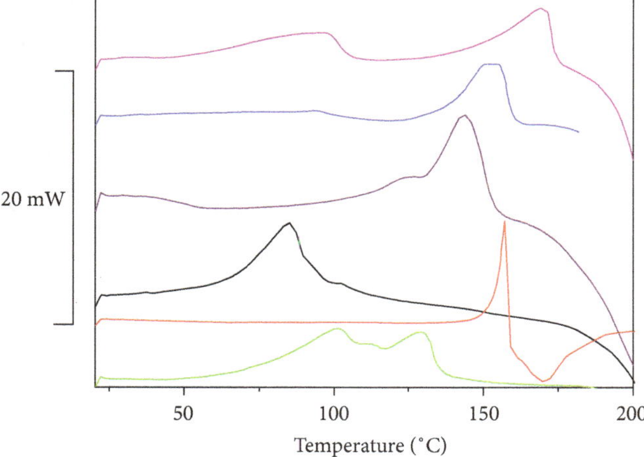

FIGURE 3: DSC curves of OMP (red) and prepared salts (OMP-K (violet), OMP-Na (blue), OMP-Mg (pink), OMP-Li (green), OMP-Ca (black)). The curves are shifted for clarity.

to 110°C and the second one from 120°C to 160°C. Both steps are in the accordance with the release of two water molecules [5, 6, 8, 23–25]. In OMP-Mg, the loss in mass in two steps was also observed, from 85°C to 105°C and from 135°C to 170°C which is in accordance with the release of two water molecules [12]. In OMP-Ca, the continuous loss in mass was observed from 40°C to 150°C (loss of three water molecules), while OMP-Li experienced the continuous loss of mass from 70°C to 150°C (also in accordance with the loss of three water molecules). Thermal analysis showed that all prepared omeprazole salts were in hydrate form.

3.2. Differential Scanning Calorimetry. Characteristic graphs of DSC measurements OMP and its salts are shown in Figure 3. DSC of omeprazole displays only one melting point at 150°C after which decomposition occurs, whereas the DSC curves of omeprazole salts are more complex. In the case of

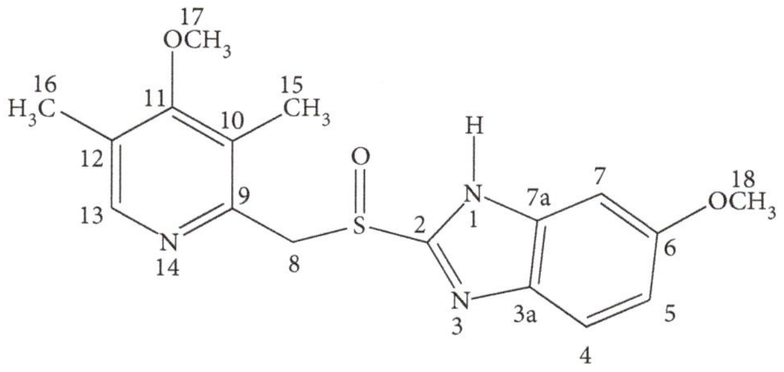

FIGURE 4: OMP with numbered carbon and nitrogen atoms.

OMP-Mg, two melting points were measured (at 100°C and 170°C) followed by decomposition. OMP-Na also has two melting points (at 100°C and 150°C) again followed by decomposition. The result is in agreement with the thermogravimetric measurements and results from literature [23] where the OMP-Na form A was measured. Potassium salt has only one melting point at 145°C followed by decomposition, whereas OMP-Li experiences three melting points (at 100°C, 110°C, and 125°C) followed by decomposition. Similarly, OMP-Ca has also three melting points (at 58°C, 90°C, and 100°C) followed by decomposition.

The correlation of results obtained by TGA and DSC is very high and were thus used for clarifying spectroscopic measurements such as DVS and hot-stage in combination with in situ.

3.3. NMR. In order to additionally characterize solid OMP and its salt, we measured ^{1}H MAS and ^{13}C CP-MAS NMR spectra. For assignment of the ^{13}C spectra, we used atom numbering shown in Figure 4. We focused on the expected differences mainly in ^{13}C spectra between various types of OMP (Table 1). Comparison of the proton ^{1}H MAS NMR spectra of solid samples of OMP and all prepared salts revealed that the OMP sample comprises the signal at 13.5 ppm (Figure 5). The signal at this chemical shift is assigned to a proton on the nitrogen atom N1, which is involved in a hydrogen bond. This type of proton is missing in all almost identical proton spectra of OMP salts. That may indicate that salt formation removes the proton attached to N1 nitrogen.

It is evident from Figure 6 that one of the most prominent differences between omeprazole and its salts in the ^{13}C CP-MAS NMR spectra of solid samples is in the region between 90 ppm and 100 ppm, where carbon atom C7 resonates. The rest of OMP ^{13}C spectra were assigned following the data from literature [26]. All spectra presented in Figure 6 correspond to a tautomeric form of 6-metoxy-omeprazole. In contrast to omeprazole where the corresponding C7 ^{13}C chemical shift is located at 92.0 ppm, the other samples of OMP salts show this signal at 98.0 ppm. The differences in signal chemical shifts indicate that in samples of OMP salts certain changes appear in the vicinity of the carbon atom C7, such as the absence of the proton on a nitrogen atom N1.

Another prominent difference between OMP and its salts can be found in the region spanned between 45 ppm and 65 ppm. One peak found in the omeprazole spectrum is

Table 1: Carbon chemical shifts (ppm) of OMP and its salts.

C atom	OMP	OMP-Na	OMP-K	OMP-Li	OMP-Mg	OMP-Ca
C15	8.96	10.43	11.79	11.61	11.90	11.09
C16	12.60	12.34	13.34	14.40	13.00	13.54
C8, C17, C18	57.81	53.75	52.78	53.35	52.55	52.39
	—	58.48	59.22	60.23	58.79	54.71
	—	60.74	—	—	—	60.05
	—	—	—	—	—	61.79
C7	92.02	99.02	98.40	97.16	97.21	96.59
	—	—	—	—	—	98.67
C5	112.94	111.31	108.39	113.21	109.31	110.29
		114.60	117.08	117.53	116.78	111.70
C4	121.85	125.11	124.32	126.82	124.44	116.60
C10, C12	126.00	127.41	126.94	129.49	127.95	125.82
	—	—	—	—	—	127.56
	—	—	—	—	—	128.98
C3a–C7a	136.00	139.30	139.41	139.62	137.20	137.67
						139.44
C9–C13	150.13	147.33	146.42	147.18	141.85	144.56
	—	150.71	149.02	—	147.15	146.59
	—	154.30	150.59	—	149.46	147.78
	—	—	—	—	—	149.47
	—	—	—	—	—	151.29
C2–C6	158.05	158.24	153.32	155.48	153.66	153.63
	—	—	—	157.18	156.59	157.868
C11	164.41	164.51	163.05	165.29	164.01	159.90
	—	—	164.12	—	—	164.22

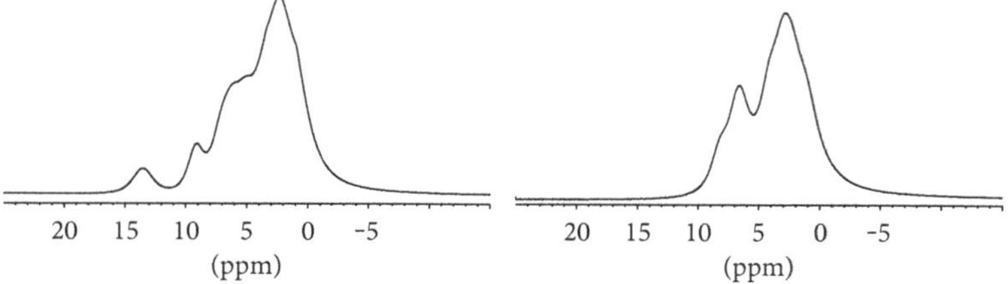

Figure 5: Left spectrum: ^{1}H MAS spectrum of OMP; right spectrum: ^{1}H MAS spectrum of OMP-Na.

evidently multiplied in the salt spectra. This splitting may be connected to the changes in the environment of carbon atoms C8, C17, and C18 of OMP during salt formation, when the cation comes in the vicinity of an OMP molecule.

3.4. Vibrational Analysis. To get a better view on the assignment of OMP vibrational spectrum (Figure 7), the OMP molecule is divided into four fragments. The first fragment is a substituted pyridine ring, the second one is a benzimidazole fragment, and the third one is a C-SO-C bridge. The fourth fragment consists of various methoxy groups. Each and every one of the fragments aforementioned is already assigned in literature [27–30]. Although the band frequencies of fragments may slightly differ from one found in the vibrational spectra, they serve as a relevant guide for band assignment in the spectra of OMP. Band frequencies of band in infrared and Raman spectrum of OMP are tabulated and described in Table 2.

Considering that prior analyses have already determined that the absence of the NH group is the main factor in salt formation, most of the attention will be given to extract vibrations connected to this group. NH stretching in infrared spectrum is usually visible as a medium intensity band in the range between 3500 and 3200 cm^{-1}. However, the NH group from the benzimidazole part of the molecule is a strong proton donor, which may participate in hydrogen bonds with various types of proton acceptors which means that NH stretching frequency would redshift with respect to the

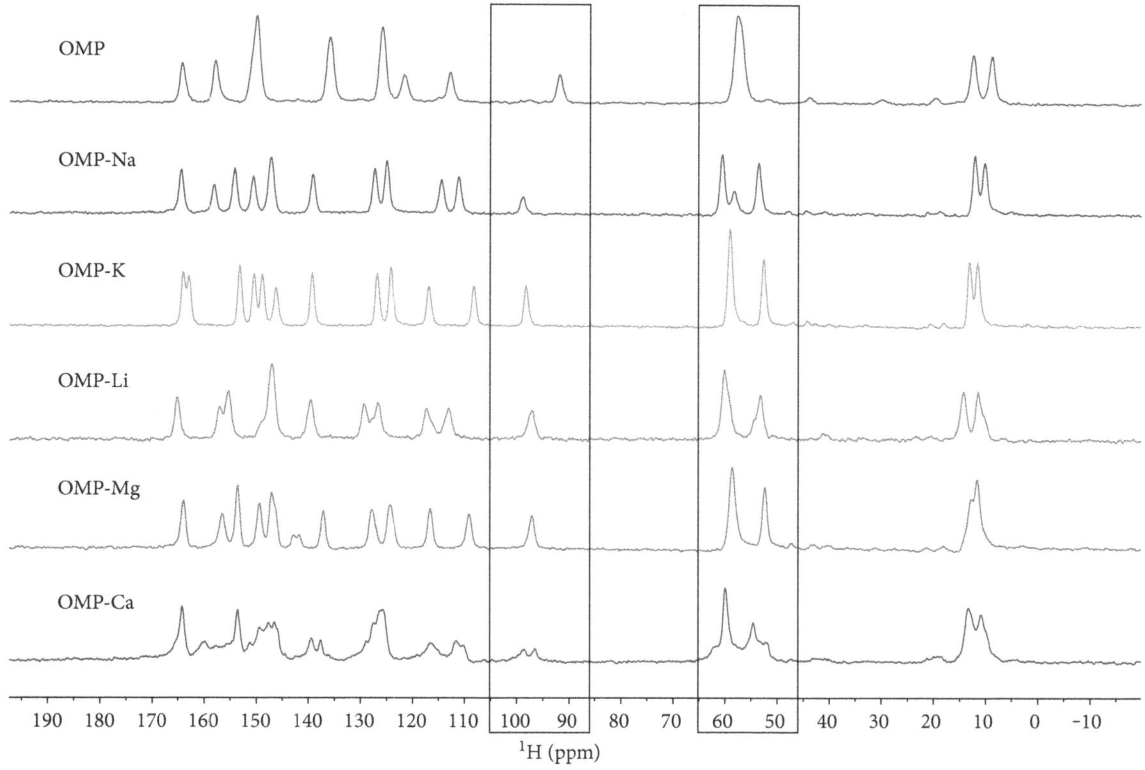

FIGURE 6: Comparison of ^{13}C CP-MAS NMR spectra of solid samples of OMP and OMP salts.

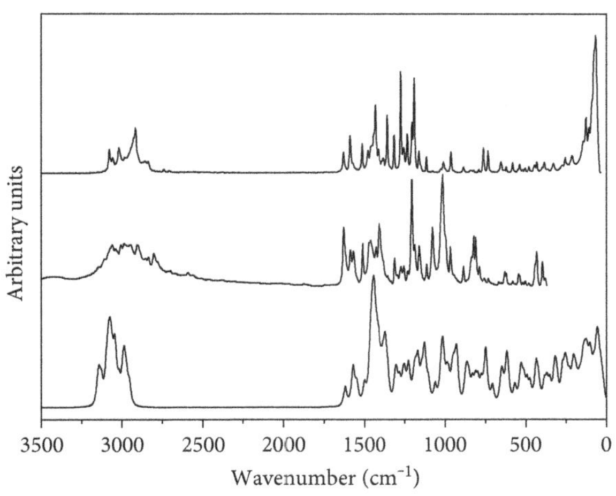

FIGURE 7: Raman (upper), infrared (middle), and calculated (lower) spectrum of a solid-state OMP.

strength of the established hydrogen bond. A formation of hydrogen bonds also broadens the NH vibrational band. Therefore, the NH stretching vibration can be found in the infrared spectrum of OMP as a broad band near 2930 cm^{-1}. Another characteristic band for NH group is NH deformation, which is located near 1587 cm^{-1}. This type of vibration is rarely strictly localized. In the case of OMP, it is coupled with C-C stretching and CH deformations. The coupling reduces the applicability of this type of vibrations for confirmation of the presence of pure OMP.

For the complete assignment of the vibrational spectra of OMP, we used characteristic spectra of particular fragments, which can be found in the molecule. Assignment of key vibrations was checked by calculation (Table 3). Figure 8 presents the most significant vibrations in the calculated spectra of OMP.

By general comparison of infrared and Raman spectra of all five prepared salts, we can conclude that they are structurally very similar with a few exceptions.

Raman and infrared spectra of OMP and all 5 prepared salts in solid state are shown in Figure 9.

Tables with assigned bands for OMP in Raman and infrared spectra were used to support the determination of specificity of individual spectra of salts (Tables 4 and 5).

Although the vibrational spectra of OMP and its salts are very similar, particular differences can be used for the accurate determination of the sample composition. Since all salts are in the form of crystal hydrates, it is that the NH stretching band will be masked by the broad OH stretching band of hydrating water. However, OMP has specific Raman bands of medium and medium-weak intensity at 1630 cm^{-1} (CH deformation, CC stretching) and 1512 cm^{-1} (CH + NH deformation) that are absent in the spectra of salts. On the other hand, all five prepared salts have characteristic bands of medium intensity at 3006 cm^{-1}–2980 cm^{-1} (CH stretching (BI, P)) and at 1364 cm^{-1} (C4H + C5H deformation).

Bands of the highest intensity in the range between 1300 and 1100 cm^{-1} are partly overlapped; therefore, the identification of specific salts is unlikely just by analysing the

TABLE 2: Peak frequencies of bands in infrared and Raman spectrum of OMP.

IR (cm^{-1})	Raman (cm^{-1})	Assignation
~3070 (shift)	3076	CH stretching (BI, P)
3057	3055	CH stretching (BI, P)
	3016	CH stretching (BI, P)
2984	2981	CH_3 (M) stretching asymmetric
2972	~2968 (sh)	CH_3 (M) stretching asymmetric
2944	2954	CH_3 (M) stretching asymmetric
~2930	Covered	NH stretching vibration
2904	2915	CH_3 stretching asymmetric
2854	2857	CH_3 (M) stretching symmetric
2835	2837	CH_3 stretching symmetric
1629	1629	QS (P), CH deformation, CC stretching (P, BI)
1587	1588	CH + NH deformation, CC stretching (BI)
1568	1569	QS (P), CH deformation, CC stretching (P, BI)
1512	1514	CH + NH deformation (BI)
1470 (sh)	1477	SCS (P), CH (B), NH deformation, N3C2 stretching
1462	1460	CH_3 deformation (M), NH + CH deformation (BI)
1428	1432	SCS (P), CH_3 deformation symmetric
1410	1412	NH + CH deformation (BI), CH deformation (P)
1356	1356	NH + C4H + C5H deformation (BI)
1312	1314	CH deformation, R-O-CH_3 stretching
1293		R-O-CH_3 stretching
1273	1273	NH + CH deformation symmetric (BI)
1253	1255	NH + CH deformation symmetric (BI)
1231	1232	NH + CH deformation asymmetric (BI)
1206	1203	NH + CH deformation symmetric (BI)
1188	1190	NH + CH deformation symmetric (BI)
1160	1160	CH deformation asymmetric
1113	1113	CH deformation asymmetric
1077	1076	NH deformation (BI), CH_3 (M)
1014	1009	S=O stretching vibration
967	963	CH deformation (BI)
885	886	CH wagging
835	842	CH3 rocking of methoxy group
822	823	QIPB (P)
810	790	CH3 rocking ip
787		CH3 rocking ip
759	762	CH wagging (IM)
732	733	CH wagging (B)
667	653	NH wagging (BI)
631	635	S-C stretching sulfoxide chain
621	620	QIPB (P), ring deformation
582	582	CH + NH wagging (BI)

TABLE 2: Continued.

IR (cm^{-1})	Raman (cm^{-1})	Assignation
546	546 (sh)	oop ring deformation
535	538	oop ring deformation
518	519	oop ring deformation
502	504	oop ring deformation
478	479	oop ring deformation
	444	R-O-CH3 deformation (M)
	429	Ring deformation (BI)
	382	QOOPB (P)
	327	CH wagging asymmetric (BI)
	254	R-O-CH3 deformation (M)
	212	CH wagging
	142	Phonon vibration
	124	Phonon vibration
	104	Phonon vibration
	87	Phonon vibration
	72	Phonon vibration
	66	Phonon vibration

BI: benzimidazole; P: pyridine; M: methoxy group; IM: imidazole; QS: quadrant stretch; SCS: semi-circle stretch; oop: out-of-plane; ip: in-plane; QIPB: quadrant in-plane bend; QOOPB: quadrant out-of-plane bend.

TABLE 3: Calculated stretching and deformation modes of OMP polar groups.

Group	Frequency (cm^{-1})
C-OCH3 (pyridine)	1423, 1181, 1013
C-OCH3	1442, 1060, 988
N=C-C-CH3	1547, 1411, 1250
HN-C-S	3015, 1570, 1411, 1182
N=C-S	1411, 1182
S=O	978, 618, 422, 381

aforementioned wavenumber range. By looking at infrared spectra, we come to the similar conclusion as in the case of Raman spectra. OMP spectrum clearly differs from the spectra of salts by the appearance of medium intensity bands at 1514 and 1356 cm^{-1} (CH + NH deformation (BI) and NH + C4H + C5H deformation (BI)), which are absent in salts. On the other hand, salts have a very specific peak at 1364 cm^{-1} (C4H + C5H deformation (BI)) of weaker intensity, which is absent in the spectrum of OMP. For specific distinction of each individual salt, we have to use the complementarity property of infrared and Raman spectroscopy with a comparison of more characteristic bands to determine specific OMP salt.

Specific OMP-K spectrum is a strong and narrow band at 3600 cm^{-1} in infrared spectrum (loosely hydrogen-bonded OH, OH stretching vibration of absorbed water molecules) and a medium intense band at 1400 cm^{-1} (CH deformation).

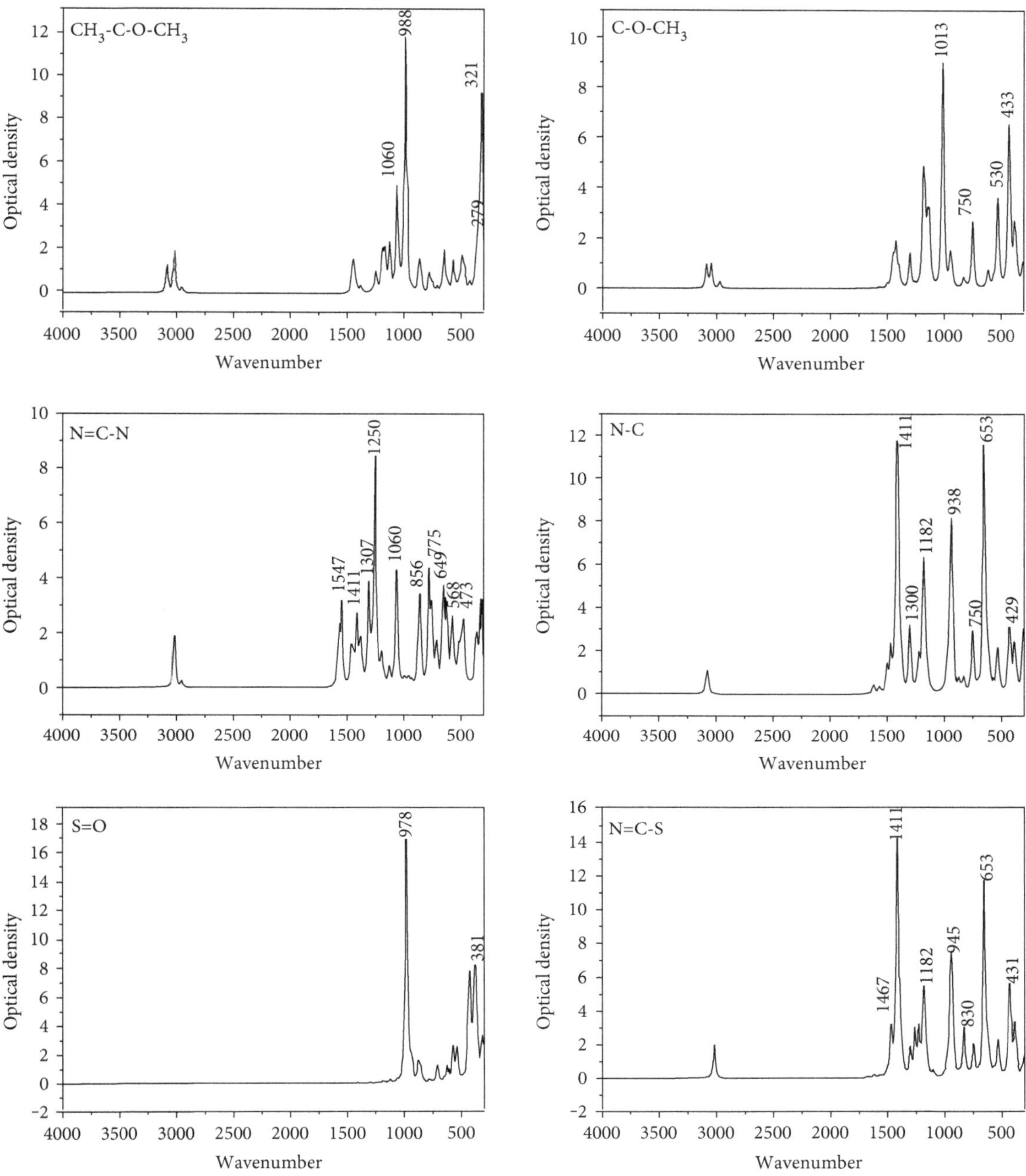

FIGURE 8: Calculated bands of selected omeprazole functional groups. These bands are taken from the calculation of the optimised OMP molecule. The whole calculated spectrum is presented in Figure 7.

The latter one can be used for identification of the presence of OMP-K salt. OMP-Na is characterized in infrared spectrum by an especially intense and for OH stretching at 3600–3100 cm^{-1} unusually narrow band. Raman spectrum bands at 1242 cm^{-1} (CH deformation) and at 655 cm^{-1} (S-C stretching) can be used for identification. Raman spectrum of OMP-Ca salts exhibits medium weak intensity band at 990 cm^{-1} (S=O stretching) that can be used for salt characterization, while in the infrared spectrum characteristic bands at 1396 cm^{-1} and 1023 cm^{-1} (–CH deformation / S=O stretching) individualize this salt.

In the same manner, OMP-Li is best described by Raman bands with strong and medium intensity which are located at 1190 cm^{-1} and 628 cm^{-1} (CH deformation, S-C stretching). Infrared spectrum medium intensity band at 1386 cm^{-1} can be used for salt determination.

Infrared spectrum of OMP-Mg contains less specific information. The most applicable are bands at 1410, 1230, and 637 cm^{-1}. Raman spectrum contains partially overlapped band of medium intensity at 2928 cm^{-1} (CH stretching) and a very intense band at 1253 cm^{-1} (CH symmetric deformation). There are also two highly specific bands at 1002 and

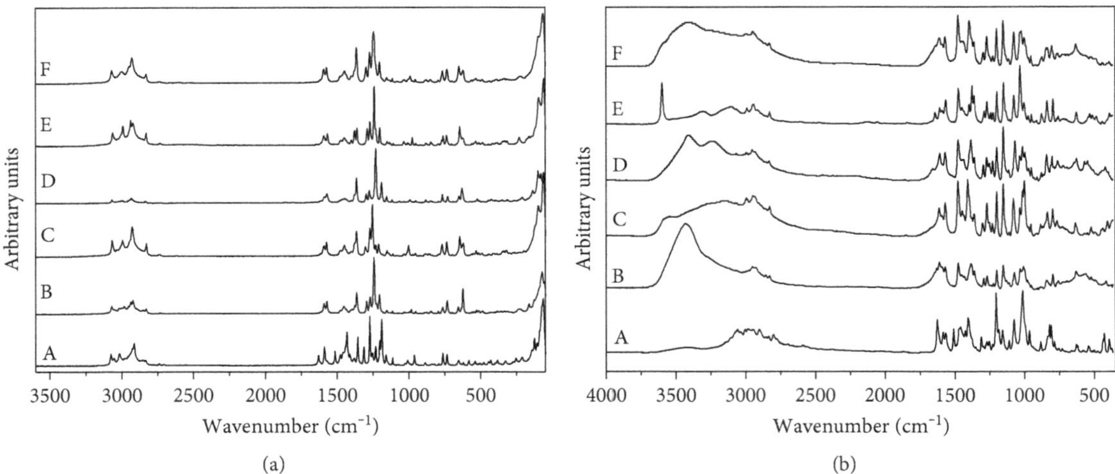

Figure 9: Raman (a) and infrared (b) spectra of OMP and prepared salts in solid state. (A) OMP, (B) OMP-Na, (C) OMP-Mg, (D) OMP-Li, (E) OMP-K, and (F) OMP-Ca.

Table 4: Assignment of characteristic vibrations in infrared spectra of OMP and its salts.

Vibration	OMP	OMP-Li	OMP-Na	OMP-K	OMP-Mg	OMP-Ca
O-H, stretching, H-bonded	—	3416 3241	3434	3316 3109	3543 3156	3406
O-H stretching, free	—	—	—	3600	—	—
CH, stretching (BI, P)	~3070	3063	3063	3062	3063	3063
CH_3 (M), stretching	2944	2935	2930	2947	2930	2947
NH, stretching	~2930	—	—	—	—	—
QS (P), CH deformation, CC stretching (P, BI)	1629	1608	1613	1608	1613	1613
CH + NH deformation, CC stretching (BI)	1587	—	—	—	—	—
CC stretching (BI)	—	1591	1591	1591	1591	1591
QS (P), CH deformation, CC stretching (P, BI)	1568	1571	1570	1566	1571	1571
CH + NH deformation (BI)	1512	—	—	—	—	—
SCS (P), CH (B), NH deformation, N3C2 stretching	1470 (shift)	1477	1477	1475	1477	1477
NH + CH deformation (BI), CH deformation (P)	1410	—	—	—	—	—
CH deformation (P)	—	1386	1384	1400	1410	1396
NH + CH deformation symmetric (BI)	1273	—	—	—	—	—
CH deformation (BI)	—	1270	1270	1270	1270	1270
NH + CH deformation symmetric (BI)	1253	—	—	—	—	—
CH deformation (BI)	—	1252	1242	1242	1252	1252
NH + CH deformation symmetric (BI)	1231	—	—	—	—	—
CH deformation (BI)	—	1228	1225	1225	1230	1224
NH + CH deformation symmetric (BI)	1206	—	—	—	—	—
CH deformation (BI)	—	1200	1200	1200	1200	1200
NH + CH deformation symmetric (BI)	1188	—	—	—	—	—
S=O stretching vibration	1014	1035 1015 998	1031 1008 —	1031 — 1003	1034 1015 1002	1031 1023 1002
CH deformation (BI)	967	950	950	950	950	950
S-C stretching	631 621	628 —	633 —	630 —	637 —	634 —

Table 5: Description and comparison of specific vibrations in Raman spectra of OMP and its salts.

Vibration	OMP	OMP-Li	OMP-Na	OMP-K	OMP-Mg	OMP-Ca
O-H, stretching, H-bond	—	—	—	—	—	—
O-H, stretching, free	—	—	—	—	—	—
CH, stretching (BI, P)	3076	3068	3070	3062	3066	3069
	3016	2994	2981	2991	2995	3000
CH_3 stretching, symmetric	2837	2834	2830	2830	2830	2830
NH. stretching	Overlapped	—	—	—	—	—
QS (P), CH deformation, CC stretching (P, BI)	1629	1589	1592	1592	1592	1592
CH + NH deformation, CC stretching (BI)	1588	1570	1572	1565	1572	1572
CH + NH deformation (BI)	1514	—	—	—	—	—
SCS (P), CH_3 deformation symmetric	1432	1450	1454	1450	1450	1446
NH + C4H + C5H deformation (BI)	1356	—	—	—	—	—
CH deformation (BI)	—	1364	1364	1364	1364	1364
NH + CH deformation symmetric (BI)	1273	—	—	—	—	—
CH deformation (BI)	—	1274	1273	1270	1272	1273
NH + CH deformation symmetric (BI)	1255	—	—	—	—	—
CH deformation (BI)	—	1230	1242	1241	1253	1243
NH + CH deformation symmetric (BI)	1232	—	—	—	—	—
CH deformation (BI)	—	1190	1225	1229	1231	1229
NH + CH deformation symmetric (BI)	1203	—	—	—	—	—
CH deformation (BI)	—	—	1206	1201	1211	1203
NH + CH deformation symmetric (BI)	1190	—	—	—	—	—
S=O stretching vibration	1009	992	985	1006	1002	990
CH deformation (BI)	963	954	954	954	954	954
S-C stretching	653	650	655	644	644	649
	635	628	623	620	620	620

644 cm^{-1}, but they are of weaker intensity and therefore would be poorly visible in the multicomponent mixture.

4. Conclusions

NMR and XRPD methods were used to confirm OMP salt formation by spectra/diffractogram analyses and comparisons with patent data. By comparing ^1H MAS NMR spectra of OMP and salts, the absence of a signal at 13.5 ppm in the latter has been attributed to the deprotonation of nitrogen in salts. Furthermore, the comparison of ^{13}C CP-MAS NMR spectra of solid samples of OMP and its salts showed the shift of signal at 92.0 ppm which is assigned to the atom C7 in OMP to 98.0 ppm in salt spectra. This shift may be initiated by structural changes in the vicinity of the C7 atom. Sodium, magnesium, and calcium salts were additionally compared and confirmed by XRPD technique (patent data comparison). Potassium and lithium salts had their first description by using the XRPD technique.

Thermal techniques (TGA and DSC) were used to test the solid-state properties of prepared salts. The results showed consistent phase transitions in all salts with the exception of OMP base which stayed unchanged during the whole cycle up to the melting point. We showed that all salts exhibit hydrate form formation.

We assigned infrared and Raman spectra by the application of already assigned model molecules and quantum calculation of vibrational frequencies. It has been confirmed that salt formations occur by hydrogen cation cleavage from N-H bond on benzimidazole ring (absence of bands on 1514 and 582 cm^{-1} in Raman spectra of salts) and electron delocalisation in sulfoxide chain (bands in the region between 1009 cm^{-1} and 1016 cm^{-1} and between 660 and 600 cm^{-1}). We found several distinct bands which are characteristic for a particular formation of different OMP salts. Therefore, we have shown that both infrared and Raman spectroscopy are relevant techniques to determine the presence of particular salts in more complex mixtures.

References

[1] M. Hickey and M. Peterson, "Novel omeprazole forms and related methods," Patent: US 2006/0160783 A1, 2006.

[2] T. Lind, C. Cederberg, G. Ekenved, U. Haglund, and L. Olbe, "Effect of omeprazole-a gastric proton pump inhibitor-on pentagastrin stimulated acid secretion in man," *Gut*, vol. 24, no. 4, pp. 270–276, 1983.

[3] M. J. Kendall, "Review article: esomeprazole - the first proton pump inhibitor to be developed as an isomer," *Alimentary Pharmacology & Therapeutics*, vol. 17, Supplement 1, pp. 1–4, 2003.

[4] A. A. Al-Badr, "Omeprazole," in *Profiles of drug substances, excipients and related methodology*, H. G. Brittain, Ed., vol. 35, pp. 151–262, Elsevier, Amsterdam, 2008.

[5] A. Gustavsson, K. Kjellbom, and I. Ymén, "Omeprazole sodium salt," Patent: US 6.207.188 B1, 2001.

[6] M. Reddy, S. Eswaraiah, V. V. N. K. V. Raju et al., "Crystalline form C of omeprazole sodium and the related process of its preparation, a crystalline form D of omeprazole sodium and the related process of its preparation, and a process for preparation of crystalline form a of omeprazole sodium," Patent: US 2004/0224987, 2004.

[7] L. Vranicar Savanovic, Z. Ham, and J. Rzen, "Crystalline solvate of omeprazole sodium," Patent: US 2009/0221646, 2009.

[8] N. Markovic, S. Agntovic-Kustrin, B. Glass, and C. A. Prestidge, "Physical and thermal characterisation of chiral omeprazole sodium salts," *Journal of Pharmaceutical and Biomedical Analysis*, vol. 42, no. 1, pp. 25–31, 2006.

[9] N. Furuyama, S. Hasegawa, T. Hamaura et al., "Evaluation of solid dispersions on a molecular level by the Raman mapping technique," *International Journal of Pharmaceutics*, vol. 361, no. 1-2, pp. 12–18, 2008.

[10] E. Widjaja and R. K. Hong Seah, "Application of Raman microscopy and band-target entropy minimization to identify minor components in model pharmaceutical tablets," *Journal of Pharmaceutical and Biomedical Analysis*, vol. 46, no. 2, pp. 274–281, 2008.

[11] D. Clark and S. Šašić, "Chemical images: technical approaches and issues," *Cytometry Part A*, vol. 69A, no. 8, pp. 815–824, 2006.

[12] A. E. Brändström, "Omeprazole salts," Patent: EP 0 124 495 B1, 1987.

[13] N. H. Milac, A. Copar, B. Podobnik et al., "Crystalline form of omeprazole," Patent: US 7.553.856 B2, 2009.

[14] N. Kumar, T. Sharma, and B. Vijayaraghavan, "Novel amorphous form of omeprazole salts," Patent: WO 01/87831 A2, 2002.

[15] G. Kresse and J. Hafner, "*Ab initio* molecular dynamics for liquid metals," *Physical Review B*, vol. 47, p. 558, 1993.

[16] G. Kresse and J. Furthmuller, "Efficient iterative schemes for *ab initio* total-energy calculations using a plane-wave basis set," *Physical Review B*, vol. 54, article 11169, 1996.

[17] G. Kresse and D. Joubert, "From ultrasoft pseudopotentials to the projector augmented-wave method," *Physical Review B*, vol. 59, p. 1758, 1999.

[18] J. P. Perdew, K. Burke, and M. Ernzerhof, "Generalized gradient approximation made simple," *Physical Review Letters*, vol. 77, p. 3865, 1996, Erratum Phys. Rev. Lett., vol. 78, pp. 1396, 1997.

[19] H. J. Monkhorst and J. D. Pack, "Special points for Brillouin-zone integrations," *Physical Review B*, vol. 13, p. 5188, 1976.

[20] P. M. Bhatt and G. R. Desiraju, "Tautomeric polymorphism in omeprazole," *Chemical Communications*, vol. 20, pp. 2057–2059, 2007.

[21] D. M. Wood and A. Zunger, "A new method for diagonalising large matrices," *Journal of Physics A: Mathematical and General*, vol. 18, no. 9, p. 1343, 1985.

[22] S. Baroni, S. de Gironcoli, and A. del Corso, "Phonons and related crystal properties from density-functional perturbation theory," *Reviews of Modern Physics*, vol. 73, p. 515, 2001.

[23] F. S. Murakami, A. P. Cruz, R. N. Pereira, B. R. Valente, and M. A. S. Silva, "Development and validation of a RP-HPLC method to quantify omeprazole in delayed release tablets," *Journal of Liquid Chromatography & Related Technologies*, vol. 30, no. 1, pp. 113–121, 2007.

[24] S. Martindale, *The Complete Drug Reference*, Pharmaceutical Press, London, Thomson Micromedex, Greenwood Village, 33 edition, 2002.

[25] M. Dellagreca, M. R. Iesce, L. Previtera, M. Rubino, F. Temussi, and M. Brigante, "Degradation of lansoprazole and omeprazole in the aquatic environment," *Chemosphere*, vol. 63, no. 7, pp. 1087–1093, 2006.

[26] R. M. Claramunt, C. López, and J. Elguero, "The structure of *omeprazole* in the solid state: a ^{13}C and ^{15}N NMR/CPMAS study," *ARKIVOC*, vol. 2006, no. 5, pp. 5–11, 2006.

[27] M. A. Palafox, "Scaling factors for the prediction of vibrational spectra. I. Benzene molecule," *International Journal of Quantum Chemistry*, vol. 77, no. 3, pp. 661–684, 2000.

[28] M. A. Morsy, M. A. Al-Khalidi, and A. Suwaiyan, "Normal vibrational mode analysis and assignment of benzimidazole by ab initio and density functional calculations and polarized infrared and Raman spectroscopy," *The Journal of Physical Chemistry A*, vol. 106, no. 40, pp. 9196–9203, 2002.

[29] P. J. Larkin, *IR and Raman Spectroscopy Principles and Spectral Interpretation*, Elsevier, Waltham, MA, USA, 2011.

[30] K. C. Medhi, "Vibrational spectra and thermodynamic functions of 2-methoxypyridine and 2-methoxy-d_3-pyridine," *Bulletin of the Chemical Society of Japan*, vol. 57, no. 1, pp. 261–266, 1984.

A Derivative Spectrometric Method for Hydroquinone Determination in the Presence of Kojic Acid, Glycolic Acid, and Ascorbic Acid

Zenovia Moldovan, Dana Elena Popa, Iulia Gabriela David, Mihaela Buleandra, and Irinel Adriana Badea

Department of Analytical Chemistry, Faculty of Chemistry, University of Bucharest, 4-12 Regina Elisabeta Av., District 3, 030018 Bucharest, Romania

Correspondence should be addressed to Dana Elena Popa; dana_lena1978@yahoo.com

Academic Editor: Jose S. Camara

A new, simple, and sensitive spectrometric method was developed for hydroquinone (HQ) determination in the presence of other depigmenting agents (kojic acid (KA), glycolic acid (GA), and ascorbic acid (AA)), commonly introduced in skin lightening products. The method is based on the oxidation of the depigmenting agents by potassium dichromate in sulfuric acid medium and subsequent measurement of the amplitude of the first-order derivative absorption spectrum at 268 nm. By applying the zero-crossing method, at this wavelength, the oxidation products of KA, AA, and GA do not interfere in the indirect determination of HQ. Beer's law was obeyed in the range of 0.22–22 $\mu g \cdot mL^{-1}$ HQ, with a detection limit of 0.07 $\mu g \cdot mL^{-1}$. The developed method was applied with good results for the first time to the rapid determination of HQ in binary, ternary, and quaternary mixtures, thus proving that it could represent an effective tool for various skin lightening products analyses.

1. Introduction

Human skin contains melanocytes (cells located at the base of the epidermis), which produce melanin (a dark macromolecular vital pigment) by a combination of enzymatically catalyzed chemical reactions. This process is named melanogenesis and it intensifies after exposure to UVB radiation, causing the skin to visibly tan. The aim of the melanogenesis is to protect the hypodermis from the DNA photodamage. The first step consists of tyrosine oxidation to dopaquinone catalyzed by tyrosinase [1]. The abnormal accumulation of melanin induces melasma, a chronic skin disorder that results in brownish facial pigmentation. Taking into account that tyrosinase is the enzyme responsible for the melanin synthesis tyrosinase inhibitors are used as whitening or anti-hyperpigment agents due to their ability to suppress dermal-melanin production [2].

Hydroquinone (HQ) is considered to be one of the strongest inhibitors of melanin production and for more than 25 years it has been established as the most effective ingredient for treating melasma [3]. However, its long-term application has numerous adverse effects, including irritative dermatitis, melanocyte destruction, contact dermatitis, and ochronosis. Matsumoto et al. [4] published recently a study related to the risk of systemic effects of HQ when using skin lightening cosmetics containing it. The adverse effects of HQ are transitory below the 3.0% level (2% HQ is the maximum concentration permitted by the United States Food and Drug Administration (USFDA), whereas 4.0% HQ formulations are only available by prescription [5]). Concentrations of HQ above 5.0% could cause local irritation [6] and even persistent hypopigmentation named leukoderma [7]. Therefore, over the past years HQ has become a controversial skin-care agent for topical use. It should be mentioned that the whitening products containing HQ have been banned in many countries because of concerns related to cancer risk [8]. However, in order to minimize the risk of side effects, many clinical studies reported partial or total replacement of HQ in various cosmetics. The medical literature data also report that combination therapy is more effective than single agent use

[9]. As a result, a series of dermatological creams contain binary, ternary, or quaternary mixtures of HQ and other tyrosinase inhibitors such as glycolic acid (GA), kojic acid (KA), and ascorbic acid (AA). The addition of this last one, which is a well-known antioxidant, enhances the stability of HQ as it can be easily oxidized (even in a tube) and become ineffective [10].

Taking into consideration both benefits and risks of using HQ-containing cosmetics, the quantitative determination of the HQ level in bleaching creams is imperative. For this purpose, many studies on HQ determination in different cosmetics are reported. The employed analytical methods are based on the specific properties of HQ exploited by chromatographic (HPLC [11–14]), capillary electrochromatographic [15], voltammetric [16–19], and spectrometric techniques [20–31]. The advantages of the spectrometric techniques consist in the fact that they use accessible and simpler equipment, have shorter analysis time, and are cheaper than the chromatographic techniques. The use of UV-Vis spectrometry has enhanced rapidly over the last few years. Some of the advantages of these methods are precision, short analysis time, and less reagents consumption [20]. The spectrometric determinations of HQ in cosmetic products were based on direct measurement of UV absorbance of HQ [21, 22], UV derivative spectrometry [23], spectrometric ratio difference method [24], the successive ratio subtraction coupled with constant multiplication UV spectrometry [25], fluorescence spectroscopy [26], or the absorbance measurement of the product resulting in a redox reaction between HQ and specific reagents. For instance, trace levels of HQ were determined by UV absorption measurements after its oxidation to p-benzoquinone (BQ) by oxygen in the presence of ammonium meta-vanadate as an oxidizing catalyst [27]. The catalytic oxidation of HQ to BQ by $KMnO_4$ in alkaline medium was also used for the spectrometric determination of HQ [28]. Ammonium molybdate (Mo(VI)) was used to oxidize HQ in acidic medium and the resulting molybdenum (V) was spectrometrically monitorized [29]. The inhibitory effect of HQ on the oxidation of an organic reagent (Rhodamine B) was used for HQ determination by a kinetic spectrometric method [30]. UV spectrometric investigations of the HQ polymerization in the presence of Cr(VI) were also recently reported [31].

In the present work a simple, accurate, and precise first-order derivative spectrometric method was proposed for the first time to quantify HQ in the presence of other depigmenting agents, namely, KA, GA, and AA, commonly present in cosmetic products. The method is based on the oxidation of HQ by $K_2Cr_2O_7$ in sulfuric acid medium and subsequent absorbance measurement of the first-order derivative spectrum of the oxidation product (BQ) at 268 nm.

2. Materials and Methods

2.1. Chemicals. All chemicals were of analytical reagent grade and were purchased from Sigma-Aldrich. Deionized-distilled water was used throughout the experiments.

Aqueous stock solutions of HQ, KA, GA, and AA (10^{-2} mol·L^{-1}) were freshly prepared and used to obtain the working standard solutions. When not used, the solutions were stored in the refrigerator. A 5 mol·L^{-1} H_2SO_4 solution was obtained by diluting concentrated sulfuric acid (98%, 1.84 g·mL^{-1}). Working solutions of 10^{-3} and 5×10^{-3} mol·L^{-1} $K_2Cr_2O_7$ resulted after dilution of a 10^{-1} mol·L^{-1} $K_2Cr_2O_7$ stock solution. Eppendorf vary-pipettes (10–100; 100–1000; and 500–2500 μL) were employed to deliver accurate volumes.

2.1.1. Depigmenting Agents in Acid Medium. An aliquot (1 mL) of 10^{-3} mol·L^{-1} solution of HQ, AA, KA, or GA was transferred into a 5 mL volumetric flask. After adding 5 mol·L^{-1} H_2SO_4 (1 mL), the mixture was brought to the mark with distilled water and homogenized. Then, the absorbance spectrum was recorded against water as reference. The derivative spectra were plotted with a 2 nm interval from the zero-order spectra of the individual analyzed solutions.

2.1.2. Mixtures of Depigmenting Agents and Potassium Dichromate in Acid Medium. An aliquot (1 mL) of 10^{-3} mol·L^{-1} solution of HQ, KA, GA, or AA, 1 mL of 5 mol·L^{-1} H_2SO_4, and a known volume of 10^{-3} or 5×10^{-3} mol·L^{-1} $K_2Cr_2O_7$ (added in small excess, to obtain a weak yellow colored solution) were transferred into a 5 mL volumetric flask and diluted to the mark. For each solution the absorption spectrum was recorded in the range 215–400 nm, against water as reference. This was necessary due to the fact that between 325 and 400 nm only $K_2Cr_2O_7$ is absorbed, thus being mandatory the removal of the $K_2Cr_2O_7$ excess influence. For each mixture, a corresponding blank solution was prepared by transferring into a 5 mL calibrated flask 1 mL of 5 mol·L^{-1} H_2SO_4 and a known volume of 10^{-3} or 5×10^{-3} mol·L^{-1} $K_2Cr_2O_7$ to obtain a final aqueous solution having the concentration equal to that of the unreacted $K_2Cr_2O_7$ (deduced from its absorbance at 350 nm; see the details given in Section 3.1.). The spectrum of the analyte oxidation product was obtained by subtracting the blank spectrum from the spectrum recorded for the analyte in the presence of a small excess of the oxidant ($K_2Cr_2O_7$).

Binary, ternary, and quaternary mixtures of HQ and other dermatological active agents (KA, GA, and AA) in acidic medium and in the presence of $K_2Cr_2O_7$ were prepared in the same manner as the individual depigmenting agents. The differences between absorbencies of the mixtures and the corresponding blank solutions were also obtained.

2.2. Apparatus. Absorbance measurements were performed in the 215–400 nm wavelength range on a UV-VIS spectrometer (V-530 Jasco-Japan), with fully integrated PC running Spectra Manager software, equipped with quartz cells of 1.00 cm. Suitable settings were slit width 1 cm and scan speed 100 nm·min^{-1}.

3. Results and Discussions

3.1. Zero- and First-Order Derivative UV Spectrometric Studies. HQ, KA, GA, and AA absorb UV radiation having char-

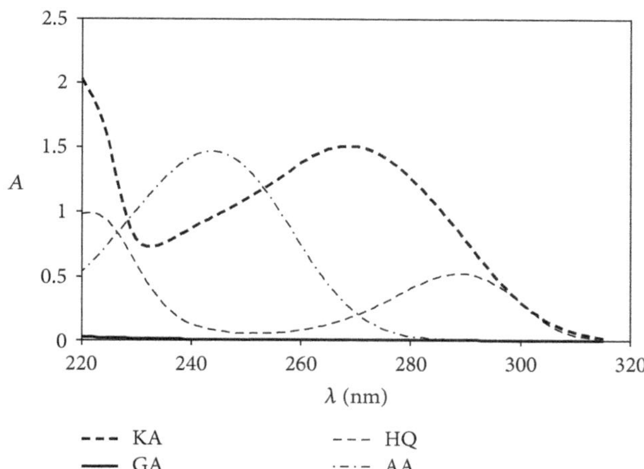

Figure 1: UV absorption spectra of HQ, KA, AA, and GA; every compound has the same concentration (2×10^{-4} mol·L^{-1}).

acteristic absorption spectra (Figure 1). As it can be observed, the spectra of HQ and KA overlap significantly. Moreover, in the given experimental conditions, the calculated molar absorptivity (ε_{HQ} = 2600 L·mol^{-1}·cm^{-1} at $\lambda_{max,HQ}$ = 289 nm) of HQ is smaller than that of AA (ε_{AA} = 7350 L·mol^{-1}·cm^{-1} at $\lambda_{max,AA}$ = 244 nm) and of KA (ε_{KA} = 7500 L·mol^{-1}·cm^{-1} at $\lambda_{max,KA}$ = 269 nm).

By applying the zero-crossing method, the first- and second-order derivative spectra of HQ, KA, and AA do not permit the determination of HQ in ternary mixtures due to the fact that there is no wavelength where only HQ presents a measurable signal (Figures 2(a) and 2(b)).

It was observed that HQ presents an analytical signal at 303 nm in its third-order derivative spectrum, while the same order derivative spectra of the other tested compounds intersect the abscissa (Figure 2(c)). However, the analytical signal of HQ is very small, even at a concentration of 2×10^{-4} mol·L^{-1} (22 μg·mL^{-1}), which leads to a very low sensitivity.

Under these circumstances, a new methodology was established. This one considers the capacity to be oxidized of the above-mentioned compounds when $K_2Cr_2O_7$ in acidic medium is used as oxidizing agent. Thus, another series of UV spectra were recorded for the individual active compounds in the presence of $K_2Cr_2O_7$ in sulfuric acid medium. As depicted in Figure 3 the oxidation product of HQ in presence of the $K_2Cr_2O_7$ excess (spectrum (1)) exhibits an absorption band in the same wavelength region (325–400 nm) as the $K_2Cr_2O_7$ solution (spectrum (2)). It must be mentioned that the UV absorption spectrum of BQ (not shown) presents a characteristic absorption band in the range 225–250 nm similar to the spectrum of the HQ oxidation product resulting from the reaction of HQ with $K_2Cr_2O_7$ in sulfuric acid medium (spectrum (2)). In order to obtain only the spectrum of the HQ oxidation product (spectrum (3)), the spectral subtraction of spectrum (2) from spectrum (1) was performed. Each analyzed mixture was prepared in the presence of a known $K_2Cr_2O_7$ excess.

The concentration of the unreacted $K_2Cr_2O_7$ was deduced from the absorbance of the mixture at 350 nm, where only dichromate ion is absorbed. To achieve this aim a calibration curve was accomplished using $K_2Cr_2O_7$ solutions with different concentrations (2×10^{-5}–2×10^{-4} mol·L^{-1}) in 1 mol·L^{-1} H_2SO_4. Using the equation of the regression line ($A_{350\,nm} = 0.4875c - 0.0002$; $R^2 = 0.9994$; c is the dichromate concentration) the blank concentration (having the same absorbance at 350 nm as the mixture of the analytes in presence of dichromate excess) is calculated.

The UV spectra of the investigated compounds in the absence and in the presence of $K_2Cr_2O_7$ differ significantly in the case of HQ and AA (Figure 4), whereas for KA only a small decrease of the absorption band intensity is observed in the presence of $K_2Cr_2O_7$ in sulfuric acid medium, suggesting that KA is not oxidized in these conditions.

The redox reactions between HQ, GA [32], AA [33], and $K_2Cr_2O_7$ in sulfuric acid medium are shown in Scheme 1.

By overlaying the spectra of the reaction products of HQ, KA, GA, AA, and $K_2Cr_2O_7$, the first one cannot be determined in presence of the others (Figure 5(a)). Although at the first sight one may consider that HQ can be determined in the presence of GA and AA, it must be mentioned that in real samples the GA and AA concentrations are even five times higher than the HQ concentration, their contribution becoming significant.

Applying the first-order derivative, the analytical signal attributed to the oxidation product of HQ (benzoquinone (BQ)) could be used for the indirect determination of HQ in the presence of the other ingredients, at the zero-crossing of KA oxidation product (268 nm), where the amplitudes of the first-order derivative spectra of AA and GA oxidation products are also zero (Figure 5(b)).

3.2. Optimization of the Working Parameters. In order to optimize the working conditions of the proposed method, the influence of the sulfuric acid concentration at constant HQ and dichromate contents was studied. It was observed that the analytical signal increases with increasing the concentration of H_2SO_4 up to 1 mol·L^{-1}; then it remains almost constant. Further experiments were made on samples prepared in 1 mol·L^{-1} H_2SO_4. The stability of the reaction product between HQ and dichromate was monitored by spectrometry in the time range 0–1800 sec. The measurements made at 240 nm (the wavelength corresponding to the maximum absorbance of the HQ oxidation product in the presence of dichromate) indicated that the absorbance was stable within the tested period.

3.3. Analytical Parameters of the Indirect Spectrometric Method Developed for HQ Quantitative Determination. Using the above optimized spectrophotometric method developed for the indirect determination of HQ (via its oxidation in the presence of dichromate) a linear relationship was obtained between $dA/d\lambda$ (at 268 nm) and the HQ concentration. The parameters of the calibration curve, obtained by linear square regression of the results, are given in Table 1.

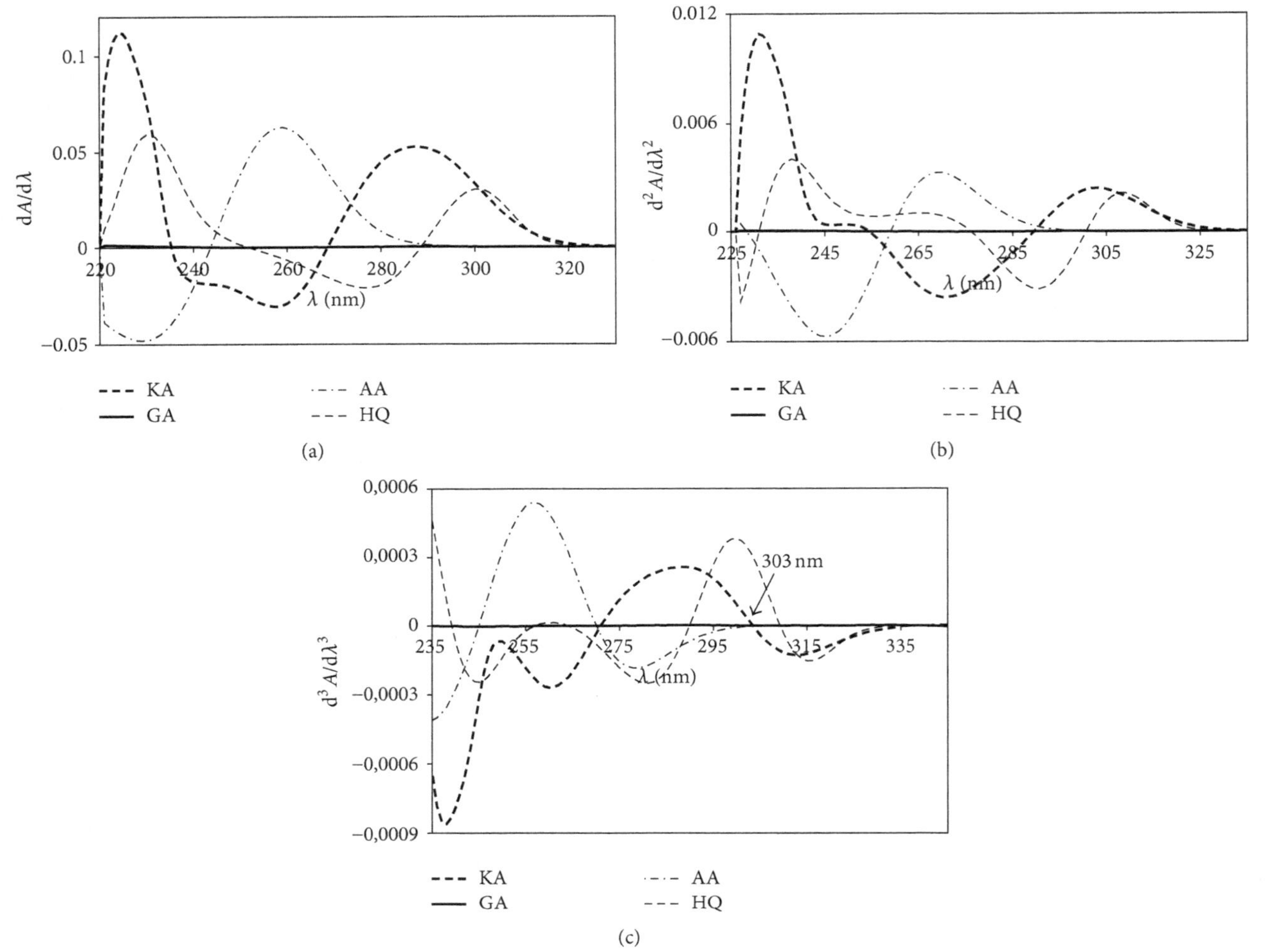

FIGURE 2: (a) First-order, (b) second-order, and (c) third-order derivative spectra of HQ, KA, AA, and GA; every compound has the same concentration (2×10^{-4} mol·L^{-1}).

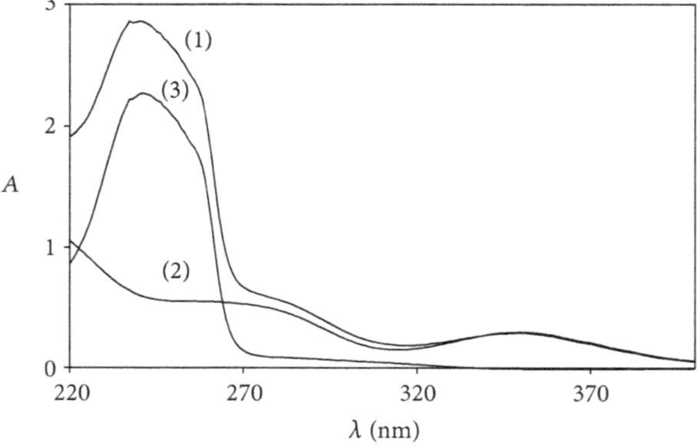

FIGURE 3: UV spectra of (1) 2×10^{-4} mol·L^{-1} HQ in the presence of 2×10^{-4} mol·L^{-1} $K_2Cr_2O_7$ (in excess), in 1 mol·L^{-1} sulfuric acid; (2) $K_2Cr_2O_7$ at a level of concentration corresponding to those unreacted in solution (1); (3) = (1) − (2).

The limit of quantification (LOQ) was determined by the analysis of samples with known concentration of HQ and by establishing the minimum level at which the analyte can be quantified with acceptable accuracy and precision. The limit of detection (LOD) was considered as the signal to noise ratio of 3 : 1 [34].

A comparison with other reported methods shows that the proposed spectrometric method is more sensitive, with

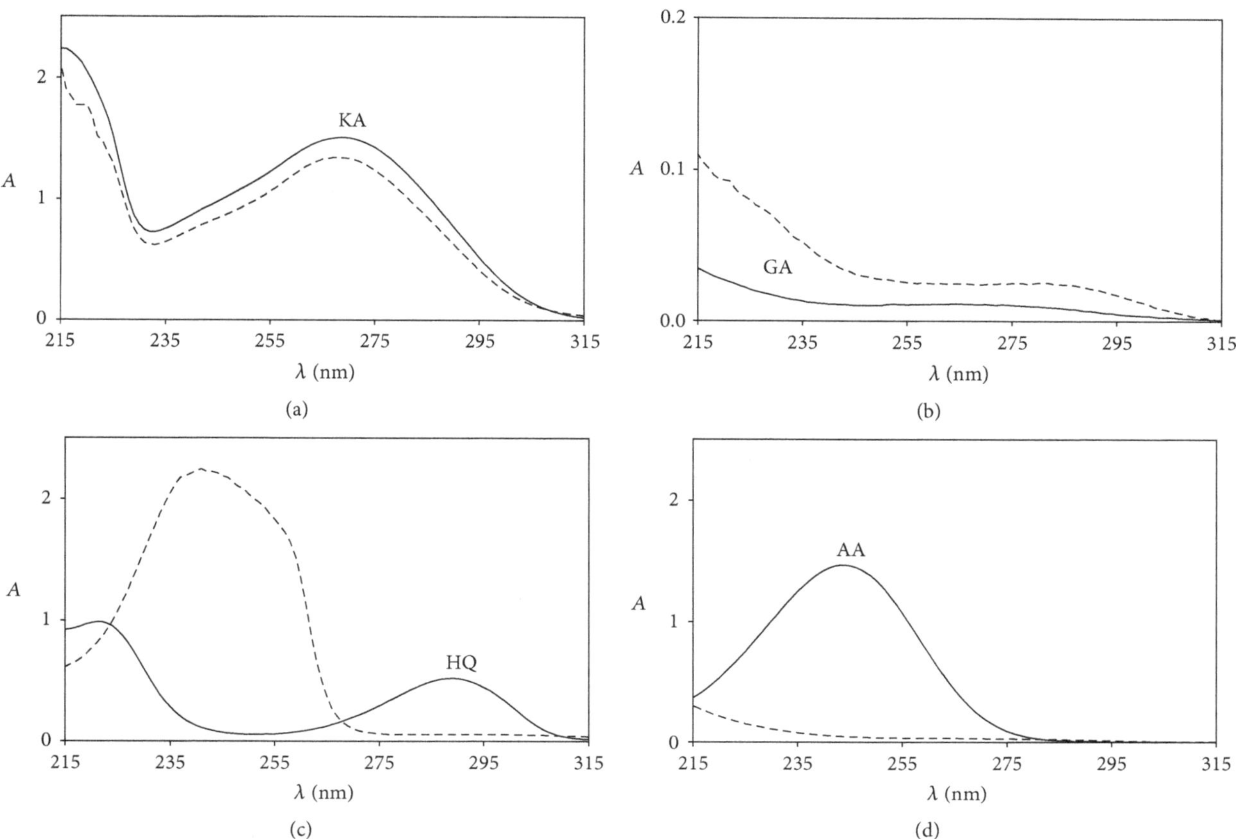

FIGURE 4: Overlaid spectra of the studied compounds in the absence (solid lines) and in the presence (dashed line) of $K_2Cr_2O_7$ in 1 mol·L^{-1} sulfuric acid; every compound has the same concentration (2×10^{-4} mol·L^{-1}).

TABLE 1: Analytical parameters of the first-order derivative spectrometric method for the indirect HQ determination.

Parameter	Value
Linear range, μg·mL^{-1}	0.22–22
Intercept, a	0.0018
Intercept standard deviation, s_a	0.00032
Slope, b	0.0031
Slope standard deviation, s_b	0.00003
Determination coefficient, R^2	0.9994
LOD, μg·mL^{-1}	0.07
LOQ, μg·mL^{-1}	0.22

a larger linear range of 0.22–22 μg·mL^{-1}. At the same time, the proposed method is less time consuming, no heating is required, and it is eco-friendly (no use of organic solvents) and inexpensive (Table 2).

3.4. Precision and Accuracy. The precision and the accuracy of the proposed spectrometric method were obtained using solutions containing HQ at three different concentration levels, within the established linear range. The results presented in Table 3 show high accuracy (estimated by the percent recovery (R%), between 99.39% and 100.15%) and precision (estimated by the means of relative standard deviation (RSD%), between 1.34% and 2.60%) of the obtained results.

The obtained percent recovery values lie within the accepted limits for these concentration levels [35].

3.5. Application of the Developed Method to the Determination of HQ in Binary, Ternary, and Quaternary Mixtures. The literature data report various combinations of different topical agents for melasma treatment [36]. Hydroquinone is generally the main component of the formulations [10]. It is usually combined with glycolic acid, kojic acid, and ascorbic acid resulting in binary (4% HQ + 2% GA; 4% HQ + 10% GA; 2% HQ + 2% KA) [37, 38], (4% HQ + 10% AA) [39]; ternary (2% HQ + 10% GA + 2–4% KA) [40], (4% HQ + 2.5% AA + 0.75% KA) [41]; or quaternary (2% HQ + GA + KA + AA) [42] mixtures.

As it can be seen, the four ingredients are found in the dermatological formulations in variable mixtures and concentrations (HQ $\leq$ 4%; AA $\leq$ 10%; KA $\leq$ 2%; GA $\leq$ 10%). In order to determine how the possible interfering compounds affect the HQ quantitation, binary mixtures were prepared, the concentration of the other active compounds

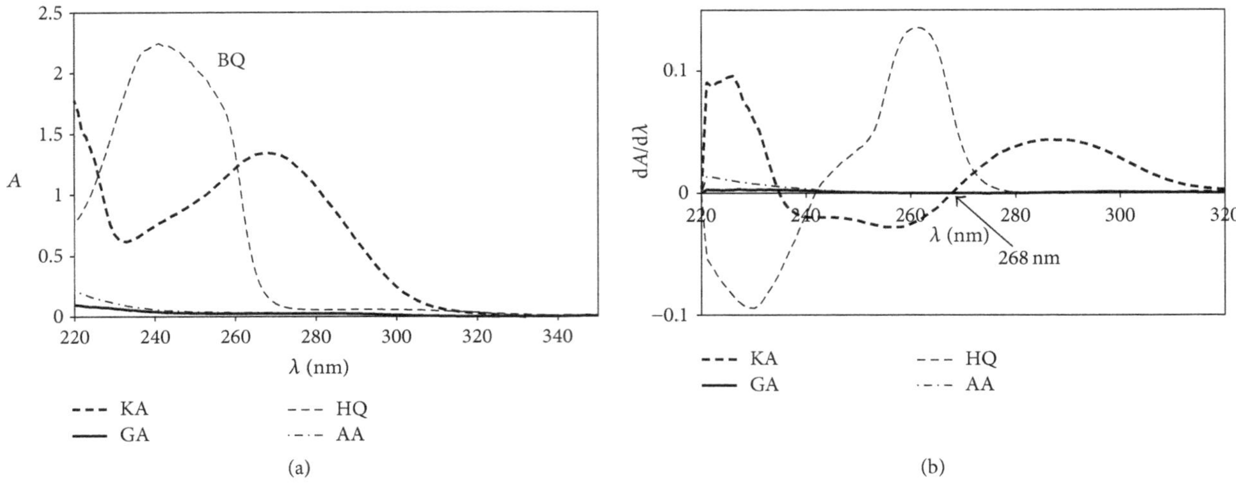

FIGURE 5: Overlaid (a) zero-order and (b) first-order derivative spectra of HQ, KA, GA, and AA in the presence of $K_2Cr_2O_7$ in $1\,mol{\cdot}L^{-1}$ sulfuric acid; every compound has the same concentration ($2 \times 10^{-4}\,mol{\cdot}L^{-1}$).

TABLE 2: Comparison of the proposed method with the reported spectrometric methods for HQ determination.

Analytical signal/working conditions	Linear range, $\mu g{\cdot}mL^{-1}$	Application	Ref.
$A_{302\,nm}$/HQ in H_2SO_4	2–12	Body lotions	[22]
$A_{293\,nm}$/HQ in methanolic solution	10–50	Cosmetic creams	[21]
$A_{293\,nm}$/HQ + O_2 + NH_4VO_3 in 2-propanol : water 1 : 1	0.025–2	Cosmetic creams	[27]
$dA/d\lambda_{302\,nm}$/HQ in H_2SO_4	10–26	Cosmetic creams	[23]
$A_{580\,nm}$/HQ + $(NH_4)_2MoO_4$ + 10% H_2SO_4; 20-minute reaction time, 100°C	10–100	Skin whitening formulations	[29]
$A_{610\,nm}$/HQ + $KMnO_4$ in alkaline medium; 30–35-minute reaction time	1–26	Pharmaceuticals	[28]
$dA/d\lambda_{302\,nm}$/HQ in the presence of KA, AA, and GA + $K_2Cr_2O_7$ in H_2SO_4, instantaneous reaction; room temperature	0.22–22	Synthetic mixtures	Present work

being smaller than, equal to, and higher than the HQ concentration.

The recipes of the dermatological formulations that contain all the four active ingredients do not contain information about the KA, GA, and AA concentrations. Therefore, in the present study, the ternary mixtures contain all the four components at the lowest concentration level (percent ratio, HQ : KA : GA : AA = 1 : 1 : 1 : 1) and at the highest accepted concentration level (percent ratio, HQ : KA : GA : AA = 2 : 1 : 5 : 5).

The determination of HQ in combination with the mentioned active ingredients was studied by applying the proposed indirect derivative spectrometric method. Different combinations of HQ and the other depigmenting compounds and the analytical results for HQ determination are given in Tables 4 and 5. The percent recovery values (R%) summarized in Table 3 are very close to 100% (between 99.20 and 101.60%), a fact that reveals the accuracy of the proposed method in the determination of HQ in binary mixtures.

TABLE 3: The precision and accuracy of the results obtained by the proposed spectrometric method; n -number of independent measurements; $n = 5$; SD: standard deviation.

HQ, $\mu g{\cdot}mL^{-1}$		RSD, %	R, %
Considered	Found ± SD		
2.2	2.19 ± 0.05	2.31	99.39
5.5	5.51 ± 0.07	1.34	100.15
11	10.99 ± 0.29	2.60	99.88

In the case of ternary and quaternary mixtures, HQ can be determined by the proposed spectrometric method with good recovery values, the standard deviations supporting also the fact that the results are reliable and comparable.

All the experimental data obtained during this study led to the results presented throughout this paper and these revealed that the developed spectrometric method is a useful

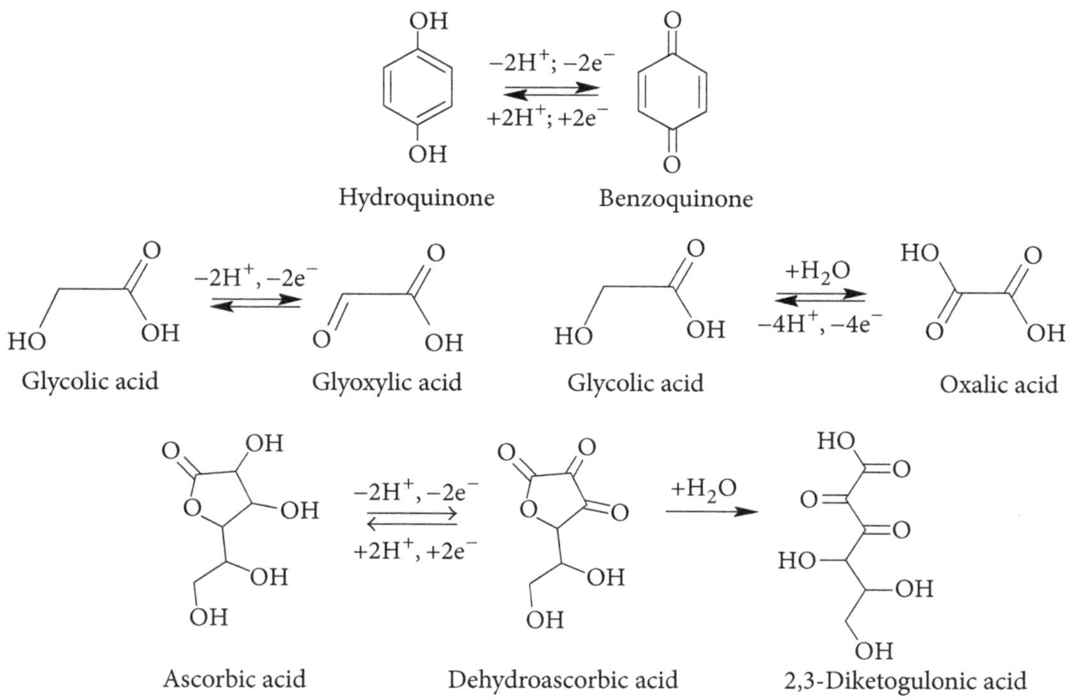

Scheme 1

Table 4: Results of the HQ determination in synthetic binary mixtures by applying the proposed spectrometric method; $n = 3$.

Mixtures	Percent ratio, HQ : KA/AA/GA	HQ concentration, $\mu g \cdot mL^{-1}$ Considered	Found ± SD	R%
HQ + KA	4 : 1	4	4.04 ± 0.08	101.06
	2 : 1	4	4.04 ± 0.05	101.06
	1 : 1	4	4.06 ± 0.07	101.33
	1 : 1.5	4	3.96 ± 0.05	99.47
	1 : 2	4	3.97 ± 0.04	99.20
HQ + AA	2 : 1	4	3.99 ± 0.06	99.73
	1 : 1	4	3.98 ± 0.05	99.47
	1 : 2	4	4.02 ± 0.04	100.53
	1 : 2.5	4	4.00 ± 0.06	100.00
	1 : 5	4	4.04 ± 0.07	101.06
HQ + GA	1 : 1	4	4.04 ± 0.05	101.06
	1 : 2	4	4.05 ± 0.06	101.17
	1 : 3	4	4.03 ± 0.07	100.80
	1 : 4	4	4.02 ± 0.06	100.53
	1 : 5	4	4.06 ± 0.08	101.60

analytical tool for HQ quantitation in the presence of GA, KA, and AA from real samples containing varied mixtures of the mentioned compounds.

4. Conclusions

The new proposed method for the hydroquinone determination in the presence of other dermatologic active ingredients (kojic acid, ascorbic acid, and glycolic acid) is cheap and simple, and it is not time consuming. The used reaction system contains an oxidant ($K_2Cr_2O_7$) that does not necessitate any additional steps. Moreover, the method does not contain variables which influence the reliability of the results. The accuracy, the precision, and the results obtained analyzing the synthetic mixtures recommend the spectrometric procedure for skin depigmenting products analysis and control by means of hydroquinone level monitoring, in binary, ternary, or quaternary mixtures with kojic acid, glycolic acid, and ascorbic acid. The application of the spectrometric method

TABLE 5: Results of the HQ determination in ternary and quaternary synthetic mixtures by applying the proposed spectrometric method; $n = 3$.

Percent ratio, HQ : GA : KA : AA	HQ concentration, μg·mL^{-1} Considered	Found ± SD	R%
1 : 5 : 1 : 0	2	1.85 ± 0.08	92.55
1 : 5 : 2 : 0	2	1.87 ± 0.06	93.62
4 : 0.75 : 0 : 2.5	8	8.34 ± 0.35	104.26
1 : 1 : 1 : 1	2	1.83 ± 0.05	91.49
2 : 5 : 1 : 5	4	4.19 ± 0.05	104.79

on synthetic mixtures represents the first step for further researches on real samples.

Competing Interests

The authors declare that they have no competing interests.

References

[1] R. Uchida, S. Ishikawa, and H. Tomoda, "Inhibition of tyrosinase activity and melanine pigmentation by 2-hydroxytyrosol," *Acta Pharmaceutica Sinica B*, vol. 4, no. 2, pp. 141–145, 2014.

[2] V. M. Sheth and A. G. Pandya, "Melasma: a comprehensive update: part I," *Journal of the American Academy of Dermatology*, vol. 65, no. 4, pp. 689–697, 2011.

[3] N. C. Dlova, S. H. Hamed, J. Tsoka-Gwegweni, and A. Grobler, "Skin lightening practices: an epidemiological study of South African women of African and Indian ancestries," *British Journal of Dermatology*, vol. 173, no. 2, pp. 2–9, 2015.

[4] M. Matsumoto, H. Todo, T. Akiyama et al., "Risk assessment of skin lightening cosmetics containing hydroquinone," *Regulatory Toxicology and Pharmacology*, vol. 81, pp. 128–135, 2016.

[5] United States Food and Drug Administration, Hydroquinone studies under the National Toxicology Program (NTP), 2010, http://www.fda.gov/AboutFDA/CentersOffices/OfficeofMedicalProductsandTobacco/CDER/ucm203112.htm.

[6] P. L. García, M. I. R. M. Santoro, A. K. Singh, and E. R. M. Kedor-Hackmann, "Determination of optimum wavelength and derivative order in spectrophotometry for quantitation of hydroquinone in creams," *Revista Brasileira de Ciencias Farmaceuticas/Brazilian Journal of Pharmaceutical Sciences*, vol. 43, no. 3, pp. 397–404, 2007.

[7] Y.-H. Lin, Y.-H. Yang, and S.-M. Wu, "Experimental design and capillary electrophoresis for simultaneous analysis of arbutin, kojic acid and hydroquinone in cosmetics," *Journal of Pharmaceutical and Biomedical Analysis*, vol. 44, no. 1, pp. 279–282, 2007.

[8] M. M. C. G. Peters, T. W. Jones, T. J. Monks, and S. S. Lau, "Cytotoxicity and cell-proliferation induced by the nephrocarcinogen hydroquinone and its nephrotoxic metabolite 2,3,5-(tris-glutathion-S-yl)hydroquinone," *Carcinogenesis*, vol. 18, no. 12, pp. 2393–2401, 1997.

[9] M. Rendon, M. Berneburg, I. Arellano, and M. Picardo, "Treatment of melasma," *Journal of the American Academy of Dermatology*, vol. 54, no. 5, pp. S272–S281, 2006.

[10] A. D. Katsambas and A. J. Stratigos, "Depigmenting and bleaching agents: coping with hyperpigmentation," *Clinics in Dermatology*, vol. 19, no. 4, pp. 483–488, 2001.

[11] S. Siddique, Z. Parveen, Z. Ali, and M. Zaheer, "Qualitative and quantitative estimation of hydroquinone in skin whitening cosmetics," *Journal of Cosmetics, Dermatological Sciences and Applications*, vol. 2, no. 3, pp. 224–228, 2012.

[12] D. Amponsah, R. Voegborlo, and G. E. Sebiawu, "Determination of amount of hydroquinone in some selected skin-lightening creams sold in the Ghanaian market," *International Journal of Scientific & Engineering Research*, vol. 5, no. 6, pp. 544–550, 2014.

[13] M. H. Gbetoh and M. Amyot, "Mercury, hydroquinone and clobetasol propionate in skin lightening products in West Africa and Canada," *Environmental Research*, vol. 150, pp. 403–410, 2016.

[14] A. M. Htet, E. E. Thin, M. M. Saw, and S. Win, "Chemical analysis of hydroquinone and retinoic acid contents in facial whitening creams," *Asian Journal of Pharmaceutical Sciences*, vol. 11, no. 1, pp. 89–90, 2016.

[15] C. Desiderio, L. Ossicini, and S. Fanali, "Analysis of hydroquinone and some of its ethers by using capillary electrochromatography," *Journal of Chromatography A*, vol. 887, no. 1-2, pp. 489–496, 2000.

[16] M. Buleandra, A. A. Rabinca, C. Mihailciuc et al., "Screen-printed Prussian Blue modified electrode for simultaneous detection of hydroquinone and catechol," *Sensors and Actuators, B: Chemical*, vol. 203, pp. 824–832, 2014.

[17] M. Aragó, C. Ariño, À. Dago, J. M. Díaz-Cruz, and M. Esteban, "Simultaneous determination of hydroquinone, catechol and resorcinol by voltammetry using graphene screen-printed electrodes and partial least squares calibration," *Talanta*, vol. 160, pp. 138–143, 2016.

[18] G. Xu, B. Li, and X. Luo, "Carbon nanotube doped poly(3,4-ethylenedioxythiophene) for the electrocatalytic oxidation and detection of hydroquinone," *Sensors and Actuators B: Chemical*, vol. 176, pp. 69–74, 2013.

[19] J. Tashkhourian, M. Daneshi, F. Nami-Ana, M. Behbahani, and A. Bagheri, "Simultaneous determination of hydroquinone and catechol at gold nanoparticles mesoporous silica modified carbon paste electrode," *Journal of Hazardous Materials*, vol. 318, pp. 117–124, 2016.

[20] M. R. Siddiqui, Z. A. AlOthman, and N. Rahman, "Analytical techniques in pharmaceutical analysis: a review," *Arabian Journal of Chemistry*, 2013.

[21] P. O. Odumosu and T. O. Ekwe, "Identification and spectrophotometric determination of hydroquinone levels in some cosmetic creams," *African Journal of Pharmacy and Pharmacology*, vol. 4, no. 5, pp. 231–234, 2010.

[22] T. E. Kipngetich, M. Hillary, and M. Shadrack, "UV-VIS analysis and determination of hydroquinone in body lotions and creams sold in retail outlets in Baraton, Kenya," *Baraton Interdisplinary Research Journal*, vol. 3, no. 1, pp. 23–28, 2013.

[23] P. L. García, M. I. R. M. Santoro, A. K. Singh, and E. R. M. Kedor-Hackmann, "Determination of optimum wavelength and derivative order in spectrophotometry for quantitation of hydroquinone in creams," *Brazilian Journal of Pharmaceutical Sciences*, vol. 43, no. 3, pp. 397–404, 2007.

[24] E. S. Elzanfaly, A. S. Saad, and A.-E. B. Abd-Elaleem, "Simultaneous determination of retinoic acid and hydroquinone in skin ointment using spectrophotometric technique (ratio difference

method)," *Saudi Pharmaceutical Journal*, vol. 20, no. 3, pp. 249–253, 2012.

[25] M. R. Elghobashy, L. I. Bebawy, R. F. Shokry, and S. S. Abbas, "Successive ratio subtraction coupled with constant multiplication spectrophotometric method for determination of hydroquinone in complex mixture with its degradation products, tretinoin and methyl paraben," *Spectrochimica Acta Part A: Molecular and Biomolecular Spectroscopy*, vol. 157, pp. 116–123, 2016.

[26] H. Chen, Z. Lin, H. Tang, T. Wu, and C. Tan, "Quantitative analysis of dihydroxybenzenes in complex water samples using excitation-emission matrix fluorescence spectroscopy and second-order calibration," *Journal of Spectroscopy*, vol. 2014, Article ID 412039, 7 pages, 2014.

[27] S. Uddin, A. Rauf, T. G. Kazi, H. I. Afridi, and G. Lutfullah, "Highly sensitive spectrometric method for determination of hydroquinone in skin lightening creams: application in cosmetics," *International Journal of Cosmetic Science*, vol. 33, no. 2, pp. 132–137, 2011.

[28] B. B. Qassim and H. S. Omaish, "Development of FIA system for the spectrophotometric determination of hydroquinone in pure material and pharmaceutical formulations," *Journal of Chemical and Pharmaceutical Research*, vol. 6, no. 3, pp. 1548–1559, 2014.

[29] A. F. Seliem and H. M. Khalil, "Sensitive spectrophotometric method for determination of hydroquinone in some common cosmetics in Najran region in K.S.A," *Ultra Chemistry*, vol. 9, no. 2, pp. 221–228, 2013.

[30] Y. Ni, Z. Xia, and S. Kokot, "A kinetic spectrophotometric method for simultaneous determination of phenol and its three derivatives with the aid of artificial neural network," *Journal of Hazardous Materials*, vol. 192, no. 2, pp. 722–729, 2011.

[31] C.-Y. Cheng, Y.-T. Chan, Y.-M. Tzou, K.-Y. Chen, and Y.-T. Liu, "Spectroscopic investigations of the oxidative polymerization of hydroquinone in the presence of hexavalent chromium," *Journal of Spectroscopy*, vol. 2016, Article ID 7958351, 8 pages, 2016.

[32] L. R. Goldfrank and N. E. Flomenbaum, "Toxic alcohols," in *Goldfrank's Toxicologic Emergencies*, L. R. Goldfrank, N. E. Flomenbaum, N. A. Lewin et al., Eds., pp. 1049–1060, Appleton and Lange, Stamford, Conn, USA, 6th edition, 1998.

[33] H. Borsook, H. W. Davenport, C. E. P. Jeffreys, and R. C. Warner, "The oxidation of ascorbic acid and its reduction in vitro and in vivo," *The Journal of Biological Chemistry*, vol. 117, no. 1, pp. 237–279, 1937.

[34] A. Shrivastava and V. Gupta, "Methods for the determination of limit of detection and limit of quantitation of the analytical methods," *Chronicles of Young Scientists*, vol. 2, no. 1, pp. 21–25, 2011.

[35] G. Latimer, *Official Methods of Analysis of AOAC International*, AOAC, Gaithersburg, Md, USA, 2012.

[36] D. Bandyopadhyay, "Topical treatment of melasma," *Indian Journal of Dermatology*, vol. 54, no. 4, pp. 303–309, 2009.

[37] C. B. Lynde, J. N. Kraft, and C. W. Lynde, "Topical treatments for melasma and postinflammatory hyperpigmentation," *Skin Therapy Letter*, vol. 11, no. 9, pp. 1–6, 2006.

[38] A. J. Stratigos and A. D. Katsambas, "Optimal management of recalcitrant disorders of hyperpigmentation in dark-skinned patients," *American Journal of Clinical Dermatology*, vol. 5, no. 3, pp. 161–168, 2004.

[39] S. Bruce and J. Watson, "Evaluation of a prescription strength 4% hydroquinone/10% L-ascorbic acid treatment system for normal to oily skin," *Journal of Drugs in Dermatology*, vol. 10, no. 12, pp. 1455–1461, 2011.

[40] J. T. E. Lim, "Treatment of melasma using kojic acid in a gel containing hydroquinone and glycolic acid," *Dermatologic Surgery*, vol. 25, no. 4, pp. 282–284, 1999.

[41] R. C. Monteiro, B. N. Kishore, R. M. Bhat, D. Sukumar, J. Martis, and H. K. Ganesh, "A comparative study of the efficacy of 4% hydroquinone vs 0.75% Kojic acid cream in the treatment of facial melasma," *Indian Journal of Dermatology*, vol. 58, no. 2, p. 157, 2013.

[42] http://www.amazon.com/Procelain-Skin-Whitening-Hydroquinone-Hyperpigmentation/dp/B009JJA49U.

8

Determination of Benzylpenicillin Potassium Residues in Duck Meat Using Surface Enhanced Raman Spectroscopy with Au Nanoparticles

Yijie Peng, Muhua Liu, Jinhui Zhao, Haichao Yuan, Yao Li, Jinjiang Tao, and Hongqing Guo

Optics-Electrics Application of Biomaterials Lab, College of Engineering, Jiangxi Agricultural University, Nanchang, Jiangxi 330045, China

Correspondence should be addressed to Jinhui Zhao; zjhxiaocao@sina.com

Academic Editor: Muhammad Tahir

A new method using surface enhanced Raman spectroscopy (SERS) with Au nanoparticles was established for the rapid detection of benzylpenicillin potassium (PG) residues in duck meat. Au nanoparticles were used as SERS enhancement substrate, and the maximum absorption peak of Au nanoparticles using the UV-Vis spectrophotometer was 548 nm. In the research, the SERS spectra of PG solutions and PG duck meat extract as well as their vibrational assignment were analyzed. The effects of Au nanoparticles addition, sample addition, NaCl solution addition, and adsorption time on the SERS intensities of PG duck meat extract were discussed. It is revealed that a good linearity can be obtained between the SERS intensities at 993 cm^{-1} and the PG residues concentrations (0.5~15.0 $mg{\cdot}L^{-1}$) detected in duck meat extract. The linear equation was $Y = 831.68X + 1997.1$, and the determination coefficient was 0.9553. The determination coefficient of PG in duck meat extract between the actual values and the predictive values was 0.9757, and the root mean square error (RMSEP) was 0.6496 mg/L. The recovery rate of PG in duck meat extract was 90~121%. The results showed that the method using SERS with Au nanoparticles could pave a new way for the rapid detection of PG residues in duck meat.

1. Introduction

Penicillin is a type of β-lactam antimicrobial drugs and, as one of the most output and the most widely used antibiotics, has been successfully applied to treat bacterial infections in both human and animals, such as gastrointestinal, urethra, and respiratory tract infections [1, 2]. Because of its high solubility, benzylpenicillin potassium (PG) is one of the most used penicillins and has been regarded as the best choice of treating duck diseases [2–4]. The penicillin residues are strictly controlled and limited in China, with a maximum residue limit (MRL) of penicillin residues in animal foods of 50 $\mu g{\cdot}L^{-1}$, according to the regulation of China's Ministry of Agriculture [5]. If the amount of penicillin is not rationally used in the duck-raising process, violation residues in duck meat will be caused and the penicillin residues in duck meat would endanger human's health.

Currently, the main methods reported for the detection of penicillin residues in animal foods include the high-performance liquid chromatography (HPLC) [1, 6], microbial method [7], enzyme-linked immunosorbent assay [8], and electrochemical immunosensor [5]. The advantages of these techniques are the sensitivity and practicality; however, these methods are also suffering several drawbacks, such as complex pretreatment and unsatisfactory cost-effectiveness. Therefore, it is very urgent to establish a new method for the simple, rapid, and accurate detection of penicillin residues in duck meat.

Surface enhanced Raman spectroscopy (SERS) is an interesting phenomenon that some molecules and functional groups were adsorbed onto the roughened surface of a suitable metal, such as gold, silver, or copper, and the intensities of their Raman signals can be greatly enhanced. This detection technique, which possesses many advantages such as convenient operation, high accuracy rate, fast testing velocity, and the portable instrument, has been widely applied in the biological engineering, medical, food detection, and other research works [9–11]. Di Anibal et al. adopted a screening tool to detect Sudan I dye in culinary spices using SERS and multivariate analysis [12]. Tang et al. applied the alkaline silver colloid as SERS enhancement substrate to detect the melamine in milk [13]. Zhu used SERS to detect the nitrofuran antibiotics residues in chicken, fish, and shrimp meat [14]. In this research, Au nanoparticles were used as the SERS enhancement substrate and PG in duck meat extract was used as the research object, and a new method using SERS with Au nanoparticles was established for the rapid detection of PG residues in duck meat.

2. Materials and Methods

2.1. Reagents and Equipment. Duck was purchased from the vegetable market of Jiangxi Agricultural University. PG standard substance (99.8%) was purchased from standard substances network of China. Tetrachloroaurate trihydrate ($HAuCL_4 \cdot 3H_2O$, $M = 393.83$, its Au $\geq$ 49.0%) was purchased from Sigma-Aldrich. Trisodium citrate, sodium chloride, acetonitrile, and hexane were of analytical grade. Ultrapure water was also used.

QE65000 portable Raman spectrometer (Ocean Optics Co., Ltd.), T6 series UV-Vis spectrophotometer (Beijing Purkinje General Instrument Co., Ltd.), laboratory ultrapure water machine (Kertong Water Co., Ltd.), wire coil heater (Beijing Yongxing Instrument Co., Ltd.), K-50B ultrasonic cleaner (Hefei Kinnic Machinery Co., Ltd.), FA1004B electronic balance (accuracy of 0.1 mg, Shanghai Flat Instrument Co., Ltd.), VORTEX-5 vortex mixer (Haimen, Jiangsu Province, Kylin-Bell Lab Instruments Co., Ltd.), HSC-24B termovap sample concentrator (Tianjin City Heng Austrian Science and Technology Development Co., Ltd.), JW-1024 low speed centrifuge (Anhui Jiaven Enqipment Industry Co., Ltd.), and quartz sampling bottle (Beijing Cheng Teng Equipment Co., Ltd.) were used.

2.2. Experimental Methods

2.2.1. Pretreatment of Duck Meat Extract. The duck meat extract [6, 15]: 5 g of duck meat mud, 2 g of anhydrous sodium sulfate, and 25 mL acetonitrile were added into 50 mL polypropylene centrifuge tube. The mixture was oscillated for 1 min, followed by the ultrasonication treatment for 5 min. After centrifuging the mixture at 4200 r/min for 10 min, the supernatant of the mixture was removed. The above processes were repeated twice, and the supernatants were mixed. 4 mL of the supernatants with 8 mL of hexane saturated with acetonitrile was mixed, oscillated for 1 min, and centrifuged at 4200 r/min for 8 min. The hexane layer of the mixed solutions was removed, and the residual solutions were dried with nitrogen. The residue was dissolved in 4 mL ultrapure water, and the solutions were filtrated through a 0.45 μm filter membrane to remove lipids. The duck meat extract was obtained and stored under the environment of 4°C.

2.2.2. Synthesis of Au Nanoparticles. Au nanoparticles [16, 17]: 3 mL (1%) of auric chloride acid and 47 mL of ultrapure water were mixed in a 100 mL beaker, and the mixed solutions were heated with wire coil heater. After the boiling had commenced, 2 mL (1%) of trisodium citrate solution was added rapidly into the above boiling solution under a glass rod stirring. After 5 min, the color of the mixed solution was changed to coffee. The synthesized Au nanoparticles were stored at the room temperature.

2.2.3. Standard Solution and Sample Solution. PG standard solution (100 $mg \cdot L^{-1}$): 10 mg PG was dissolved in 100 mL duck meat extract. Sodium chloride standard solution (0.1 $mg \cdot L^{-1}$ NaCl): 0.585 g of sodium chloride was dissolved in 100 mL ultrapure water.

The PG sample solution: the different volumes of PG standard solution were put into 10 mL centrifuge tube and the capacity was fixed to 5 mL using duck meat extract. Ultimately, the different mass concentrations of PG duck meat extract samples (0.5, 1.0, 2.0, 4.0, 6.0, 7.0, 9.0, 10.0, 12.0, and 15.0 $mg \cdot L^{-1}$) were obtained through the above method.

1 mL of Au nanoparticles, 20 μL of sample solution (0.5~15.0 $mg \cdot L^{-1}$), and 100 μL of NaCl solution (0.1 $mg \cdot L^{-1}$) were sequentially added into the quartz sampling bottle. The mixture was well mixed, and then the SERS spectra were collected when the adsorption reaction lasted for 5 min.

2.3. Parameter Settings of Instruments. Raman spectrometer parameter settings: the spectrometer's power and the laser wavelength were 700 mW and 785 nm, respectively. The integration time of 10 s, the average integral number of 2, and the smoothness of 1 were applied.

UV-Vis spectrophotometer parameter settings: the display range and the scanning range were 0~2 s and 400~700 nm, respectively. The interval of 2 nm and the spectrophotometric mode of Abs were selected. Finally, the absorption spectra were collected under the fast scan speed.

3. Results and Discussion

3.1. UV-Visible Absorption Spectra of Au Nanoparticles and SERS Assignment of PG. In the formation of sodium citrate reduction of auric chloride acid, the sodium citrate, which was not directly reacted with the boiling solution of auric chloride acid, was firstly decomposed into the stable acetoacetate and formic acid, and then Au nanoparticles were produced via the oxidation reaction between formic acid and auric chloride acid [18]. The sodium citrate was decomposed rapidly with electromagnetic heating, and the reactive substances were fused quickly [18]. Consequently, the smaller particle size and even distribution of Au nanoparticles could be synthetized. 500 μL of Au nanoparticles was diluted with 2 mL ultrapure water, and the UV-Visible absorption spectra were collected. As shown in Figure 1, the maximum absorption peak of Au nanoparticles was 548 nm and the half-peak width was about 79 nm. The results showed that the particle size of Au nanoparticle was somehow single and distributed uniformly [19].

The SERS spectra and structural formulas of PG were shown in Figure 2, and it can be seen that PG was mainly composed of the β-lactam ring, the thiazolidine ring, and the acyl side chain of the substituted benzene ring. As seen

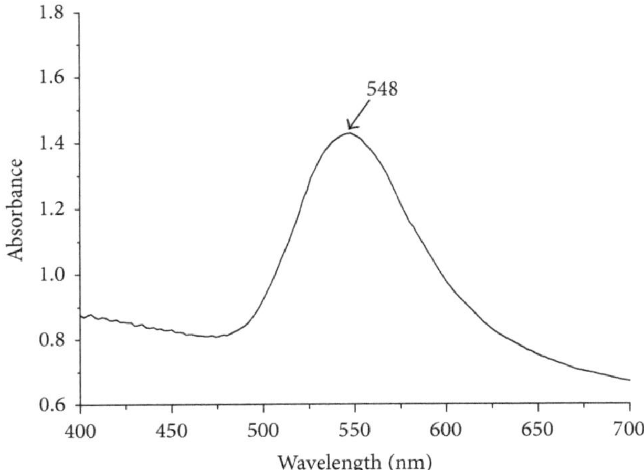

FIGURE 1: The UV-Visible absorption spectra of Au nanoparticles.

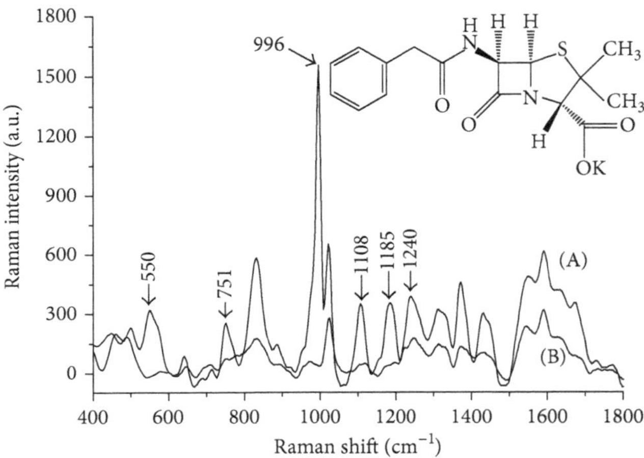

FIGURE 2: The SERS spectra and structural formulas of the PG: (A) Au nanoparticles + PG (7 mg/L), (B) Au nanoparticles.

from curves (A) and (B) in Figure 2, a visible difference was observed in the SERS spectra of Au nanoparticles and Au nanoparticles + PG at 550~1240 cm^{-1}, which indicates that some interaction was produced between PG and Au nanoparticles. As shown from curve (A) in Figure 2, the SERS peak at 550 cm^{-1} might result from the deformation vibration of thiazole ring and benzene ring, the SERS speak at 751 cm^{-1} might be ascribed to the CC stretching vibration, and the SERS peak at 996 cm^{-1} might be ascribed to the stretching vibration of benzene ring and the breathing vibration of triangle ring. The SERS peak at 1108 cm^{-1} was possibly due to CH_3 rocking vibration of plane and C-C-C bending vibration, the SERS peak at 1185 cm^{-1} was possibly due to CN stretching vibration and CH bending vibration of thiazole ring, and the SERS peak at 1240 cm^{-1} was possibly due to CH bending vibration and C-OH stretching vibration from curve (A) [20–23].

3.2. SERS Spectra of PG Solutions and PG Duck Meat Extract. The negative charge of Au nanoparticles resulted from many anions of citrate self-assembled on Au nanoparticles surface. Because of electrostatic repulsion of Au nanoparticles, Au nanoparticles remained stable in aqueous solution [24]. When the NaCl solution was added, the negative charge of Au nanoparticles was neutralized by the reaction with Na^+. The neutralizing reaction led to the disappearance of Au nanoparticles electrostatic repulsion and the aggregation of Au nanoparticles. As a result, the signal of resonance light scattering was enhanced. As shown in Figure 3(a), when the NaCl solution was added into the mixture of Au nanoparticles and PG solution (7 mg·L^{-1}), the SERS signal of the mixture was further enhanced. In the comparison of curves (A) and (C) in Figure 3, the SERS peak of PG solution at 993 cm^{-1} was blue shifted about 3 cm^{-1}. It can be also seen that the SERS peaks of PG solution at 1492~1674 cm^{-1} were excited from curves (A) and (B). The SERS peak at 1492 cm^{-1} might result from the CH_3 bending vibration. The SERS peak at 1563 cm^{-1} might be attributed to double-peak vibration of degenerate ring, and the SERS peak at 1674 cm^{-1} might be ascribed to C=O stretching vibration of peptide bond, NH and CH bending vibration [20–23].

Some constituents of duck meat, such as protein and fat, could significantly interfere with the SERS signal of PG. Therefore, the acetonitrile was firstly used as the extraction solvent to precipitate the protein in duck meat. Secondly, the hexane saturated with acetonitrile was used to remove fat and other impurities. The remaining acetonitrile was dried with nitrogen, and the residue was dissolved in ultrapure water. Finally, the relatively pure duck meat extract was obtained. Au nanoparticles + duck meat extract + NaCl, Au nanoparticles + PG duck meat extract (7 mg·L^{-1}) + NaCl, and Au nanoparticles + PG (7 mg·L^{-1}) + NaCl were shown in Figure 3(b), and we could see that some SERS peaks of duck meat extract were the same with the SERS peaks of PG duck meat extract, such as 717 cm^{-1}, 1024 cm^{-1}, 1244 cm^{-1}, and 1369 cm^{-1}. The SERS spectra of PG duck meat extract presented the peaks at 993 cm^{-1} and 1492 cm^{-1}, while the SERS spectra of duck meat extract did not present peaks at 993 cm^{-1} and 1492 cm^{-1}. Meanwhile, the two SERS peaks of PG duck meat extract anastomosed with the SERS peaks of PG solution, which rendered a realistic basis for the detection of PG residues in duck meat.

3.3. Effect of Au Nanoparticles Addition on the SERS Intensity of PG Duck Meat Extract. The SERS effect is caused by the electromagnetic effect and the chemical effect, and the measured molecules are adsorbed onto the roughened surface of metal substrate. However, not all of the adsorbed molecules can produce the SERS effect on the surface of metal substrate. Most of the enhancement effect results from the portion of the adsorbed molecules on the positive surface of metal substrate. Therefore, the volume concentrations of Au nanoparticles are critical for the enhancement of SERS intensities [25, 26]. In this research, the volumes of PG duck meat extract (7 mg·L^{-1}) and the NaCl solution (1 mol/L) were fixed to 50 μL and 100 μL, respectively, and then the SERS spectra of the mixture with the addition of 0.5, 0.7, 1, 1.2, and 1.5 mL of Au nanoparticles were collected, respectively.

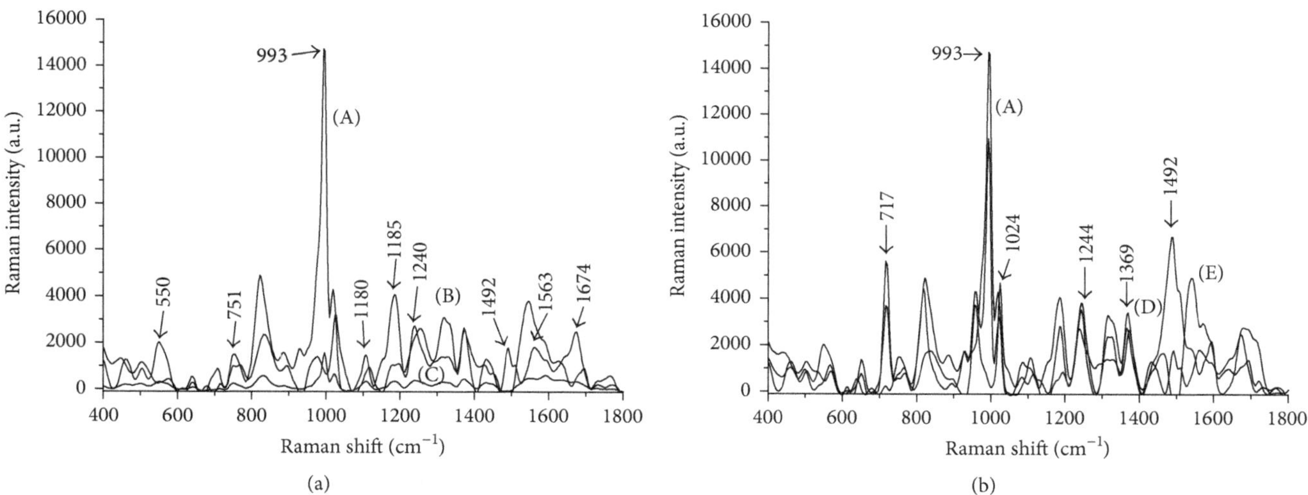

FIGURE 3: The SERS spectra of the PG solutions (a) and PG duck extract (b): (A) Au nanoparticles + PG (7 mg/L) + NaCl; (B) Au nanoparticles + NaCl; (C) Au nanoparticles + PG (7 mg·L^{-1}); (D) Au nanoparticles + PG duck meat extract + NaCl; (E) Au nanoparticles + duck meat extract + NaCl.

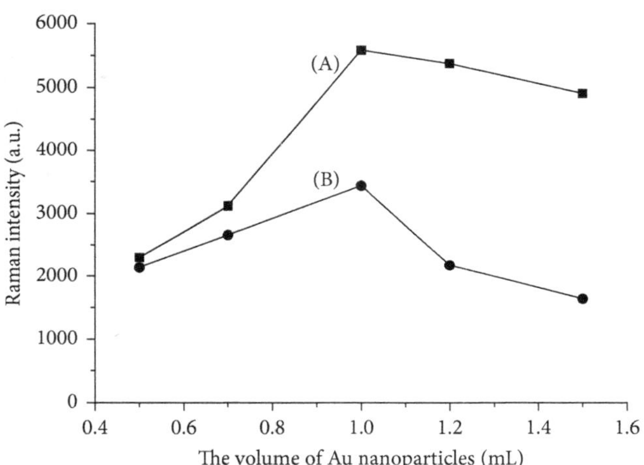

FIGURE 4: Effect of the Au nanoparticles addition on the SERS intensity of PG duck meat extract: (A) 993 cm^{-1}, (B) 1492 cm^{-1}.

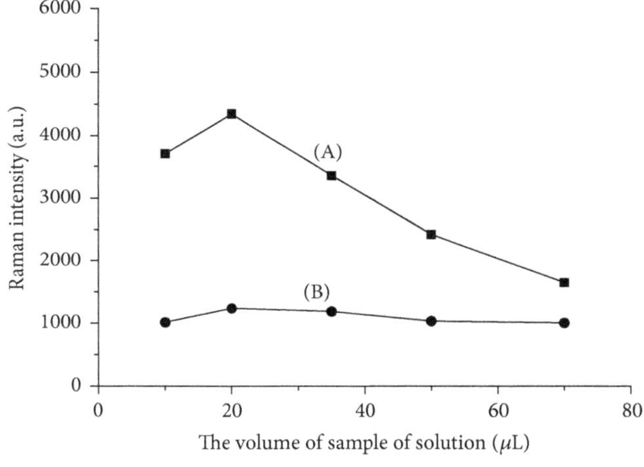

FIGURE 5: Effect of sample addition on the SERS intensity of PG duck meat extract: (A) 993 cm^{-1}, (B) 1492 cm^{-1}.

As shown in Figure 4, the SERS intensities at 993 cm^{-1} and 1492 cm^{-1} were continuously strengthened with the increases of Au nanoparticles volume. However, when the volumes of Au nanoparticles exceeded 1 mL, the SERS intensities decreased gradually. It is speculated that Au nanoparticles firstly adsorbed PG and the adsorption effect enhanced the Raman signal with the increases of Au nanoparticles volumes. However, when the volumes of Au nanoparticles increased to a certain value, the neutralization of Na^+ to Au nanoparticles could cover up the part of the PG adsorption effect and decreased the SERS intensities. Therefore, the Au nanoparticles addition was determined as 1 mL in the following experiment.

3.4. Effect of Sample Addition on the SERS Intensity of PG Duck Meat Extract. Owing to the portion of adsorbed molecules on the positive surface of metal substrate, the volume concentration of the sample solution (7 mg/L) could impact the SERS intensity of PG duck meat extract. When Au nanoparticles and the NaCl solution were fixed, respectively, in this paper, the SERS spectra of the different volumes of the sample solutions (10, 20, 35, 50, and 70 μL) were collected and analyzed. As seen from Figure 5, when the volumes of the sample solutions increased from 10 μL to 20 μL, the SERS intensities at 993 cm^{-1} and 1492 cm^{-1} were enhanced. However, when the volumes were more than 20 μL, the SERS intensities decreased gradually. Therefore, the sample addition was selected as 20 μL in the following experiment.

3.5. Effect of NaCl Addition on the SERS Intensity of PG Duck Meat Extract. The larger particle size of aggregate, which was caused by the adsorbed PG on the surface of Au

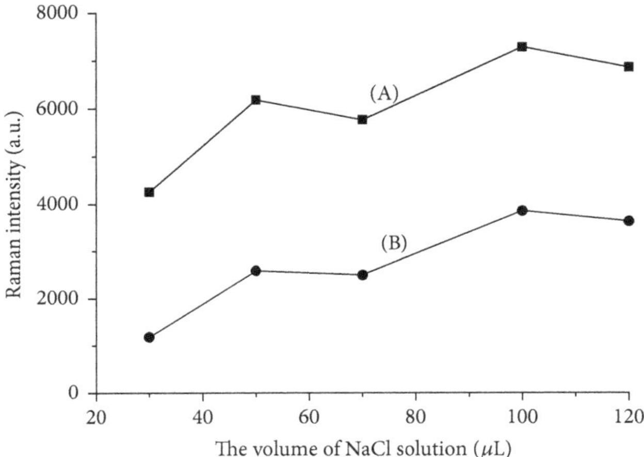

Figure 6: Effect of NaCl addition on the SERS intensity of PG duck meat extract: (A) 993 cm^{-1}, (B) 1492 cm^{-1}.

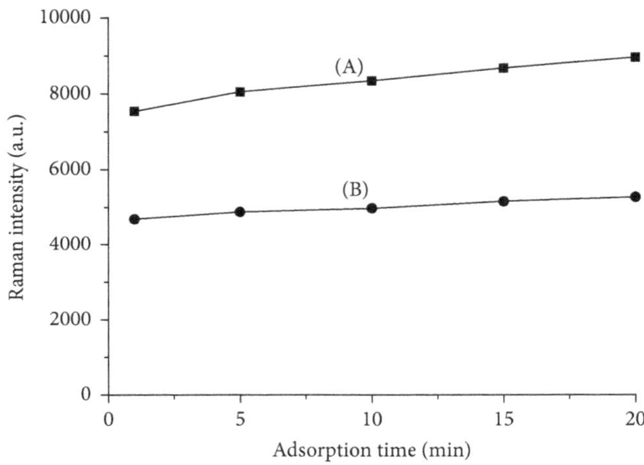

Figure 7: Effect of adsorption time on the SERS intensity of PG duck meat extract: (A) 993 cm^{-1}, (B) 1492 cm^{-1}.

nanoparticles, would result in the local plasma resonance between PG and Au nanoparticles and the enhancement of SERS intensities [27]. When the NaCl solution was added into the mixture of Au nanoparticles + PG duck meat extract, Au nanoparticles would gather rapidly and the color of the mixture was changed from coffee to blue. The particle size of the above Au nanoparticles became bigger, which is manifested as the enhancement of the resonance light scattering, and the SERS intensities of PG duck meat extract were further enhanced. In order to investigate the effects of different volume concentrations of the NaCl solution on the SERS intensities of the mixture, 1 mL of Au nanoparticles and 20 μL of PG duck meat extract were firstly added into the quartz sampling bottle in turn. And, then, different volumes of the NaCl solution (30, 50, 70, 100, and 120 μL) were added, respectively, and their SERS spectra were collected. When the NaCl addition was 100 μL, the SERS intensities at 993 cm^{-1} and 1492 cm^{-1} reached the maximum value in Figure 6. The reason was probably that the proper volume of NaCl solution could have the effect of the activator, but the excess volume of the NaCl solution made the mixture produce the coagulation and resulted in the SERS intensities weakening [28]. The experimental results indicated that 100 μL of the NaCl solution was the optimum addition, so the NaCl addition was selected as 100 μL in the following experiment.

3.6. Effect of Adsorption Time on the SERS Intensity of PG Duck Meat Extract. When the volumes of Au nanoparticles, PG duck meat extract, and the NaCl solution were fixed, respectively, the adsorption time had some influences on the enhancement effect of SERS intensities. After 1 mL of Au nanoparticles, 20 μL of PG duck meat extract and 100 μL of the NaCl solution were mixed together, Au nanoparticles would gather, and then the aggregation of Au nanoparticles would produce the SERS spectra. When the adsorption times of the mixture were 1, 5, 10, 15, and 20 min, respectively, the SERS spectra of the mixture in different adsorption times were collected. As shown from curve (A) in Figure 7, the SERS intensities at 993 cm^{-1} were of larger increasing range at 1~5 min. Although the SERS intensities had some enhancement effect at 5~20 min, the enhancement extent was relatively stable. As seen from curve (B) in Figure 7, the SERS intensities at 1492 cm^{-1} were enhanced gradually at 1~20 min, but the enhancement extent did not change obviously. The above results might be because the PG molecules in duck meat extract could not be completely adsorbed on the surface of Au nanoparticles before 5 min, so the SERS intensities at 993 cm^{-1} and 1492 cm^{-1} were continuously enhanced. After 5 min, the PG molecules might be completely adsorbed on the surface of Au nanoparticles. Therefore, the SERS spectra were collected after the reaction for 5 min.

3.7. Prediction and Analysis Model. The proposed method based on SERS with Au nanoparticles was employed to determine PG residues in duck meat. The different mass concentrations of PG duck meat extract samples were prepared and the SERS spectra were collected under the optimum conditions. The research showed that when the concentration range of PG in duck meat extract was 0.5~15.0 mg·L^{-1}, the SERS intensities at 993 cm^{-1} and 1492 cm^{-1} showed the enhancement trend with the increasement of PG concentrations. When the mass concentration of PG in duck meat extract was lower than 4 mg·L^{-1}, the SERS peak at 1492 cm^{-1} was hardly observed. However, the SERS peak at 993 cm^{-1} could be observed under the same conditions when the mass concentration of PG in duck meat extract was 0.5 mg·L^{-1}. Therefore, the PG duck meat extract samples (the concentration range of 0.5~15.0 mg·L^{-1}) and the SERS intensities at 993 cm^{-1} were utilized for the quantitative analysis.

The calibration curve between the mass concentrations of PG in duck meat extract and the SERS intensities at 993 cm^{-1} was established using 6 samples, and 4 of the remaining samples were used to verify the accuracy of the calibration curve. As shown in Figure 8 and Table 1, when the concentration range of PG in duck meat extract

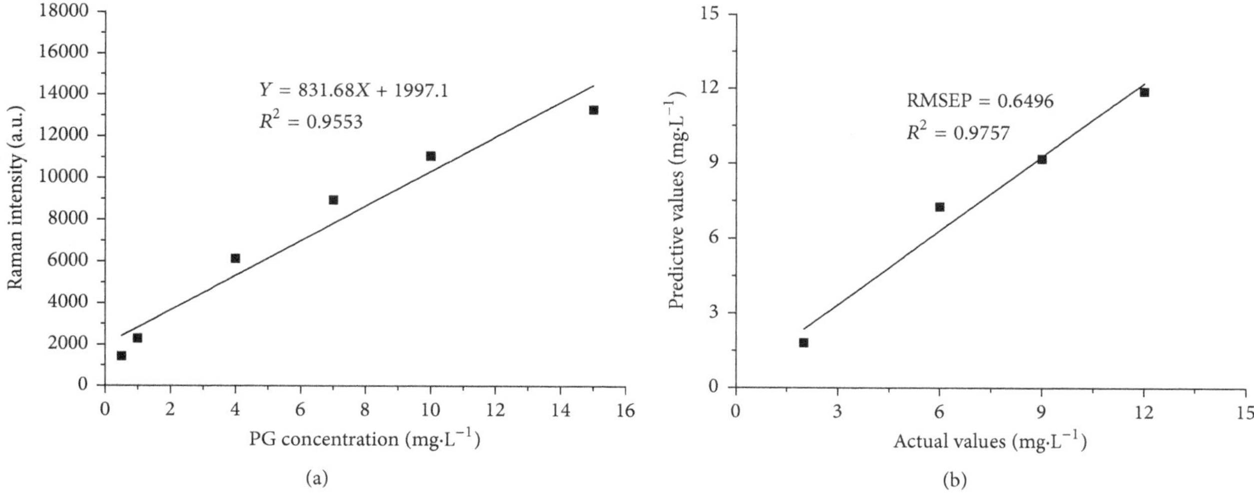

FIGURE 8: Linear equation of the PG duck extract (a) and the relationship of the prediction samples (b).

TABLE 1: Determined results of the recovery rate.

Sample number	Concentration of PG in duck meat extract/(mg·L^{-1})	Recovery amount/(mg·L^{-1})	Recovery rate/%
1	2	1.81	90.3
2	6	7.27	121.1
3	9	9.19	102.2
4	12	11.91	99.3

was 0.5~15.0 mg·L^{-1}, there was a good linear relationship between the concentrations of PG in duck meat extract and the SERS intensities at 993 cm^{-1}. The linear equation was $Y = 831.68X + 1997.1$, and the determination coefficient was 0.9553. The determination coefficient of PG in duck meat extract between the actual values and the predictive values was 0.9757, and the root mean square error (RMSEP) was 0.6496 mg·L^{-1}. The recovery rate of PG in duck meat extract was 90~121%. The result showed that the proposed method was feasible and relatively reliable using SERS with Au nanoparticles for the rapid detection of PG residues in duck meat.

4. Conclusions

Firstly, the absorption spectra of Au nanoparticles, the SERS spectra of PG solution, and PG duck meat extract and their vibrational assignment were analyzed, and these analyses could provide the realistic basis for the detection of PG residues in duck meat. Secondly, the effects of Au nanoparticles addition, sample addition, NaCl solution addition, and adsorption time on the SERS intensities of PG duck meat extract were discussed, respectively, and their optimal experimental conditions were determined, respectively. Finally, the SERS spectra of 10 samples with different mass concentrations of PG in duck meat extract were collected under the optimal conditions, and the model of PG residues in duck meat extract was established and analyzed. The result showed that a good linearity was obtained between the SERS intensities at 993 cm^{-1} and the concentrations of PG in duck meat extract in the range of 0.5~15 mg·L^{-1}, $Y = 831.68X + 1997.1$, and the determination coefficient was 0.9553. The determination coefficient of PG in duck meat extract between the actual values and the predictive values was 0.9757, the root mean square error (RMSEP) was 0.6496 mg·L^{-1}, and the recovery rate was 90~121%. The experimental results indicated that it was feasible that SERS with Au nanoparticles was used for the rapid detection of PG residues in duck meat.

Competing Interests

The authors declare that they have no competing interests.

Acknowledgments

This research was supported by External Science and Technology Cooperation Plan of Jiangxi Province, China (20132BDH80005), and National Natural Science Foundation of China (31101295). Additional support for this research was provided by the Science and Technology Support Project of Jiangxi Province, China (2012BBG70058).

References

[1] C. J. Liu, H. Wang, Z. X. Du, Y. B. Jiang, J. H. Shan, and Y. H. Cai, "Simultaneous determination of 2 penicillins and their 5 major metabolites in bovine muscle by ultra performance liquid chromatography tandem mass spectrometry," *Chinese Journal of Analytical chemistry*, vol. 36, no. 5, pp. 617–622, 2011.

[2] Y. Q. Liu and P. S. Ma, "Determination of diffusion coefficient of benzylpenicillin potassium in aqueous solution," *Chemical Industry and Engineering*, vol. 15, no. 3, pp. 6–14, 1998.

[3] M. H. Han, C. Q. Gu, X. Y. Hu et al., "Identification and virulence of four enterococcus strains isolated from ducks,"

Chinese Journal of Veterinary Science, vol. 27, no. 6, pp. 821–829, 2007.

[4] X. L. Cai, Q. Wei, Y. S. Cui et al., "Serotype identification and drug sensitivity test of 22 strains of riemerella anatipes-tifersis isolated from ducks in Zhejiang province," *Acta Agriculturae Zhejiangensis*, vol. 19, no. 1, pp. 46–49, 2007.

[5] X. J. Li, L. Li, J. Ma, Y. Xue, and G. H. Wu, "Determination of residues penicillin in chicken muscle by an electrochemical immunosensor," *Chines Journal of Applied Chemistry*, vol. 26, no. 6, pp. 716–720, 2009.

[6] C. Wang and X. Wang, "Determination of 5 penicillin residues in pork by high performance liquid chromatography with pre-column derivatization," *Chinese Journal of Analytical Chemistry*, vol. 29, no. 7, pp. 779–781, 2001.

[7] K. Y. Zhang, D. J. Wang, Z. H. Yuan, and S. X. Fan, "Determination of Ampicillin residues in chicken meat and swineport," *Chinese Journal of Veterinary Science*, vol. 24, no. 5, pp. 470–472, 2004.

[8] K. Jiang, Y. P. Chen, Y. F. Jin, J. F. Huang, X. Z. Chen, and D. L. Zhang, "Application of enzyme-linked immunosorbent assay in detecting β-lactam antibiotic residues in dairy products," *China Dairy Industry*, vol. 38, no. 1, pp. 51–54, 2010.

[9] B. S. Jia, W. L.-J. Ha-si, X. Lin, F. Yang, S. Lin, and Z. W. Lv, "Preparation of raman surface-enhangced scattering substrate and its application," *Journal of Food Safety and Quality*, vol. 5, no. 5, pp. 1490–1494, 2014.

[10] Y. Wang, W. S. Chen, Z. F. Hua et al., "Diagnosis of esophageal tissue based on surface-enhanced Raman spectroscopy," *Scientia Sinica Vitae*, vol. 43, no. 6, pp. 525–532, 2013.

[11] M.-H. Zhou, C.-Y. Liao, Z.-Y. Ren, H.-M. Fan, and J.-T. Bai, "Bioimaging technologies based on surface-enhanced Raman spectroscopy and their applications," *Chinese Optics*, vol. 6, no. 5, pp. 633–642, 2013.

[12] C. V. Di Anibal, L. F. Marsal, M. P. Callao, and I. Ruisánchez, "Surface Enhanced Raman Spectroscopy (SERS) and multivariate analysis as a screening tool for detecting Sudan I dye in culinary spices," *Spectrochimica Acta—Part A: Molecular and Biomolecular Spectroscopy*, vol. 87, pp. 135–141, 2012.

[13] J.-Q. Tang, C. Tian, C.-Y. Zeng, and S.-Q. Man, "Alkaline silver colloid for surface enhanced Raman scattering and application to detection of melamine doped milk," *Spectroscopy and Spectral Analysis*, vol. 33, no. 3, pp. 709–713, 2013.

[14] X. Y. Zhu, *Studies on the Detection of Nitrofuran Antibiotics by Surface-Enhanced Raman Scattering*, Jiangnan University, Wuxi, China, 2013.

[15] X. M. Li, Y. Z. Cao, J. J. Zhang, J. Mu, L. Wang, and H. Q. Wang, "Determination of 9 penicillins in fugu and eel by high pressure liquid chromatography-electrospray tandem mass spectrometry," *Journal of Food Safety and Quality*, vol. 4, no. 4, pp. 1165–1172, 2013.

[16] P. Liu, B. Y. He, and Y. F. Huang, "The rapid diagnostic stripe of the candida albicans in cow mastitis," China, 201310674116. 9, 2013.

[17] W. Ji, *Studies on the detection of chloramphenicol and its sodium succinate by surface-enhanced raman scattering [M.S. thesis]*, Jiangnan University, Jiangsu, China, 2013.

[18] L. Wang, S. H. Huang, Z. F. Huang et al., "Rapid synthesis of gold colloids assisted by microwave irradiation and SERS activity characterized," *The Journal of Light Scattering*, vol. 26, no. 1, pp. 13–18, 2014.

[19] Q. Zhang and Y. X. Weng, "A possible mechanism for the reduces fluorescence quantum efficiency of bacteriochlorophyll a molecules adsorbed on the gold nanoparticles," *The Journal of Light Scattering*, vol. 27, no. 1, pp. 1–8, 2015.

[20] X. Y. Qin, H. M. Sui, R. Li et al., "Study on surface enhanced raman spectroscopy of benzylpenicillin sodium drug molecule," *Spectroscopy and Spectral Analysis*, vol. 34, no. 10, pp. 253–254, 2014.

[21] X. F. Qian, Y. Shi, J. T. Zhao, and W. Shi, "Studies of penicillins by micro raman spectroscopy," *Spectroscopy and Spectral Analysis*, vol. 22, no. 1, pp. 51–53, 2002.

[22] Q. Q. Kang and G. M. Zhou, "Study on amoxicillin and the interaction with DNA by spectrometric methods," *The Journal of Light Scattering*, vol. 23, no. 4, pp. 368–375, 2011.

[23] W. Z. Liang, *The Application of Raman Spectroscopy in Antibiotic Drug*, Tianjin University, Tianjin, China, 2012.

[24] Y. N. Zhang, *Synthesis and Interaction with DNA of Polypyridine Copper Complexes*, Henan Normal University, Henan, China, 2013.

[25] Z. Y. Zhu, R. A. Gu, and T. H. Lu, *Application of Raman spectroscopy in Chemistry*, Northeastern University Press, Shenyang, China, 1998.

[26] X. G. Yang and Q. L. Wu, *Raman Spectroscopy Analysis and Application*, National Defense Industry Press, Beijing, China, 2008.

[27] Y. Q. He, S. P. Liu, Z. F. Liu, X. L. Hu, and Q. M. Lu, "Resonance rayleigh scattering spectral method for the determination of kanamycin with gold nanoparticle as probe," *Acta Chimica Sinica*, vol. 63, no. 11, pp. 997–1002, 2005.

[28] L. Lin, *The Qualitative and Quantitative Analysis of Pesticide Residue in Tea Using Surface-Enhanced Raman Spectroscopy (SERS)*, Jiangxi Agricultural University, Nanchang, China, 2014.

9

Simultaneous Quantification of Paracetamol and Caffeine in Powder Blends for Tableting by NIR-Chemometry

Dana Maria Muntean,[1] Cristian Alecu,[2] and Ioan Tomuta[1]

[1] *Department of Pharmaceutical Technology and Biopharmacy, University of Medicine and Pharmacy Iuliu Hatieganu Cluj-Napoca, 41 V. Babes, 400012 Cluj-Napoca, Romania*
[2] *S.C. Laropharm S.R.L., 145A Alexandriei, 70000 Bragadiru, Romania*

Correspondence should be addressed to Ioan Tomuta; tomutaioan@umfcluj.ro

Academic Editor: Rizwan Hasan Khan

Near-infrared spectroscopy (NIRS) is a technique widely used for rapid and nondestructive analysis of solid samples. A method for simultaneous analysis of the two components of paracetamol and caffeine from powder blends has been developed by using chemometry with near-infrared spectroscopy (NIRS). The method development was performed on samples containing 80, 90, 100, 110, and 120% active pharmaceutical ingredients, and near-infrared spectroscopy (NIRS) spectra of prepared powder blends were recorded and analyzed in order to develop models for the prediction of drug content. Many calibration models were applied in order to perform quantitative determination of drug content in powder, and choosing the appropriate number of factors (principal components) proved to be of highly importance for a PLS chemometric calibration. Once the methods were developed, they were validated in terms of trueness, precision, and accuracy. The results obtained by NIRS methods were compared with those obtained by HPLC reference method, and no significant differences were found. Therefore, the NIR chemometry methods were proved to be a suitable tool for predicting chemical properties of powder blends and for simultaneous determination of active pharmaceutical ingredients.

1. Introduction

Near-infrared spectroscopy (NIRS) has been proved to remain a powerful analytical tool for analyzing a vast variety of samples from petrochemical, food, agricultural, and pharmaceutical industries [1]. During the last decade, it has experienced a significant increase for quantitative pharmaceutical tests. The quantitative analytical strategies concerning pharmaceuticals, such as HPLC, GC, and UV spectrophotometry, usually require dissolving the samples, separating them, and determining their ingredients, being destructive. Compared to traditional analytical methods, near-infrared spectroscopy (NIRS) has been proved to be a more versatile and fast-growing analytical tool not only in the pharmaceutical sciences but also in the industry. The main advantages of this technique are that it is noninvasive and nondestructive, has no sample preparation required, large number of molecules which could be quantified, and is very fast due to the high frequency of spectrum acquisition [2].

In recent years, NIRS methods are starting to become popular in the pharmaceutical area. NIRS needs chemometric analysis of data in order to be used as quantitative technique. Chemometric methods such as three linear regression modeling methods, principal component regression (PCR), partial least squares (PLS) [3], or one nonlinear regression model based on artificial neural networks (ANN) [4, 5] are used [6, 7]. For the determination of the concentrations' chemical compounds in pharmaceutical field using near-infrared spectroscopy, PCR and PLS models are intensively used [6–8].

In order to increase patient compliance, the combination of two or more active compounds in the same commercial preparation may be used. The manufacturing process of tablets containing two or more active compounds, such us fixed-dose combination tablets, involves the following unit operations: dispersing, granulation, mixing, tableting, and coating. Each of the unit operations may have huge influence on the quality of the final product. For example, powder mixing is an essential unit operation for manufacturing fixed-dose combination, because inadequate mixing process conducts to poor quality of the final product due to low blend uniformity that is critical to ensure compliant content uniformity per united dose [9]. An important goal for NIRS analysis in pharmaceutical industry is a direct evaluation of powder blends during the mixing process as on-line, in-line, or at-line testing [2, 10].

Paracetamol (N-acetyl-p-aminophenol, acetaminophen) is a long-established substance, being one of the most extensively employed drugs in the world. For patients with sensitivity to aspirin, it is a noncarcinogenic and effective substitute and it is accepted to be a suitable drug for the relief of pain and fever in adults and children [11]. Due to its intensive use for therapeutic purposes, the quick methods for its determination during quality control are of great importance. In last years, many methods for paracetamol determination have been reported, such as chromatographic, spectrofluorimetric, chemiluminiscent, spectrophotometric, and electrochemical techniques [12].

Caffeine (3,7-dihydro-1,3,7-trimethyl-1H-purine-2,6-dione) is an alkaloid N-methyl derivative of xanthine that is broadly distributed in natural products, commonly used in beverages. Its consumption has many physiological effects, such as gastric acid secretion, diuresis, and stimulation of the central nervous system [13]. It is used in combination with nonsteroidal anti-inflamatory drugs in analgesic formulations or in combination with ergotamine in migraine treatment. The special literature reveals various methods for caffeine analyzing, including spectrophotometric [14, 15] and chromatographic [16] ones. These methods are generally expensive, time consuming, and complicated.

HPLC methods are widely used for powder blend uniformity evaluation, due to good selectivity, specificity, and linear range. However, this technique requires sample preparation and chromatographic separation of the analytes, so it takes hours and therefore can be done only offline. Monitoring the blend uniformity in the mixing steps of tablet manufacturing is considered to be an important goal for PAT concept and can be done only by a direct analysis. Once calibrated and validated, NIRS methods seem to be the best analytical options, due to short analysis time and low cost per analysis in contrast with HPLC analysis methods.

In this paper, we explored the applications of chemometry on near-infrared spectroscopy (NIRS) for the quantitative analysis of paracetamol and caffeine, in powder blend for tableting, predicting their concentrations simultaneously, without any processing of the sample.

2. Materials and Method

2.1. Materials. All the raw material powders including paracetamol (Novacyl, France) and anhydrous caffeine (BASF, Germany) as active compounds and lactose (Meggle, Germany), cornstarch (Roquette, France), colloidal silicon dioxide—Aerosil (Rohm Pharma Polymers, Germany), polyvinyl-pyrrolidone (BASF, Germany), talcum (IMERYS Luzenac, France), and magnesium stearate (Union Derivan S.A, Spain) as excipients were pharmaceutical grade.

Table 1: Composition of calibration and validation samples.

Concentration level	1 80%	2 90%	3 100%	4 110%	5 120%
Paracetamol	38.71	43.55	48.39	53.23	58.06
Caffeine	3.84	4.35	4.84	5.32	5.81
	Tablets composition (mg/tablet)				
Paracetamol	240.0	270.0	300.0	330.0	360.0
Caffeine	24.0	27.0	30.0	33.0	36.0
Lactose	146.0	113.0	80.0	47.0	14.0
Cornstarch					
Colloidal silicon dioxide					
PVP	210.0	210.0	210.0	210.0	210.0
Talcum					
Magnesium stearate					
Tablet weight	620.0	620.0	620.0	620.0	620.0

Calibration samples: levels 1, 2, 3, 4, and 5. Validation samples: levels 2, 3, and 4.

2.2. Sample Preparation. Pharmaceutical industries frequently use wet granulation in order to convert fine cohesive powders into dense and round granules. The granules are produced by vigorous mixing of a wet-powdered mixture composed of active compounds, some excipients, and binder. The overall purpose of this operation is to obtain a final product with improved characteristics, such as better flowability and compressibility. Other benefits were obtained using wet granulation; the distribution of the drug in the final product, as well as the dissolution properties of tablets may be improved.

For calibration and validation purpose, powder blends for tablets were prepared as presented in Table 1.

Paracetamol, caffeine, lactose, some cornstarch, and PVP (which was previously dissolved in 4.8 ml distilled water) were mixed. The wet mixture was passed through a sieve. It was left to dry at room temperature until the next day; when it was weighted, the amount of remaining excipients to be added was adjusted according to the weighted mixture. The powder blend was passed through a 800 μm sieve.

The mixture composition was designed for a tablet weight of approximately 620 mg and a usual amount of active ingredients of 300 mg paracetamol (48.38%, w/w) and 30 mg caffeine (4.84%, w/w). This formulation will be further considered as 100% active content formulation.

2.3. Experimental Design for Model Development. A calibration set was built using a full factorial experimental design with two factors and two levels, using Modde 11.0 Software (Umetrics, Sweden) to build and analyze the experimental plans. Each sample in the calibration set contained two components (paracetamol, caffeine); each component was taken at five concentration levels (80, 90, 100, 110, and 120% reported to the theoretical amount). The composition of calibration set samples is presented in a full factorial matrix of experimental plan, in Table 2.

TABLE 2: Experimental design matrix for calibration set.

Exp name	Run order	X_1	X_2	Exp name	Run order	X_1	X_2
N1	9	38.71	3.87	N15	13	58.06	4.84
N2	26	43.55	3.87	N16	25	38.71	5.32
N3	16	48.39	3.87	N17	24	43.55	5.32
N4	21	53.23	3.87	N18	17	48.39	5.32
N5	11	58.06	3.87	**N19***	**4**	**53.23**	**5.32**
N6	15	38.71	4.35	N20	7	58.06	5.32
N7*	**8**	**43.55**	**4.35**	N21	28	38.71	5.81
N8	6	48.39	4.35	N22	1	43.55	5.81
N9	27	53.23	4.35	N23	3	48.39	5.81
N10	23	58.06	4.35	N24	18	53.23	5.81
N11	14	38.71	4.84	N25	22	58.06	5.81
N12	20	43.55	4.84	N26	12	48.39	4.84
N13*	**2**	**48.39**	**4.84**	N27	19	48.39	4.84
N14	10	53.23	4.84	N28	5	58.06	4.84

X_1—paracetamol concentration (mg/tablet); X_2—caffeine concentration (mg/tablet). *Validation samples.

2.4. Recording of NIR Spectra. Near-infrared spectra of powder blends were recorded using a Fourier-transform NIRS analyzer (Antaris II, ThermoElectron Scientific, USA) in reflectance sampling configuration, equipped with an indium gallium arsenide (InGaAs) detector. Since the powder samples are not homogeneous, the device is equipped with a system for the rotation of samples during the measurements, so that obtained spectrum is representative for the sample and ensures the reproducibility of measurements. Each reflectance spectrum was recorded using OMNIC software (Termo Scientific, USA) by integrating 32 scans, over the range of 11000 to 4000 cm^{-1}, with a resolution of 8 cm^{-1}.

2.5. HPLC as Reference Method. After the collection of all NIR spectra from each individual powder blend, high-performance liquid chromatography (HPLC) analysis was performed for reference. Weighted powder samples were dissolved in methanol, in a 25 ml volumetric flask. The flask was vibrated in an ultrasonic bath (Transsonic T700, Germany) until complete dissolution. 5 ml were transferred to a centrifuging tube and were centrifuged (Sigma 2-16, Sartorius, Germany) for 5 min at 5000 rpm. One milliliter of the resulting supernatant was pipetted into a 10 ml volumetric flask and diluted to volume with water-acetonitrile (75:25, v/v).

Separately, 10 mg of paracetamol and 10 mg of caffeine were accurately weighted (using 0,01 mg analytical balance) into a 10 ml volumetric flask, and the same operation as described above was carried out to prepare the standard solution for calibration curve.

The samples were then analyzed by HPLC with UV detection. The HPLC system was a 1100 series model (Agilent Technologies, USA) consisted in a binary pump, an autosampler, a column thermostat, and a UV detector. Separation was carried out on a Gemini C18 (100×3.00 mm; 3 μm) column, with a mobile phase consisting of acetate buffer-acetonitrile (88 : 12, v/v) and a flow rate of 0.7 ml min^{-1}. Detection was performed at 245 nm for paracetamol (retention time at 1.9 min) and 275 nm for caffeine (retention time at 2.9 min). The calibration curves used for paracetamol and caffeine determination were in the range of 40–200 μg ml^{-1} and 4–20 μg ml^{-1}, respectively.

2.6. Data Processing. NIR spectra recorded for multivariate calibration models were previously processed using several established methods: first derivative, second derivative, standard normal variate (SNV), multiplicative scattering correction (MSC), straight line subtraction (SLS), minimum maxim normalization (MMN), in order to construct the calibration models. The partial least square (PLS) regression was conducted using multivariate analysis Opus Quant software (Bruker Optics, Germany).

This software allows validation of the chemometric multivariate calibration by the "full cross-validation." According to this procedure, iterative calibrations were performed by removing in turn each standard from the training set and then predicting the excluded sample with that calibration [17].

2.7. Method Validation. Once a calibration is developed and favourable predictions are expected, they must be validated to be accepted for routine use. Independent sets of samples are needed for external validation. There are several validation parameters that must be determined in order to be consistent with the recommendations of the International Conference of Harmonization (ICH) and with other guidelines: accuracy, precision (repeatability and intermediate precision), linearity, and range of application. The validation was performed according to the strategy proposed by Hubert et al., [18, 19] and adapted by the authors for NIR methods [10, 22, 23].

Validation of NIR methods for both paracetamol and caffeine assays was performed considering 90%, 100%, and 110% active compound content (formulations N7, N13, and N23). Four replicates were prepared for each formulation, in three different days, resulting a 36-sample validation set. In order to see which model fits the best, linearity and accuracy profiles were computed and compared, considering ICH Q2 guideline requirements.

3. Results and Discussion

Development of chemometric multivariate calibration models means to calculate the calibration parameters of the obtained data after the analysis of the NIR spectral calibration set. To do this, various methods of pretreatment of spectra in combination with selecting different spectral regions and different methods of regression analysis may be used. The entire spectrum and selecting certain spectral regions containing strong absorption bands in combination with different pretreatment methods of single spectra as standard normal variate (SNV), first derivative (FD), multiplicative scattering correction (MSC), straight line subtraction (SLS), minimum maximum normalization (MMN), or combined (FD + SNV, FD + MSC, and FD + SLS) were tested for building a calibration model. Once the calibration model has been

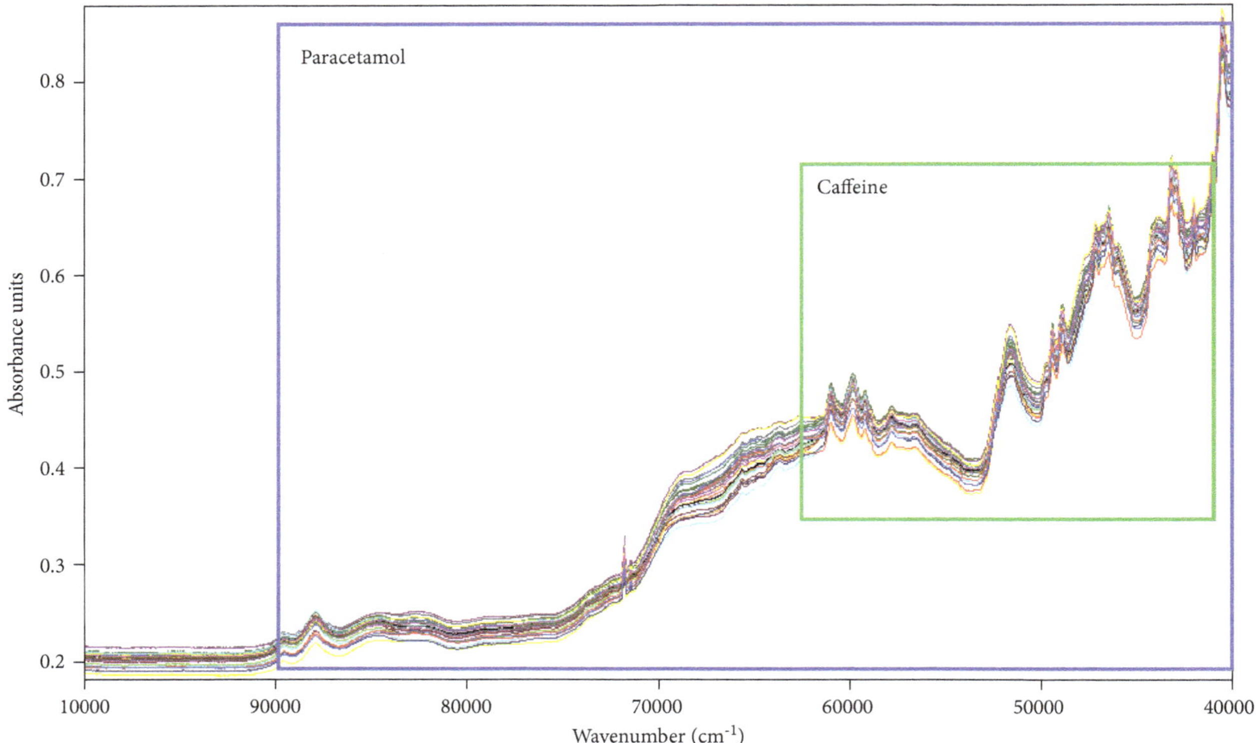

FIGURE 1: Reflectance spectrum of powder mixture recorded at a resolution of 8 cm^{-1} for a calibration set. The highlighted blue area defines the spectral ranges selected for the paracetamol, and the green area defines the spectral ranges selected for the caffeine.

developed, the capacity of prediction was tested on the test samples used during development.

3.1. Spectral Investigation. The NIR reflectance spectrum of the calibration set is presented in Figure 1. The calibration model for powder blends was built after recording three spectra of each formulation. Overall, 84 spectra were recorded and analyzed. NIR spectra of powder blends contain both chemical and physical information, so robust calibration models must be developed. Therefore, preprocessing methods and wavelength selection ranges should be carefully chosen to extract the chemical information that is mainly correlated with the API concentration [9]. In the spectral region with wavelengths less than 4000 cm^{-1}, the amount of radiation reaching the detector is reduced, background noise is high, and therefore it was eliminated from the analysis. Eliminating these spectral regions was carried out by other researchers too, who found similar results [20, 21]. Thus, model development for paracetamol assay was done using the spectral regions 9000–4000 cm^{-1}. In the case of caffeine, the spectral regions selected for model development and future analysis were 6200–4100 cm^{-1}.

3.2. Model Development and Validation for Active Compounds. Many calibration models were applied in order to perform quantitative determination of drug content in powder. Choosing the appropriate number of factors (principal components) is highly important for a PLS chemometric calibration. The number of factors for the experimental data obtained in calibration has to be chosen so that to avoid "over fitting." There were different methods proposed in order to select the optimum number of factors [22, 23]. Root mean square error of prediction (RMSEP) was used as a diagnostic test to assess the predictive ability of the model. Evaluation of prediction models was also made by plotting the known concentrations depending on the estimated concentrations. A good correlation coefficient R^2 between real and estimated concentrations was obtained for the studied components at optimized models, indicating good predictive ability of the models. The RMSEP and R^2 values obtained for optimized models for paracetamol are presented in Table 3 and for caffeine in Table 4.

In case that PLS algorithms are used for the development of methods, it is known that the method of spectral data pretreatment and the number of factors (components) are critical parameters. Selecting the optimal number of factors was performed using the criterion of Haaland and Thomas [24–26].

As presented in Table 3, the preprocessing based methods resulted in better prediction ability models. No preprocessing method showed high values for RMSECV, in the comparison of the preprocessing methods. For paracetamol, the models based on spectral range 9000–4000 cm^{-1} showed good ability of prediction when using 3–6 PLS factors. The authors chose the model d (SNV) showing low RMSECV and good R^2. For caffeine, the selected spectral range was 6200–4100 cm^{-1}, and the lowest RMSEP and best R^2 were obtained for the model c (SLS) as can be seen in Table 3.

Table 3: Statistical parameters and number of PLS factors for different models proposed for paracetamol and caffeine assays in powder blends.

					Paracetamol						
Model	a	b	c	**d***	e	f	g	h	i	j	k
Pretreatment	None	COE	SLS	**SNV**	mMN	MSC	FD	SD	FS+SLS	FD+SVN	FD+MSC
Spectral range selected (cm^{-1})						9000-4000					
Number of PLS factors	2	5	6	**5**	5	5	5	5	5	3	3
R^2	0.978	0.997	0.997	**0.997**	0.998	0.998	0.997	0.997	0.997	0.998	0.997
RMSECV (%)	0.966	0.334	0.311	**0.274**	0.261	0.259	0.328	0.363	0.326	0.302	0.337
Bias (%)	−0.0075	0.00088	0.00055	**0.00044**	−0.0029	0.00047	−0.0093	0.0010	−0.0039	0.0098	0.0048
					Caffeine						
Model	a	b	**c***	d	e	f	g	h	i	J	k
Pretreatment	None	COE	**SLS**	SNV	mMN	MSC	FD	SD	FS+SLS	FD+SVN	FD+MSC
Spectral range selected (cm^{-1})						6200-4100					
Number of PLS factors	14	14	**12**	11	11	11	10	8	11	9	9
R^2	0.958	0.96	**0.962**	0.958	0.941	0.945	0.946	0.881	0.952	0.939	0.940
RMSEP	0.137	0.132	**0.131**	0.137	0.163	0.156	0.155	0.23	0.146	0.165	0.164
Bias (%)	−0.0019	−0.0020	**−0.0085**	−0.0036	−0.0061	−0.0054	−0.0046	−0.0029	−0.0029	−0.00407	−0.0045

*Selected model for validation of the methods.

Table 4: Validation results of NIR method for paracetamol and caffeine assays in powder blends.

Concentration level (paracetamol)	Mean paracetamol content (mg/tablet)	Trueness		Precision		Accuracy	
		Relative bias (%)	Recovery (%)	Repeatability (RSD %)	Intermediate precision (RSD %)	Relative tolerance limits (%)	Tolerance limits (mg/tablet)
				Paracetamol			
43.55	43.90	0.805	100.8	0.384	0.393	[−0.11, 1.72]	[43.50, 44.30]
48.39	48.44	0.101	100.1	0.188	0.185	[−0.32, 0.52]	[48.64, 48.64]
53.23	53.07	−0.307	99.69	0.187	0.163	[−0.68, 0.06]	[53.26, 53.26]
				Caffeine			
4.355	4.437	1.880	101.88	1.34	3.19	[−6.26, 9.92]	[4.08, 4.82]
4.839	4.808	−0.659	99.34	2.49	2.79	[−7.53, 6.21]	[4.48, 5.14]
5.323	5.137	−1.592	98.41	1.83	1.63	[−5.28, 2.01]	[4.95, 5.33]

Validation of NIR methods for both paracetamol and caffeine assays was performed considering 90%, 100%, and 110% active compound content (formulations N7, N13, and N23). The linear profile, as seen in Figure 2, representing measured concentration versus theoretical concentration of paracetamol and caffeine, respectively, showed a good correlation coefficient and slope, confirming the linearity for the proposed model.

The β-expectation tolerance limits are included within the ±5% acceptance limits. Based on the data presented in Table 3 and Figure 3, we concluded that model d is best fitted for paracetamol and model c is best fitted for caffeine assay, in powder blends for tableting. Based on ICH Q2 and EMA guidelines, we validated the following characteristics: linearity, trueness, precision, and accuracy, as shown in Table 4.

The precision of the methods was evaluated by calculating two parameters: repeatability and intermediate precision at three concentration levels. Both parameters had satisfactory values for all concentration levels. The best recovery was obtained for the 100% paracetamol content, while best intermediate precision was obtained for the 110% paracetamol content. As for caffeine, the best results in terms of precision were obtained for the 90% active compound. In terms of accuracy, the β-expectation tolerance limits are fully included within the ±5% acceptance limits for both substances, so it can be concluded that the proposed models will provide results with adequate accuracy for both paracetamol and caffeine assays, whatever the active content of the powder blends is within the concentration limits studied.

According to data presented in Figures 2, 3, and Table 4, the NIR-chemometric methods using model d for paracetamol and model c for caffeine have satisfactory accuracy and linearity profile. It can be concluded that the

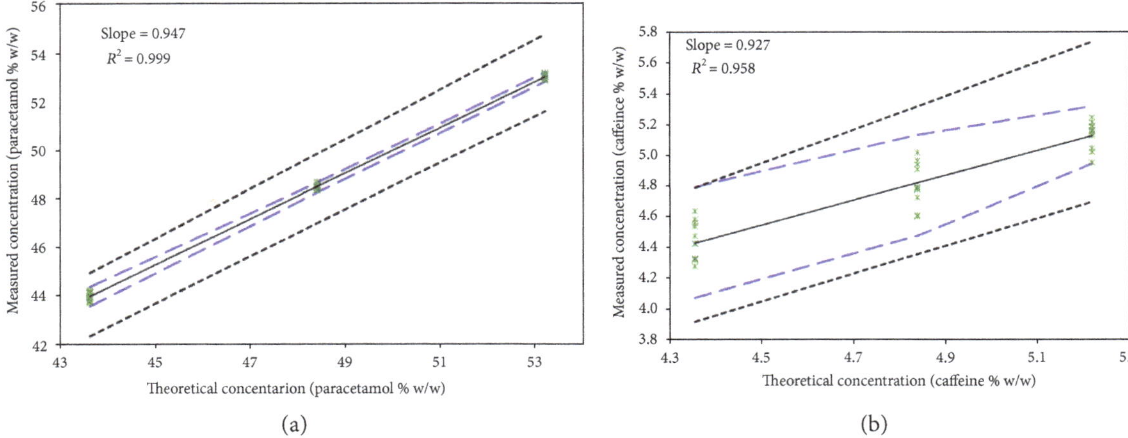

FIGURE 2: Linearity profile of the NIR—chemometic methods for paracetamol and caffeine determination. (a) paracetamol; (b) caffeine.

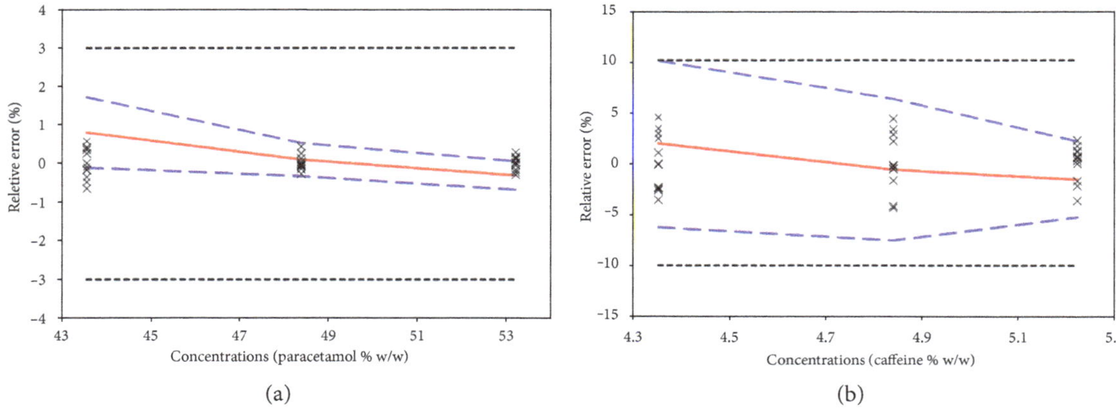

FIGURE 3: Accuracy profile of the NIR—chemometic methods for paracetamol and caffeine determination. (a) paracetamol; (b) caffeine.

TABLE 5: Paracetamol and caffeine determination by NIR-validated method and HPLC reference method.

API	Paracetamol		Caffeine	
Method	NIR	HPLC	NIR	HPLC
Theoretic concentration (mg)	48.39	48.39	4.839	4.839
Found (mg)	48.43	48.82	4.810	4.674
Recovery (%)	100.08	100.89	99.40	96.59
CV	0.17	1.19	2.15	4.53
t value	1.186		1.865	
p value	0.236		0.0915	

proposed NIR-chemometric method is suitable for paracetamol and caffeine assays in powder blends within the concentration range of 43.55–53.23% for the paracetamol and 4.35–5.32 for the caffeine.

3.3. Application of the Method. Once the NIR methods for the determination of paracetamol and caffeine were validated, they were applied for the active content determination in six control powder blends containing 48.39% w/w paracetamol and 4.84% w/w caffeine. The reference HPLC method has been also used for the active content assay in the same samples. The results obtained by the NIR method were compared with the values obtained by the reference HPLC method, in terms of the active content recovery, as shown in Table 5.

As can be seen, the recovery was quite similar for both methods. Also, Student's t-test was used to compare the two methods. The results did not show any statistical difference ($p > 0.05$) between the results predicted by NIR and HPLC which are used as the reference method.

4. Conclusions

The use of chemometry on near-infrared spectroscopy (NIRS) was explored for nondestructive quantitative analysis of two components in powder blends. The two components were determined simultaneously using pretreated spectra together with chemometry and PLS multivariate calibration. The models were validated in terms of trueness, precision, and accuracy, for an active content of 90, 100, and 110% for paracetamol and caffeine. Good statistical indicators were obtained. Furthermore, it was proved that the proposed methods are suitable for active pharmaceutical ingredient determination, as the results obtained are similar with those obtained by HPLC, which are used as the reference method.

Considering the results presented in this work, the NIR chemometry methods proved to be a suitable tool for predicting the chemical concentration of powder blends during preparation of fixed-dose combination tablets with paracetamol and caffeine. These methods can be used for on-line, in-line, or at-line monitoring of the blend uniformity in the mixing steps of the manufacturing process of tablets, with considerable saving in time and money in comparison with HPLC analysis.

Acknowledgments

This work was supported by a grant of the Romanian National Authority for Scientific Research and Innovation, CNCS-UEFISCDI, Project no. PN-III-P2-2.1-BG-2016-0201.

References

[1] Y. Dou, Y. Sun, Y. Ren, P. Ju, and Y. Ren, "Simultaneous non-destructive determination of two component of combined paracetamol and amantadine hydrochloride in tablets and powder by NIR spectroscopy and artificial neural networks," *Journal of Pharmaceutical and Biomedical Analysis*, vol. 37, no. 3, pp. 543–549, 2005.

[2] M. Jamrogiewicz, "Application of the near-infrared spectroscopy in the pharmaceutical technology," *Journal of Pharmaceutical and Biomedical Analysis*, vol. 66, no. 1, pp. 1–10, 2012.

[3] R. G. Brereton, "Introduction to multivariate calibration in analytical chemistry," *Analyst*, vol. 125, no. 11, pp. 2125–2154, 2000.

[4] M. Blanco, J. Coello, H. Iturriaga, S. Maspoch, and M. Porcel, "Simultaneous enzymatic spectrophotometric determination of ethanol and methanol by use of artificial neural networks for calibration," *Analytica Chimica Acta*, vol. 398, no. 1, pp. 83–92, 1999.

[5] A. Safavi, H. Abdollahi, and M. R. Hormezi Nezhad, "Artificial neural networks for simultaneous spectrophotometric differential kinetic determination of Co(II) and V(IV)," *Talanta*, vol. 59, no. 3, pp. 515–523, 2003.

[6] T. Wu, H. Chen, Z. Lin, and C. Tan, "Identification and quantitation of melamine in milk by near-infrared spectroscopy and chemometrics," *Journal of Spectroscopy*, vol. 2016, Article ID 6184987, 8 pages, 2016.

[7] L. Xu, S. M. Yan, C. B. Cai et al., "Nonlinear multivariate calibration of shelf life of preserved eggs (Pidan) by near infrared spectroscopy: stacked least squares support vector machine with ensemble preprocessing," *Journal of Spectroscopy*, vol. 2013, Article ID 797302, 7 pages, 2013.

[8] J. Irudayaraj and J. Tewari, "Simultaneous monitoring of organic acids and sugars in fresh and processed apple juice by Fourier transform infrared-attenuated total reflection spectroscopy," *Applied Spectroscopy*, vol. 57, no. 12, pp. 1599–1604, 2003.

[9] L. Martínez, A. Peinado, and L. Liesum, "In-line quantification of two active ingredients in a batch blending process by near-infrared spectroscopy: influence of physical presentation of the sample," *International Journal of Pharmaceutics*, vol. 451, no. 1-2, pp. 67–75, 2013.

[10] C. Sîrbu, I. Tomuta, L. Vonica, M. Achim, L. L. Rus, and E. Dinte, "Quantitative characterization of powder blends for tablets with indapamide by near-infrared spectroscopy and chemometry," *Farmácia*, vol. 62, no. 1, pp. 48–57, 2014.

[11] R. N. Goyal, V. K. Gupta, M. Oyama, and N. Bachheti, "Differential pulse voltammetric determination of paracetamol at nanogold modified indium tin oxide electrode," *Electrochemistry Communications*, vol. 7, no. 8, pp. 803–807, 2005.

[12] B. C. Lourencao, R. A. Medeiros, R. C. Rocha-Filho, L. H. Mazo, and O. Fatibello-Filho, "Simultaneous voltammetric determination of paracetamol and caffeine in pharmaceutical formulations using a boron-doped diamond electrode," *Talanta*, vol. 78, no. 3, pp. 748–752, 2009.

[13] N. Spataru, B. V. Sarada, D. A. Tryk, and A. Fujishima, "Anodic voltammetry of xanthine, theophylline, theobromine and caffeine at conductive diamond electrodes and its analytical application," *Electroanalysis*, vol. 14, no. 11, pp. 721–728, 2002.

[14] M. Z. Ding and J. K. Zou, "Centrifugal extraction procedure for the ultraviolet spectrophotometric determination of caffeine in beverages," *Chinese Journal of Analytical Chemistry*, vol. 36, no. 3, pp. 381–384, 2008.

[15] Y. Yamauchi and A. NakamuraI. Kohno, M. Kitai, K. Hatanaka, and T. Tanimoto, "Simple and rapid UV spectrophotometry of caffeine in tea coupled with sample pre-treatment using a cartridge column filled with polyvinylpolypyrrolidone (PVPP)," *Chemical & Pharmaceutical Bulletin*, vol. 56, no. 2, pp. 185–188, 2008.

[16] Y. Yamauchi and A. NakamuraI. Kohno, K. Hatanaka, M. Kitai, and T. Tamimoto, "Quasi-flow injection analysis for rapid determination of caffeine in tea using the sample pre-treatment method with a cartridge column filled with polyvinylpolypyrrolidone," *Journal of Chromatography. A*, vol. 1177, no. 4, pp. 190–194, 2008.

[17] A. Espinosa-Mansilla, F. Salinas, and I. De Orbe Paya, "Simultaneous determination of sulfadiazine, doxycycline, furaltadone and trimethoprim by partial last squares multivariate calibration," *Analytica Chimica Acta*, vol. 313, no. 1-2, pp. 103–112, 1995.

[18] P. Hubert, J. J. Nguyen-Huu, B. Boulanger et al., "Harmonization of strategies for the validation of quantitative analytical procedures: a SFSTP proposal-part II," *Journal of Pharmaceutical and Biomedical Analysis*, vol. 45, no. 1, pp. 70–81, 2007.

[19] P. Hubert, J. J. Nguyen-Huu, B. Boulanger et al., "Harmonization of strategies for the validation of quantitative analytical procedures: a SFSTP proposal-part I," *Journal of Pharmaceutical and Biomedical Analysis*, vol. 36, no. 3, pp. 579–586, 2004.

[20] Z. Xiaobo, Z. Jiewen, M. J. W. Povey, M. Holmes, and M. Hanpin, "Variables selection methods in near-infrared spectroscopy," *Analytica Chimica Acta*, vol. 667, no. 1-2, pp. 14–32, 2010.

[21] J. L. Ramirez, M. K. Bellamy, and R. J. Romañach, "A novel method for analyzing thick tablets by near infrared spectroscopy," *AAPS PharmSciTech*, vol. 2, no. 3, pp. 15–24, 2001.

[22] I. Tomuta, L. Rus, R. Iovanov, and L. L. Rus, "High-throughput NIR-chemometric methods for determination of drug content and pharmaceutical properties of indapamide tablets," *Journal of Pharmaceutical and Biomedical Analysis*, vol. 84, pp. 285–292, 2013.

[23] I. Tomuta, R. Iovanov, E. Bodoki, and L. Vonica, "Development and validation of NIR-chemometric methods for chemical and pharmaceutical characterization of meloxicam tablets," *Drug Development and Industrial Pharmacy*, vol. 40, no. 4, pp. 549–559, 2014.

[24] C. P. Meza, M. A. Santos, and R. J. Romanach, "Quantitation of drug content in a low dosage formulation by transmission near infrared spectroscopy," *AAPS PharmSciTech*, vol. 7, no. 1, article 29, pp. E1–E9, 2006.

[25] A. Porfire, I. Tomuta, L. Tefas, S. E. Leucuta, and M. Achim, "Simultaneous quantification of l-α-phosphatidylcoline and cholesterol in liposomes using near infrared spectrometry and chemometrics," *Journal of Pharmaceutical and Biomedical Analysis*, vol. 63, pp. 87–94, 2012.

[26] A. Porfire, D. Muntean, M. Achim, L. Vlase, and I. Tomuta, "Simultaneous quantification of simvastatin and excipients inliposomes using near infrared spectroscopy and chemometry," *Journal of Pharmaceutical and Biomedical Analysis*, vol. 107, pp. 40–49, 2015.

10

Ultraviolet Spectroscopy Used to Fingerprint Five Wild-Grown Edible Mushrooms (Boletaceae) Collected from Yunnan, China

Yan Li,[1,2,3] Ji Zhang,[1,2] Tao Li,[4] Tianwei Yang,[5] Yuanzhong Wang,[1,2] and Honggao Liu[5]

[1]*Institute of Medicinal Plants, Yunnan Academy of Agricultural Sciences, Kunming 650200, China*
[2]*Yunnan Technical Center for Quality of Chinese Materia Medica, Kunming 650200, China*
[3]*College of Traditional Chinese Medicine, Yunnan University of Traditional Chinese Medicine, Kunming 650500, China*
[4]*College of Resources and Environment, Yuxi Normal University, Yuxi 653100, China*
[5]*College of Agronomy and Biotechnology, Yunnan Agricultural University, Kunming 650201, China*

Correspondence should be addressed to Yuanzhong Wang; boletus@126.com and Honggao Liu; honggaoliu@126.com

Academic Editor: Khalique Ahmed

Nowadays, wild-grown edible mushrooms which are natural, nutritious, and healthy get more and more popular by large consumers. In this paper, UV spectra of different Boletaceae mushrooms with the aid of partial least squares discriminant analysis (PLS-DA) and hierarchical cluster analysis (HCA) were shown to be a practical and rapid method for discrimination purpose. The specimens of *Boletus edulis*, *Boletus ferrugineus*, *Boletus tomentipes*, *Leccinum rugosiceps*, and *Xerocomus* sp. were described based on the UV spectra. From the results, all the specimens were characterized by strong absorption at the wavelengths of 274 and 284 nm and showed the shoulder at 296 nm. However, changes could be seen in the peak heights at the same wavelength for different samples. After analyzing by chemometrics, visual discrimination among samples was presented and the relationships among them were also obtained. This study showed that UV spectroscopy combined with chemometrics methods could be used successfully as a simple and effective approach for characterization of these five wild-grown edible mushrooms at species and genus levels. Meanwhile, this rapid and simple methodology could also provide reference for the discrimination of edible mushrooms.

1. Introduction

Wild-grown edible mushrooms are considered as healthy food sources and have long been attracting a great deal of interest by humankind for the natural and nutritional effects [1]. Some species of them are consumed as a delicacy and constitute an increasing share in the world diet [2]. The consumption of wild-grown edible mushrooms is 5.6 kg of fresh product per household yearly in Czech Republic while higher value is found to be 20–24 kg in China [3, 4]. Moreover, these mushrooms are not only an important source of revenue for rural economies but also a substantial economic resource in several regions of the world [5].

In the daily life, many edible mushrooms are usually sliced and dried after collecting for better storage and sales. However, since the market demands for wild-grown edible mushrooms increased, some unscrupulous traders sell inferior mushrooms for profiteering such as mixing different species of dried mushroom slices even toxic ones which lead to unfair competition [6]. In a previous study, Dentinger and Suz [7] analyzed 15 pieces of dried Chinese porcini from a single commercial packet purchased in London and identified three species of mushrooms that have never been formally described by science until now. It suggested that insufficient knowledge of the porcini species contained within food products may pose a health concern. Therefore, for economical, biodiversity-related reasons and so forth, it is often important to discriminate the species of wild-grown edible mushrooms accurately.

Classical mushroom discrimination is according to careful observation of microscopic and macroscopic morphological characters. However, these methods require trained and experienced people. At present, some analytical techniques for discrimination of edible mushrooms have been published including high-performance liquid chromatography (HPLC), gas chromatography-mass spectrometry (GC-MS), infrared (IR) spectroscopy, nuclear magnetic resonance (NMR) spectroscopy, and DNA sequence analyses [8–12]. Since these techniques provide a nonselective signal, the aid of appropriate chemometrics methods is also necessary for interpreting them [13, 14]. Nevertheless, some imperfections about these methods have been reported. For example, IR needed the experienced technicians and it was hard to model. GC-MS which was expensive could be used to analyze the substance which had the low boiling point and good thermal stability, merely. The ideal chemical technique for the discrimination of mushroom species would provide rapid and accurate analysis. Recently, ultraviolet (UV) spectrometry that reflect the comprehensive fuzz information of samples should be accessible as discrimination tool in diverse research fields, such as the analysis of foods [15], herb medicines [16], automotive window tints [17], for the advantages like rapidity, simplicity, and low cost. Additionally, Li et al. [18] distinguished wild and cultured *Macrocybe gigantea* with different storage times based on UV spectroscopy combined with multivariate analyses. As reported by Yang et al. [19] UV spectra fingerprints in combination with chemometrics methods could be used to discriminate different parts of edible mushrooms. The rapid and reliable method of UV spectroscopy has shown huge potential for the analysis of edible mushrooms

Mushrooms in the family Boletaceae with 50 genera and 800 species, which are mainly characterized by soft fleshy context, are important groups in the macrofungi of basidiomycete [20, 21]. They are wildly collected and consumed in the main production areas such as eastern Asia, Europe, and North America [22, 23]. In southwestern China, Yunnan Province which is mild and rainy in summer and autumn providing ideal conditions for fungal growth is one of the most important centers for producing, consuming, and trading Boletaceae mushrooms [24]. *Boletus edulis* which is one of the most famous delicious edible mushrooms in the world is widely liked by people [25]. Other species such as *Boletus tomentipes* are generally trading on the market during summer and autumn [26]. Apart from flavor and taste, the fruiting bodies of the wild-grown edible mushrooms in this family are considered sources of proteins, amino acids, vitamins, and carbohydrates, as well as minerals and antioxidants [1, 26]. In addition to edibleness, many medicinal properties in these mushrooms, such as antioxidation, antitumor, and antibiotic effect, have been claimed [27, 28].

In this study, an analytical method for the discrimination of five wild-grown edible mushrooms (Boletaceae) by UV spectroscopy was established and verified. All the spectroscopic data were analyzed by partial least squares discriminant analysis (PLS-DA) and hierarchical cluster analysis (HCA) that aimed to distinguish different specimens and find correlations among these species. The results represented a detailed report on the differentiation of tested Boletaceae species which may provide a utility of methodology for discriminating wild-grown edible mushrooms rapidly and accurately.

2. Experiment

2.1. Materials. For the real samples used in this study, the fruiting bodies of wild Boletaceae including five species were collected from Yunnan Province during the collection season (June to September) in 2011. All the samples were authenticated by Dr. Honggao Liu, College of Agronomy and Biotechnology, Yunnan Agricultural University, and preserved in the specimen room of this university. The sample information is listed in Table 1.

2.2. Apparatus. Apparatus were as follows: UV-2550 UV-Vis Spectrophotometer (Shimadzu, Japan); DFT-100 type grinder (Zhejiang Wenling City Linda Machinery Company, China); SY3200-T type ultrasonic washer (Shanghai Shengyuan Ultrasonic Equipment Company, China); 100-mesh stainless steel sieve (Beijing Zhongxi Tai'an Technology Service Company, China); and AR1140 Electronic Analytical Balance (NJ, USA).

2.3. Sample Preparations. All the samples were cleaned up and dried at the temperature of 50°C before analysis. Then all of them were ground to fine powder and passed through a 100-mesh stainless steel sieve. The sieved powders were stored in the labeled Ziploc bags at room temperature until further analysis. In each mushroom species, the mixed mushroom samples ($n = 7$) of the same collection site were used to perform the experiment. 0.1 g of each mixed powdered sample and 10.0 mL chloroform (analytical grade) were put into a 25 mL colorimetric tube and extracted by ultrasonication for 30 min. The extracts were filtered and kept as stock solutions for testing.

2.4. Data Acquisition and Chemometrics Methods. Each stock solution was analyzed by UV-2550 UV-Vis spectrophotometer at 0.2 nm sampling interval and 1.0 nm slit width. Scans were collected over a range of 190–400 nm and each sample was measured in triplicate. The UV spectra were treated by the three groups of average and second derivative, in order to eliminate the solvent interference and increase accuracy of spectra. The number of points for derivative was five and the spectral data were mean centered during the analyzed.

The absorption readings obtained over the spectral points of all the samples were converted into a data matrix using Excel 2007 (Microsoft, USA) with the spectral points as variables represented by columns and the corresponding spectral absorption measurements represented by rows. Then the raw spectral data of all the samples were analyzed by partial least squares discriminant analysis (PLS-DA) and hierarchical cluster analysis (HCA) to evaluate the relationships in terms of similarity or dissimilarity among groups of multivariate data. The two statistical analyses were performed by using SIMCA-P$^+$ 11.5 (Umetrics, Umeå, Sweden) and SPSS

TABLE 1: Information of all the samples.

Number	Species	Site of collection
1	*Leccinum rugosiceps* (Peck) Singer	Wuhua District, Kunming, Yunnan
2	*Leccinum rugosiceps* (Peck) Singer	Qujing, Yunnan
3	*Leccinum rugosiceps* (Peck) Singer	Pubei, Yimen, Yunnan
4	*Boletus ferrugineus* Schaeff.	Wuding, Yunnan
5	*Boletus ferrugineus* Schaeff.	Qujing, Yunnan
6	*Boletus ferrugineus* Schaeff.	Wuhua District, Kunming, Yunnan
7	*Xerocomus* sp.	Nanhua, Chuxiong, Yunnan
8	*Xerocomus* sp.	Yao'an, Chuxiong, Yunnan
9	*Xerocomus* sp.	Qujing, Yunnan
10	*Boletus tomentipes* Earle	Fuliangpeng, Eshan, Yunnan
11	*Boletus tomentipes* Earle	Pu'er, Yunnan
12	*Boletus tomentipes* Earle	Xiaojie, Eshan, Yunnan
13	*Boletus tomentipes* Earle	Qianchang, Yao'an, Yunnan
14	*Boletus tomentipes* Earle	Tianshentang, Nanhua, Yunnan
15	*Boletus tomentipes* Earle	Tongchang, Yimen, Yunnan
16	*Boletus tomentipes* Earle	Shaqiao, Nanhua, Yunnan
17	*Boletus tomentipes* Earle	Pu'er, Yunnan
18	*Boletus edulis* Bull.	Baofeng, Jinning, Yunnan
19	*Boletus edulis* Bull.	Yulu, Nanhua, Yunnan
20	*Boletus edulis* Bull.	Pubei, Yimen, Yunnan
21	*Boletus edulis* Bull.	Longchuan, Nanhua, Yunnan
22	*Boletus edulis* Bull.	Qianchang, Yao'an, Yunnan
23	*Boletus edulis* Bull.	Qianchang, Yao'an, Yunnan
24	*Boletus edulis* Bull.	Shaqiao, Nanhua, Yunnan

20.0 (IBM Corp., Armonk, USA), respectively. Data were visualized by using the two statistical approaches.

3. Results and Discussion

3.1. Selection of Extraction Solvent. Every powdered mushroom sample was taken out to form the mixed sample in order to select the extraction solvent. A total of four different extraction solvents (petroleum ether, chloroform, absolute ethanol, and 0.5 mol/L NaOH) were used and all reagents were of analytical grade. The number of the absorption peaks was used to validate the most appropriate extraction solvent. As shown in the result in Figure 1, the number of the absorption peaks about the chloroform extract is the highest of all the extracts while others have only one or two absorption peaks. It suggested that chloroform extract may include more constituent information about the mushroom samples to reflect their characteristics and chloroform could be the most appropriate extraction solvent.

3.2. UV Spectra of the Wild Edible Boletaceae Mushrooms. The UV spectra of all the specimens are shown in Figure 2. On account of the detection range of the UV-Vis spectrophotometer, the wavelengths of absorption peaks are arranged from 235 to 400 nm for the sake of avoiding the spectral noise. It shows that all the specimens have higher overlap rate from 235 to 335 nm than that of other wavelengths. Every sample has some characteristic absorption peaks to reveal its fingerprint feature. In Figure 2, it indicated that some chemical components appeared to be very similar among these five species of Boletaceae mushrooms because all the specimens are characterized by strong absorption at the wavelengths of 274 and 284 nm and show the shoulder at 296 nm. However, changes can be seen in the peak heights at the same wavelength for different samples and in the ratios between absorbance values at different wavelengths for the same sample. The UV absorption bands of the presented specimens were usually associated with the presence of different chromophores exemplified in conjugated systems as well as other UV-absorbing systems [29]. To a certain degree, when the substance was in high concentration, the corresponding absorbance was high, too [30]. It suggested that the contents of chemical compositions in different species of mushrooms were variable. This result was in agreement with the reports in previous literatures that the accumulation of chemical composition may be affected by the mushroom species [31, 32]. These differences may be used to discriminate the Boletaceae specimens.

3.3. Partial Least Squares Discriminant Analysis. PLS-DA is a well-established chemometric approach for supervised analyses based on a PLS model in which the dependent variable

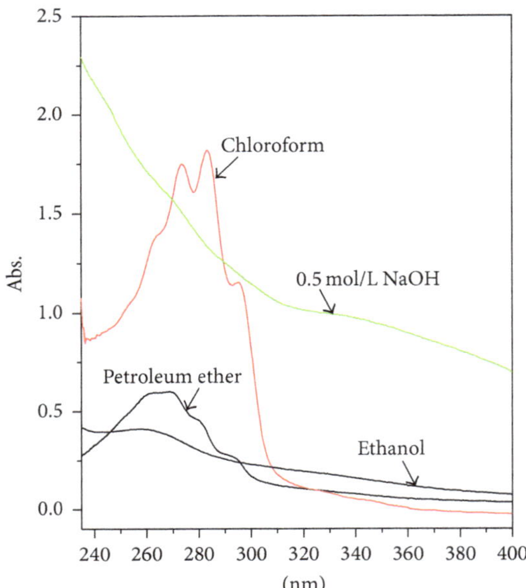

Figure 1: UV spectra of different extraction solvents.

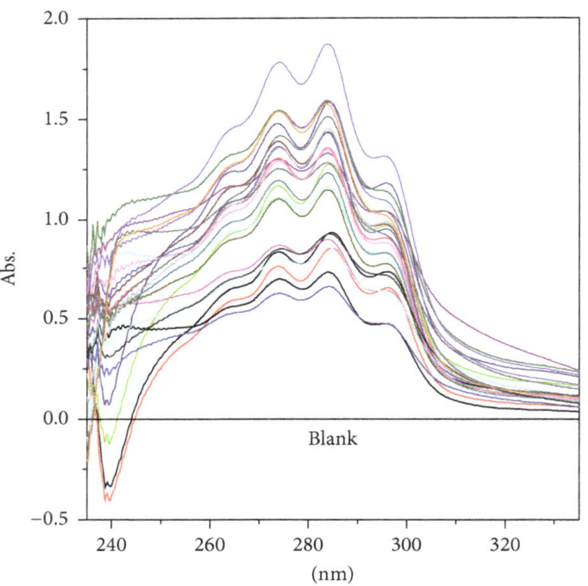

Figure 2: UV spectra fingerprints of the wild edible mushrooms (Boletaceae).

(Y block) represents class membership [33]. This method was used as a representative technique to discriminate the mushroom samples according to their species in this study. The first four principal components (PCs) of PLS-DA could explain 85% of the total variance. The score plot of the sample data with 95% confidence ellipses is shown in Figure 3. Clear separation of the five species of wild-grown Boletaceae mushrooms is observed in the two-dimensional diagram. The mushroom samples which belong to the same species could form cluster and be distinct with other species. PC 1 is determined mainly by negative scores for the samples of

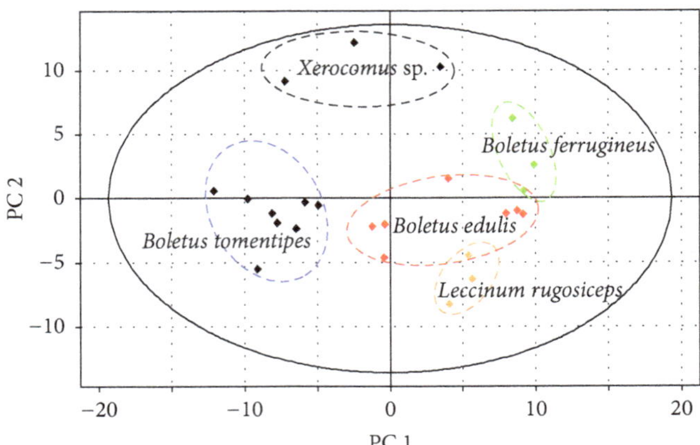

Figure 3: PLS-DA score plot based on UV spectra of five species of wild-grown Boletaceae mushrooms.

B. tomentipes while the strongest positive scores on PC 2 are found for samples which belong to *Xerocomus* sp. All the samples could be distinguished into five classes. Additionally, the individual differences among samples of *B. edulis* seem to be relatively obvious because these samples are distributed dispersedly. Similarly, clear differences are also visible in the individuals of *Xerocomus* sp. What is more, the species in the same genus tend to cluster together. As a result, 24 tested samples were classified entirely as their groups by PLS-DA.

According to the spectrographic PLS-DA analysis, a series of scores (variable importance for the projections, or VIPs) were computed to assess the contribution of absorbance to these dimensions. A variable which has the VIP score greater than 1.0 is usually considered important for the discrimination whereas variables with VIPs smaller than 1.0 are less important [34]. Figure 4 shows the contribution of information from each individual variable to the overall samples separation by PLS-DA. As shown in Figure 4, a total of 159 variables have the VIPs greater than 1.0 and about 50% of them are in the region of 270–300 nm. It showed that the absorption of the wavelength from 270 to 300 nm was likely to be considered as main factor for discrimination of all the mushrooms. Moreover, in this region, the absorption in the wavelength of 272.4, 274.6, 278, 279.4, 280.2, 280.4, 282.2, 283.2, 284, 285, 285.2, 289.8, 291, 292.2, 296, 296.2, 296.4, 297.8, and 298 nm makes a relatively great contribution because the VIPs of these variables are greater than 2.0 (Table 2).

From the corresponding loading plot (Figure 5), the contribution of different selected variables (VIP > 1.0 and marked by hollow red square) on the PCs has been recognized. The loadings of variables which made a relatively great contribution (VIPs > 2.0) are shown in Table 2. PC 1 plays a significant role in discriminating *B. tomentipes* and *Xerocomus* sp. from other samples. Variables in the wavelength of 280.2, 289.8, and 291 nm have contributions to PC 1, which separate samples number 10 to 17 from the others, with *B. tomentipes* having a negative loading value. The most relevant variables for discriminating *Xerocomus* sp. from other mushrooms are in the wavelength of 278, 279.4, 280.4,

TABLE 2: VIPs and loadings of selected variables of PLS-DA.

Var ID/nm	VIP	Loading 1	Loading 2
272.4	2.27903	0.10182	−0.034039
274.6	2.13471	0.0953719	−0.048547
278	2.48458	−0.111003	0.0341494
279.4	2.05394	−0.091763	0.0367178
280.2	2.2159	−0.098999	−0.024235
280.4	2.22486	−0.099399	0.0190163
282.2	2.24673	0.100376	0.0255896
283.2	2.17515	0.0971786	−0.015926
283.8	2.02639	0.0905325	−0.002524
284	2.07926	0.0928945	−0.02318
285	2.28531	0.1021	−0.017045
285.2	2.1474	0.0959388	−0.005978
289.8	2.46534	−0.110143	−0.039376
291	2.15336	−0.096205	−0.006975
292.2	2.45586	−0.10972	0.0165164
296	2.18898	0.0977962	−0.083971
296.2	2.08193	0.0930139	−0.03437
296.4	2.26702	0.101283	0.0613701
297.8	2.1033	0.0939687	−0.005287
298	2.18613	0.097669	0.0402933

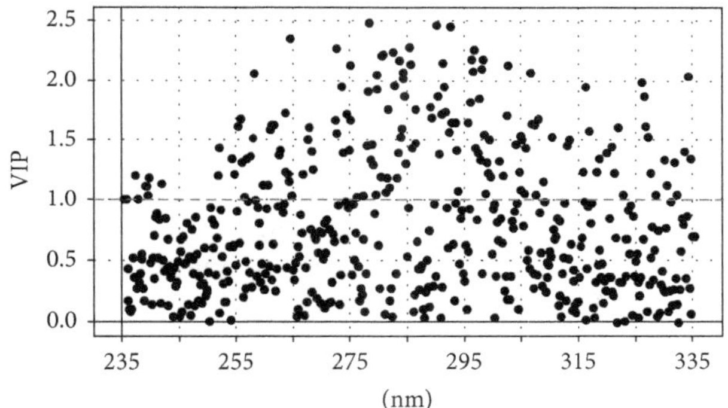

FIGURE 4: Variable Importance for the Projection (VIP) plot of absorbance for the contribution to sample separation from PLS-DA.

and 292.2 nm. In addition, the absorption of wavelength of 282.2, 296.4, and 298 nm plays the discriminating roles for *B. ferrugineus* from other species while *L. rugosiceps* and *B. edulis* are in the same quadrant where the absorbance of wavelength of 272.4, 274.6, 283.2, 283.8, 284, 285, 285.2, 296, 296.2, and 297.8 nm contributes to separating these two species from the other samples.

3.4. Hierarchical Cluster Analysis. HCA is an unsupervised pattern recognition method for clustering samples according to the similarities among them [35]. It was used based on the first four PCs of PLS-DA to classify the samples into groups using the single linkage method for cluster building and the distance between clusters was computed by the squared Euclidean method. As shown in the dendrogram (Figure 6), all the samples have the correct cluster at species level. Overall, all the five species are divided into three main fractions when the distance is 19 and that apparently did reflect interdependent relationships occurring among them. In the first case up to three subfractions can be recognized that relate to *B. ferrugineus* (samples number 4, 5, and 6), *B. tomentipes* (samples number 10 to 17), and *B. edulis* (samples number 18 to 24), respectively, which belonged to genus *Boletus*, family Boletaceae [36, 37]. As a consequence, these three species mushrooms had the correct cluster at genus and family levels and corresponded with the fungal classification. Three samples—samples number 7 to 9—are similar and join to form the second cluster. They are all classified as *Xerocomus* sp. [20]. The last one also contains three mushroom samples. Combined with the information in Table 1, these samples were *L. rugosiceps* and pertained to genus *Leccinum*, family Boletaceae [20, 37].

Generally, the different species of edible mushrooms were discriminated based on the morphological characteristics and macroscopic color reactions. However some literatures suggested that these two methods relied on experience and subjective factors [8, 38]. In addition, the color reactions of the mushrooms may be affected by the environment, climate, agrotype, growing season, and physiological status of the fruiting bodies and this may cause the deviation of subjective judgment [38]. In this study, discrimination of five wild-grown edible mushrooms (Boletaceae) profiled by UV spectroscopy analysis combined with chemometrics allowed for the digitalization of these sample properties providing a novel approach for objective annotation of different edible mushroom (Boletaceae) attributes such as species. In contrast, this analysis was efficient, rapid, and reliable, as based on the chloroform extracts.

PLS-DA was a good tool that could provide an overall look at the initial differences in UV spectra and it was possible to show clear differences in chemical components among the five species of wild-grown edible Boletaceae mushrooms. It suggested that this method could be used to differentiate mushroom specimens according to their species. However, although PLS-DA showed clear separation of the mushrooms, better clustering could be observed in the HCA dendrogram (Figure 6). All the specimens could be distinguished accurately at species level and the relationship among them has been also presented.

Obviously, although there were differences among the five species of wild-grown edible mushrooms, some similar constituents between *B. edulis* and *B. tomentipes* could be reflected on account of these two species samples joined together at first. More interestingly, according to the results of dendrogram, the interspecific differences between *B. edulis* and *B. tomentipes* could be the smallest in *B. edulis*, *B. ferrugineus*, and *B. tomentipes* even though these three species belonged to the same genus. This may be related to the genetic stability and variability of different species of Boletaceae mushrooms during the long-term evolution process. With regard to the level of genus, *Boletus* may have similar relationship with *Xerocomus* based on the chemical analysis

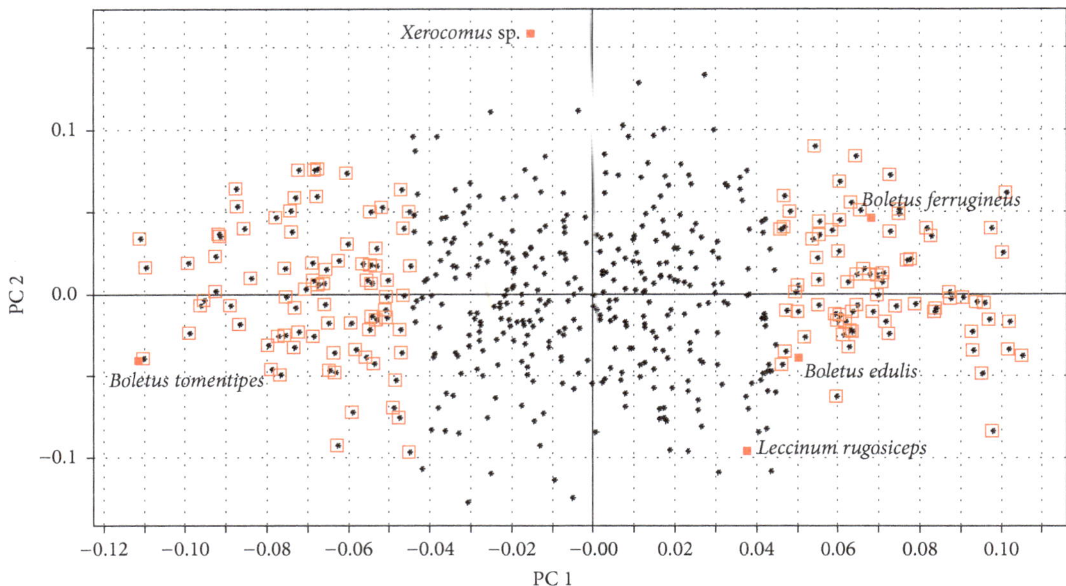

FIGURE 5: Loading plot generated from the PLS-DA model of the mushroom samples.

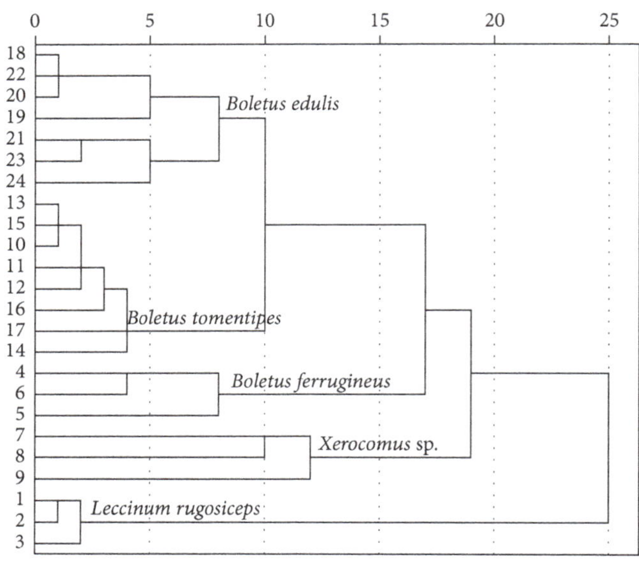

FIGURE 6: Dendrogram resulting from hierarchical cluster analysis.

because these two genera could cluster together firstly among the three genera studied in this study. This was consistent with the previous study that some mycologists have incorporated *Xerocomus* into *Boletus* [39]. Additionally, *Boletus* was clearly different from *Leccinum* that the samples which belonged to the same genus could be clustered to differentiate with the other one. Indeed, a recent paper has demonstrated this result by molecular analyses that explained this phenomenon in a genetic way [40]. On the other hand, this method for chemical analysis verified the consequence of previous study. It could provide a new way to classify the edible mushrooms.

4. Conclusions

This paper described the combination of UV spectroscopy and chemometrics as a rapid discrimination approach of wild-grown Boletaceae mushrooms. The results demonstrated that obvious differences could be found in the whole chemical components based on the chloroform extracts of *B. edulis*, *B. ferrugineus*, *B. tomentipes*, *L. rugosiceps*, and *Xerocomus* sp. All the samples could be distinguished accurately at species and genus levels and the relationships among them have been presented. On the other hand, this study could offer an economical, effective, and useful discrimination approach of wild-grown edible mushrooms.

Competing Interests

The authors declare that there is no conflict of interests regarding the publication of this paper.

Acknowledgments

This work was sponsored by the National Natural Science Foundation of China (31660591, 21667031) and the Science Foundation of the Yunnan Province Department of Education (2016ZZX106).

References

[1] X.-M. Wang, J. Zhang, L.-H. Wu et al., "A mini-review of chemical composition and nutritional value of edible wild-grown mushroom from China," *Food Chemistry*, vol. 151, pp. 279–285, 2014.

[2] P. Kalač, "Chemical composition and nutritional value of European species of wild growing mushrooms: a review," *Food Chemistry*, vol. 113, no. 1, pp. 9–16, 2009.

[3] L. Šišák, "The importance of mushroom picking as compared to forest berries in the Czech Republic," *Mykologický Sborník*, vol. 84, no. 3, pp. 78–83, 2007.

[4] D. Zhang, T. Y. Gao, P. Ma, Y. Luo, and P. C. Su, "Bioaccumulation of heavy metal in wild growing mushrooms from Liangshan Yi nationality autonomous prefecture, China," *Wuhan University Journal of Natural Sciences*, vol. 13, no. 3, pp. 267–272, 2008.

[5] M. De Román and E. Boa, "Collection, marketing and cultivation of edible fungi in Spain," *Micologia Aplicada International*, vol. 16, no. 2, pp. 25–33, 2004.

[6] Y. M. Shi, G. Liu, Y. L. Sun, S. X. Wei, C. Q. Yan, and X. J. He, "Identifition of *Tricholoma matsutake* (S. Ito et Imai) Sing and *Agaricus Blazei* Murrill using Fourier transform infrared spectroscopy and hierarchical cluster analysis," *The Journal of Light Scattering*, vol. 22, pp. 171–174, 2010.

[7] B. T. M. Dentinger and L. M. Suz, "What's for dinner? Undescribed species of porcini in a commercial packet," *PeerJ*, vol. 2, article e570, 2014.

[8] G. Liu, J.-H. Liu, A.-M. Yang, Q. Dong, and D.-S. Song, "Identification of edible mushrooms by Fourier transform infrared spectroscopy," *Spectroscopy and Spectral Analysis*, vol. 24, no. 8, pp. 941–945, 2004.

[9] J. Zhang, X. Zhong, S. Li, G. Zhang, and X. Liu, "Metabolic characterization of natural and cultured *Ophicordyceps sinensis* from different origins by ^{1}H NMR spectroscopy," *Journal of Pharmaceutical and Biomedical Analysis*, vol. 115, pp. 395–401, 2015.

[10] L. Barros, M. Dueñas, I. C. F. R. Ferreira, P. Baptista, and C. Santos-Buelga, "Phenolic acids determination by HPLC-DAD-ESI/MS in sixteen different Portuguese wild mushrooms species," *Food and Chemical Toxicology*, vol. 47, no. 6, pp. 1076–1079, 2009.

[11] R. Malheiro, P. Guedes de Pinho, S. Soares, A. César da Silva Ferreira, and P. Baptista, "Volatile biomarkers for wild mushrooms species discrimination," *Food Research International*, vol. 54, no. 1, pp. 186–194, 2013.

[12] G. Wu, B. Feng, J. P. Xu et al., "Molecular phylogenetic analyses redefine seven major clades and reveal 22 new generic clades in the fungal family *Boletaceae*," *Fungal Diversity*, vol. 69, no. 1, pp. 93–115, 2014.

[13] M. Y. Shen, M. Y. Xie, S. P. Nie, Y. Q. Wan, and J. H. Xie, "Discrimination of different *Ganoderma* species and their region based on GC-MS profiles of sterols and pattern recognition techniques," *Analytical Letters*, vol. 44, no. 5, pp. 863–873, 2011.

[14] I. Marekov, S. Momchilova, B. Grung, and B. Nikolova-Damyanova, "Fatty acid composition of wild mushroom species of order Agaricales-Examination by gas chromatography-mass spectrometry and chemometrics," *Journal of Chromatography B*, vol. 910, pp. 54–60, 2012.

[15] P. H. G. D. Diniz, M. F. Barbosa, K. D. T. de Melo Milanez, M. F. Pistonesi, and M. C. U. de Araújo, "Using UV-Vis spectroscopy for simultaneous geographical and varietal classification of tea infusions simulating a home-made tea cup," *Food Chemistry*, vol. 192, pp. 374–379, 2016.

[16] H. A. Gad, S. H. El-Ahmady, M. I. Abou-Shoer, and M. M. Al-Azizi, "A modern approach to the authentication and quality assessment of thyme using UV spectroscopy and chemometric analysis," *Phytochemical Analysis*, vol. 24, no. 6, pp. 520–526, 2013.

[17] K. J. van der Pal, M. Maric, W. Van Bronswijk, and S. W. Lewis, "Ultraviolet-visible spectroscopic characterisation of automotive window tints for forensic purposes," *Analytical Methods*, vol. 7, no. 13, pp. 5391–5395, 2015.

[18] Y. Li, J. Zhang, H. Liu, H. Jin, Y. Wang, and T. Li, "Discrimination of storage periods for *Macrocybe gigantea* (Massee) using UV spectral fingerprints," *Czech Journal of Food Sciences*, vol. 33, no. 5, pp. 441–448, 2016.

[19] T. W. Yang, B. K. Cui, J. Zhang et al., "Identification of different parts of edible bolete mushrooms by UV fingerprint," *Mycosyst*, vol. 33, no. 2, pp. 262–272, 2014.

[20] X. L. Mao, *The Microfungi in China*, Henan Science and Technology Press, Henan, China, 1st edition, 2000.

[21] G.-L. Wang, S.-Q. Wu, and Q.-F. Wu, "Separation, purification and identification of acidic polysaccharide fraction extracted from *Boletus edulis* and its influence on mouse lymphocyte proliferation in vitro," *Journal of Chemical and Pharmaceutical Research*, vol. 5, no. 12, pp. 431–437, 2013.

[22] N. Sitta and M. Floriani, "Nationalization and globalization trends in the wild mushroom commerce of Italy with emphasis on porcini (*Boletus edulis* and allied species)," *Economic Botany*, vol. 62, no. 3, pp. 307–322, 2008.

[23] L. Sun, M. Yang, X. Bai, and Y. Zhuang, "Effects of different cooking methods on nutritional characteristics of *Boletus aereus*," *Advanced Materials Research*, vol. 634, no. 1, pp. 1474–1480, 2013.

[24] F. K. Zhu, L. Qu, W. X. Fan, M. Y. Qiao, H. L. Hao, and X. J. Wang, "Assessment of heavy metals in some wild edible mushrooms collected from Yunnan Province, China," *Environmental Monitoring and Assessment*, vol. 179, no. 1–4, pp. 191–199, 2011.

[25] M. Bovi, L. Cenci, M. Perduca et al., "BEL β-trefoil: a novel lectin with antineoplastic properties in king bolete (*Boletus edulis*) mushrooms," *Glycobiology*, vol. 23, no. 5, pp. 578–592, 2013.

[26] T. Li, Y. Wang, J. Zhang, Y. Zhao, and H. Liu, "Trace element content of *Boletus tomentipes* mushroom collected from Yunnan, China," *Food Chemistry*, vol. 127, no. 4, pp. 1828–1830, 2011.

[27] R. Gu, "Analysis about the condition of extraction and purification of polysaccharide in *Boletus edulis*," *Farm Machinery*, no. 20, pp. 150–151, 2011.

[28] W. Radzki, A. Sławińska, E. Jabłońska-Ryś, and W. Gustaw, "Antioxidant capacity and polyphenolic content of dried wild edible mushrooms from Poland," *International Journal of Medicinal Mushrooms*, vol. 16, no. 1, pp. 65–75, 2014.

[29] Y. E. Zeng and L. Zhang, *Instrumental Analysis*, Science Press, Beijing, China, 5th edition, 2010.

[30] K.-L. Wei, Z.-Y. Wen, X. Wu, Z.-W. Zhang, and T.-L. Zeng, "Research advances in water quality monitoring technology based on UV-Vis spectrum analysis," *Spectroscopy and Spectral Analysis*, vol. 31, no. 4, pp. 1074–1077, 2011.

[31] I. Akata, B. Ergönül, and F. Kalyoncu, "Chemical compositions and antioxidant activities of 16 wild edible mushroom species grown in Anatolia," *International Journal of Pharmacology*, vol. 8, no. 2, pp. 134–138, 2012.

[32] Z. Zhang, Z. Gu, Y. Yang, S. Zhou, Y. F. Liu, and J. Q. Tang, "Evaluation of the umami taste of three species of dried wild edible fungi," *Food Scence*, vol. 34, no. 21, pp. 51–54, 2013.

[33] S. Wold, M. Sjöström, and L. Eriksson, "PLS-regression: a basic tool of chemometrics," *Chemometrics and Intelligent Laboratory Systems*, vol. 58, no. 2, pp. 109–130, 2001.

[34] N. Shetty, M. H. Olesen, R. Gislum, L. C. Deleuran, and B. Boelt, "Use of partial least squares discriminant analysis on visible-near infrared multispectral image data to examine germination ability and germ length in spinach seeds," *Journal of Chemometrics*, vol. 26, no. 8-9, pp. 462–466, 2012.

[35] C. Sârbu, R. D. Naşcu-Briciu, A. Kot-Wasik, S. Gorinstein, A. Wasik, and J. Namieśnik, "Classification and fingerprinting of kiwi and pomelo fruits by multivariate analysis of chromatographic and spectroscopic data," *Food Chemistry*, vol. 130, no. 4, pp. 994–1002, 2012.

[36] X. L. Wu, X. L. Mao, G. E. Tuli et al., *Medicinal Fungi of China*, Science Press, Beijing, China, 1st edition, 2013.

[37] Index Fungorum, http://www.indexfungorum.org/Names/Names.asp.

[38] H. L. Wei, H. B. Li, L. L. Wang et al., "Molecular recognition of species in *Boletus* sect. *Appendiculati*," *Mycosyst*, vol. 33, no. 2, pp. 242–253, 2014.

[39] P. M. Kirk, P. F. Cannon, D. W. Minter, and J. A. Stalpers, *Ainsworth & Bisby's Dictionary of the Fungi*, CAB International, Wallingford, UK, 10th edition, 2008.

[40] T. Riviere, A. G. Diedhiou, M. Diabate et al., "Genetic diversity of ectomycorrhizal Basidiomycetes from African and Indian tropical rain forests," *Mycorrhiza*, vol. 17, no. 5, pp. 415–428, 2007.

Laser-Induced Breakdown Spectroscopy Applied on Liquid Films: Effects of the Sample Thickness and the Laser Energy on the Signal Intensity and Stability

Violeta Lazic and Massimiliano Ciaffi

Department of FSN-TECFIS-DIM, ENEA, Via E. Fermi 45, 00044 Frascati, Italy

Correspondence should be addressed to Violeta Lazic; violeta.lazic@enea.it

Academic Editor: Nikša Krstulović

Droplets of organic liquids on aluminum substrate were probed by an Nd:YAG laser, both in a steady state and during rotation at speeds 18–150 rpm. Rotation transforms the droplet into film, which estimated thickness at high speeds was below 3 μm and 20 μm for diesel and peanut oil, respectively. Line intensities from the liquid (C I) and the support (Al I) material were tracked as a function of the film thickness and the laser energy. By film thinning, the line intensities from liquid sample were enhanced up to a factor 100x; simultaneously, the LIBS signal fluctuations were reduced 5–10 times with respect to the steady droplet. In certain experimental conditions, the line intensities from the support material become very weak with respect to the C I line, indicating an efficient screening of the substrate by highly excited plasma from the liquid layer. At a fixed rotation speed, there is a laser energy threshold, dependent on the liquid thickness, above which the LIBS signal becomes stable. Here, we discuss the relative processes and optimization of the experimental conditions for the LIBS measurements frome one laser shot to another.

1. Introduction

Laser-induced breakdown spectroscopy (LIBS) provides information about elemental composition of the sample volume involved in the plasma [1–3]. LIBS could be applied for analysis of materials in solid, liquid, and gaseous state, as well as of aerosols [4–6] and of particles suspended in liquid media [7, 8].

Rapid chemical analysis of liquids by LIBS [9] has many potential applications; among them, there are monitoring of waters [10–13], of quality in pharmaceutical or food industry [14–16], medical diagnostics [17, 18], and control of industrial processes or mechanical equipment [19, 20]. Laser-induced plasma on or inside liquids generates intense shockwaves, bubbling, and expulsion of droplets, which together with the laser energy losses due to liquid vaporization reduce the available energy for the plasma excitation. LIBS measurements on liquids usually have low analytical sensitivity, which could be improved by proper experimental approach and sample preparation, as reviewed in [9, 21] and updated in [22].

Excluding some complex methods for preparation of liquid samples that eliminate advantages of LIBS in terms of simplicity, rapidity, and possibility for in-field measurements, on evaporating liquids (like waters and alcohols), an intense LIBS signal could be obtained by probing the residues left on a solid substrate [13, 14, 18, 23]. The sample drying should be performed close to room temperature in order to prevent losses of volatile components. This process is time-consuming, it leaves not uniform residue distribution, and it could not be applied on liquids like oils.

Recently, another simple approach for LIBS measurements of liquids has been reported and it regards the signal generation on a thin liquid film placed over an absorbing substrate [22, 24]. Here, the breakdown initiates on the substrate and the plasma atomizes and ionizes the liquid layer. In this way, the laser-induced splashes, the plasma formation threshold, and the matrix effect (very strong in case of a direct plasma formation on liquids [9]) are importantly reduced. The LIBS signal strongly gains in intensity and so in analytical sensitivity.

Xiu et al. [24] prepared a thin oil film of arbitrary thickness on aluminium substrate by spreading manually the liquid. The plasma distribution in terms of intensity, temperature, and electron density was studied comparatively on dry aluminium and the same was covered with the oil film. It was noticed that on bare aluminium, the plasma remains close to the target surface while in presence of a thin liquid film, the maximum plasma emissivity is shifted away from the surface for a 1.5–2 mm. Vapor confinement by the initial liquid layer, later transformed to the plume, produces the plasma temperature and the electron density higher than in the case of the uncovered substrate.

Well controlled and reproducible oil film thickness could be obtained by rotating substrate at adjustable speed or by electrowetting on dielectric with controlled voltage in case of water solution [22]; the latter is impossible to thin by rotation due to high cohesive forces. This is the only work that reports changes of the LIBS signal with thickness of a liquid film. It was found that progressive thinning of a liquid film first eliminates the laser-induced splashes than the aerosol expulsion, and up to a certain thickness, it increases the emission lines from elements present in the liquid. However, the systematic studies of influence of liquid film height over substrate on the LIBS signal have not been yet performed although such sampling method, simple and applicable also for small sample volumes, could lead to a huge increase in the detection sensitivity [22].

For comparison, when analyzing water solutions in form of thin liquid sheet in air, it was found that the LIBS line intensities increase up to the sheet thickness of 20 μm [25], and this was explained by the largest liquid-laser interaction volume. For major sheet thicknesses, the LIBS signal decreases due to larger energy losses into mechanical effects and liquid evaporation. For a fixed liquid jet thickness, there is an optimal laser energy for the LIBS signal strength and the further energy growth reduces both the line intensities and their shot-to-shot stabilities [26].

On the other hand, it is well known that the efficiency of wet laser ablation (LA), particularly important for medicine and nanoparticle production, depends on a liquid type and its height above the solid target. In [27], by using KrF laser, the highest ablation rate of silica covered with water was obtained for the liquid layer of 1.1 mm, corresponding to the strongest local pressure induced by the laser. In LA of a bone covered by 1 mm thick water layer, it was observed that increase of the laser fluency above a certain level does not increase the ablation rate because of the plasma shielding [28]. For LA of aluminium by an Nd:YAG laser at 1064 nm, the optimum water thickness was of 3 mm, producing a 28-fold increase in crater volume compared to ablation in air [29]. In LA of zirconia by fs pulses, the highest ablation rate in presence of acetone was achieved for the liquid height between 0.2 and 0.7 mm [30]. At fixed experimental conditions, the film thickness for the maximum target ablation depends on the liquid properties [31].

There are also several works that explain laser-induced detachment of a thin liquid film from a solid-absorbing target [32]. Below the threshold for the plasma formation, the laser pulse rapidly rises temperature at the liquid-solid interface, creating a high-pressure vapor layer that expels the thin film from the surface. However, in scientific literature, there is a lack of knowledge relative to the plasma formation involving a liquid film over absorbing substrate and how the optical emission depends on the experimental conditions.

In the present work, we study LIBS signal behavior as a function of the estimated liquid film thickness over aluminium substrate and influence of the laser pulse energy. The examined liquids are diesel and peanut oil, chosen due to very different kinematic viscosities, which determine the liquid film thickness after the substrate rotation at a certain speed. The line intensities and their shot-to-shot fluctuations were considered for both Al I and C I emissions relative to the substrate material and the liquid, respectively.

2. Experimental

2.1. Instruments and Layout. Figure 1 shows the experimental layout, where the plasma was generated by an Nd:YAG laser (Quantel, CFR Ultra) emitting 6.5 ns long pulses at 1064 nm. The laser beam, without or after passing a beam splitter that transmits about 34% of the incident energy, was focused perpendicularly to the sample by means of a plano-convex quartz lens with focal length $f = 100$ mm. The spot diameter on the sample, placed slightly above the focal plane, was of about 0.42 mm, and it was determined on silica wafer. The LIBS measurements were performed with the incident laser energy between 18 mJ and 190 mJ, corresponding to the irradiance 2–21 GW/cm^2.

The plasma emission was collected at angle of about 60° from the target plane by two quartz lenses (focal lengths 100 mm and 70 mm) and brought to a spectrometer (Jobin-Yvon 550), used here with grating 1200 gr/mm, by a fiber bundle arranged at the exit into 0.1 mm × 12.2 mm array. At the spectrometer's output, an ICCD (Andor iStar DH734) was used for the LIBS signal detection. The acquisition gate and delay from the laser pulse were adjusted by a delay generator (Quantum Composer 9600+) to values of 2 μs and 1.2 μs, respectively. The measurements were performed at the laser repetition rate limited to 2 Hz.

Images of the plasma and of the liquid droplet/film were taken by a commercial photographic camera. The contact area between the liquid and the support material was determined from dimensionally calibrated photographs with help of ImageJ software (free source). The average liquid thickness on the substrate was calculated from the measured area occupied by the droplet and the known, deposited sample volume.

2.2. Samples and Preparation. Liquids used in the experiment are a commercial diesel fuel and a refined edible peanut oil. These two liquids were chosen due to very different kinematic viscosities (see Table 1), which produce large differences in the liquid film thicknesses on the substrate.

As a substrate, we used an aluminum disk with diameter of 25 mm, matching the spin coater. If not specified differently in the text, aluminium surface was polished by a sandpaper 400 grits/mm but for some measurements, we also tested the surface polished with 800 grits/mm. Before placing a liquid

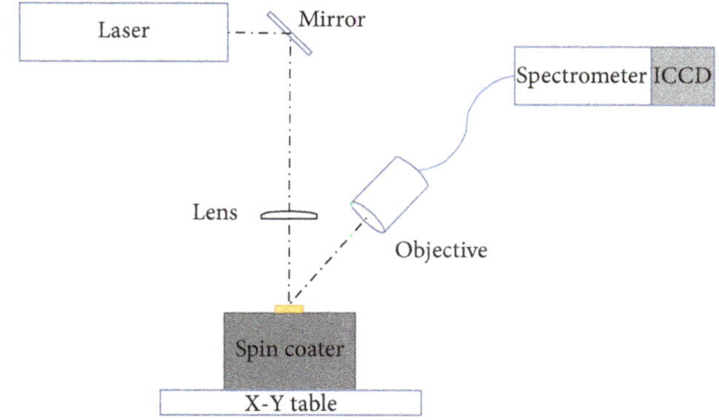

Figure 1: Experimental layout.

droplet on the substrate, the same was washed in a water flux, dried, and then washed by a high-purity methanol.

Liquid sample placed on the substrate had precisely controlled volume (±1%), delivered by an autoclavable pipette (Labgene Scientific). A droplet placed at the disc center was spread into a film by rotating the substrate on a spin coater (Laurell Technologies, KL-SCI-20), which speed can be adjusted between 18 and 150 rpm. After turning on the spin coater and before starting the LIBS measurements or stopping the rotation to photograph the film surface, we waited at least 30 s to stabilize first the rotation speed and successively the liquid film thickness.

Although the LIBS measurements could be performed on a single droplet of a few microliters, here, we used larger sample volumes, sufficient to cover completely the aluminium disk. At low rotation speed (<25 rpm), the excess of the liquid accumulates at the support edges; otherwise, it escapes from the surface. The sampling was performed during the rotation, by registering separately the spectra produced by 10 laser pulses. Between the two LIBS measurements, the whole holder was shifted horizontally for 2 mm in order to avoid development of the craters on support material, and another 10 μL droplet was added, waiting again 30 s to stabilize the liquid film thickness.

3. Results

3.1. Effects of the Rotation Speed. On the steady aluminum support, a 20 μL droplet has the average thickness over the contact area of about 140 μm and 320 μm for diesel and peanut oil, respectively (Table 1). Due to a meniscus shape of the droplet, in the central part where the laser beam was focused, the local thickness is about twice larger than the average value. On diesel droplets, we studied the LIBS signal behavior as a function of the rotation speed at two different laser energies, namely, 23 mJ and 70 mJ. In absence of rotation, the laser pulse, even at the low energy (23 mJ), produces splashes that reach the nearby optical elements, so it was necessary to clean them after each laser shot. The LIBS spectrum from the steady droplet shows rather weak spectral lines (Figure 2(a)) from both the liquid (C I) and the support material (Al I and Si I). After switching on the spin coater at its minimum speed (18 rpm), the splashes disappear but there is some aerosol spraying, which deposition on the nearby optical components was visible after a few tenths of the laser pulses. For this reason, at low rotation speed, the lenses were cleaned after each 10 laser shots. At 18 rpm, the LIBS signal is about 20 times more intense than in the case of the steady droplet (Figure 2(b)). Further increase of the rotation speed leads to disappearance of the aerosol, and for the speed of 40 rpm or higher, the optical elements did not require periodical cleaning. At the rotation speed of 100 rpm, where the estimated liquid film thickness was in order of a few μm (Table 1), the spectral lines are about 100 times more intense compared to the steady droplet. Here, also the emission lines from Mn II and Fe II around 260 nm were clearly observed, indicating the method's capability to detect also minor sample constituents even by applying a relatively low laser energy.

Figure 3 shows the average intensities of C I (247.8 nm) and Al I (256.8 nm) lines and their relative standard deviations (RSDs) as a function of the rotation speed for pulse energies of 23 mJ and 70 mJ. By using the lower pulse energy, the line intensities from both C I and Al I steadily grow in range 25–90 rpm, and then the saturation occurs. At this point, the liquid film is probably fully evaporated locally and efficiently atomized by the laser pulse; thus, its further thinning does not contribute to the signal intensity. RSD of the characteristic lines rapidly decays with disappearance of the splashes, that is, passing from a thick droplet to a liquid film. However, RSD of C I line keeps at high values, around 0.5, up to the speed of 90 rpm where the LIBS signal saturates. For the faster rotation, the RSD oscillates around value of 0.2. At the higher laser energy (70 mJ), the LIBS signal growth is much slower than in the previous case, and C I line intensity becomes almost saturated starting from about 50 rpm. For the same speed, the RSD of the C I line drops down and then oscillates around value of 0.1. This value of RSD was obtained also for Al I lines from the rotating support uncovered by a liquid, and we hypothesize that the vibrations present during rotation keep the LIBS signal fluctuations at this limit. From the obtained results here, we might conclude that there exists a threshold thickness of liquid film below which the laser pulse produces a full local sample evaporation and a more stable LIBS signal. This threshold thickness and the LIBS signal stability relative to the liquid increase with the laser energy. From Figures 3(a) and 3(b), we might note that the ratio of C I and Al I line intensities is higher at higher pulse energy, and this will be discussed later.

In analyzing behavior of the Al I at 256.8 nm, some cautions must be taken as this line is resonant, although much less intense than the nearby line at 257.5 nm, and a self-absorption might occur. The experimentally measured ratio of these two lines for the data shown in Figure 3 was in range 0.58–0.59 while their predicted ratio from the NIST database should be about 0.36 according to the formula valid for the plasma in local thermal equilibrium:

$$I^{ki} = a \cdot N \frac{g_k A_{ki} e^{-((E_k)/k)T}}{U(T)}, \tag{1}$$

where a is constant depending on experimental conditions; g_k and A_{ki} are the level degeneracy and the tabulated transition probability, respectively; E_k is energy of the upper

Table 1: Physical properties and the estimated droplet/film thicknesses for diesel and peanut oil.

	Diesel	Peanut oil
Surface tension at 20°C (mN/m)	25.8 [33]	10.0 [34]
Kinematic viscosity (mm^2/s) at 40°C	2–4*	37 [35]
Average thickness of 20 μL droplet (μm)	141 ± 10	320 ± 20
Av. thickness of droplet 5 μL at 18 rpm (μm)	19 ± 4	102 ± 10
Av. thickness of liquid film at 100 rpm (μm)**	≤3.0	≤20

* means specified by the supplier. ** means measured for droplet volumes of 0.2 μL and 5 μL for diesel and oil, respectively.

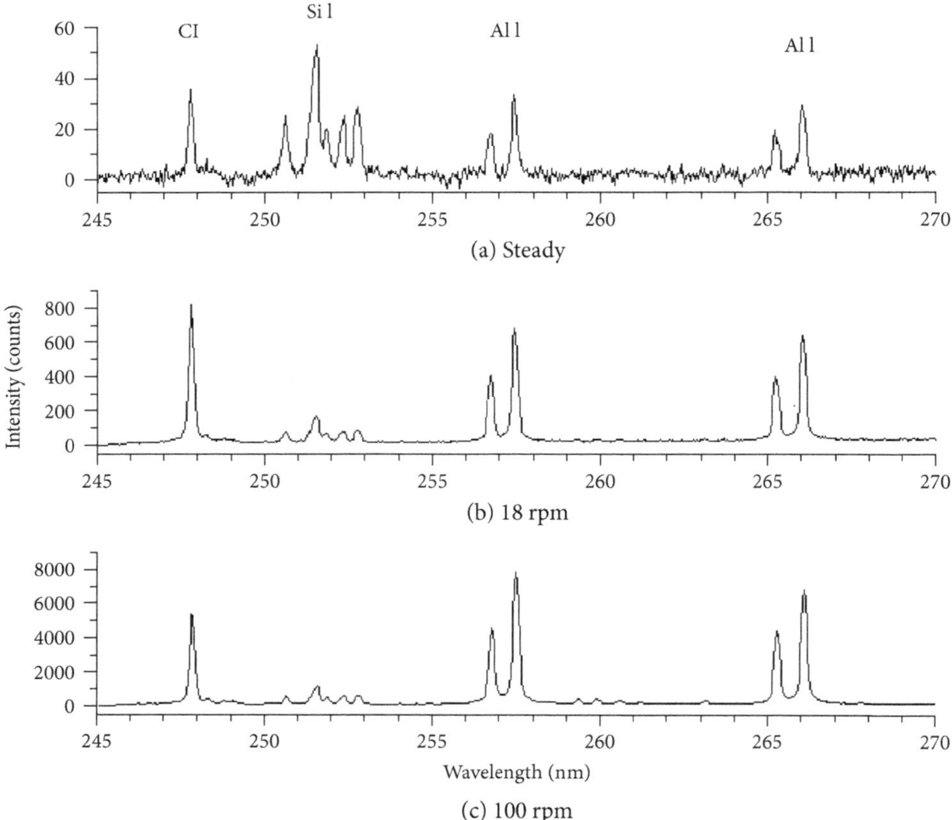

Figure 2: LIBS spectra from diesel, averaged over 10 laser pulses with energy of 23 mJ. Steady droplet (a) and support rotated at 18 rpm (b) and 100 rpm (c).

transition level, $U(T)$ is the temperature dependent partition function of the considered species with density N in the plasma. The low measured ratio of the two Al I (doublet) means that the more intense transition (257.5) is self-absorbed. We do not know if and how much the less intense line is self-absorbed but the stable line intensity ratio from the doublet in different experimental conditions indicates that behavior of the chosen Al I line (256.8 nm) in Figure 3 is not caused by the line saturation.

We performed the analog measurements on peanut oil, which a droplet leaves much thicker liquid film with respect to diesel (Table 1), by applying the laser energy of 70 mJ. Also in this case, the intensity of C I line becomes more stable close to the rotation speed of 50 rpm (Figure 4) but the corresponding RSD is about three times higher than in the case of diesel (see Figure 3(d)). For the steady droplet, the RSD is much higher than the same one measured on diesel, and it decreases below 1.0 only at speed of about 35 rpm.

In the following, we attempted to estimate the average liquid film thickness as a function of the rotation speed. To this aim, we placed on the substrate a droplet with well-controlled volume V (μL) and then measured the area A (mm^2) occupied by the droplet/liquid film after performing dimensional image calibration. The average liquid thickness d (μm) was then calculated as $d = V(\mu L)/A(mm^2)$. In absence of rotation, the error in estimating A is low as the liquid is clearly encircled also when photographing the sample perpendicularly to the support (Figure 5(a)). If the liquid reaches borders of the support (Figure 5(b)), the estimated average film thickness could be higher than the real one as the losses of liquid volume

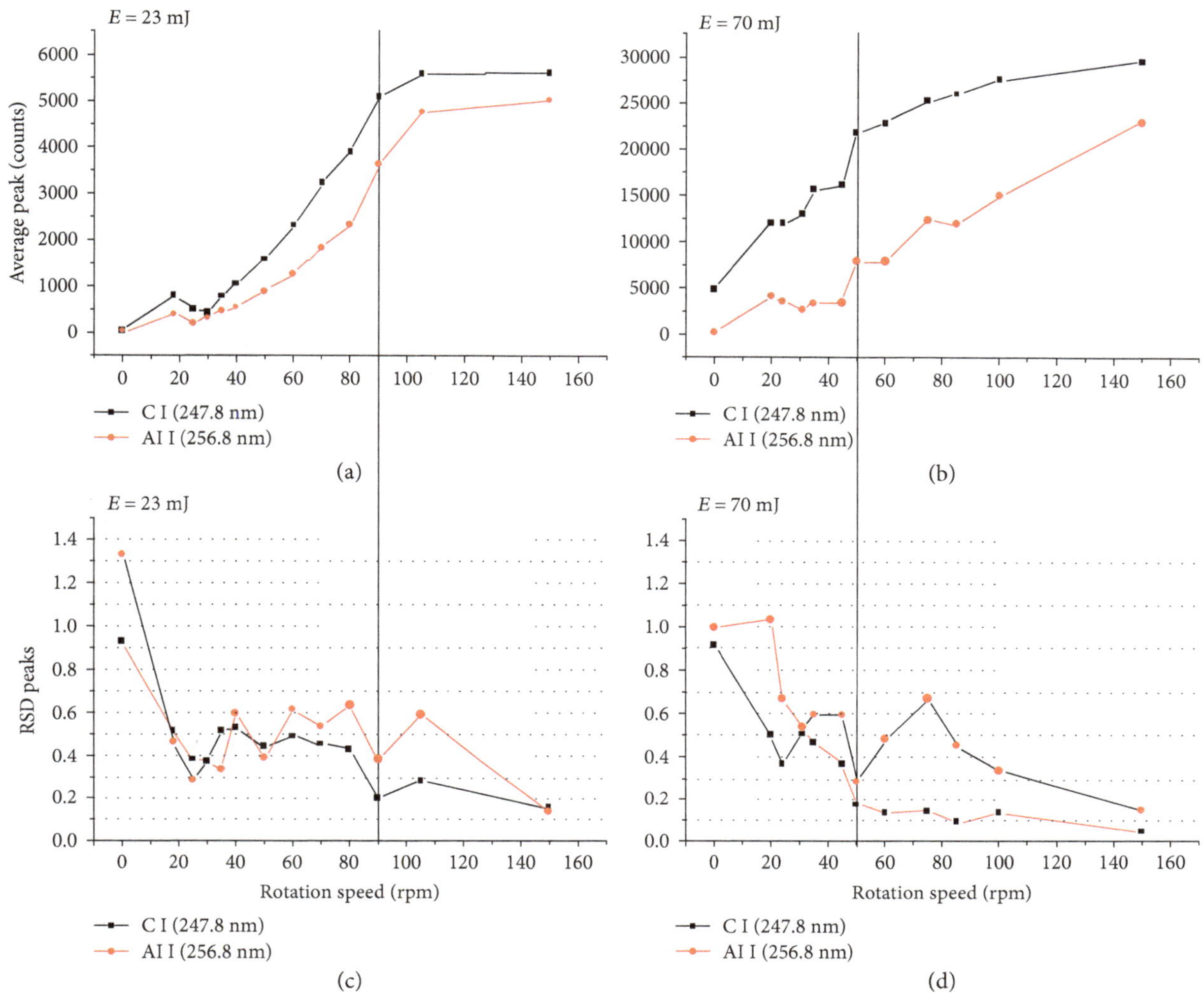

FIGURE 3: Average peak intensity of C I and Al I lines from the diesel droplet, measured over 10 pulses, as a function of the rotation speed for laser energies of 23 mJ (a) and 70 mJ (b); the corresponding RSD is shown in (c) and (d), respectively.

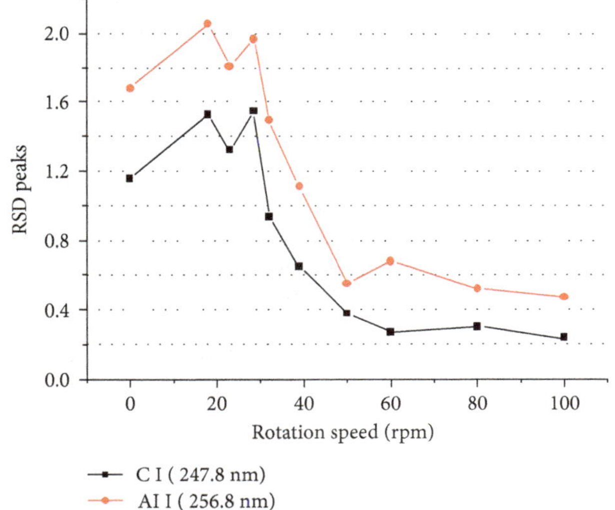

FIGURE 4: RSD of C I and Al I lines from 20 μL droplet of peanut oil as a function of the rotation speed; the laser energy is 70 mJ.

from the substrate might occur. By applying high rotation speeds, the shape of area occupied by liquid becomes irregular, introducing larger errors in its circle, and it was necessary to take the photographs angularly—this also increases the error due to the perspective (Figure 5(c)).

Initially, we attempted to determine the film thickness of 20 μL oil droplet but the liquid reaches the borders of the support already at the rotation speed of 18 rpm (Figure 5(b)). Further increase of the rotation speed causes losses of the liquid, visibly sputtered away from the support, and to a consequent overestimation of the average film thickness, based on the known deposited liquid volume. For this reason, the measurements of the oil film thickness were performed also for the droplet of 5 μL, on the substrate polished with sandpaper of 400 grits/mm or 800 grits/mm (Figure 6). For the substrate polishing by 400 grits/mm, at the rotation speed of 18 rpm the droplet of 20 μL has the average thickness similar to that of the smaller droplet (5 μL). However, by the applied method here, the estimated film thickness at larger rotation speed differs significantly in the two cases due to the liquid volume loss that affects calculations of the average

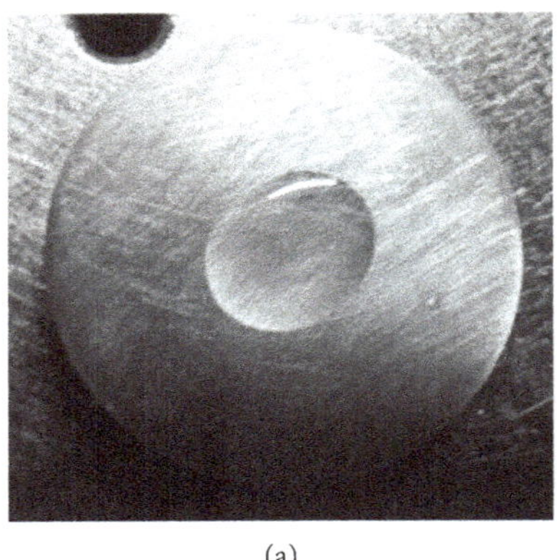
(a)

(b)

(c)

FIGURE 5: Photos of 20 μL oil droplet on (a) steady support; after rotating at 18 rpm (b) and 45 rpm (c).

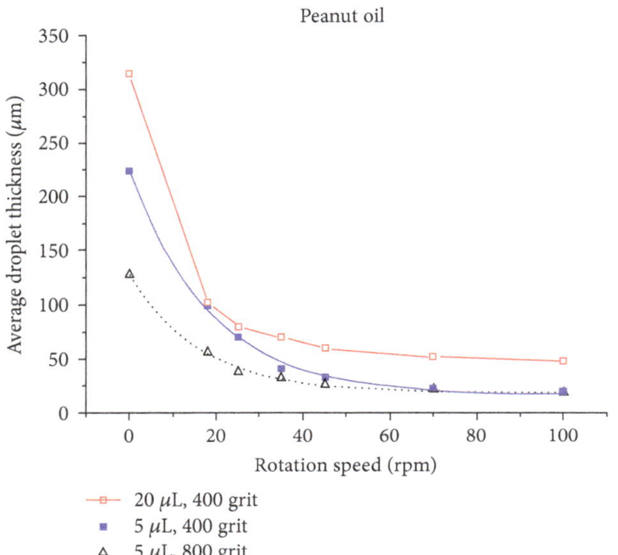

FIGURE 6: Average droplet thickness of peanut oil with volume of 20 μL or 5 μL, placed on the support polished with sandpaper of 400 grits/mm and 800 grits/mm. For 5 μL droplets, the data are fitted exponentially.

film thickness. On the substrate polished by the finer sandpaper (800 grits/mm), the average liquid height is importantly lower than for the rougher polishing except at high rotation speeds, where the estimated thicknesses converge to about 20 μm. Here, the film height also tends to a constant value at lower rotation speed compared to the rougher substrate, where the exponential decay coefficients are 15.5 rpm^{-1} and 18 rmp^{-1}, respectively. Comparing the Figures 4 and 6, we might conclude that on peanut oil, the LIBS signal excited by 70 mJ pulses becomes stable for the thickness below 50 μm.

A steady diesel droplet with volume of 20 μL has the thickness about twice lower than the oil droplet, and this explains the much lower RSD of the C I line (Figure 3(d)). Due to a very low kinematic viscosity, in rotation at the minimum speed (18 rpm), the large diesel droplet already spreads over the whole substrate and partially leaves it, thus compromising the calculations of the average area occupied by the liquid and so, of the film thickness. For the droplet of 5 μL, we estimated the thickness of 19 μm at rotation speed of 18 rpm. To measure the area occupied by diesel droplet at high rotation speeds, it was necessary to reduce its volume to 0.2 μL only; otherwise, the liquid volume was partially lost from the support disc. In this case, the estimated liquid thickness reaches the stable value of about 3 μm at speed of only 35 rpm. Rapid thinning of diesel droplet with the rotation speed explains a drop of the RSD relative to C I line below value of 0.5 already at 18 rpm; on the oil, this value of RSD was achieved at the speeds above 40 rpm. Stability of the C I line intensity from one laser shot to another, much better on diesel than on oil (Figures 3(c) and 3(d) and Figure 4), could be attributed to large differences in the liquid film thickness.

3.2. Influence of Laser Energy on the LIBS Signal. Influence of pulse energy on the LIBS signal was studied at constant rotation speed of 40 rpm, on both diesel and peanut oil; for the latter, we used the support polished with the abrasive papers with 400 or 800 grits/mm. At the chosen rotation speed, the estimated thickness of diesel film was of about 3 μm while peanut oil film was thick of about 40 μm and 27 μm for the polishing by 400 and 800 grits/mm, respectively. For the rougher polishing and the oil sample, by increasing laser energy, the C I line intensity starts to grow from about 50 mJ. At the pulse energy of 190 mJ, the C I peak is about 60 times higher than at the minimum energy used here (Figure 7(a)). Simultaneously, the growth of the Al I emission intensity is much slower compared to the C I line, so the line intensity ratio I(Al)/I(C) rapidly drops from 2.0, reaching values in range 0.1–0.2 only (Figure 7(b)). This means that starting from a certain laser energy, here around 30 mJ, the plasma produced on the liquid film effectively screens the support material from ablation and further energy increase mainly contributes to excitation of the species originating from the liquid. This fact is very important for reducing influence of the substrate's constituents on the spectra from the analyzed liquid sample. Figure 8 shows the comparative spectra from peanut oil obtained at two very different pulse energies.

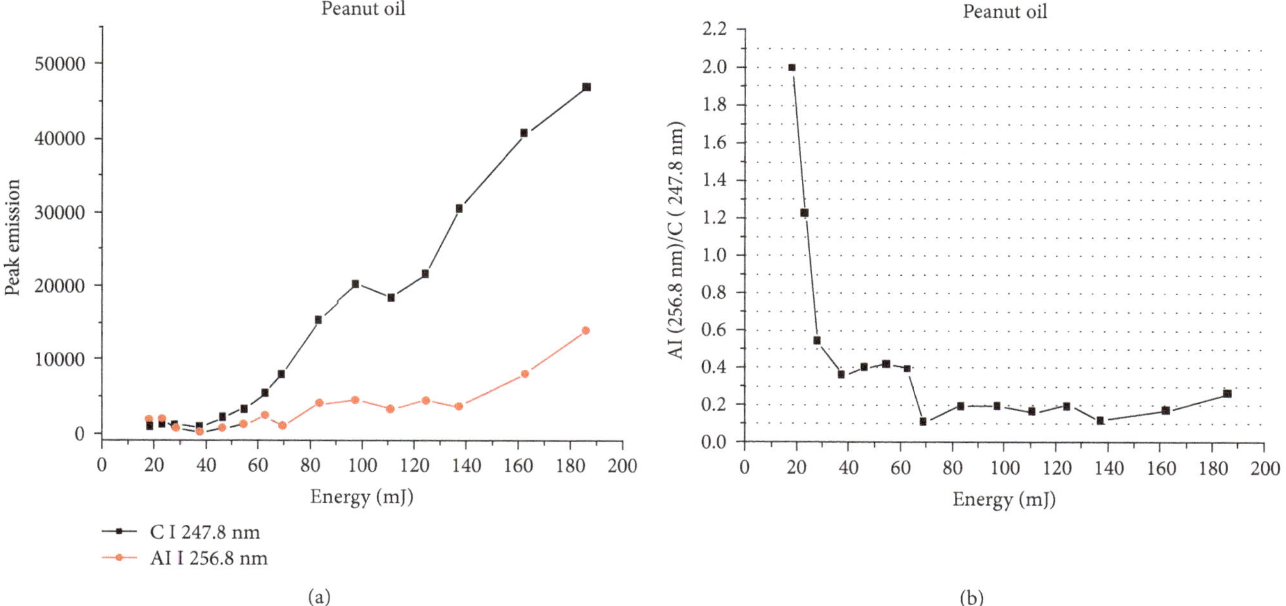

FIGURE 7: (a) Peak emission of C I and Al I lines averaged over 10 pulses as a function of the laser energy. (b) The corresponding line intensity ratio. The sample is peanut oil rotated at speed of 40 rpm.

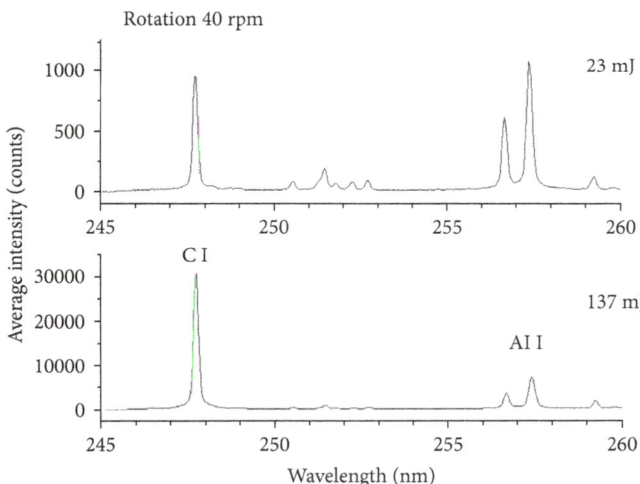

FIGURE 8: LIBS spectra from peanut oil on aluminum substrate (polishing 400 grits/mm) averaged over 10 laser pulses with energies of 23 mJ (top) and 137 mJ (bottom); the rotation speed is 40 rpm.

On peanut oil, starting from the pulse energy of about 40 mJ, the RSD of C I line steadily decreases from value above 1.1 or 1.3 to 0.2 (Figure 9(a)). On the rougher substrate, the signal stability is initially worse than for the finer polishing, and the stable RSD is reached at higher pulse energy (125 mJ compared to 110 mJ). Further increase of the laser energy corresponds to a stable RSD of C I peak in the spectrum except at the highest energy applied here (190 mJ), where some other instabilities seem to occur. At the same time, the RSD of Al peak fluctuates at high values, between 0.7 and 1.6 (not shown). On diesel film, about 10 times thinner than the oil, the intensity of C I line is much more stable from one shot to another, and an approximately constant RSD ≈ 0.1 is reached at the laser energy of about 80 mJ. From these results, it is clear that there is a laser energy threshold E_{st} above which the LIBS signal from liquid film becomes stable. Basing on the previously estimated liquid film thickness and the data shown in Figure 9(a), we found a linear relation between the film thickness and the energy threshold E_{st} (Figure 9(b)). We hypothesize that above the threshold energy E_{st}, the liquid layer locally detaches from the surface and becomes fully evaporated, resulting in a stable emission intensity from one laser shot to another and independently on a strongly variable support ablation. Further increase of the laser energy mainly contributes to the excitation of the atomized species from the liquid, without affecting importantly the support ablation (see Figure 7).

3.3. Optimization of the Experimental Conditions. Previously, we showed that the LIBS signal from an analyzed organic liquid improves in intensity and stability by reducing the liquid film thickness, achieved here by the substrate rotation. Thinning of the liquid film on substrate also reduces the laser energy requirements for obtaining an intense stable signal, and this is very important if a laser source with limited performances is used.

Unfortunately, increase of the rotation speed above a certain value does not bring significant improvements of the LIBS signal as the minimum achievable film thickness depends on kinematic viscosity of liquid, which is a measure of resistance to flow and shear. Kinematic viscosity η of oils decreases exponentially with temperature, and it is described by the Arrhenius equation:

$$\eta = A_1 e^{E_a/RT}, \tag{2}$$

where A_1 and E_a are constants specific for the oil, R is the universal gas constant, and T is the absolute temperature.

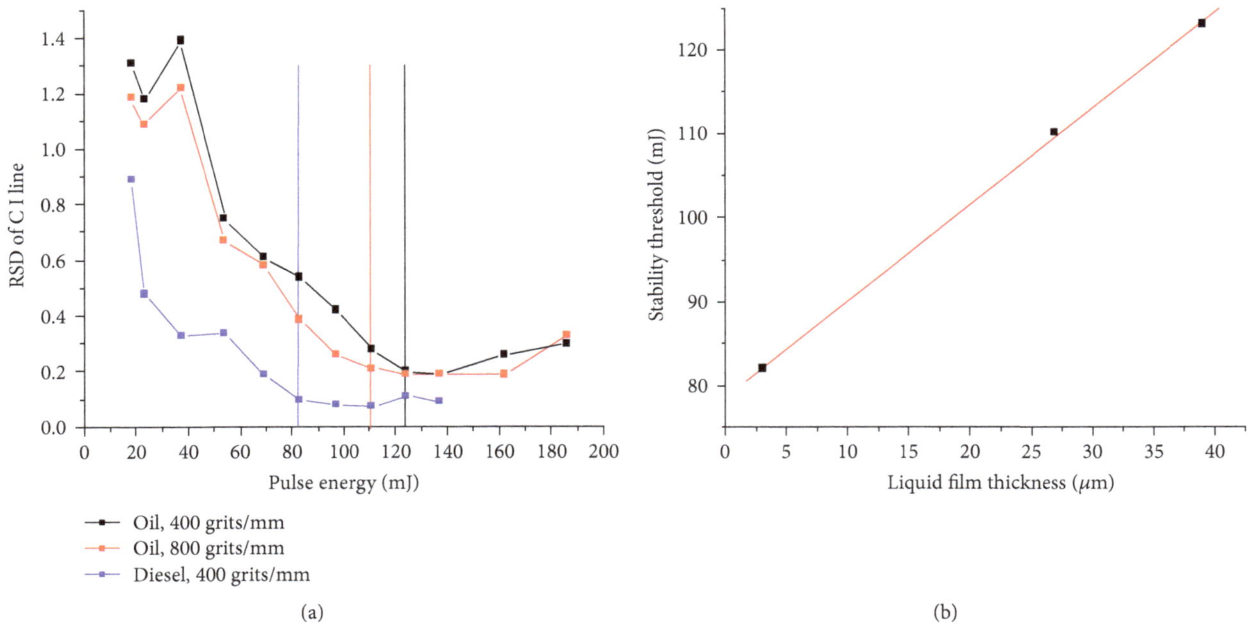

Figure 9: (a) RSD of C I line as a function of the laser energy at rotation speed of 40 rpm, measured on diesel (blue) and on peanut oil for two polishing grades of the substrate; the energies where the signal becomes stable are indicated with vertical lines. (b) The threshold energy for stable C I line intensity as a function of the estimated liquid film thickness.

Table 2: Kinematic viscosities of diesel [36], sunflower oil [36], and peanut oil [35] at different temperatures.

Temperature (°C)	Kinematic viscosity (mm^2/s)		
	Diesel	Sunflower oil	Peanut oil
10	5.4	118.7	133
20	4.15	73.4	82
30	3.3	48.5	54
40	2.7	33.8	37
50	2.3	24.5	27

Table 2 shows kinematic viscosity at different temperatures for diesel and sunflower oil, published in [36], and the known or extrapolated values by (2) for refined peanut oil [35]. It is evident that close to room conditions, the kinematic viscosity of organic liquids rapidly reduces with raising the temperature. From the practical point of view, this means that a more efficient thinning of a liquid film could be achieved by keeping the sample above room temperature, for example, at 30°C. Anyway, in order to have well-repeatable LIBS measurements, it would be opportune to keep the liquid sample at a stable temperature or to adjust the rotation speed in a way to maintain the wanted liquid film thickness.

If the rotation speed of the motor employed is limited to lower values than here (150 rpm), the film thickness at a certain and not high rotation speed could be further reduced by finer polishing of the substrate. However, due to a simultaneous liquid atomization and the substrate ablation, for detecting the trace elements in liquids, it is necessary to use a support material free of the same, for example, silica wafer [22]. In the case of a metallic substrate, its polishing should not transfer impurities to the material. For example, we compared the spectra from the bare aluminium substrate (not covered by a liquid) after polishing it by a sandpaper or by Al_2O_3 particles. In the first case, by sweeping the monochromator over different wavelength ranges, we detected Cr, Cu, and Zr lines in the spectra. These lines were not present in the LIBS spectra taken on the same substrate polished by Al_2O_3 particles, and this clearly indicates a transfer of impurities from the sandpaper grains to the substrate surface.

4. Conclusions

A droplet of a viscous liquid could be efficiently reduced to a thin film by rotating the support on which it is placed. In this way, the average liquid's height was reduced here from about 140 μm to 3 μm and from about 320 μm to 20 μm for droplets of diesel and peanut oil, respectively. The minimum film thickness that could be achieved at high rotation speeds is roughly proportional to the kinematic viscosity of liquid, much lower for diesel than for peanut oil.

Thinning of a liquid droplet/film eliminates unwanted laser-induced splashes and spraying during LIBS measurements, thus minimizing losses of the sample volume and a necessity to clean the nearby optical elements. Increase of the support rotation speed enhances the intensities of the spectral lines from liquid up to a factor 100 until the saturation is reached, corresponding to an almost constant liquid film thickness. The saturation speed lowers with the applied laser pulse energy.

Besides an impressive increase in the LIBS signal with thinning of a liquid film, shot-to-shot fluctuations of the line intensities originated from the liquid are reduced many times. A better signal stability corresponds to a thinner liquid film and to a higher pulse energy.

Raising the laser energy above a certain threshold produces a full local atomization/ionization of the liquid layer and reduces abruptly the LIBS signal fluctuations. Simultaneously, intensities of the lines originating from liquid are largely enhanced while those generated by the support ablation remain relatively weak, and this reduces the substrate interference in the spectra from the liquid sample. The threshold laser energy increases with the liquid thickness, which could be reduced also by a finer substrate polishing.

The proposed simple method for liquid analysis, together with the presented parameter optimizations here (energy, film thickness), allows to obtain a very intense and a relatively stable LIBS signal from small liquid volumes. Further studies should regard the analytical sensibility of the method and the effects of substrate material on the plasma formation.

References

[1] D. W. Hahn and N. Omenetto, "Laser-induced breakdown spectroscopy (LIBS), part I: review of basic diagnostics and plasma-particle interactions: still-challenging issues within the analytical plasma community," *Applied Spectroscopy*, vol. 64, pp. 335A–366A, 2010.

[2] D. W. Hahn and N. Omenetto, "Laser-induced breakdown spectroscopy (LIBS), part II: review of instrumental and methodological approaches to material analysis and applications to different fields," *Applied Spectroscopy*, vol. 66, pp. 347–419, 2012.

[3] R. Fantoni, L. Caneve, F. Colao, L. Fornarini, V. Lazic, and V. Spizzichino, "Methodologies for laboratory laser induced breakdown spectroscopy semi-quantitative and quantitative analysis—a review," *Spectrochimica Acta Part B: Atomic Spectroscopy*, vol. 63, pp. 1097–1108, 2008.

[4] D. W. Hahn, "Laser-induced breakdown spectroscopy for sizing and elemental analysis of discrete aerosol particles," *Applied Physics Letters*, vol. 72, pp. 2960–2962, 1998.

[5] B. Hettinger, V. Hohreiter, M. Swingle, and D. W. Hahn, "Laser-induced breakdown spectroscopy for ambient air particulate monitoring: correlation of total and speciated aerosol particle counts," *Applied Spectroscopy*, vol. 60, pp. 237–245, 2006.

[6] L. A. Álvarez-Trujillo, V. Lazic, J. Moros, and J. J. Laserna, "Standoff monitoring of aqueous aerosols using nanosecond laser-induced breakdown spectroscopy: droplet size and matrix effects," *Applied Optics*, vol. 56, pp. 3773–3782, 2017.

[7] T. Bundschuh, T. U. Wagner, and R. Koster, "Laser-induced breakdown detection (LIBD) for the highly sensitive quantification of aquatic colloids. Part I: principle of LIBD and mathematical model," *Particle & Particle Systems Characterization*, vol. 22, pp. 172–180, 2005.

[8] T. Kovalchuk-Kogan, V. Bulatov, and I. Schechter, "Optical breakdown in liquid suspensions and its analytical applications," *Advances in Chemistry*, vol. 2015, Article ID 463874, 21 pages, 2015.

[9] V. Lazic, "LIBS analysis of liquids and of materials inside liquids," in *Laser-Induced Breakdown Spectroscopy*, S. Mussazzi and U. Perini, Eds., vol. 624, Springer-Verlag, Berlin, Germany, 2014, Chap. 8.

[10] M. Yao, J. Lin, M. Liu, and Y. Xu, "Detection of chromium in wastewater from refuse incineration power plant near Poyang Lake by laser induced breakdown spectroscopy," *Applied Optics*, vol. 51, pp. 1552–1557, 2012.

[11] K. Rifai, S. Laville, F. Vidal, M. Sabsabi, and M. Chakera, "Quantitative analysis of metallic traces in water-based liquids by UV-IR double-pulse laser-induced breakdown spectroscopy," *Journal of Analytical Atomic Spectrometry*, vol. 27, pp. 276–283, 2012.

[12] S. T. Jarvinen, S. Saari, J. Keskinen, and J. Toivonen, "Detection of Ni, Pb and Zn in water using electrodynamic single-particle levitation and laser-induced breakdown spectroscopy," *Spectrochimica Acta Part B: Atomic Spectroscopy*, vol. 99, pp. 9–14, 2014.

[13] N. E. Schmidt and S. R. Goode, "Analysis of aqueous solutions by laser-induced breakdown spectroscopy of ion exchange membranes," *Applied Spectroscopy*, vol. 56, pp. 370–374, 2002.

[14] L. St-Onge, E. Kwong, M. Sabsabi, and E. B. Vadas, "Rapid analysis of liquid formulations containing sodium chloride using laser-induced breakdown spectroscopy," *Journal of Pharmaceutical and Biomedical Analysis*, vol. 36, pp. 277–284, 2004.

[15] Y. G. Mbesse Kongbonga, H. Ghalila, M. Boyomo Onana, and Z. Ben Lakhdar, "Classification of vegetable oils based on their concentration of saturated fatty acids using laser induced breakdown spectroscopy (LIBS)," *Food Chemistry*, vol. 147, pp. 327–331, 2014.

[16] Z. Abdel-Salam, J. Al Sharnoubi, and M. A. Harith, "Qualitative evaluation of maternal milk and commercial infant formulas via LIBS," *Talanta*, vol. 115, pp. 422–426, 2013.

[17] N. H. Cheung and E. S. Yeung, "Distribution of sodium and potassium within individual human erythrocytes by pulsed-laser vaporization in a sheath flow," *Analytical Chemistry*, vol. 66, pp. 929–936, 1994.

[18] N. Melikechi, H. Ding, S. Rock, O. Marcano, and D. Connolly, "Laser-induced breakdown spectroscopy of whole blood and other liquid organic compounds," *Proceedings of the SPIE*, vol. 6863, article 68630O, 2008.

[19] N. K. Rai and A. K. Rai, "LIBS—an efficient approach for the determination of Cr in industrial wastewater," *Journal of Hazardous Materials*, vol. 150, no. 3, pp. 835–838, 2008.

[20] L. Zheng, F. Cao, J. Xiu et al., "On the performance of laser-induced breakdown spectroscopy for direct determination of trace metals in lubricating oils, Spectrochim," *Spectrochimica Acta Part B: Atomic Spectroscopy*, vol. 99, pp. 1–8, 2014.

[21] V. Lazic and S. Jovićević, "Laser induced breakdown spectroscopy inside liquids: processes and analytical aspects," *Spectrochimica Acta Part B: Atomic Spectroscopy*, vol. 101, pp. 288–311, 2014.

[22] V. Lazic, R. Fantoni, A. Palucci, and M. Ciaffi, "Sample preparation for repeated measurements on a single liquid droplet using laser-induced breakdown spectroscopy," *Applied Spectroscopy*, vol. 71, pp. 670–677, 2017.

[23] D. Bae, S.-H. Nama, S.-H. Han, J. Yoo, and Y. Lee, "Spreading a water droplet on the laser-patterned silicon wafer substrate for surface-enhanced laser-induced breakdown spectroscopy," *Spectrochimica Acta Part B: Atomic Spectroscopy*, vol. 113, pp. 70–78, 2015.

[24] J.-S. Xiu, X.-S. Bai, V. Motto-Ros, and J. Yu, "Characteristics of indirect laser-induced plasma from a thin film of oil on a metallic substrate," *Frontiers of Physics*, vol. 10, pp. 231–239, 2015.

[25] H. Ohba, M. Saeki, I. Wakaida, R. Tanabe, and Y. Ito, "Effect of liquid-sheet thickness on detection sensitivity for laser-induced breakdown spectroscopy of aqueous solution," *Optics Express*, vol. 22, pp. 24478–24490, 2014.

[26] N. K. Rai and A. K. Rai, "LIBS—an efficient approach for the determination of Cr in industrial wastewater," *Journal of Hazardous Materials*, vol. 150, pp. 835–838, 2008.

[27] S. Zhu, Y. F. Lu, and M. H. Hong, "Laser ablation of solid substrates in a water-confined environment," *Applied Physics Letters*, vol. 79, 2001.

[28] H. W. Kang, H. Lee, S. Chen, and A. J. Welch, "Enhancement of bovine bone ablation assisted by a transparent liquid layer on a target surface," *IEEE Journal of Quantum Electronics*, vol. 42, pp. 633–642, 2006.

[29] N. Krstulović, S. Shannon, R. Stefanuik, and C. Fanara, "Underwater-laser drilling of aluminum," *The International Journal of Advanced Manufacturing Technology*, vol. 69, pp. 1765–1773, 2013.

[30] N. Bärch, A. Gatti, and S. Barrcikowski, "Improving laser ablation of zirconia by liquid films: multiple influence of liquids on surface machining and nanoparticle generation," *Journal of Laser Micro/Nanoengineering*, vol. 4, pp. 66–70, 2009.

[31] D. Kim, B. Oha, and H. Lee, "Effect of liquid film on near-threshold laser ablation of a solid surface," *Applied Surface Science*, vol. 222, pp. 138–147, 2004.

[32] P. Frank, J. Graf, F. Lang, J. Boneberg, and P. Leiderer, "Laser-induced film ejection at interfaces: comparison of the dynamics of liquid and solid films," *Applied Physics A*, vol. 101, pp. 7–11, 2010.

[33] A. B. Chhetri and K. C. Watts, "Surface tensions of petro-diesel, canola, jatropha and soapnut biodiesel fuels at elevated temperatures and pressures," *Fuel*, vol. 104, pp. 704–710, 2013.

[34] N. Siddiqui and A. Ahmad, "A study on viscosity, surface tension and volume flow rate of some edible and medicinal oils," *International Journal of Science, Environment and Technology*, vol. 2, pp. 1318–1326, 2013.

[35] T. W. Ryan III, L. G. Dodge, and T. J. Callahan, "The effects of vegetable oil properties on injection and combustion in two different diesel engines," *Journal of the American Oil Chemists Society*, vol. 61, pp. 1610–1619, 1984.

[36] B. Esteban, J.-R. Riba, G. Baquero, A. Rius, and R. Puig, "Temperature dependence of density and viscosity of vegetable oils," *Biomass and Bioenergy*, vol. 42, pp. 164–171, 2012.

12

Investigation by Raman Spectroscopy of the Decomposition Process of HKUST-1 upon Exposure to Air

Michela Todaro,[1,2] Antonino Alessi,[1,3] Luisa Sciortino,[1] Simonpietro Agnello,[1] Marco Cannas,[1] Franco Mario Gelardi,[1] and Gianpiero Buscarino[1]

[1] *Dipartimento di Fisica e Chimica, Università di Palermo, 90123 Palermo, Italy*
[2] *Dipartimento di Fisica e Astronomia, Università di Catania, 95123 Catania, Italy*
[3] *Laboratoire H. Curien, UMR CNRS 5516, Université de Lyon, 42000 Saint-Etienne, France*

Correspondence should be addressed to Gianpiero Buscarino; gianpiero.buscarino@unipa.it

Academic Editor: Artem E. Masunov

We report an experimental investigation by Raman spectroscopy of the decomposition process of Metal-Organic Framework (MOF) HKUST-1 upon exposure to air moisture ($T = 300$ K, 70% relative humidity). The data collected here are compared with the indications obtained from a model of the process of decomposition of this material proposed in literature. In agreement with that model, the reported Raman measurements indicate that for exposure times longer than 20 days relevant irreversible processes take place, which are related to the occurrence of the hydrolysis of Cu-O bonds. These processes induce small but detectable variations of the peak positions and intensities of the main Raman bands of the material, which can be related to Cu-Cu, Cu-O, and O-C-O stretching modes. The critical analyses of these changes have permitted us to obtain a more detailed description of the process of decomposition taking place in HKUST-1 upon interaction with moisture. Furthermore, the reported Raman data give further strong support to the recently proposed model of decomposition of HKUST-1, contributing significantly to the development of a complete picture of the properties of this considerable deleterious effect.

1. Introduction

Metal-Organic Frameworks (MOFs) are a new class of hybrid compounds constituted by combination of metallic clusters and organic linkers [1–4]. The formation of a crystalline porous matrix is achieved by reacting these two basic elements in appropriate synthesis conditions [1–4]. The possibility to adapt crystalline structures to a specific request, in terms of appropriate choice of metallic centers, linkers length, pore dimensions, and high surface area, makes MOFs very interesting in a wide range of application fields as catalysis and biomedical field as drug delivery systems, chemical sensors, gas storage, and so on [1–4].

One of the most promising MOFs from the point of view of fundamental and applied research is HKUST-1, synthesized for the first time in 1999 [5]. It is a reference material for adsorption of small molecules and gases and selective capture and transporting of a large variety of gases. In particular, several recent studies have been focused on the hydrogen, methane, and carbon dioxide storage in order to use alternative clean energy sources and with the intent to eliminate greenhouse gases from the atmosphere [6–9].

The outstanding adsorption properties of HKUST-1 are strictly related to its structural characteristics. Indeed, the cubic crystalline structure of this material, whose chemical composition is $Cu_3(BTC)_2(H_2O)_3$, is composed of two Cu^{2+} ions linked together by four carboxylate groups to form a paddle-wheel unit. Each carboxylate bridge is part of 1,3,5-benzene tricarboxylate (BTC) linker molecule. A picture of this structure is reported in Figure 1.

The crystal structure involves large cavities with diameter of about 9 Å and small pockets with diameter of about 6 Å. The principal adsorption sites of polar molecules as NH_3 and H_2O are those located on axial binding site of Cu^{2+} ion, called open metal sites [5].

The good affinity of HKUST-1 with respect to water molecules has been extensively investigated in the past. Indeed, it is well known that the adsorption of water molecules

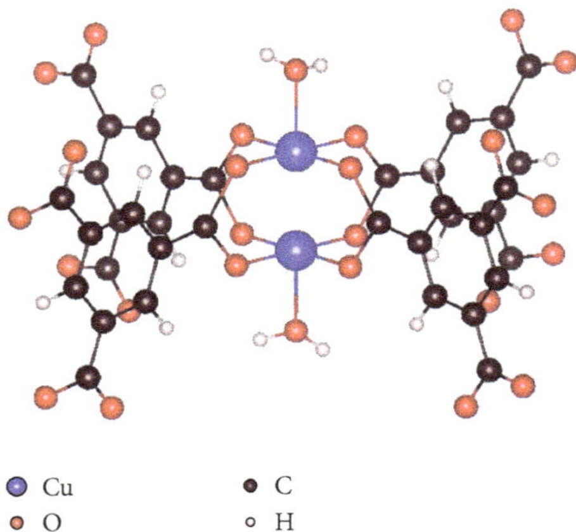

Figure 1: Schematic representation of the paddle-wheel unit of as synthesized HKUST-1. Each Cu^{2+} ion is coordinated with four carboxylate groups, belonging to four 1,3,5-benzene tricarboxylate (BTC) linker molecules and one water molecule.

on open metal sites actually takes place within few minutes of exposure to moisture [10]. This feature has been already taken into account in industrial applications because, for instance, the absorption of some gases is assisted by the presence of water molecules in matrix [11, 12].

In spite of this, one of the crucial aspects to consider is the stability of HKUST-1 with respect to a prolonged exposure to moisture, which can affect the crystalline matrix [13, 14]. Indeed, the range of applications of a MOF is significantly reduced if the integrity of the crystalline matrix is compromised as a consequence of such treatments. In fact, some studies have pointed out that HKUST-1 is not much stable with respect to moisture. In particular, it has been shown that when the material is exposed to moisture, it adsorbs a lot of water molecules until they reach the liquid state into the cavities. When this stage is obtained, water becomes able to hydrolyze the Cu-O bonds, breaking the crystalline network of the material. Recently, the effects of the interaction of HKUST-1 with air moisture ($T = 300$ K, 70% relative humidity) for a long period of time have been investigated in detail by electron paramagnetic resonance (EPR), X-ray diffraction (XRD), scanning electron microscopy (SEM), and surface analysis techniques. It has been shown that the decomposition process of HKUST-1 upon exposure to moisture takes place through three different stages. The first stage, with duration of about 20 days, is characterized by the adsorption of a large amount of water molecules by the material. Water molecules gradually fill the cavities of the material pushing towards the Cu^{2+} ions. The authors [15] pointed out by XRD that during this first stage the material undergoes a swelling, but no irreversible structural changes take place in the material. The decomposition process of the crystalline matrix actually starts for times longer than 20 days, when the effects of the hydrolysis of the Cu-O on the structure of the material become more and more evident on increasing the duration of exposure to air [15]. In particular, during the second stage of the process of decomposition, which takes place for exposure times from 20 days to 50 days, about 35% of the paddle-wheels become hydrolyzed and consequently one of the four carboxylate bridges involved in these paddle-wheels is detached. Finally, in the third stage of the process of decomposition, which takes place for exposure times longer than 50 days, a second detaching of a carboxylate bridge takes place in the same paddle-wheels already affected in the previous stage of the process.

In order to shed new light on the fundamental issue concerning the structural effects induced by hydrolysis on carboxylate MOFs, here we present an experimental investigation performed by Raman spectroscopy focused on the process of decomposition induced in HKUST-1 upon exposure to air moisture ($T = 300$ K, 70% relative humidity) and showing the potentialities of the technique to deepen the understanding of the underneath process in terms of molecular groups changes.

2. Materials and Methods

Commercial HKUST-1 in powder form was purchased from Sigma-Aldrich as Basolite C300. A sample of about 10 mg of HKUST-1 was exposed to air at $T = 300$ K and relative humidity (RH) of 70% for different times. We have monitored the vibrational properties of the sample of HKUST-1 with Raman spectroscopy from about 3 hours up to 165 days of exposure to air. Furthermore, we have performed a measurement on the activated sample of HKUST-1.

The Raman measurements were acquired in the region from 150 to 1700 cm^{-1} with a Bruker *Senterra* μ-Raman spectrometer equipped with a diode laser working at $\lambda = 532$ nm. We have performed measurements both at high resolution (3–5 cm^{-1}) and at low resolution (9–15 cm^{-1}). The advantage of the low resolution mode is that it gives spectra with significantly higher signal-to-noise ratio with respect to those obtained in high resolution mode. The nominal laser power was 0.2 mW to avoid sample modifications. For statistical reasons, each sample was measured in three different points and the resulting spectra were averaged.

The spectrum of the activated sample of HKUST-1 was acquired with a Horiba LabRam HR-Evolution μ-Raman spectrometer with a 532 nm laser and 1 mW nominal laser power, at high resolution (4 cm^{-1}). No spectral distortions were induced by laser power.

The sample was put inside a high pressure Linkam THMS600PS cell at 150°C (heating ramp 100 C/min) in atmospheric pressure and monitored starting from 10 minutes up to 1 hour, recording Raman spectra.

All the recorded spectra were subtracted by a baseline and subsequently normalized with respect to the intensity of the band at ~1000 cm^{-1}. We have chosen this peak for the normalization because it is well known that it is a very stable signal as it is related to the symmetric stretching of the C=C bonds in the rigid benzene rings.

3. Results

The high resolution Raman spectrum of the sample of activated HKUST-1 and that acquired for the sample after

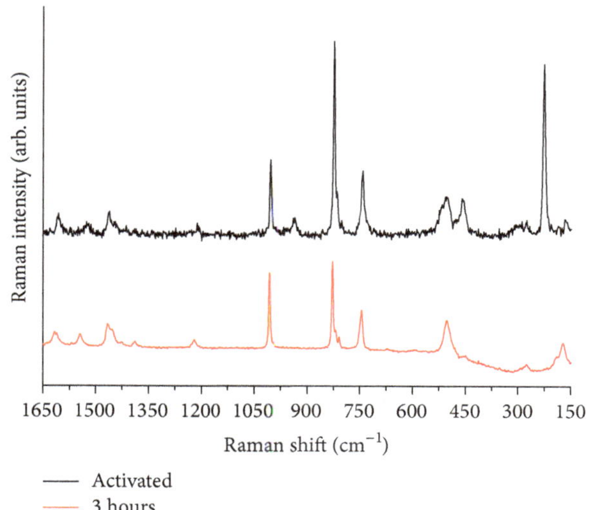

FIGURE 2: High resolution Raman spectra of the HKUST-1 sample activated (black line) and after about 3 hours of exposure to air moisture at $T = 300$ K (red line). Spectra are vertically shifted, for viewing purposes, after the normalization described in the text.

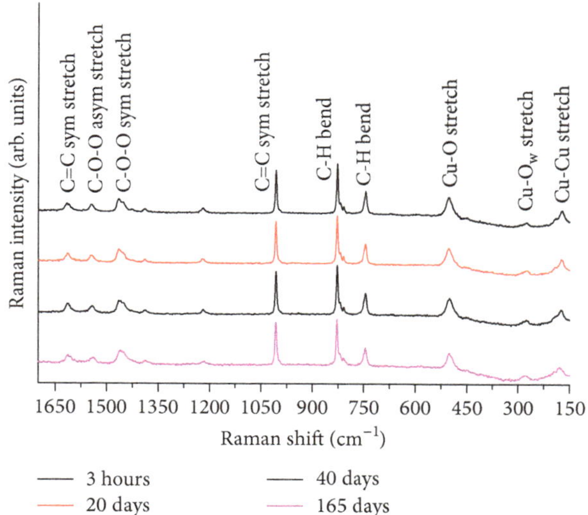

FIGURE 3: High resolution Raman spectra of HKUST-1 acquired after about 3 hours and 20, 40, and 165 days of exposure to air at $T = 300$ K. Spectra are vertically shifted, for viewing purposes, after the normalization described in the text.

about 3 hours of exposure to air ($T = 300$ K, RH 70%) are reported in Figure 2.

The two spectra show the same spectroscopic characteristics with respect to those reported in literature for activated and hydrated HKUST-1 [16, 17]. In particular, the intense peak at about 230 cm^{-1} evident in the spectrum of activated sample is replaced by two bands in that of the sample exposed to air for 3 hours, in line with results reported by other authors [16].

Regarding the spectrum of the sample exposed to air for 3 hours, between 150 and 600 cm^{-1}, the peaks of vibrational modes involving Cu^{2+} ions are evident. More precisely, the doublet at 173–191 cm^{-1} and the peak at 276 cm^{-1} are assigned in literature to a stretching mode involving Cu-Cu dimer and Cu-O_w, where O_w indicates the oxygen of the water molecule adsorbed on Cu^{2+} ion, respectively [16–19]. The doublet at 449–502 cm^{-1}, whose first component is barely detectable, is related to Cu-O stretching modes involving oxygen atoms of carboxylate bridges [16, 17].

In the central region from 700 to 1100 cm^{-1}, it is possible to recognize the peaks related to the vibrational modes of benzene rings. In particular, at 745 cm^{-1} and at 828 cm^{-1}, we observe the C-H out-of-plane bending modes of rings, whereas at 1006 cm^{-1} the symmetric stretching mode of C=C is well evident in the Raman spectrum [16–19]. In the region from 1400 to 1700 cm^{-1}, two vibrational modes of carboxylate bridges are present. The first feature at 1460 cm^{-1} corresponds to the symmetric stretching of O-C-O, whereas the peak at 1544 cm^{-1} corresponds to the asymmetric stretching of O-C-O [16]. At 1616 cm^{-1}, the C=C symmetric stretching of benzene ring is recognized [16, 17].

In Figure 3, a comparison of the Raman spectra obtained for the HKUST-1 exposed to air moisture ($T = 300$ K, RH 70%) for different times is reported.

As shown in Figure 3, the Raman spectrum obtained for the sample exposed to air for about 3 hours contains essentially the same peaks as the spectra obtained after longer exposure to air up to 165 days. This indicates that no new vibrations are induced by the decomposition process of the material in the investigated wavenumbers range. In order to obtain a more detailed description of the variations induced in the Raman spectra of HKUST-1 during the exposure of the material to air, we have estimated the intensity and the position of all the relevant peaks present in the Raman spectra. For this analysis we have taken advantages of the high signal-to-noise ratio of the spectra obtained in the low resolution mode.

In Figure 4, the variations induced by exposure of HKUST-1 to air in the intensity and in the position of the peak at about 174 cm^{-1} attributed to a mode involving the Cu-Cu stretching, are presented. A comparison among some indicative spectra in the region of this peak is also reported in Figure 4(b). As shown, we observe that both the amplitude and the peak position of this band remain almost constant up to about 20 days of exposure to moisture. At variance, Figure 4(a) suggests that for longer exposures the peak position undergoes a blue shift and correspondingly the peak amplitude decreases. However, by inspection of the spectra of Figure 4(b), it is easy to recognize that actually the situation is more complex. In fact, it emerges that in correspondence between the longest exposure times the peak exhibits at least two well distinguishable components, both peaked at larger wavenumbers with respect to the position observed after 3 hours of exposure of the material to air moisture.

Similarly, in Figure 5, the changes induced in the peak at about 502 cm^{-1} related to Cu-O stretching are reported. As shown, in this case we have found that this feature remains stable up to about 20 days of exposure to air; thereafter both

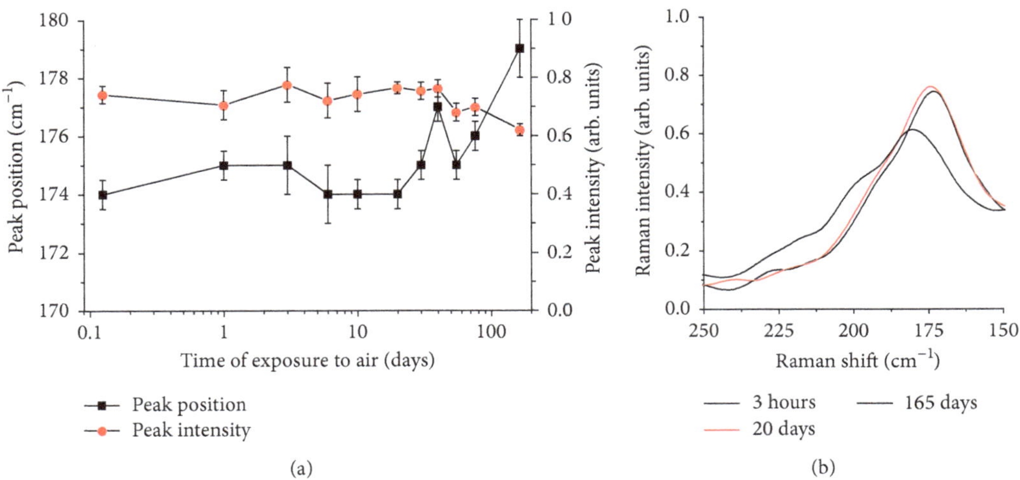

FIGURE 4: (a) Position (black squares) and intensity (red circles) of the Raman peak at about 174 cm^{-1} for HKUST-1 as a function of time of exposure to air moisture. (b) Raman spectra in the region of the peak at about 174 cm^{-1} acquired for HKUST-1 after 3 hours, 20 days, and 165 days of exposure to air moisture.

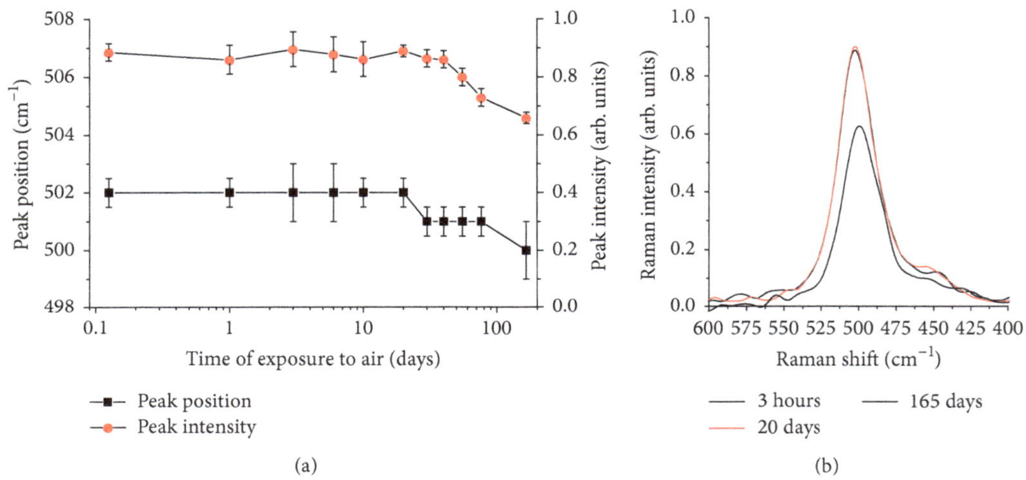

FIGURE 5: (a) Position (black squares) and intensity (red circles) of the Raman peak at about 502 cm^{-1} for HKUST-1 as a function of time of exposure to air moisture. (b) Raman spectra in the region of the peak at about 502 cm^{-1} acquired for the HKUST-1 after 3 hours, 20 days, and 165 days of exposure to air moisture.

the amplitude and the position of the peak gradually change. The former reduces from 0.88 to 0.64 for exposure to air up to 165 days, whereas the latter undergoes a red shift of about 2 cm^{-1}. This shift is small but well reliable, as it falls above the experimental uncertainty, as it is evident from Figure 5(a). The amplitude difference of the peak at about 502 cm^{-1} between spectra at 20 and 165 days of exposure to moisture is well recognizable in Figure 5(b).

Finally, in Figure 6, the data concerning the peak at about 1615 cm^{-1} are summarized. In this case we observe that the peak position remains essentially unchanged during the first 20 days of exposure of the material to air, whereas it undergoes a red shift for longer exposure times. The maximum shift observed for this band is of about 2 cm^{-1}. At variance, no relevant changes of the peak intensity take place during 165 days of exposure of the material to air moisture. Features similar to those here described were also found for the peaks at 1460 cm^{-1} and at 1544 cm^{-1} due to the symmetric and asymmetric stretching of O-C-O groups. All the other peaks not mentioned above were found to be essentially not affected by the exposure of the material to air moisture up to 165 days and will not be considered further.

4. Discussion

The main purpose of the present investigation is to better understand the processes of decomposition taking place in the carboxylate MOF HKUST-1 upon exposure to air moisture. Since in a recent work [15] a model has been proposed for this process, here we expect to find further support to

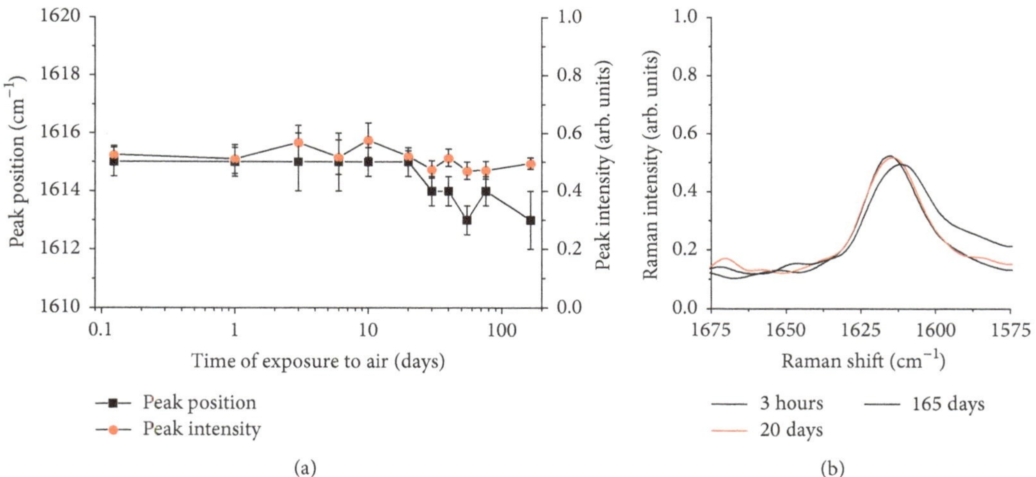

FIGURE 6: (a) Position (black squares) and intensity (red circles) of the Raman peak at about 1615 cm^{-1} for HKUST-1 as a function of time of exposure to air moisture. (b) Raman spectra in the region of the peak at about 1615 cm^{-1} acquired for HKUST-1 after 3 hours, 20 days, and 165 days of exposure to air moisture.

that model and to obtain a more detailed knowledge of the process. This goal is actually achieved by the overall data presented in the previous section and discussed in detail in the following.

The proposed model indicates that, during the first 20 days of exposure of HKUST-1 to air moisture, no irreversible processes take place. In fact, during this first stage, large quantities of water molecules are absorbed by the material, just filling the volume of the cavities. In agreement with the expectation that no irreversible changes take place during the first 20 days of exposure, the Raman data reported here indicate no relevant changes affecting the peaks falling within the range of investigated wavenumbers. Model [15] also suggests that, during this first stage of the process, the electronic energy levels of the Cu^{2+} ions significantly change and that the interaction between the two Cu^{2+} ions within the paddle-wheels becomes stronger, as a consequence of the interaction of these ions with the large quantities of water filling the cavities and pushing in all directions. This effect is actually not caught by the Raman spectroscopy in the present investigation, as no change in the properties of the Cu-Cu vibration is recognized. The reason for this failure is probably due to the fact that the two Cu^{2+} are very far away with respect to each other and consequently their bond strength is expected to be very weak. In fact, the estimated distance between them is of about 2.64 Å [16]. Since the variations of the properties of this bond induced by exposure of the material to air moisture are expected to be small perturbations of the pristine small value, they presumably fall below the detection limit of our Raman equipment.

On the basis of previous reports [15, 18, 19], it is known that upon exposure of HKUST-1 to air moisture for times between 20 and 40 days the irreversible process of hydrolysis of the Cu-O bonds is active. As a consequence of this process one of the four carboxylate bridges is detached in about 35% of the paddle-wheels of the material, generating a paramagnetic defect named $E_1'(Cu)$ center, which involves a Cu ion in +1 oxidation state with reduced coordination and a Cu^{2+} ion with the pristine coordination but modified environment [15]. Finally, the last stage of the process of decomposition of HKUST-1 starts for exposure times longer than 50 days and it involves a further detaching of a second carboxylate bridge in the same paddle-wheels already affected in the previous stage of the process [15]. Such final structure has been attributed to a paramagnetic defect named $E_2'(Cu)$ center. The occurrence of relevant irreversible modification of the material for exposure times longer than 20 days is actually confirmed by the reported Raman data. As shown in Figures 4, 5, and 6, some of the Raman bands suffer detectable peak shifts for such exposure times. More in detail, the blue shift of the peak related to the Cu-Cu bonds shown in Figure 4 indicates that the structure of the dimer significantly changes after prolonged exposure to air moisture. In addition, the fact that this peak exhibits two distinguishable components after 165 days of exposure is very interesting, as it indicates that the paddle-wheels of the material are of two different types, with different structures, and characterized by well distinguishable Cu-Cu vibrations. The presence of two structures presumably corresponds to the hydrolyzed and nonhydrolyzed paddle-wheels, within the pristine structure. On the basis of the model [15], the relative weight of these two components is about 35% and 65%, respectively. These quantities are roughly consistent with the relative amplitude of the two components shown in Figure 4 in the range from 150 cm^{-1} to 200 cm^{-1}. By careful inspection of the peak at about 174 cm^{-1} for the spectra obtained for HKUST-1 exposed to air moisture for about 3 hours and 20 days shown in Figure 4(b), a very small shoulder is recognized at about 178 cm^{-1}. This feature is presumably due to a small fraction of preexisting defective paddle-wheels essentially generated during the synthesis of the material. Concerning the properties of the Cu-O bonds, together with a notable red shift, it also suffers a considerable reduction of the peak intensity for exposure times longer than 20 days. This interesting result agrees with the model, indicating that after hydrolysis many Cu^{+1} ions suffer a reduction of their coordination number with respect to the pristine structure.

Finally, we note that since the hydrolysis process involves the oxygen of the carboxylate bridge, one expects also to observe a change of the spectroscopic properties of the peaks at 1460 cm^{-1} and at 1544 cm^{-1}, corresponding to the symmetric and asymmetric stretching of O-C-O groups. Also this expectation has been fulfilled in our present investigation. As a matter of fact, we have observed a gradual red shift of these bands for long exposure times of the material to air. It is worth noticing that, in agreement with the model of decomposition proposed previously [15], we have observed no change of the intensities of these two bands. This is due to the decomposition process of the material that does not change the number of carboxylate bridges but just the properties of their pristine coordination with the Cu^{2+} ions. We have found similar changes in the peak at 1615 cm^{-1} due to the stretching mode of C=C bonds of the benzene ring. This is not surprising because such vibrations actually take place in the aromatic ring of the linker and consequently the modifications occurring in the carboxylate bridges are expected to affect indirectly also the C=C vibration properties. Also, in this case, no change in the number of the C=C bonds is expected. Accordingly, as shown in Figure 6, there is no change of the intensity of the peak at 1615 cm^{-1} is recognized.

5. Conclusions

We report an experimental investigation by Raman spectroscopy of the decomposition process induced by exposure of the MOF HKUST-1 to air moisture. Our data indicate the occurrence of relevant structural processes taking place in the material for exposure times longer than 20 days. In agreement with previous reports [15] these processes are related to hydrolysis which affects a significant fraction of the Cu-O bonds of the crystal, irreversibly reducing the crystalline order of HKUST-1. The coexistence of two types of paddle-wheels with different structures, corresponding to hydrolyzed and nonhydrolyzed paddle-wheels, predicted by a previous study [15], has been confirmed by detecting a splitting of the Raman peak attributed to the Cu-Cu vibration in two well distinguishable components. The predicted change of the coordination number of a fraction of Cu ions in the material during the decomposition process [15] is also confirmed by the observation of a reduction of the intensity of the Raman peak at 502 cm^{-1}, attributed to Cu-O stretching. Summarizing, the overall data reported here support and extend the model of the process of decomposition of HKUST-1 upon exposure to air recently proposed, contributing to reach a complete and reliable description of this technologically relevant process.

Competing Interests

The authors declare that there are no competing interests regarding the publication of this paper.

Acknowledgments

The authors thank the people of the LAMP group (http://www.unipa.it/lamp/) at the Department of Physics and Chemistry of the University of Palermo for useful discussions. The CHAB laboratories at ATeN center (University of Palermo, http://www.chab.center/home/, Med-CHHAB, PONa3_00273) are acknowledged for the LabRam spectrometer and Linkam equipments use.

References

[1] H. Furukawa, K. E. Cordova, M. O'Keeffe, and O. M. Yaghi, "The chemistry and applications of metal-organic frameworks," *Science*, vol. 341, no. 6149, Article ID 1230444, pp. 1–12, 2013.

[2] J. L. C. Rowsell and O. M. Yaghi, "Metal-organic frameworks: a new class of porous materials," *Microporous and Mesoporous Materials*, vol. 73, no. 1-2, pp. 3–14, 2004.

[3] J. Canivet, A. Fateeva, Y. Guo, B. Coasne, and D. Farrusseng, "Water adsorption in MOFs: fundamentals and applications," *Chemical Society Reviews*, vol. 43, no. 16, pp. 5594–5617, 2014.

[4] L. Sciortino, A. Alessi, F. Messina, G. Buscarino, and F. M. Gelardi, "Structure of the FeBTC metal-organic framework: a model based on the local environment study," *The Journal of Physical Chemistry C*, vol. 119, no. 14, pp. 7826–7830, 2015.

[5] S. S.-Y. Chui, S. M.-F. Lo, J. P. H. Charmant, A. G. Orpen, and I. D. Williams, "A chemically functionalizable nanoporous material $[Cu_3(TMA)_2(H2O)_3]_n$," *Science*, vol. 283, no. 5405, pp. 1148–1150, 1999.

[6] J.-R. Li, R. J. Kuppler, and H.-C. Zhou, "Selective gas adsorption and separation in metal-organic frameworks," *Chemical Society Reviews*, vol. 38, no. 5, pp. 1477–1504, 2009.

[7] K.-S. Lin, A. K. Adhikari, C.-N. Ku, C.-L. Chiang, and H. Kuo, "Synthesis and characterization of porous HKUST-1 metal organic frameworks for hydrogen storage," *International Journal of Hydrogen Energy*, vol. 37, no. 18, pp. 13865–13871, 2012.

[8] Y. Liu, C. M. Brown, D. A. Neumann, V. K. Peterson, and C. J. Kepert, "Inelastic neutron scattering of H_2 adsorbed in HKUST-1," *Journal of Alloys and Compounds*, vol. 446-447, pp. 385–388, 2007.

[9] J. Getzschmann, I. Senkovska, D. Wallacher et al., "Methane storage mechanism in the metal-organic framework $Cu_3(btc)_2$: an *in situ* neutron diffraction study," *Microporous and Mesoporous Materials*, vol. 136, no. 1–3, pp. 50–58, 2010.

[10] F. Gul-E-Noor, D. Michel, H. Krautscheid, J. Haase, and M. Bertmer, "Time dependent water uptake in $Cu_3(btc)_2$ MOF: identification of different water adsorption states by ^1H MAS NMR," *Microporous and Mesoporous Materials*, vol. 180, pp. 8–13, 2013.

[11] N. C. Burtch, H. Jasuja, and K. S. Walton, "Water stability and adsorption in metal-organic frameworks," *Chemical Reviews*, vol. 114, no. 20, pp. 10575–10612, 2014.

[12] A. Özgür Yazaydin, A. I. Benin, S. A. Faheem et al., "Enhanced CO_2 adsorption in metal-organic frameworks via occupation of open-metal sites by coordinated water molecules," *Chemistry of Materials*, vol. 21, no. 8, pp. 1425–1430, 2009.

[13] J. B. Decoste, G. W. Peterson, B. J. Schindler, K. L. Killops, M. A. Browe, and J. J. Mahle, "The effect of water adsorption on the structure of the carboxylate containing metal-organic frameworks Cu-BTC, Mg-MOF-74, and UiO-66," *Journal of Materials Chemistry A*, vol. 1, no. 38, pp. 11922–11932, 2013.

[14] Z. Liang, M. Marshall, and A. L. Chaffee, "CO_2 adsorption-based separation by metal organic framework (Cu-BTC) versus zeolite (13X)," *Energy & Fuels*, vol. 23, no. 5, pp. 2785–2789, 2009.

[15] M. Todaro, G. Buscarino, L. Sciortino et al., "Decomposition process of carboxylate MOF HKUST-1 unveiled at the atomic scale level," *The Journal of Physical Chemistry C*, vol. 120, no. 23, pp. 12879–12889, 2016.

[16] C. Prestipino, L. Regli, J. G. Vitillo et al., "Local structure of framework Cu(II) in HKUST-1 metallorganic framework: spectroscopic characterization upon activation and interaction with adsorbates," *Chemistry of Materials*, vol. 18, no. 5, pp. 1337–1346, 2006.

[17] N. R. Dhumal, M. P. Singh, J. A. Anderson, J. Kiefer, and H. J. Kim, "Molecular interactions of a Cu-based metal-organic framework with a confined imidazolium-based ionic liquid: a combined density functional theory and experimental vibrational spectroscopy study," *Journal of Physical Chemistry C*, vol. 120, no. 6, pp. 3295–3304, 2016.

[18] K. Tan, N. Nijem, P. Canepa et al., "Stability and hydrolyzation of metal organic frameworks with paddle-wheel SBUs upon hydration," *Chemistry of Materials*, vol. 24, no. 16, pp. 3153–3167, 2012.

[19] K. Tan, N. Nijem, Y. Gao et al., "Water interactions in metal organic frameworks," *CrystEngComm*, vol. 17, no. 2, pp. 247–260, 2015.

13

Influence of Lead on the Interpretation of Bone Samples with Laser-Induced Breakdown Spectroscopy

Abdolhamed Shahedi,[1] Esmaeil Eslami,[1] and Mohammad Reza Nourani[2]

[1] *Department of Physics, Iran University of Science and Technology, Narmak, Tehran 16846-13114, Iran*
[2] *Tissue Engineering Division, Biotechnology Research Center, Baqiyatallah University of Medical Sciences, Tehran, Iran*

Correspondence should be addressed to Esmaeil Eslami; eeslami@iust.ac.ir

Academic Editor: Yu Shang

This study is devoted to tracing and identifying the elements available in bone sample using Laser-Induced Breakdown Spectroscopy (LIBS). The bone samples were prepared from the thigh of laboratory rats, which consumed 325.29 g/mol lead acetate having 4 mM concentration in specified time duration. About 76 atomic lines have been analyzed and we found that the dominant elements are Ca I, Ca II, Mg I, Mg II, Fe I, and Fe II. Temperature curve and bar graph were drawn to compare bone elements of group B which consumed lead with normal group, group A, in the same laboratory conditions. Plasma parameters including plasma temperature and electron density were determined by considering Local Thermodynamic Equilibrium (LTE) condition in the plasma. An inverse relationship has been detected between lead absorption and elements like Calcium and Magnesium absorption comparing elemental values for both the groups.

1. Introduction

Nowadays, an increasing spread of polluting industries, the pollution caused by toxic metals, and the hazards on human health are crystal clear to everyone. Thus detecting these pollutants and their effects on biological organisms is of great importance. Laser-Induced Breakdown Spectroscopy (LIBS) method has attracted much attention of many researchers due to its properties, unique capabilities such as low cost and minimal time duration of preparing the sample, prompt and simultaneous analysis of multielements, capability of being used in all three states of materials, and its nondestructive and noncontact nature [1, 2]. This method has been successfully used to identify corpses, bones, and humans' fossils or mummies remaining from centuries. It determines characteristic backgrounds like age, sex, race, and stature of bodies [3]. Clinical application of Laser-Induced Breakdown Spectroscopy to the analysis of teeth and dental materials used the pulsed lasers for controlled material ablation. This method is suggested as an alternative to mechanical drilling for the removal of caries and in tooth modification. Spectral analysis of the ablated plasma can be exploited to monitor precisely the laser drilling process in vivo and in real time [4, 5]. LIBS procedure proposes a new technique for diagnosing the presence or occurrence of cancer accurately and safely. It is now apparent that the malignant tumors can be predicted via detection of Calcium, Potassium, and Copper in body [6]. The method was also used to detect the elements of newborn heart muscles by reviewing elements such as Potassium, Calcium, and Sodium [7]. Analyzing biological seminal fluid [8], detecting pathogenic bacteria and pathogens [9], and so forth are among other applications of this method.

The presented technique depends on the interaction between the laser and the soft and hard biological surfaces in order to induce plasma. The plasma formation results from the matrix effects initiated by the absorption of the laser energy by the ablated material [2]. When the plasma expands in an ambient air, relaxation of the plasma plume by different processes—one of which is the recombination of the free electrons with the positive ions—occurs [10]. This recombination is a radiative process, which provides a continuum emission in addition to the bremsstrahlung. Another radiative process is the atomic emissions due to transitions between the different energy levels of an element from

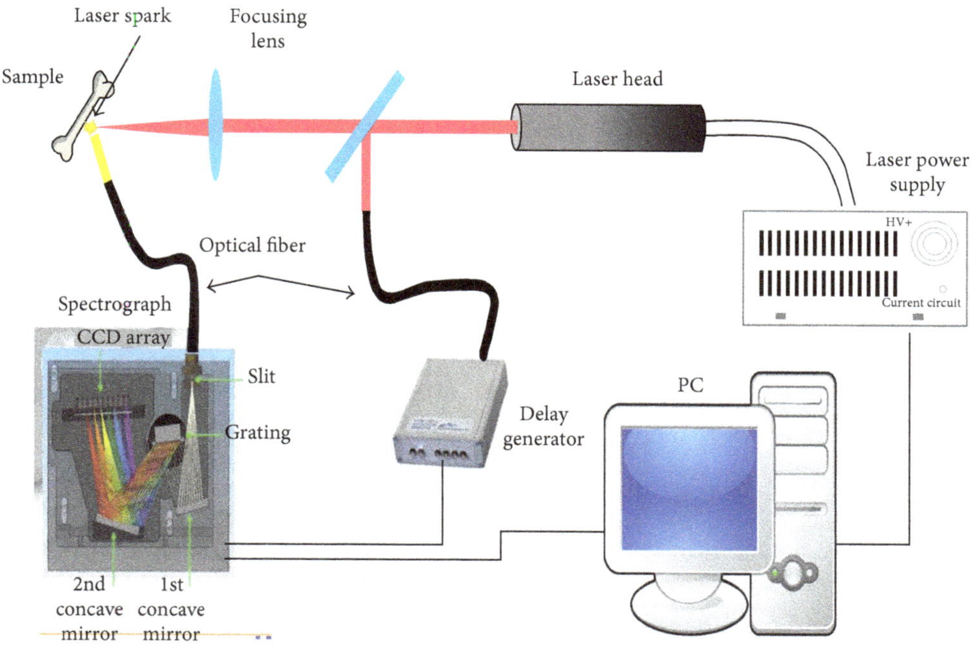

FIGURE 1: Schematic of the LIBS setup.

which spectrometric measurement can provide qualitative information on the atomic composition of the investigated biological samples. In addition, LIBS can also be used to provide information about the relative quantitative elemental composition of the samples. This is achieved by investigating the ratio between the intensity levels of the atomic emission lines: high concentrations of an element yield a higher intensity of the characteristic elemental emission spectra [11, 12].

In this study, we attempted to employ this method to differentiate between two groups of rates based mainly on the statistical analysis of the collected spectra. These spectra intrinsically possess information of the different concentration levels of the elements found in both the groups.

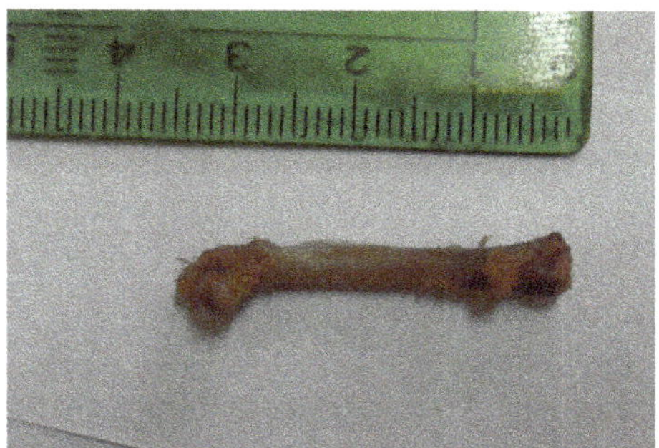

FIGURE 2: An example of rat's bone.

2. Methods and Materials

2.1. Test Layout. The experimental setup is schematically shown in Figure 1. It consists of a Nd:Yag laser having 10 ns pulse width at 1064 nm wavelength, 2 Hz pulse repetition frequency, and 160 mJ energy. To eliminate bremsstrahlung radiation in early stage when plasma is generated, and to have atomic spectrum with the highest ratio of signal to noise, different time intervals in plasma radiation are required. A delay time between laser pulse and data recording by photodiode was about 10 μs. As illustrated in Figure 1, a part of laser light is reflected toward photodiode by a half-silvered mirror (reflects about 4% of light) which lies in a 45-degree angle in line with light and delay contributor starts the camera by sending a pulse. Laser beam is focused on the surface of the sample located at the focal length of the converging lens with $f = 10$ cm. An optical fiber locates at 45 degrees with beams in optimal distance from the location where plasma was generated (to avoid saturation) and transmits the light resulting from the plasma to the 2D intensified CCD (Andor iStar, DH720) through a Czerny-Turner spectrograph (Chromex, 500 IS) in the wavelength range of 115–920 nm. LIBS spectra were collected from 10 different spots of each sample; 10 spectra were collected from each spot. These spectra enable us to analyze the sample quantitatively and quantitatively. To obtain ratio of signal to high noise, best conditions such as delay time and optical field of view were optimized.

2.2. Sample Preparation. Female rats with average age of four weeks with same race and conditions were chosen for preparing bone samples. Ten rats were divided into two equal groups. Group A received 250 mL distilled water including lead salts (lead acetate 325.29 g/mol) with 4 mM concentration daily and group B received only 250 mL distilled water with no additives daily for 50 days. Most portion of lead is absorbed in bones [13]. So the left legs of the rats were split up and cleaned. Since almost all portions of lead injected to

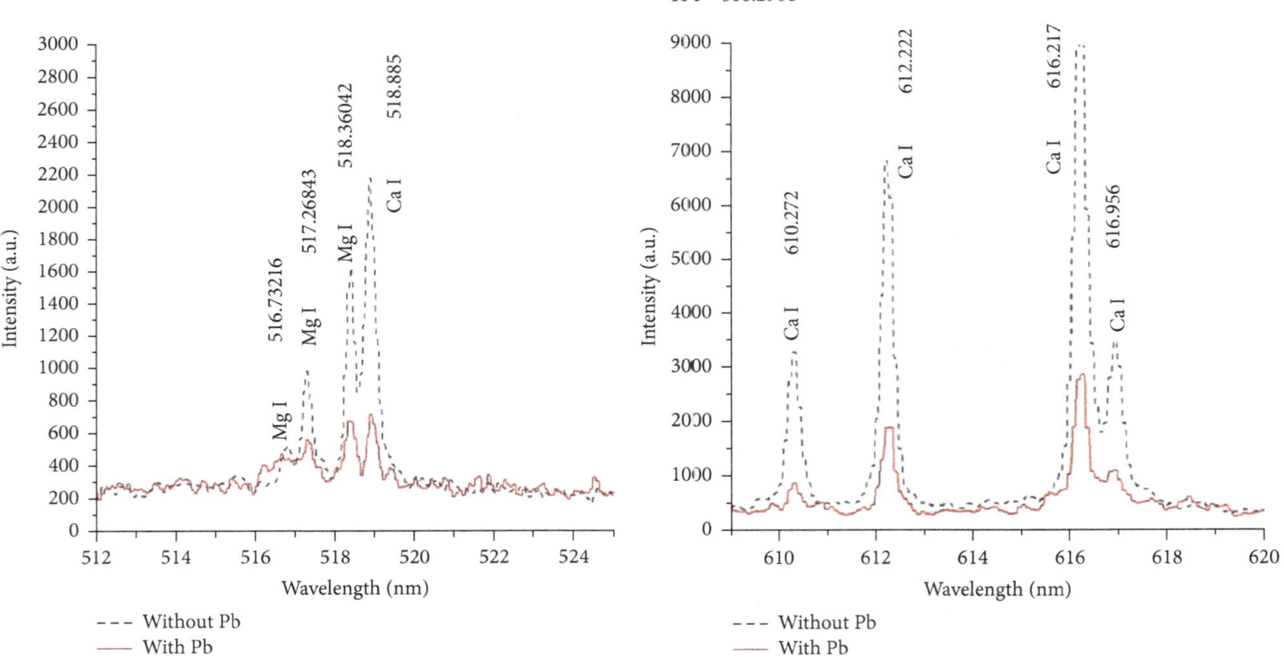

FIGURE 3: Typical mean LIBS spectra in both groups.

the body were absorbed in bone, they influence absorption of other elements in bone according to the procedure to prepare for laser pulse irradiation (Figure 2).

3. Plasma Characteristics

The LIBS spectra collected from the samples were processed from the raw data using the ORIGIN software. Figure 3 shows the mean spectra (mean value and standard deviation) of 10 bone spectra of both groups. To exemplify, the most prominent peaks existing in the spectra of both samples were used to show the standard deviation of all of the 10 spot-measurements. The elements with prominent emissions common to both bone types, found under the measurement conditions of this study, were Ca, Mg, and Fe. The LIBS spectra of group B exhibit more prominent atomic spectral lines than those of group A. The identification of the atomic lines is based on [14, 15]. As can be seen in this figure a significant decrease of Mg and Ca can be observed due to lead consumption. Note that during all experiments an intense line on Pb I at 405.78 nm was observed. As illustrated in Figure 4, its intensity is increased with lead consummation in bone. However, our analysis is mainly focused on the effect of Pd on other elements.

3.1. Plasma Temperature. Plasma temperatures are often determined from the measurement of ratios of the intensities of neutral to neutral lines, usually for the same element. The line intensities are combined with the Boltzmann equation to determine

$$\ln\left(\frac{I\lambda_{ki}}{A_{ki}g_k}\right) = \ln\left(\frac{hc}{4\pi}\frac{N_0}{U(T)}\right) - \frac{E_k}{KT_{\text{exc}}}, \tag{1}$$

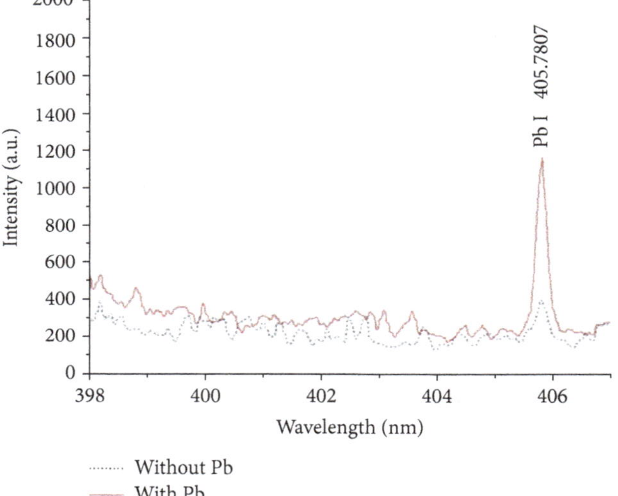

FIGURE 4: Typical lead spectrum at 405.78 nm.

where I is the intensity of the spectral line, λ_{ki} is the wavelength of the transition from level k to i, A_{ki} is the transition probability, g_k is the statistical weight, E_k is the energy value of higher level, and T_{exc} is the excitation temperature. Thus, a plot of the above equation versus the energy of the upper level E_j yields a straight line called Boltzmann plot. Its slope equals $-(KT_{\text{exc}})^{-1}$ [16]. In this work, the plasma temperature was determined by using the total intensity of the different strong lines as Fe, Mg, and Ca in the visible region. Figure 5 shows the plasma temperature

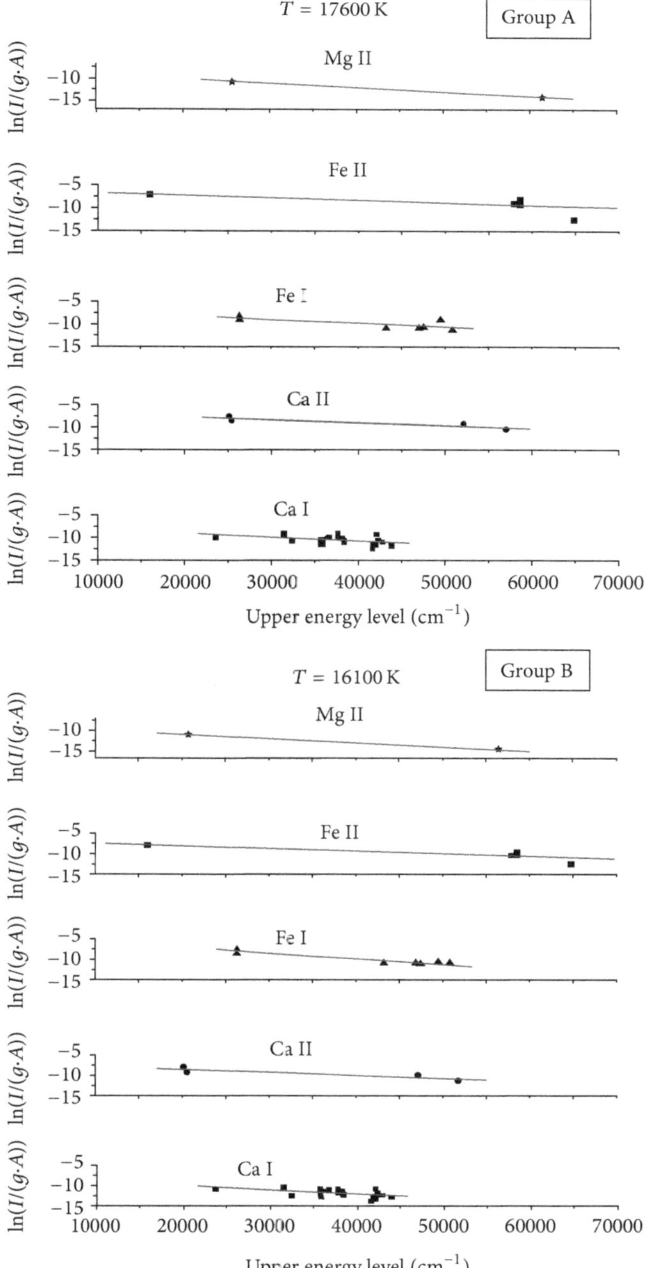

FIGURE 5: Boltzmann plots for different elements present in group A and group B.

obtained for both the groups. The results are summarized in Table 1.

3.2. Electron Density. Plasma near the LTE condition could be characterized by a single temperature describing the distribution of species in energy levels, the population of ionization stages, or the kinetic energy of electrons, ions, and atoms. In this case, the excitation temperature would be the same as the plasma temperature. According to the Saha-Boltzmann equation the electron density can be derived from the intensity ratio of two lines corresponding to different ionization stages of the same element [17, 18]:

$$n_e = \frac{2\left(2\pi m_e kT\right)^{3/2}}{h^3}\frac{I^{\mathrm{I}}_{mn}A_{jk}g^{\mathrm{II}}_j}{I^{\mathrm{II}}_{jk}A_{mn}g^{\mathrm{I}}_m}e^{-(E_{\mathrm{ion}}+E^{\mathrm{II}}_j-E^{\mathrm{I}}_m)/KT}, \quad (2)$$

where E_m and E_j are the upper level energies of neutral and single ionized transitions. E_{ion} is the ionization energy and n_e is the electron density. In practice, thermodynamic equilibrium is really hard to achieve in the whole area while it can exist in fine regions of plasma which are referred to as Local Thermodynamic Equilibrium (LTE). This condition is normally reached during adequate number of collisions where the energy is distributed more equally between colliding particles with the same masses. Thus, LTE may exist at high electron density limit given by McWhirter criterion [19, 20]:

$$n_e\left(\mathrm{cm}^3\right) \geq 1.6\times10^{12}\left[T_{\mathrm{exc}}(\mathrm{K})\right]^{1/2}\left[\Delta E\,(\mathrm{eV})\right]^3, \quad (3)$$

where ΔE is the energy difference between the two levels [21]. Table 1 shows an estimation of the electron density for both groups. The electron density was calculated using the above standard relation. We based our estimate on the Ca line at 452.69 nm; its spectroscopic parameters are tabulated in [19]. An equality expression related to the electron density was used to estimate whether LTE is likely to prevail in our measurements. The value 4.5×10^{15} (cm^{-3}), which, according to [12, 19], is needed for LTE to exist, is well below the values encountered in our experiments. This suggests that the analytical measurements in our study were most likely carried out under LTE conditions.

TABLE 1: Plasma temperature and electron density for both groups.

Sample	Plasma temperature (K)	Electron density (cm^{-3})
With Pb	17613 ± (14%)	9×10^{15}
Without Pb	16140 ± (18%)	10×10^{15}

4. Qualitative Analysis of Elements

The following are the dominant elements available in both samples. The elements with prominent emissions common to both bone types are Ca, Mg, and Fe. The content of these elements in bone is illustrated in Figure 6 for both groups.

Previous study shows that the Calcium (Ca I, Ca II) is the main source for the consumption of lead in bone [22]. This effect is clearly observed in Figure 3. Although the lead concentration is augmented itself, the portion of lead is consumed via chemical reactions to reduce Calcium. The results for Magnesium (Mg I, Mg II) are similar to Calcium. Magnesium like Calcium competes with lead for absorption in bone. So as expected, Mg I and Mg II decreased in sample of group A [23]. There is irregular reduction process for absorption in Fe I and Fe II. Unlike Ca and Mg, Iron element increased in some recorded wavelengths when the rats consumed lead. This behavior was observed in the

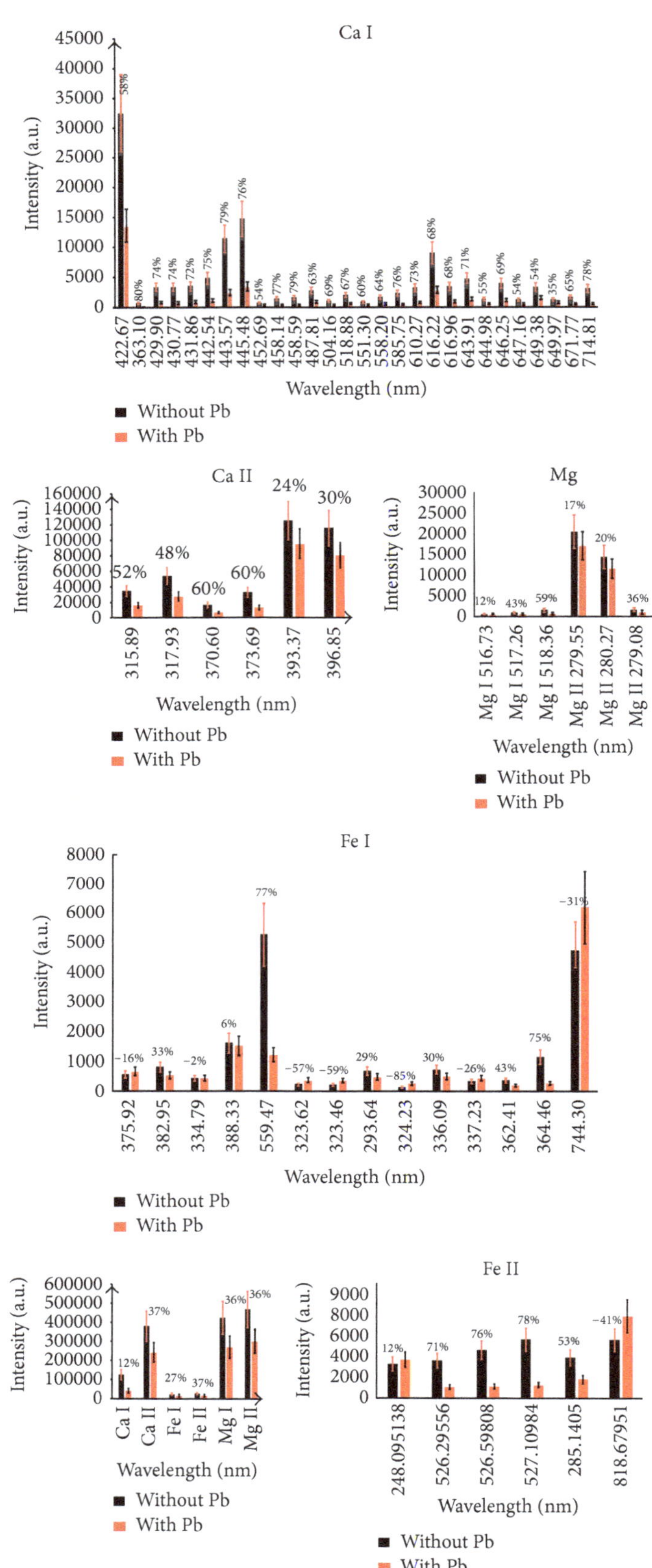

FIGURE 6: Comparison of relative intensity of different element presented in both groups.

previous work [24]. However, a decrease in Fe element at some lines is observed for group A.

In conclusion, results revealed that temperature and electron density of plasma resulting from laser beam are similar for both lead and nonlead groups and they obey Local Thermodynamic Equilibrium (LTE) conditions. The effect of its absorption in elements such as Calcium, Magnesium, and Fe had been recognized, although Laser-Induced Breakdown Spectroscopy (LIBS) method for measuring absorbed lead having such concentration in bone lacks proper performance. Qualitative comparison in value of elements in both the groups revealed that values of Ca I and Ca II and Mg I and Mg II were decreased by absorbing lead. Although there is an indirect relation for Fe concentration with absorption of lead at various wavelengths, it revealed that values of minerals such as Fe I and Fe II were globally decreased.

Competing Interests

The authors declare that they have no competing interests.

References

[1] K. Song, Y.-I. Lee, and J. Sneddon, "Applications of laser-induced breakdown spectrometry," *Applied Spectroscopy Reviews*, vol. 32, no. 3, pp. 183–235, 1997.

[2] D. A. Cremers and L. J. Radziemski, *Laser Induced Breakdown Spectroscopy*, Cambridge University Press, 2006.

[3] M. Z. Martin, N. Labbé, N. André et al., "High resolution applications of laser-induced breakdown spectroscopy for environmental and forensic applications," *Spectrochimica Acta Part B: Atomic Spectroscopy*, vol. 62, no. 12, pp. 1426–1432, 2007.

[4] O. Samek, D. C. S. Beddows, H. H. Telle, G. W. Morris, M. Liska, and J. Kaiser, "Quantitative analysis of trace metal accumulation in teeth using laser-induced breakdown spectroscopy," *Applied Physics A*, vol. 69, no. 1, supplement, pp. S179–S182, 1999.

[5] O. Samek, H. H. Telle, and D. C. S. Beddows, "Laser-induced breakdown spectroscopy: a tool for real-time, in vitro and in vivo identification of carious teeth," *BMC Oral Health*, vol. 1, no. 1, article 1, 2001.

[6] A. Kumar, F.-Y. Yueh, J. P. Singh, and S. Burgess, "Characterization of malignant tissue cells by laser-induced breakdown spectroscopy," *Applied Optics*, vol. 43, no. 28, pp. 5399–5403, 2004.

[7] H. P. de Souza, E. Munin, L. P. Alves, M. L. Redígolo, and Marcos, "Laser-induced breakdown spectroscopy in a biological tissue," in *Proceedings of the 26th Conference of ENFMC-Annals of Optics*, vol. 5, pp. 1–4, 2003.

[8] Z. Abdel-Salam and M. A. Harith, "Laser spectrochemical characterization of semen," *Talanta*, vol. 99, pp. 140–145, 2012.

[9] S. J. Rehse, H. Salimnia, and A. W. Miziolek, "Laser-induced breakdown spectroscopy (LIBS): an overview of recent progress and future potential for biomedical applications," *Journal of Medical Engineering & Technology*, vol. 36, no. 2, pp. 77–89, 2012.

[10] D. A. Cremers and R. C. Chinni, "Laser-induced breakdown spectroscopy—capabilities and limitations," *Applied Spectroscopy Reviews*, vol. 44, no. 6, pp. 457–506, 2009.

[11] A. K. Myakalwar, S. Sreedhar, I. Barman et al., "Laser-induced breakdown spectroscopy-based investigation and classification of pharmaceutical tablets using multivariate chemometric analysis," *Talanta*, vol. 87, no. 1, pp. 53–59, 2011.

[12] O. Samek, M. Liska, J. Kaiser et al., "Laser ablation for mineral analysis in the human body: integration of LIFS with LIBS," in *Biomedical Sensors, Fibers, and Optical Delivery Systems*, vol. 3570 of *Proceedings of SPIE*, International Society for Optics and Photonics, 1999.

[13] T. Fortune and D. I. Lurie, "Chronic low-level lead exposure affects the monoaminergic system in the mouse superior olivary complex," *Journal of Comparative Neurology*, vol. 513, no. 5, pp. 542–558, 2009.

[14] A. Zaidel, *Tables of Spectral Lines*, Springer Science & Business Media, 2013.

[15] J. E. Sansonetti, W. C. Martin, and S. Young, "Handbook of basic atomic spectroscopic data," *Journal of Physical and Chemical Reference Data*, vol. 34, no. 4, pp. 1559–2260, 2005.

[16] S. Pandhija and A. K. Rai, "In situ multielemental monitoring in coral skeleton by CF-LIBS," *Applied Physics B*, vol. 94, no. 3, pp. 545–552, 2009.

[17] D. A. Cremers and L. J. Radziemski, "Basics of the LIBS plasma," in *Handbook of Laser-Induced Breakdown Spectroscopy*, pp. 23–52, John Wiley & Sons, 2006.

[18] S. Z. Mortazavi, P. Parvin, M. R. Mousavi Pour, A. Reyhani, A. Moosakhani, and S. Moradkhani, "Time-resolved evolution of metal plasma induced by Q-switched Nd:YAG and ArF-excimer lasers," *Optics & Laser Technology*, vol. 62, pp. 32–39, 2014.

[19] H. R. Griem, *Principles of Plasma Spectroscopy*, vol. 2, Cambridge University Press, 2005.

[20] R. McWhirter, R. Huddlestone, and S. Leonard, *Plasma Diagnostic Techniques*, Academic, New York, NY, USA, 1965.

[21] A. F. M. Y. Haider and Z. H. Khan, "Determination of Ca content of coral skeleton by analyte additive method using the LIBS technique," *Optics & Laser Technology*, vol. 44, no. 6, pp. 1654–1659, 2012.

[22] J. C. Barton, M. E. Conrad, L. Harrison, and S. Nuby, "Effects of calcium on the absorption and retention of lead," *The Journal of Laboratory and Clinical Medicine*, vol. 91, no. 3, pp. 366–376, 1978.

[23] A. A. Van Barneveld and C. J. A. Van den Hamer, "Influence of Ca and Mg on the uptake and deposition of Pb and Cd in mice," *Toxicology and Applied Pharmacology*, vol. 79, no. 1, pp. 1–10, 1985.

[24] I. K. Robertson and M. Worwood, "Lead and iron absorption from rat small intestine: the effect of dietary Fe deficiency," *British Journal of Nutrition*, vol. 40, no. 2, pp. 253–260, 1978.

Spectroscopic Investigations, DFT Calculations, and Molecular Docking Studies of the Anticonvulsant (2*E*)-2-[3-(1*H*-Imidazol-1-yl)-1-phenylpropylidene]-*N*-(4-methylphenyl) hydrazinecarboxamide

Reem I. Al-Wabli,[1] Devarasu Manimaran,[2] Liji John,[2] Isaac Hubert Joe,[2] Nadia G. Haress,[1] and Mohamed I. Attia[1,3]

[1]*Department of Pharmaceutical Chemistry, College of Pharmacy, King Saud University, P.O. Box 2457, Riyadh 11451, Saudi Arabia*
[2]*Centre for Molecular and Biophysics Research, Department of Physics, Mar Ivanios College, Thiruvananthapuram, Kerala 695015, India*
[3]*Medicinal and Pharmaceutical Chemistry Department, Pharmaceutical and Drug Industries Research Division, National Research Centre (ID: 60014618), El Bohooth Street, Dokki, Giza 12622, Egypt*

Correspondence should be addressed to Isaac Hubert Joe; hubertjoe@gmail.com and Mohamed I. Attia; mattia@ksu.edu.sa

Academic Editor: Vincenza Crupi

Drug discovery for the management of neurological disorders is a challenging arena in medicinal chemistry. Vibrational spectral studies of (2*E*)-2-[3-(1*H*-imidazol-1-yl)-1-phenylpropylidene]-*N*-(4-methylphenyl)hydrazinecarboxamide ((2*E*)-IPPMP) have been recorded and analyzed to identify the functional groups and intermolecular/intramolecular interactions of the title molecule. The blue shift of the C-H stretching wavenumber reveals the presence of improper C-H⋯O hydrogen bonding. The equilibrium geometry, harmonic vibrational wavenumbers, Frontier orbital energy, and natural bond orbital analyses have been carried out using density functional theory with a B3LYP/6-311++G(d,p) level of the basis set. The vibrational modes have been unambiguously assigned using potential energy distribution analysis. The scaled wavenumbers are in good agreement with the experimental results. Natural bond orbital analysis has confirmed the intermolecular/intramolecular charge transfer interactions. HOMO-LUMO analysis was carried out to explore charge delocalization on the (2*E*)-IPPMP molecule. A molecular docking study has supported the anticonvulsant activity of the title molecule.

1. Introduction

Epilepsy is a rather neurobiological group of disorders. It has multiple origins and aspects depending upon the affected brain areas. It affects nearly 50 million people of the world's population [1–3]. The hydrazinecarboxamide derivatives have a wide spectrum of biological activities. Among these activities are anticancer and antioxidant [4], antifertility [5], antimicrobial [6, 7], anticonvulsant [8, 9], and anti-inflammatory [10] activities. The title molecule (2*E*)-2-[3-(1*H*-imidazol-1-yl)-1-phenyl-propylidene]-*N*-(4-methylphenyl)hydrazinecarboxamide ((2*E*)-IPPMP) was synthesized in our laboratory and its crystal structure was previously reported [11]. (2*E*)-IPPMP exhibited anticonvulsant activity with 83% and 50% seizure protection at a dose level of 718 μmol/kg in subcutaneous pentylenetetrazole (scPTZ) and maximal electroshock seizure (MES) screens, respectively, without any neurotoxicity [9].

Literature screening indicated that computational studies on the (2*E*)-IPPMP molecule have not yet been reported. Therefore, detailed investigations of structural properties and vibrational spectral analysis of the (2*E*)-IPPMP molecule were carried out in the present study using density functional theory (DFT) computations. Moreover, the biological

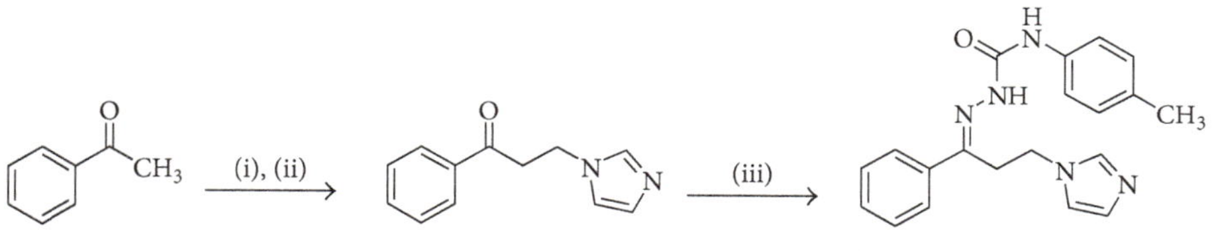

SCHEME 1: Synthesis of the title molecule (2*E*)-IPPMP. Reagents and conditions: (i) $HN(CH_3)_2 \cdot HCl$, $(CH_2O)_n$, conc. HCl, ethanol, reflux, 2 h; (ii) imidazole, water, reflux, 5 h; (iii) *N*-(4-methylphenyl)hydrazinecarboxamide, few drops of acetic acid, ethanol, RT, 18 h.

activity of the title molecule has been predicted by molecular docking analysis. It is expected that the current investigations might support the development of new potent anticonvulsant agents.

2. Experimental

2.1. General. Melting point was recorded on a Gallenkamp melting point instrument and it is uncorrected. The Fourier transform infrared spectrum of MMIMI was recorded on a Perkin Elmer RXL spectrometer (Waltham, Massachusetts, USA) in the region 4000–400 cm^{-1}, with samples in the KBr pellet method. The resolution of the spectrum was 2 cm^{-1}. The FT-Raman spectrum was measured in the range 3500–50 cm^{-1} using a Bruker RFS 100/S FT-Raman spectrophotometer (Ettlingen, Germany) with a 1064 nm Nd:YAG laser source of 100 mW power (Göttingen, Germany).

2.2. Synthesis. A solution containing *N*-(4-methylphenyl)hydrazinecarboxamide [12] (1.65 g, 10 mmol), 3-(1*H*-imidazol-1-yl)-1-phenylpropan-1-one (2.00 g, 10 mmol) [13], and a few drops of glacial acetic acid in ethanol (15 mL) was stirred at room temperature for 18 h. The reaction mixture was evaporated under reduced pressure and the residue was crystallized from ethanol to give 1.67 g (48%) of the title compound as colorless crystals (m.p. 476–478 K) which were suitable for single crystal X-ray analysis. ^{1}H and ^{13}C NMR as well as the mass spectral data of the title compound **2** are in accordance with the previously reported ones [9].

2.3. Theoretical Calculations. All DFT calculations of the (2*E*)-IPPMP molecule were performed using the Gaussian 09 program package [14] at the Becke3-Lee-Yang-Parr (B3LYP) level with a 6-311++G(d,p) basis set [15–17]. The structural parameters were computed in the gas phase as well as in the liquid phase using a polarizable continuum model (PCM) method. In order to correct the overestimations arising from some negative factors such as basis set truncation effect, neglecting electron correlations, and anharmonicity characters of the vibrational modes, the calculated wavenumbers were scaled using a uniform scaling factor of 0.9673 [18, 19]. Theoretical NMR calculation was performed on the basis of the GIAO (gauge-independent atomic orbitals) theory method using a Gaussian program.

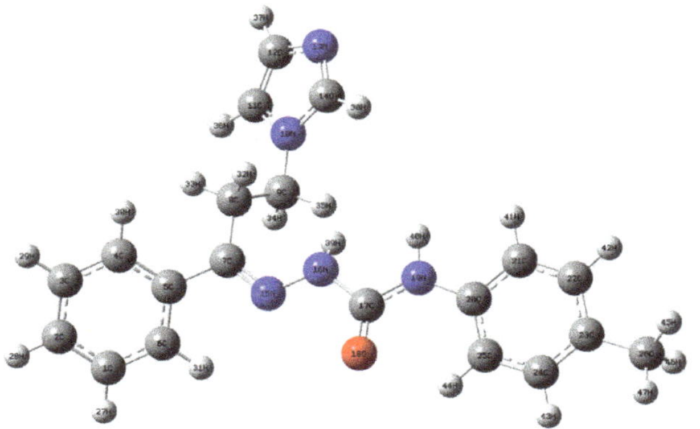

FIGURE 1: Optimized structure of (2*E*)-IPPMP.

3. Results and Discussion

3.1. Synthesis. The target compound (2*E*)-IPPMP was obtained in a three-step reaction sequence as given in Scheme 1.

3.2. Structural Geometry Analysis. The structure of the (2*E*)-IPPMP molecule was optimized using the B3LYP method with a 6-311++G(d,p) basis set. The optimized molecular structure of the isolated molecule is shown in Figure 1. The optimized geometrical parameters of the isolated (2*E*)-IPPMP molecule in the gas and water phases are given in Table 1. The calculated values were compared with the X-ray diffraction results. In the (2*E*)-IPPMP molecule the two phenyl rings are bridged by a hydrazinecarboxamide skeleton containing an imidazole ring. The C-N bond lengths C_{17}-N_{16} (1.4179 Å) and C_{17}-N_{19} (1.3792 Å) are shorter than the normal single C-N bond length 1.480 Å [20]. This discrepancy is due to the conjugation of p-type electrons of the carbonyl group and nitrogen atom, allowing the electrons to smear out along the C-N bond. In the *para*-disubstituted phenyl ring, the calculated C_{25}-H_{44} (1.0789 Å) bond length is shorter than that of the other C-C bonds; also, $O_{18} \cdots H_{44}$ distance is 2.25 Å, which is significantly shorter than the van der Waals radius (2.72 Å) [21] between O and H atoms, which indicates the possibility of C-H$\cdots$O hydrogen bonding. The elongation of the C_{20}-C_{21} (1.4004 Å) bond is due to the transfer of a lone pair of electrons from the amide nitrogen to the carbon atom. In the other phenyl ring, the C_5-H_{31}

TABLE 1: Structural geometry parameters of (2*E*)-IPPMP compared with their experimental X-ray diffraction results.

Parameter	Bond length (Å) Calc.	Exp.[a]	Solution phase	Parameter	Bond angle (°) Calc.	Exp.	Solution phase	Parameter	Dihedral angle (°) Calc.	Exp.	Solution phase
C_4-C_5	1.4012	1.386	1.402	C_2-C_1-C_6	120.31	120.52	120.25	C_6-C_1-C_2-C_3	0.16	0.21	0.3706
C_5-C_6	1.4024	1.382	1.4035	C_2-C_1-H_{27}	120.00	119.72	120.05	C_6-C_1-C_2-H_{28}	−179.76	179.84	−179.83
C_1-C_6	1.3903	1.384	1.3912	C_6-C_1-H_{27}	119.69	119.77	119.70	H_{27}-C_1-C_2-C_3	−179.89	179.84	−179.44
C_1-C_2	1.3956	1.361	1.3965	C_1-C_2-C_3	119.62	119.51	119.65	H_{27}-C_1-C_2-H_{28}	0.19	−0.13	0.36
C_2-C_3	1.3928	1.363	1.3937	C_1-C_2-C_{28}	120.20	120.24	120.17	C_2-C_1-C_6-C_5	−0.63	0.23	−0.50
C_3-C_4	1.3934	1.388	1.3942	C_3-C_2-H_{28}	120.18	120.26	120.18	C_2-C_1-C_6-H_{31}	179.22	179.80	179.97
C_5-C_7	1.4919	1.487	1.4924	C_2-C_3-C_4	120.13	120.76	120.19	H_{27}-C_1-C_6-C_5	179.42	179.79	179.31
C_8-C_7	1.5135	1.508	1.5128	C_2-C_3-H_{29}	120.15	119.63	120.18	H_{27}-C_1-C_6-C_{31}	−0.73	0.18	−0.21
C_8-C_9	1.5434	1.525	1.5428	C_4-C_3-H_{29}	119.71	119.63	119.64	C_1-C_2-C_3-C_4	0.48	0.42	0.25
C_8-H_{32}	1.0883	0.958	1.0877	C_3-C_4-C_5	120.69	120.32	120.59	C_1-C_2-C_3-H_{29}	−179.3	179.61	−179.76
C_8-H_{33}	1.0932	0.948	1.0932	C_3-C_4-H_{30}	119.24	119.89	119.12	H_{28}-C_2-C_3-C_4	−179.6	−179.60	−179.54
C_9-H_{34}	1.0907	0.970	1.0895	C_5-C_4-H_{30}	120.05	119.79	120.28	H_{28}-C_2-C_3-H_{29}	0.61	0.36	0.44
C_9-H_{35}	1.0923	0.971	1.0909	C_4-C_5-C_6	118.67	117.99	118.75	C_2-C_3-C_4-C_5	−0.65	−0.72	−0.76
N_{10}-C_9	1.4584	1.452	1.4623	C_4-C_5-C_7	120.71	120.05	120.75	C_2-C_3-C_4-H_{30}	−178.94	179.25	−179.59
N_{10}-C_{14}	1.3677	1.348	1.3628	C_6-C_5-C_7	120.61	121.06	120.49	H_{29}-C_3-C_4-C_5	179.13	179.31	179.26
N_{10}-C_{11}	1.3810	1.364	1.3800	C_1-C_6-C_5	120.57	120.95	120.57	H_{29}-C_3-C_4-H_{30}	0.84	0.71	0.42
N_{13}-C_{12}	1.3749	1.366	1.3794	C_1-C_6-H_{31}	120.54	119.58	120.27	C_3-C_4-C_5-C_6	0.18	0.75	0.62
N_{13}-C_{14}	1.3677	1.304	1.3196	C_5-C_6-H_{31}	118.89	119.58	119.15	C_3-C_4-C_5-C_7	−178.58	179.34	−178.67
C_{11}-C_{12}	1.3718	1.345	1.3713	C_5-C_7-C_8	118.92	118.08	118.76	H_{30}-C_4-C_5-C_6	178.46	179.22	179.44
C_{12}-H_{37}	1.0790	0.930	1.0792	C_5-C_7-N_{15}	115.01	115.92	115.26	H_{30}-C_4-C_5-C_7	−0.31	0.68	0.15
C_{11}-H_{36}	1.0776	0.931	1.0775	C_8-C_7-N_{15}	125.99	125.96	125.92	C_4-C_5-C_6-C_1	0.46	0.51	0.002
C_{14}-H_{38}	1.0803	0.930	1.0799	C_7-C_8-C_9	112.37	112.66	112.28	C_4-C_5-C_6-H_{31}	−179.4	179.51	179.53
C_7-N_{15}	1.2849	1.287	1.2860	C_7-C_8-H_{32}	109.46	109.95	109.63	C_7-C_5-C_6-C_1	179.22	179.58	179.30
N_{15}-N_{16}	1.4036	1.367	1.4072	C_7-C_8-H_{33}	110.05	111.01	109.68	C_7-C_5-C_6-C_{31}	−0.63	−0.39	−1.17
N_{16}-H_{39}	1.0123	0.899	1.0129	C_9-C_8-H_{32}	108.19	106.60	108.53	C_4-C_5-C_7-C_8	−36.92	−29.26	−37.14
C_{17}-N_{16}	1.4179	1.362	1.4056	C_9-C_8-H_{33}	110.37	110.76	109.71	C_4-C_5-C_7-N_{15}	146.04	152.80	145.29
C_{17}-O_{18}	1.2131	1.230	1.2222	H_{32}-C_8-H_{33}	106.18	105.53	106.85	C_6-C_5-C_7-C_8	144.33	150.64	143.58
C_{17}-N_{19}	1.3792	1.350	1.3747	C_8-C_9-N_{10}	111.95	113.52	111.52	C_6-C_5-C_7-N_{15}	−32.7	27.30	−33.99
N_{19}-H_{40}	1.0093	0.889	1.0104	C_8-C_9-H_{34}	111.61	108.87	111.63	C_5-C_7-C_8-C_9	−52.66	−68.80	−55.35
N_{19}-C_{20}	1.4143	1.419	1.4123	C_8-C_9-H_{35}	109.97	108.87	110.26	C_5-C_7-C_8-H_{32}	−172.89	49.95	−176.06
C_{20}-C_{21}	1.4005	1.384	1.4028	N_{10}-C_9-H_{34}	108.78	108.88	108.21	C_5-C_7-C_8-H_{33}	70.77	166.33	66.90
C_{21}-C_{22}	1.3895	1.381	1.3897	N_{10}-C_9-H_{35}	107.24	108.83	107.44	C_{15}-C_7-C_8-C_9	124.02	108.92	121.94
C_{22}-C_{23}	1.3984	1.380	1.4006	H_{34}-C_9-H_{35}	107.08	107.71	107.60	N_{15}-C_7-C_8-H_{32}	3.79	−15.96	1.23
C_{23}-C_{24}	1.3972	1.376	1.3974	C_9-N_{10}-C_{11}	127.03	126.90	126.69	N_{15}-C_7-C_8-H_{33}	−112.55	−132.34	−115.80
C_{24}-C_{25}	1.3931	1.384	1.3953	C_9-N_{10}-C_{14}	126.64	126.29	126.56	C_5-C_7-N_{15}-N_{16}	−177.68	179.43	−178.97
C_{20}-C_{25}	1.3987	1.384	1.3999	C_{11}-N_{10}-C_{14}	106.28	106.52	106.69	C_8-C_7-N_{15}-N_{16}	5.53	1.67	3.66
C_{21}-H_{41}	1.0864	0.929	1.0855	N_{10}-C_{11}-C_{12}	105.73	105.44	105.79	C_7-C_8-C_9-N_{10}	−174.09	−26.97	−178.38
C_{22}-H_{42}	1.0853	0.930	1.0852	N_{10}-C_{11}-H_{36}	122.10	127.32	122.03	C_7-C_8-C_9-H_{34}	63.7	66.15	60.44
C_{24}-H_{43}	1.0855	0.930	1.0855	C_{12}-C_{11}-H_{36}	132.17	124.35	132.18	C_7-C_8-C_9-H_{35}	−54.99	−176.68	−59.11
C_{25}-H_{44}	1.0790	0.929	1.0788	C_{11}-C_{12}-N_{13}	110.47	111.35	110.26	H_{32}-C_8-C_9-N_{10}	−53.12	175.97	−57.03
C_{23}-C_{26}	1.5095	1.608	1.5097	C_{11}-C_{12}-H_{37}	127.96	124.35	127.94	H_{32}-C_8-C_9-H_{34}	−175.33	62.64	−178.21
C_{26}-H_{45}	1.0928	0.960	1.0928	N_{13}-C_{12}-H_{37}	121.56	124.30	121.80	H_{32}-C_8-C_9-H_{35}	65.98	54.53	62.24
C_{26}-H_{46}	1.0926	0.960	1.0922	C_{12}-N_{13}-C_{14}	105.20	104.06	105.21	H_{33}-C_8-C_9-N_{10}	62.66	−54.53	59.38
C_{26}-H_{47}	1.0957	0.960	1.0956	N_{10}-C_{14}-N_{13}	112.32	112.62	112.05	H_{33}-C_8-C_9-H_{34}	−59.55	−54.53	−61.80

[a]Taken from [11].

TABLE 2: Second-order perturbation theory analysis of Fock matrix in natural bond orbital (NBO) basis for (2*E*)-IPPMP.

Donor (i)	Occupancy (e)	Acceptor (j)	Occupancy (e)	$E(2)^a$ kcal/mol	$E(j) - E(i)^b$ (a.u)	$E(i, j)^c$ (a.u)
$n_1(N_{10})$	1.55554	$\pi^*(C_{11}\text{-}C_{12})$	1.85684	30.62	0.29	0.087
$n_1(N_{10})$	1.55554	$\pi^*(N_{13}\text{-}C_{14})$	1.86721	46.48	0.28	0.103
$n_1(N_{13})$	1.9234	$\sigma^*(C_{11}\text{-}C_{12})$	1.98419	5.1	0.95	0.063
$n_1(N_{15})$	1.92552	$\sigma^*(C_7\text{-}C_8)$	1.97792	11.4	0.79	0.086
$n_1(N_{15})$	1.92552	$\sigma^*(N_{16}\text{-}H_{39})$	1.97686	2.02	0.77	0.036
$n_1(N_{16})$	1.79578	$\pi^*(C_7\text{-}N_{15})$	1.94162	9.06	0.36	0.052
$n_2(O_{18})$	1.83834	$\sigma^*(N_{16}\text{-}C_{17})$	1.98245	26.84	0.63	0.118
$n_2(O_{18})$	1.83834	$\sigma^*(C_{17}\text{-}N_{19})$	1.98641	24.58	0.69	0.119
$n_2(O_{18})$	1.83834	$\sigma^*(C_{25}\text{-}H_{44})$	1.97725	0.78	0.73	0.022
$n_1(N_{19})$	1.70657	$\pi^*(C_{20}\text{-}C_{25})$	1.64309	30.72	0.31	0.089
$\pi(C_2\text{-}C_3)$	1.6583	$\pi^*(C_4\text{-}C_5)$	1.64969	20.62	0.28	0.069
$\pi(C_4\text{-}C_5)$	1.64969	$\pi^*(C_1\text{-}C_6)$	1.66001	19.97	0.29	0.067
$\pi(C_{11}\text{-}C_{12})$	1.85684	$\pi^*(N_{13}\text{-}C_{14})$	1.86721	15.01	0.28	0.061
$\pi(N_{13}\text{-}C_{14})$	1.86721	$\pi^*(C_{11}\text{-}C_{12})$	1.85684	21.59	0.33	0.078
$\pi(C_{20}\text{-}C_{25})$	1.64309	$\pi^*(C_{21}\text{-}C_{22})$	1.70752	19.29	0.28	0.066
$\pi(C_{20}\text{-}C_{25})$	1.64309	$\pi^*(C_{23}\text{-}C_{24})$	1.64872	20.41	0.3	0.069
$\pi(C_{23}\text{-}C_{24})$	1.64872	$\pi^*(C_{20}\text{-}C_{25})$	1.64309	20.68	0.27	0.068
$\pi(C_{23}\text{-}C_{24})$	1.64872	$\pi^*(C_{21}\text{-}C_{22})$	1.70752	22.11	0.27	0.069
$\pi(N_{13}\text{-}C_{14})$	1.86721	$\pi^*(C_{11}\text{-}C_{12})$	1.85684	107.4	0.01	0.06
$\sigma(C_{26}\text{-}H_{46})$	1.98674	$\sigma^*(C_{23}\text{-}C_{24})$	1.64872	0.84	0.54	0.021
$\sigma(C_{26}\text{-}H_{47})$	1.97658	$\sigma^*(C_{23}\text{-}C_{24})$	1.64872	4.34	0.54	0.047
$\pi(C_{20}\text{-}C_{25})$	1.64309	$\pi^*(C_{23}\text{-}C_{24})$	1.64872	254.22	0.01	0.082
$\pi(C_{21}\text{-}C_{22})$	1.70752	$\pi^*(C_{23}\text{-}C_{24})$	1.64872	230.93	0.01	0.084

[a] $E(2)$ means energy of hyperconjugative interactions. [b] Energy difference between donor and acceptor i and j NBO. [c] $F(i, j)$ is the Fock matrix element between i and j NBO.

(1.0825 Å) bond is shorter than that of the other C-C bonds; also, $N_{15} \cdots H_{31}$ distance is 2.56 Å, which is significantly shorter than the van der Waals radius (2.75 Å) between N and H atoms, which indicates the possibility of C-H$\cdots$O hydrogen bonding. In addition, C_5-C_6 (1.4024 Å) and C_4-C_5 (1.4012 Å) bond distances on either side of the cyanide group are appreciably greater than the other C-C bonds due to the resonance effect between the cyanide group and phenyl ring. In the imidazole ring, C_{14}-N_{10} (1.3676 Å) and C_{14}-N_{13} (1.3139 Å) bonds are relatively shorter due to the lone pair interaction from the nitrogen atom. The deviation of bond lengths of N_{15}-H_{39} (1.0123 Å) and N_{19}-H_{40} (1.0092 Å) is due to the different environment of the nitrogen atom. Conjugation of the carbonyl group with the hydrazine moiety would favor planarity, but the van der Waals repulsion between H_{39} and H_{40} hinders the achievement of coplanarity. A small deviation was obtained within the calculated structural parameters in the gas and liquid phases, which were due to the solvent interactions over the solution phase calculations.

3.3. Natural Bond Orbital Analysis. Natural bond orbital (NBO) analysis was performed using the NBO 3.1 program [22] as implemented in the Gaussian 09 program package for the DFT method. The corresponding results are presented in Table 2. NBO analysis has proved to be an effective tool for chemical interpretation of hyperconjugative interaction and electron density transfer (EDT) from a filled lone pair to an unfilled antibonding orbital in the hydrogen bonding system [23–25]. The intramolecular C-H$\cdots$O hydrogen bonding is formed due to the orbital overlap between n(O) and σ^*(C-H) which results in intramolecular charge transfer (ICT) causing stabilization of H-bonded systems. This interaction results in increased electron density (ED) of the C-H antibonding orbital, which strengthens the C-H bond. NBO analysis confirms C-H$\cdots$O intramolecular hydrogen bonding formed by the orbital overlap between a lone pair $n_2(O_{18})$ and $\sigma^*(C_{25}\text{-}H_{44})$ antibonding orbital with stabilization energy of 0.78 kcal/mol. The most important interaction (n-π^*) and (n-σ^*) energies of $n_1(N_{10}) \rightarrow \pi^*(N_{13}\text{-}C_{14})$ and $n_2(O_{18}) \rightarrow \sigma^*(N_{16}\text{-}C_{17})$ are 46.48 and 26.84 kcal/mol, respectively. This larger $E(2)$ value shows ICT interactions of the molecule.

3.4. Vibrational Spectral Analysis. Computed vibrational wavenumbers and atomic displacements corresponding to the different normal modes are used to identify vibrational modes. The vibrational modes are assigned on the basis of

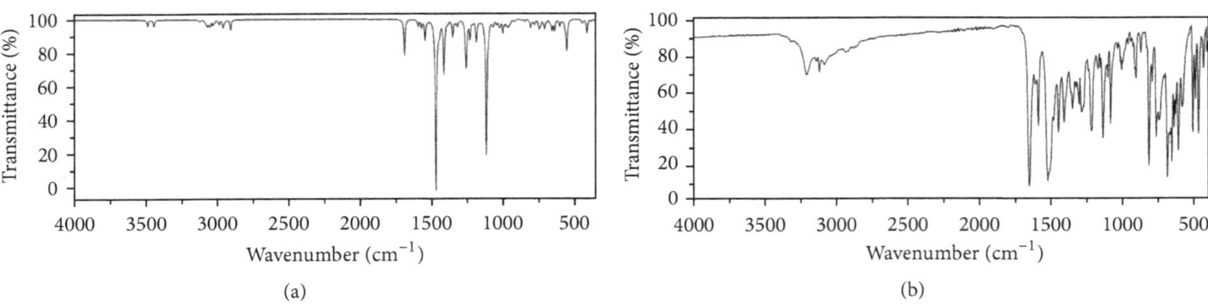

FIGURE 2: (a) Calculated (b) experiment infrared spectra of (2*E*)-IPPMP.

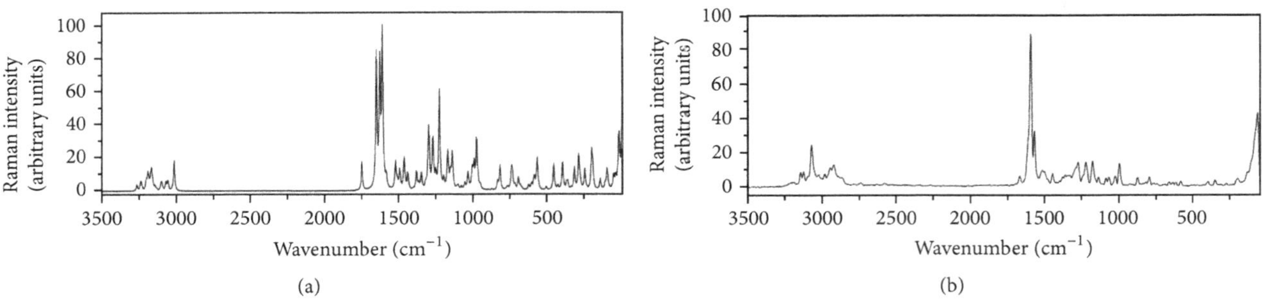

FIGURE 3: (a) Calculated (b) experiment Raman spectra of (2*E*)-IPPMP.

potential energy distribution analysis using the VEDA4 program [26]. The experimental infrared (IR) and Raman spectra are shown in Figures 2 and 3. The calculated vibrational wavenumbers, measured IR, and Raman band positions and their detailed assignments are given in Table 3.

3.4.1. Phenyl Ring Vibrations. Aromatic C-H stretching vibrations generally absorb in the region 3080–3010 cm^{-1} [27]. The observed weak IR band at 3119 cm^{-1} and Raman band at 3121 cm^{-1} correspond to aromatic C-H stretching mode. The blue shift of the C-H stretching wavenumber is due to weak intramolecular C-H···O hydrogen bonding. Aromatic C=C stretching vibrations occur in the region 1625–1430 cm^{-1} [28]. A medium IR band at 1539 cm^{-1} and a strong Raman band at 1591 cm^{-1} are observed, which correspond to aromatic C=C stretching mode. In-plane C-H deformation vibrations appear in the region 1290–1000 cm^{-1} [27]. Observed IR and Raman bands at 1288 and 1275 cm^{-1} are assigned to in-plane C-H deformation. In-plane ring deformation vibrations appear in the region 650–615 cm^{-1}. Observed IR and Raman bands at 615 and 640 cm^{-1} are assigned to in-plane ring deformation.

3.4.2. Methylene Vibrations. Asymmetric and symmetric CH_2 stretching vibrations normally appear strongly at about 2926 and 2855 cm^{-1} [29]. The Raman band at 2941 cm^{-1} is assigned to CH_2 symmetric stretching vibration. Methylene scissoring vibrations normally appear in the region 1465–1445 cm^{-1} [29]. A medium band observed in IR at 1401 cm^{-1} is attributed to methylene scissoring mode. The twisting, wagging vibrations appear in the region 1422–719 cm^{-1} [30]. The observed strong IR band at 1141 cm^{-1} and weak Raman band at 1177 cm^{-1} are assigned to CH_2 twisting modes for methylene. Wagging is observed in IR at 1352 cm^{-1} and Raman at 1361 cm^{-1}.

3.4.3. Methyl Vibrations. The asymmetric C-H stretching mode of CH_3 generally occurs at 2982–2962 cm^{-1} and CH_3 symmetric stretching at 2882–2862 cm^{-1} [29]. A weak Raman band at 2977 cm^{-1} is attributed to CH_3 symmetric stretching. Asymmetric bending vibrations of methyl groups occur in the region 1470–1450 cm^{-1} [28]. The observed weak Raman band at 1447 cm^{-1} is assigned to methyl scissoring vibration.

3.4.4. Secondary Amide Vibrations. Carbonyl stretching vibration in a secondary amide is expected in the region 1680–1630 cm^{-1} [28]. A very strong band at 1654 cm^{-1} is assigned to C=O stretching. N-H stretching vibrations generally appear in the region 3370–3170 cm^{-1}. Observed IR and Raman bands at 3141 and 3140 cm^{-1} are attributed to N-H stretching. The in-plane N-H bending vibration usually appears from 1570 to 1515 cm^{-1} [28]. The observed very strong IR band at 1523 cm^{-1} and weak Raman band at 1527 cm^{-1} are assigned to in-plane N-H bending vibration.

3.4.5. Imidazole Vibrations. Imidazole C-H stretching vibrations are expected in the region 3145–3115 cm^{-1} [31, 32]. Observed Raman bands at 3121 and 3138 cm^{-1} and IR bands at 3119 and 3138 cm^{-1} are assigned to C-H stretching mode.

TABLE 3: Vibrational assignment of (2*E*)-IPPMP based on potential energy distribution method.

Calculated wavenumber (cm^{-1})		Experimental wavenumber (cm^{-1})		Assignment with PED%
Unscaled	Scaled	FT-IR	FT-Raman	
3610	3497	3207w	—	ν N-H (92)
3565	3453	3141vw	3140vw	ν N-H (99)
3265	3163	3138vw	3138w	ν C-H imd (83)
3241	3140	3119w	3121w	ν C-H ph (99)
3202	3102		3118w	ν C-H ph (86)
3182	3082	3083vw	—	ν C-H ph (67)
3164	3066	—	3065m	ν C-H ph (27)
3105	3008	—	3018w	ν_{as} C-H (55)
3073	2977	—	2977w	ν_{sy} C-H (36)
3059	2964	—	2941w	ν_{sy} C-H (48)
3019	2925	—	2917w	ν_{sy} C-H (18)
1776	1720	1654vs	—	ν C=O (78)
1639	1588	1589m	1593vs	ν C_C ph (42)
1614	1564	—	1569m	ν C_C ph (36)
1553	1504	1523vs	1527vw	β N-H (32)
1496	1449	1449m	—	ω N-H (33)
1492	1445	—	1447vw	δ_{sci} C-H (19)
1468	1422	1401m	—	δ_{sci} C-H (79)
1399	1355	1352w	1361vw	ω C-H (51)
1382	1339	—	1338vw	ν C-C imd (20)
1327	1286	1288w	—	β C-H ph (22)
1307	1267	—	1275w	β C-H ph (12)
1263	1223	—	1222w	β C-H ph (46)
1258	1219	1218m	—	β C-H ph (56)
1209	1171	1174vw	—	β C-H ph (28)
1177	1140	1141s	1177w	t C-H (18)
1118	1084	1089m	1088vw	ν N-N (23), β C-H ph (12)
1101	1067	—	1069vw	β C-H ph (10)
1075	1041	1039vw	—	ν N-N (19)
1061	1028	1020vw	1027vw	τ CHCH (70)
1043	1011	1011w	—	β C-H imd (32)
1023	991	—	998w	ν N-C (14)
985	954	947vw	—	τ HCCH (81)
947	918	931vw	—	τ CCCH (49)
940	911	911w	—	τ HCCC (47)
917	888	876vw	875vw	β CC imd (72)
820	794	797vw	796vw	τ HCCN (55)
785	761	769w		ν C_C (14)
673	652	660s	663vw	τ CNCN (63),
652	632	—	640w	β CC ph (24)
636	617	615m	—	τ CCNC (34), β CC ph (21)
614	595	590vw	586vw	τ CCNN (17)
414	401	—	401vw	τ CCCC (29)
370	359	—	354vw	τ CCNN (14)
199	193	—	204vw	τ CCCC (23)
71	69	—	71m	τ CNCN (14), τ NNCN (19)

ν, stretching; ν_{as}, asymmetric; ν_{sy}, symmetric stretching; ph, phenyl ring; imd, imidazole ring: δ_{sci}, scissoring; t, twisting; ω, wagging; β, in-plane bending; τ, Torsion; vs, very strong; s, strong; m, medium; w, weak; vw, very weak.

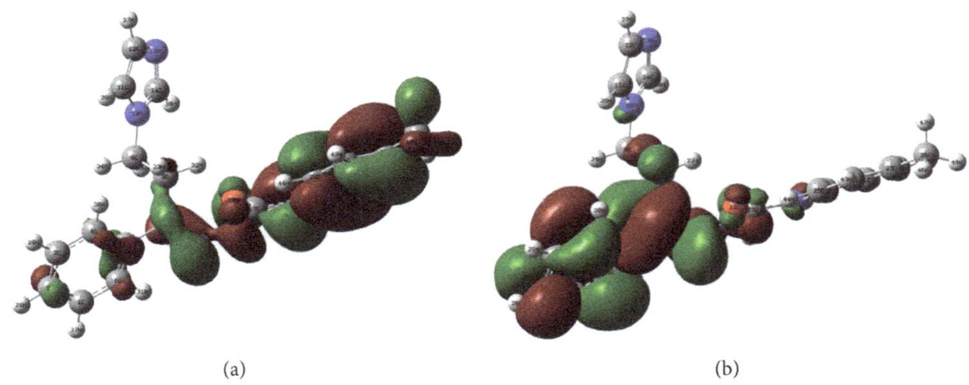

FIGURE 4: (a) HOMO and (b) LUMO plots of (2*E*)-IPPMP.

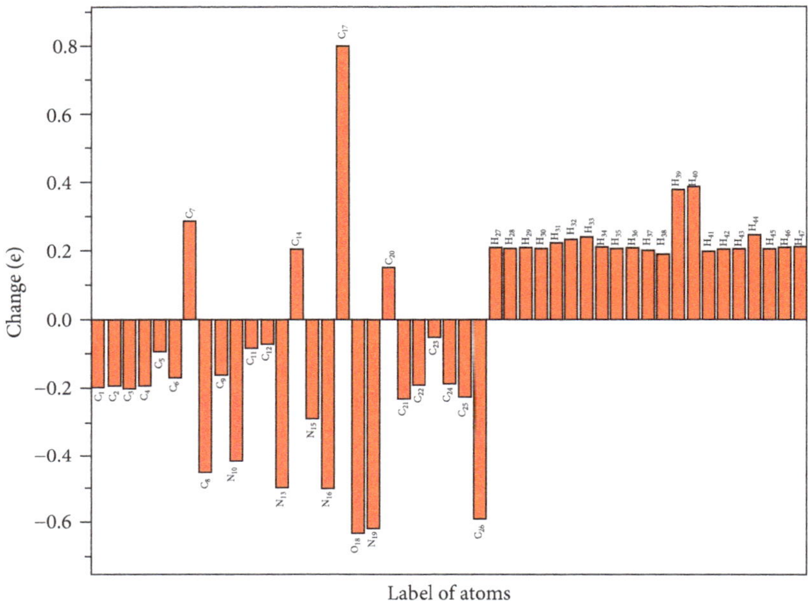

FIGURE 5: Natural charge distribution chart of the (2*E*)-IPPMP molecule.

The observed Raman band at 1338 cm^{-1} is attributed to C-C stretching mode [33].

3.4.6. Hydrazine Vibrations. In accordance with earlier reports and in agreement with the calculation, a weak band observed at 3207 cm^{-1} is assigned to hydrazine N-H stretching [27, 34]. Observed bands in IR and Raman at 1089 and 1088 cm^{-1} are assigned to N-N stretching vibration.

3.4.7. Skeletal Mode Vibrations. C-N and C-C stretching vibrations generally occur in the region 1150–850 cm^{-1} [34]. The weak IR and Raman bands observed at 1011 and 998 cm^{-1} are attributed to C-C stretching.

3.5. Frontier Molecular Orbital Energy Analysis. HOMO (highest occupied molecular orbital) and LUMO (lowest unoccupied molecular orbital) are important in defining the reactivity of a chemical species. The energy of HOMO indicates nucleophilicity and LUMO indicates electrophilicity [35]. HOMO is spread over the methyl phenyl, hydrazinecarboxamide fragment and LUMO is located on ph 2. This shows the charge transfer between the two rings through the hydrazinecarboxamide path. The HOMO (−6.61 eV) and LUMO (−1.61 eV) energies reflect the charge transfer within the molecule. The HOMO-LUMO energy gap is 4.46 eV. The frontier molecular orbital diagrams are shown in Figure 4.

3.6. Natural Population Analysis. Natural population analysis provides an effective method to calculate atomic charges and electron distribution within a molecule [36]. The net atomic charges of the (2*E*)-IPPMP molecule obtained by natural population analysis are plotted in Figure 5. All hydrogen atoms have a net positive charge. The atoms H_{40} (0.3852*e*) and H_{39} (0.3788*e*) show more positive charge than other hydrogen atoms due to their attachment with a nitrogen atom. Among the hydrogen atoms (∼0.2023*e*) of the phenyl ring, H_{44}

TABLE 4: NMR chemicals shift values of ^{13}C and ^{1}H.

^{13}C			^{1}H		
Atom	$\delta_{exp.}$	$\delta_{calc.}$	Atom	$\delta_{exp.}$	$\delta_{calc.}$
C_1	128.3	128.01	H_{27}	7.39	7.16
C_2	128.9	129.41	H_{28}	7.43	7.19
C_3	128.3	127.48	H_{29}	7.39	7.16
C_4	128.4	128.13	H_{30}	7.84	7.50
C_5	136.8	138.17	H_{31}	7.84	7.63
C_6	128.4	126.71	H_{32}	3.33	2.61
C_7	153.6	168.87	H_{33}	3.33	2.89
C_8	28.3	33.31	H_{34}	4.14	3.77
C_9	42.1	44.26	H_{35}	4.14	3.19
C_{11}	119.4	119.36	H_{36}	6.87	6.87
C_{12}	126.4	128.75	H_{37}	7.29	6.56
C_{14}	137.3	140.01	H_{38}	7.65	6.82
C_{17}	144.9	150.09	H_{39}	10.24	6.70
C_{20}	131.5	137.03	H_{40}	8.79	5.70
C_{21}	120.0	115.12	H_{41}	7.13	6.27
C_{22}	128.8	128.43	H_{42}	7.53	6.80
C_{23}	136.3	132.85	H_{43}	7.53	6.85
C_{24}	128.8	128.53	H_{44}	7.13	7.90
C_{25}	120.0	114.58	H_{45}	2.28	1.78
C_{26}	20.4	18.25	H_{46}	2.28	1.80
			H_{47}	2.28	2.08

(~0.2452*e*) shows the highest positive charge, being involved in C-H$\cdots$O intramolecular hydrogen bonding. All carbon atoms are negatively charged except C_7, C_{14}, C_{17}, and C_{20} due to their attachment with electronegative nitrogen or oxygen atoms. The atom C_{17} (0.7991*e*) shows more positive charge and N_{19} (−0.6181*e*) shows more negative charge, indicating charge delocalization in the molecule.

3.7. NMR Analysis. The scaled and experimental NMR (^{1}H and ^{13}C) chemical shift values for the (2*E*)-IPPMP molecule are presented in Table 4. The phenyl and imidazole ring carbon signals usually appear in the region 115–150 ppm. In this molecule, imidazole ring carbon signals are obtained at 119.4, 126.4, and 137.3 ppm, which are predicted at 119.36, 128.75, and 140.01 ppm, respectively.

Phenyl carbon signals were observed at 120.0, 128.4, 128.8, 128.9, 131.5, 136.3, 136.8, and 137.3 ppm, while their respective calculated ones were obtained at 115.12, 128.13, 128.53, 129.41, 137.03, 132.85, 138.17, and 140.01 ppm. A carbonyl carbon signal was seen at 144.9 ppm and its computed one was obtained at 150.09 ppm. This deviation may occur due to the presence of amide$\cdots$amide interactions in the crystalline state. On the other hand, the computed ^{1}H chemical shift values for the title molecule showed good agreement with the experimental ones (Table 4).

3.8. Molecular Docking Analysis. The title molecule (2*E*)-IPPMP was energy minimized based on the DFT method. Molecular docking was performed using AutoDock 4.2. A target protein (PDB ID: 1EOU) for antiepileptic agents was selected for the present docking analysis [37, 38]. The protein

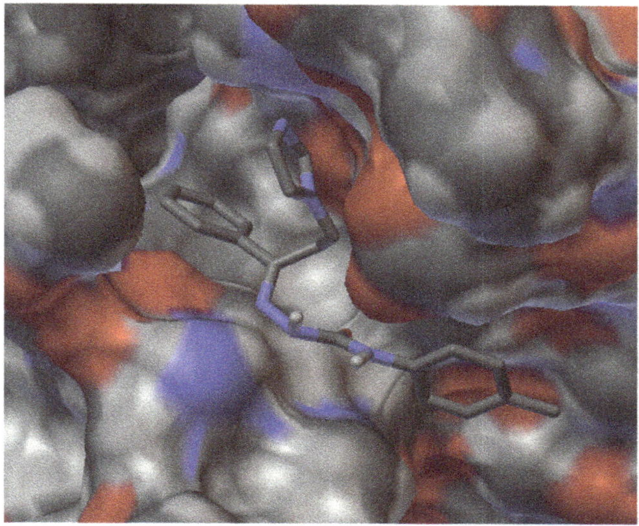

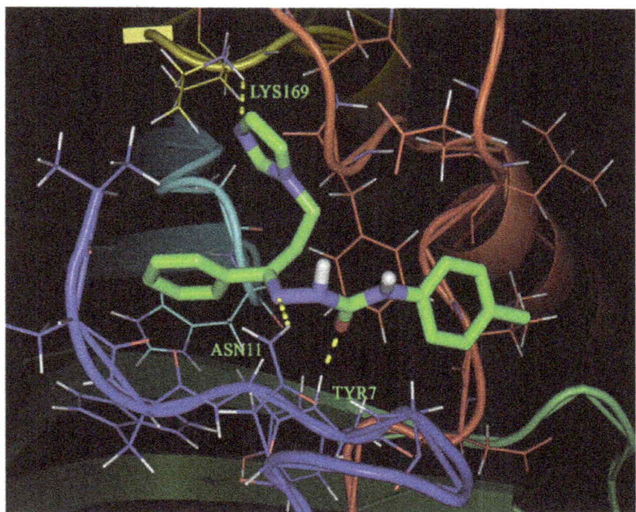

FIGURE 6: Binding pose diagrams of the (2*E*)-IPPMP molecule with its target protein.

data bank file of the target protein was downloaded from the Research Collaboratory for Structural Bioinformatics (RCSB) database, with a resolution of 2.1 Å. Protein preparation was carried out by the following steps: (i) all water molecules were removed, (ii) hydrogen atoms were added to the crystal structure, (iii) Kollman charge was added, and (iv) a previously docked inhibitor (fructose-based sugar sulfamate RWJ-37497) was removed from the protein. Rigid protein and flexible ligand dockings were carried out using the AutoDock 4.2 program package [39] with the Lamarckian genetic algorithm, applying the following protocol: trials of 100 dockings, energy evaluations of 25000000, population size of 200, mutation rate of 0.02, crossover rate of 0.8, and elitism value of 1. The docking results were evaluated by sorting the docked conformations according to their predicted binding free energy. The protein-ligand interaction complex is given in Figure 6, displaying the conformer with the best predicted binding free energy (−7.94 kcal/mol).

The amino acids ASN11, TYR7, and LYS169 in the active sites of the target protein bind with the (2*E*)-IPPMP ligand by N-H⋯O and N-H⋯N hydrogen bonds. These preliminary results support the exhibited anticonvulsant activity of the title molecule.

4. Conclusions

The geometry optimization and harmonic wavenumbers of the (2*E*)-IPPMP molecule have been performed at ground state calculations using the DFT method. FT-IR and FT-Raman measurements helped functional group identification of the title molecule. The fundamental wavenumbers are in good agreement with the theoretical results. Shifting of vibrational wavenumbers and hyperconjugative results confirmed the presence of intermolecular/intramolecular interactions in the molecule. The molecular docking results predicted the anticonvulsant activity of the (2*E*)-IPPMP molecule due to its ability to interact with a target protein (1EOU) for anticonvulsants. The results of the current study will support the development of new drug-like candidates in the anticonvulsant research area.

Competing Interests

The authors have declared that there are no competing interests.

Acknowledgments

The authors would like to extend their sincere appreciation to the Deanship of Scientific Research at King Saud University for its funding of this research through the Research Group Project no. RGP-196.

References

[1] O. K. Steinlein and D. Bertrand, "Nicotinic receptor channelopathies and epilepsy," *Pflugers Archiv—European Journal of Physiology*, vol. 460, no. 2, pp. 495–503, 2010.

[2] O. K. Steinlein, J. C. Mulley, P. Propping et al., "A missense mutation in the neuronal nicotinic acetylcholine receptor $\alpha 4$ subunit is associated with autosomal dominant nocturnal frontal lobe epilepsy," *Nature Genetics*, vol. 11, no. 2, pp. 201–203, 1995.

[3] M. De Groot, J. C. Reijneveld, E. Aronica, and J. J. Heimans, "Epilepsy in patients with a brain tumour: focal epilepsy requires focused treatment," *Brain*, vol. 135, no. 4, pp. 1002–1016, 2012.

[4] R. Gudipati, R. N. Anreddy, and S. Manda, "Synthesis, anticancer and antioxidant activities of some novel N-(benzo[d] oxa zol-2-yl)-2-(7- or 5-substituted-2-oxoindolin-3-ylidene) hydrazinecarboxamide derivatives," *Journal of Enzyme Inhibition and Medicinal Chemistry*, vol. 26, no. 6, pp. 813–818, 2011.

[5] M. K. Biyala, N. Fahmi, and R. V. Singh, "Antifertility and antimicrobial activities of palladium and platinum complexes of 6-nitro-3-(indolin-2-one) hydrazine carbothioamide and 6-nitro-3-(indolin-2-one)hydrazinecarboxamide," *Indian Journal of Chemistry—Section A: Inorganic, Physical, Theoretical and Analytical Chemistry*, vol. 45, no. 9, pp. 1999–2005, 2006.

[6] S. J. Gilani, S. A. Khan, N. Siddiqui, S. P. Verma, P. Mullick, and O. Alam, "Synthesis and in vitro antimicrobial activity of novel *N*-(6-chlorobenzo[*d*] thiazol-2-yl) hydrazine carboxamide derivatives of benzothiazole class," *Journal of Enzyme Inhibition and Medicinal Chemistry*, vol. 26, no. 3, pp. 332–340, 2011.

[7] D.-S. Guo, "Synthesis and antibacterial activities of 2-benzoyl-*N*-aryl hydrazinecarboxamides," *Chinese Journal of Organic Chemistry*, vol. 24, no. 9, pp. 1118–1121, 2004.

[8] O. Alam, P. Mullick, S. P. Verma et al., "Synthesis, anticonvulsant and toxicity screening of newer pyrimidine semicarbazone derivatives," *European Journal of Medicinal Chemistry*, vol. 45, no. 6, pp. 2467–2472, 2010.

[9] M. I. Attia, M. N. Aboul-Enein, A. A. El-Azzouny, Y. A. Maklad, and H. A. Ghabbour, "Anticonvulsant potential of certain new (2*E*)-2-[1-Aryl-3-(1*H* -imidazol-1-yl)propylidene]-*N* -(aryl/*H*)hydrazinecarboxamides," *The Scientific World Journal*, vol. 2014, Article ID 357403, 9 pages, 2014.

[10] G. S. Badu, N. Rajani, P. S. Malathy, B. Srinivas, U. Kulandaivelu, and J. V. Rao, "Synthesis, characterization and evaluation of novel *N*-(1*H*-benzimidazol-2-yl)-2-isatinylidene-hydrazinecarboxamide derivatives as anti-inflammatory agents," *Der Pharma Chemica*, vol. 2, no. 3, pp. 196–204, 2010.

[11] M. I. Attia, H. A. Ghabbour, H. W. Darwish, and H.-K. Fun, "Crystal structure of (2*E*)-2-[3-(1*H*-imidazol-1-yl)-1-phenylpropylidene]-*N*-(4-methylphenyl)hydrazinecarboxamide, $C_{20}H_{21}N_5O$," *Zeitschrift fur Kristallographie—New Crystal Structures*, vol. 229, no. 4, pp. 311–312, 2014.

[12] M. N. Aboul-Enein, A. A. El-Azzouny, M. I. Attia et al., "Design and synthesis of novel stiripentol analogues as potential anticonvulsants," *European Journal of Medicinal Chemistry*, vol. 47, no. 1, pp. 360–369, 2012.

[13] M. N. Aboul-Enein, A. A. E.-S. El-Azzouny, M. I. Attia, O. A. Saleh, and A. L. Kansoh, "Synthesis and anti-*Candida* potential of certain novel 1-[(3-substituted-3-phenyl)propyl]-1*H*-imidazoles," *Archiv der Pharmazie*, vol. 344, no. 12, pp. 794–801, 2011.

[14] M. J. Frisch, G. W. Trucks, H. B. Schlegel et al., *Gaussian 09, Revision A.02*, Gaussian, Inc, Wallingford, Conn, USA, 2009.

[15] A. D. Becke, "Density–functional thermochemistry. III. The role of exact exchange," *The Journal of Chemical Physics*, vol. 98, no. 7, pp. 5648–5652, 1993.

[16] A. D. Becke, "Density-functional exchange-energy approximation with correct asymptotic behavior," *Physical Review A*, vol. 38, no. 6, pp. 3098–3100, 1988.

[17] C. Lee, W. Yang, and R. G. Parr, "Development of the Colle-Salvetti correlation-energy formula into a functional of the electron density," *Physical Review B*, vol. 37, no. 2, pp. 785–789, 1988.

[18] A. P. Scott and L. Radom, "Harmonic vibrational frequencies: an evaluation of Hartree-Fock, Møller-Plesset, quadratic configuration interaction, density functional theory, and semiempirical scale factors," *Journal of Physical Chemistry*, vol. 100, no. 41, pp. 16502–16513, 1996.

[19] J. P. Merrick, D. Moran, and L. Radom, "An evaluation of harmonic vibrational frequency scale factors," *Journal of Physical Chemistry A*, vol. 111, no. 45, pp. 11683–11700, 2007.

[20] T. Engel, G. Drobny, and P. J. Reid, *Physical Chemistry for the Life Sciences*, Pearson Prentice Hall, New York, NY, USA, 2008.

[21] R. A. Klein, "Modified van der Waals atomic radii for hydrogen bonding based on electron density topology," *Chemical Physics Letters*, vol. 425, no. 1–3, pp. 128–133, 2006.

[22] E. D. Glendening, A. E. Reed, J. E. Carpenter, and F. Weinhold, "NBO 3.1," in *Theoretical Chemistry Institute and Department of Chemistry*, University of Wisconsin, Madison, Wis, USA, 1998.

[23] A. E. Reed, L. A. Curtiss, and F. Weinhold, "Intermolecular interactions from a natural bond orbital, donor-acceptor viewpoint," *Chemical Reviews*, vol. 88, no. 6, pp. 899–926, 1988.

[24] J. P. Foster and F. Weinhold, "Natural hybrid orbitals," *Journal of the American Chemical Society*, vol. 102, no. 24, pp. 7211–7218, 1980.

[25] F. Weinhold and C. R. Landis, *Valency and Bonding: A Natural Bond Orbital Donor-Acceptor Perspective*, Cambridge University Press, New York, NY, USA, 2005.

[26] M. H. Jamroz, "Vibrational energy distribution analysis VEDA 4," Warsaw, 2004, http://smmg.pl/software/sowtware-veda.html.

[27] G. Socrates, *Infrared Characteristic Group Frequencies*, John Wiley & Sons, New York, NY, USA, 1980.

[28] B. C. Smith, *Infrared Spectral Interpretation: A Systematic Approach*, CRC Press, New York, NY, USA, 1999.

[29] N. B. Colthup, L. H. Daly, and S. E. Wiberley, *Introduction to Infrared and Raman Spectroscopy*, Academic Press, New York, NY, USA, 1990.

[30] G. Varsanyi, *Vibrational Spectra of Benzene Derivatives*, Academic Press, New York, NY, USA, 1969.

[31] A. E. Ledesma, S. A. Brandán, J. Zinczuk, O. E. Piro, J. J. L. González, and A. B. Altabef, "Structural and vibrational study of 2-(2′-furyl)-1*H*-imidazole," *Journal of Physical Organic Chemistry*, vol. 21, no. 12, pp. 1086–1097, 2008.

[32] M. Govindarajan, A. S. Abdelhameed, A. Al-Saadi, and M. Attia, "Experimental and theoretical studies of the vibrational and electronic properties of (2*E*)-2-[3-(1*H*-imidazol-1-yl)-1-phenylpropylidene]-*N*-phenylhydrazinecarboxamide: an anticonvulsant agent," *Applied Sciences*, vol. 5, no. 4, pp. 955–972, 2015.

[33] D. N. Sathyanarayana and D. Nicholls, "Vibrational spectra of transition metal complexes of hydrazine normal coordinate analyses of hydrazine and hydrazine-d_4," *Spectrochimica Acta Part A: Molecular Spectroscopy*, vol. 34, no. 3, pp. 263–267, 1978.

[34] L. J. Bellamy, *The Infrared Spectra of Complex Molecules*, Chapman & Hall, London, UK, 1975.

[35] I. Fleming, *Molecular Orbitals and Organic Chemical Reactions*, John Wiley & Sons, Chichester, UK, 2010.

[36] A. E. Reed, R. B. Weinstock, and F. Weinhold, "Natural population analysis," *The Journal of Chemical Physics*, vol. 83, no. 2, pp. 735–746, 1985.

[37] R. Recacha, M. J. Costanzo, B. E. Maryanoff, and D. Chattopadhyay, "Crystal structure of human carbonic anhydrase II complexed with an anti-convulsant sugar sulphamate," *Biochemical Journal*, vol. 361, no. 3, pp. 437–441, 2002.

[38] H. M. Berman, J. Westbrook, Z. Feng et al., "The protein data bank," *Nucleic Acids Research*, vol. 28, no. 1, pp. 235–242, 2000.

[39] G. M. Morris, H. Ruth, W. Lindstrom et al., "Software news and updates AutoDock4 and AutoDockTools4: automated docking with selective receptor flexibility," *Journal of Computational Chemistry*, vol. 30, no. 16, pp. 2785–2791, 2009.

15

Structural Characterization of Lignin and Its Degradation Products with Spectroscopic Methods

Yao Lu,[1,2,3] Yong-Chao Lu,[4] Hong-Qin Hu,[3] Feng-Jin Xie,[3] Xian-Yong Wei,[1,3] and Xing Fan[1,3]

[1]*Key Laboratory of Coal Processing and Efficient Utilization, Ministry of Education, China University of Mining & Technology, Xuzhou 221116, China*
[2]*Advanced Analysis & Computation Center, China University of Mining & Technology, Xuzhou 221116, China*
[3]*School of Chemical Engineering and Technology, China University of Mining & Technology, Xuzhou 221116, China*
[4]*School of Basic Education Sciences, Xuzhou Medical University, Xuzhou 221004, China*

Correspondence should be addressed to Xing Fan; fanxing@cumt.edu.cn

Academic Editor: Javier Garcia-Guinea

Lignin is highly branched phenolic polymer and accounts 15–30% by weight of lignocellulosic biomass (LCBM). The acceptable molecular structure of lignin is composed with three main constituents linked by different linkages. However, the structure of lignin varies significantly according to the type of LCBM, and the composition of lignin strongly depends on the degradation process. Thus, the elucidation of structural features of lignin is important for the utilization of lignin in high efficient ways. Up to date, degradation of lignin with destructive methods is the main path for the analysis of molecular structure of lignin. Spectroscopic techniques can provide qualitative and quantitative information on functional groups and linkages of constituents in lignin as well as the degradation products. In this review, recent progresses on lignin degradation were presented and compared. Various spectroscopic methods, such as ultraviolet spectroscopy, Fourier-transformed infrared spectroscopy, Raman spectroscopy, and nuclear magnetic resonance (NMR) spectroscopy, for the characterization of structural and compositional features of lignin were summarized. Various NMR techniques, such as ^{1}H, ^{13}C, ^{19}F, and ^{31}P, as well as 2D NMR, were highlighted for the comprehensive investigation of lignin structure. Quantitative ^{13}C NMR and various 2D NMR techniques provide both qualitative and quantitative results on the detailed lignin structure and composition produced from various processes which proved to be ideal methods in practice.

1. Introduction

The main components of lignocellulosic biomass (LCBM) are cellulose, hemicellulose, and lignin. Cellulose is a polymer of glucose, accounting for 30–50 wt% of dry LCBM; hemicellulose is a mixture of heteropolymers containing various polysaccharides, such as xylan, glucuronoxylan, and glucomannan, accounting for 20–35 wt%; the mainly remaining portion with 15–30 wt% is lignin, which is a multisubstituted phenolic polymer. Lignin is the most abundant aromatic biopolymer accounting for up to 30% of the organic carbon on Earth and thus can be treated as a potential renewable feedstock for energy supplement and aromatic chemicals production [1, 2]. The annual production of lignin is more than 70 million tons [3]. The most abundant industrial lignins are produced from kraft and sulfite pulping processes in the pulp and paper industries, so-called black liquor. However, only less than 2% of the lignin produced from pulping industries was value-addedly utilized, while the rest was abandoned or burned as a low-value fuel for energy supplement [4], leading to serious waste of precious aromatic resource and environmental pollution.

Lignin is an amorphous, irregular three-dimensional, and highly branched phenolic polymer. The functions of lignin in the plant cell wall are to cover structural support, transport water and nutrients, and issue protection to prevent chemical or biological attacks, and so forth. Though the chemical structure is extremely complex, it is generally accepted that lignin is formed via irregular biosynthesis process constructed from three basic phenylpropanoid monomers,

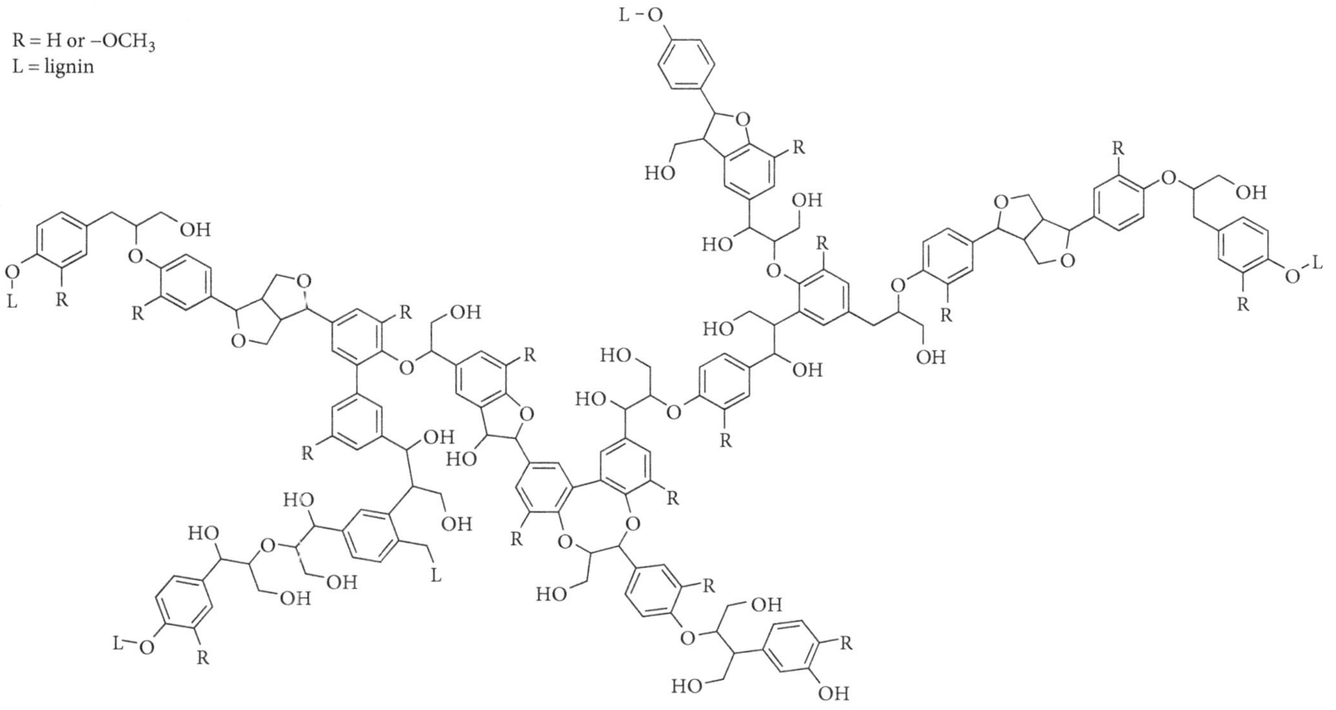

R = H or $-OCH_3$
L = lignin

Figure 1

p-hydroxyphenyl (H), guaiacyl (G), and syringyl (S) units, derived from p-coumaryl, coniferyl, and sinapyl alcoholic precursors, respectively. Figure 1 shows a typical structural model of lignin. Gymnosperms contain almost entirely G unit in lignins; dicotyledonous angiosperms contain G and S units in lignins; and all the G, S, and H units can be found in monocotyledonous lignins [5]. Other units with relatively fewer contents were also identified in the lignin of LCBM, such as ferulates and coumarates [6]. Biosynthesis of lignin is a process that monomers undergo radical coupling reactions to form racemic, cross-linked, and phenolic polymer, by which lignin content and composition may vary significantly in different LCBMs [7]. Furthermore, the structure of lignin even varies among different tissues and ages of the same individual of LCBM [8].

Typical lignin contents are 24–33% in softwoods, 19–28% in hardwoods, and 15–25% in grasses, respectively. Functional groups in lignin include methoxyl, carbonyl, carboxyl, and hydroxyl linking to aromatic or aliphatic moieties, with various amounts and proportions, leading to different compositions and structures of lignin [9]. Various linkages (see Figure 2) either in C-C or C-O type with different abundances formed in the coupling reactions involved in biosynthesis of lignin, including aryglycerol-β-ether dimer (β-O-4, 45–50%), biphenyl/dibenzodioxocin (5–5$'$, 18–25%), pino/resinol (β-5, 9–12%), diphenylethane (β-1, 7–10%), aryglycerol-α-ether dimer (α-O-4, 6–8%), phenylcoumaran (β-β', 0–3%), siaryl ether (4-O-5, 4–8%), and spirodienon.

It is difficult to draw accurate structural diagram for entire lignin by using up-to-date techniques in situ. Although relative new methods for imaging and analyzing chemical structure of lignin, such as confocal Raman scattering microscopy [10] and time-of-flight secondary ion mass spectrometry [11, 12], can provide chemical and spectral imaging of lignin for the distribution of componential units with high resolution and sensitive, these techniques are only available in several biological labs and have not been employed widely by chemical scientific groups. Up to now, the comprehensive elucidation of structural and compositional features of lignin relies on the processes for the degradation and isolation of lignin from LCBM and methods applied in the characterization of the corresponding products [2, 13]. However, in the degradation process, the original structural and compositional features of lignin may be sometimes ambiguous or even missed. Different degradation processes produce different types of lignins with various structures and compositions; furthermore, a specific analytical technique gives partial and/or limited information and is not able to provide a general picture for the entire lignin. The industrial applications of lignin are limited critically due to its complex nature and undefined chemical structure. For example, commercially purchased kraft lignins from softwoods may have different compositions as well as their structures [14]. Furthermore, the lignin-carbohydrate complex (LCC) increases the difficulty of structural analysis and isolation of lignin from LCBM [15]. The value-added utilization of lignin and its degradation products are one of the ultimate goals especially for biorefineries; therefore, the comprehensive understanding of the structure of lignin is crucial necessary, since it can provide theoretical direction on constructing and optimizing degradation processes, generating of valuable aromatic chemicals

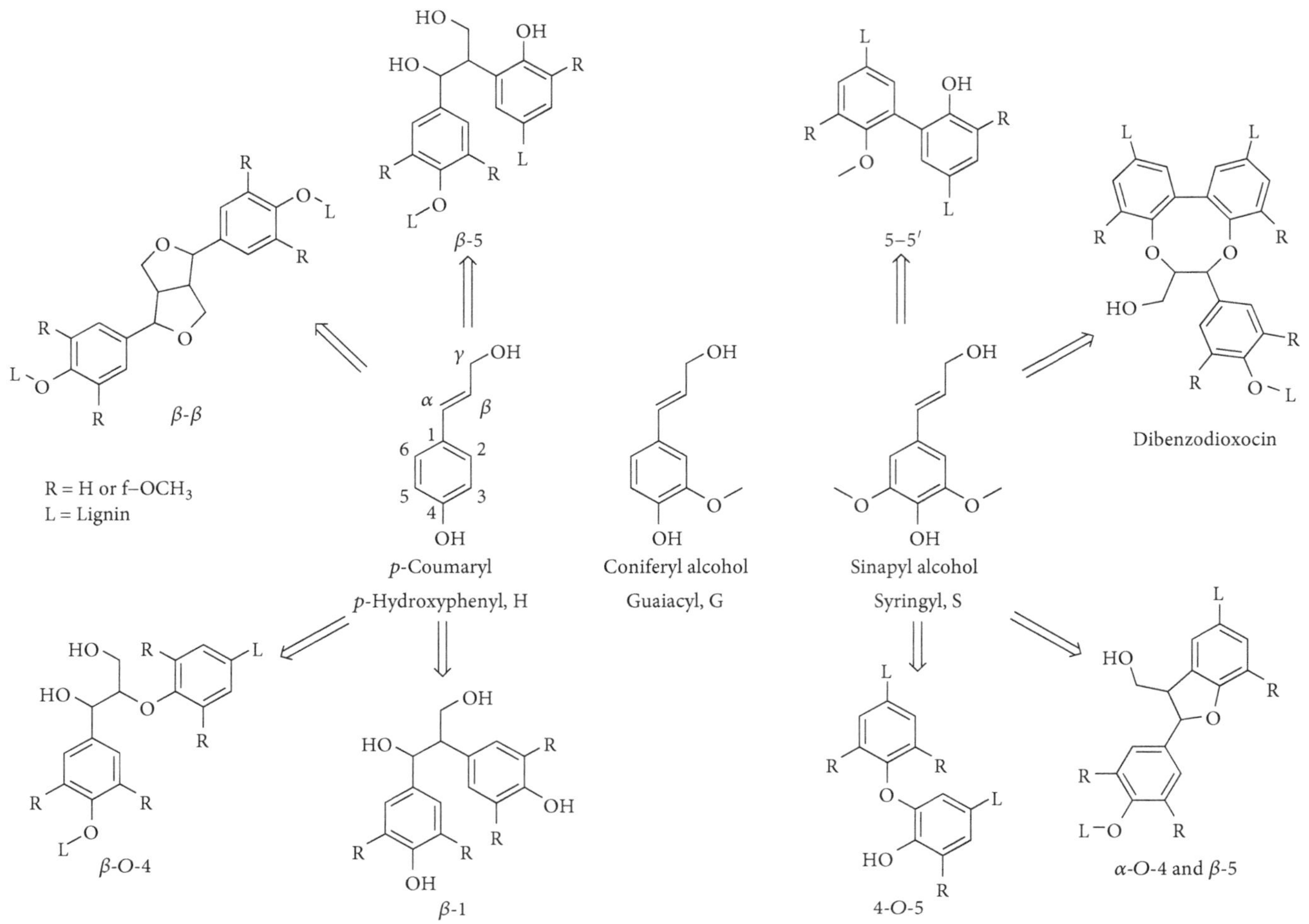

Figure 2

to act as low-molecular-mass feedstocks [16], estimating the economic viability, and so forth.

Traditionally, there are two ways to isolate lignin from other components in LCBM, so-called degradation processes: one is to extract cellulose and hemicellulose leaving most of the lignin as solid residue, and the other one is to extract lignin by using fractionation methods leaving the other components. For the former process, dilute sulfuric acid and hot water are often used to break down cellulose and hemicellulose releasing sugars and facilitating the further enzymatic hydrolysis, while leaving lignin as the main content in solid residue. For the later process, hydroxide solution, either with sodium, potassium or calcium, is used to remove lignin from LCBM samples. The degradation processes are designed to cleave the bonds between lignin and carbohydrates, leading to more or less extensive changes compared to native lignin structure. Consequently, the chemical compositional features of the resulting technical lignins, such as the relative abundance of S/G/H units, the status of side chains, and the contents of functional groups, are highly dependent on the methods and conditions used in degradation processes [17]. The most common linkages in lignin, namely, β-O-4 linkages, are relatively weak linkages and are the key target of most degradation pretreatments. Other linkages, such as β-5, β-1, β-β', 5–5′, and 4-O-5, are more complicated and difficult to be degraded. Toward the structural investigation, various lignins are produced via different degradation processes, such as milled wood lignin (MWL), acidic lignin, sulfite lignin, soda lignin, kraft lignin, organosolv lignin, cellulolytic enzyme lignin (CEL), enzymatic mild acidolysis lignin (EMAL), and lignin from thioacidolysis process [2]. In the recent years, extraction and depolymerization with ionic liquids (ILs) for the isolation or degradation of lignins were considered to be promising processes [18, 19].

For the structural and compositional elucidation of complex samples, various instrumental methods were used. For example, the chromatographic techniques coupled to mass spectrometers and high-resolution mass spectrometric techniques were used extensively in the analysis of the bio-oil, biomass, and lignin samples [20–24]. These methods concentrate on the detection of individual species basing on the chromatographic separation and high molecular resolution. However, on the other hand, spectroscopic methods, such as ultraviolet spectroscopy (UV), Fourier-transformed infrared spectroscopy (FTIR), and nuclear magnetic resonance

(NMR), concern about the analysis of the whole structure and direct detection of moieties in samples over degradation techniques [25, 26]. Detailed spectrometric information related to structural features, including functional groups, bond types, and chemical state of atoms, can be obtained. Furthermore, both of the qualitative and quantitative analyses can be carried out simultaneously.

In this review, we focused on the recent development and interesting findings on the structural investigation of lignin with spectroscopic methods over various degradation processes. Structural and compositional characters of lignin samples produced from different degradation processes were presented and compared, and developments of spectroscopic methodologies on the qualitative and quantitative elucidation of lignin structure were also summarized. The degradation processes and instrumental methods involved in the detailed and comprehensive understanding of the lignin structure were prospected.

2. Degradation Processes of Lignin

Various physical/chemical methods were carried out for the degradation and isolation of lignin. Optimization or modification of these methods was conducted on various LCBMs due to the difference in the structure of lignin. In order to facilitate further structural and/or compositional analyses or to produce high purity lignin, modified or multistep processes were usually carried out.

2.1. Milled Wood Lignin. MWL is produced via the extraction of milled sample particles from LCBM with a neutral organic solvent (e.g., 1,4-dioxane) under mild conditions to remove other components. In the extraction process, only minor changes may occur with respect to the milled sample; hence, the obtained lignin has similar property with the milled sample. Nevertheless, MWL is not considered to be a representative of the original lignin in the LCBM due to its relative low yield (based on Klason lignin).

2.2. Cellulolytic Enzyme Lignin. In order to improve the yield, CEL was developed from the extraction of enzymatically hydrolyzed MWL residue. Typically, the residual carbohydrate contents in CEL account 10–15 wt% of initial MWL sample. The structure of CEL is similar to MWL, and it is more representative of total lignin in LCBM than in MWL. CEL has commonly been used for the structural analysis of lignin in the cell wall of plants. In a recent study, cellulolytic enzyme hydrolysis was carried out prior to water/dioxane extraction of MWL to remove carbohydrates. The lignin was obtained with high yield and purity [27]. Enzymatic lignin degradation has several advantages such as mild conditions and potentially fewer inhibitors for microbes. However, the degradation of lignin in LCBM still gave a very low yield of fragmented and soluble lignin, which may due to the limitations on efficient electron transfer [28] in the process.

2.3. Sulfite, Soda, and Kraft Lignins. Sulfite, soda, and kraft lignins are by far the main technical lignins produced via industrial processes. Among them, sulfite and kraft methods are sulfur-involving processes, accounting more than 90% of the chemical pulp production worldwide [29], and soda method is sulfur-free process. In the sulfite process, water-soluble lignosulfonates are formed. Further purification is needed to remove unexpected carbohydrate impurities. This process produces the largest amount of technical lignin. However, the obtained lignin contains considerable amount of sulfur. In the soda process, lignin is dissolved in hydroxide solution and following steps including precipitation, maturation, and filtration. In the kraft process, LCBM particles are emerged in an aqueous solution containing NaOH and Na_2S. Lignins are depolymerized as water-/alkali-soluble fragments with approximately 70–75% of the hydroxyl groups become sulfonated. Industrially, kraft lignin, produced chemically from the lignin degradation in aqueous alkali, is the major constituent of black liquor (90–95%). Neither kraft lignin nor sulfite lignin is suitable for investigating the original native structures of lignins, because significant structural changes occur especially the cleavage of α-O-4 and β-O-4 linkages under the conditions of these processes. Additionally, undesirable impurities such as sulfurous compounds or carbohydrates are present in derived lignin for these fractionation processes. Currently, almost all the produced technical lignins are only high yield industrial by-products and recovered as low-value fuel. This dilemma may rely on the progress in structural characterization of lignins from various LCBMs and the further upgrading of the technical lignins targeting value-added chemical production.

2.4. Organosolv Lignin. In the organosolv process, high purity lignin and cellulose are produced at the same time with various solvents; however, no technical lignins are commercially available from this process up to now. Organosolv process typically results in more than 50% lignin removal from LCBM through cleavage of lignin-carbohydrate bonds and β-O-4 linkages. The separation of organosolv lignin can be achieved either by removing of the solvent or by precipitation with water followed by distillation. Most organosolv lignin is easily soluble in basic solutions and polar solvents, that is, ethanol or ethanol/water mixture, but will be insoluble in acidic aqueous solutions. Organosolv lignin is sulfur-free, high purity, and rich in functionality including phenolics, exhibits a narrow polydispersity, and has limited carbohydrate contamination.

The extraction conditions affect the structure of organosolv lignin, that is, severity factor (H-factor). The molecular weight of the ethanol organosolv lignin decreased within a 36–56% range with respect to the MWL with the increase of the severity. Moreover, an obvious decrease in the content of aliphatic hydroxyl groups and an increase of syringyl phenolic units and condensed phenolic structures with the increase in severity of the organosolv treatment were also observed [30]. An integrated process of hot water extraction followed by high-boiling-solvent cooking with 1,4-butanediol can fractionate bagasse vigorously into cellulose, hemicelluloses, and lignin. The organosolv lignin formed exhibited a chemical structure similar to EMAL with more newly formed phenolic OH groups [31].

2.5. Acidic Lignin. Traditionally, in the acidolysis process, lignin is extracted from LCBM sample with 1,4-dioxane containing hydrochloric acid under room temperature. The obtained lignin with high purity is considered to be a representative of the original lignin. However, a limitation of this process is that the same conditions used to hydrolyze polysaccharides also degrade the liberated monosaccharides, leading to overestimate monosaccharide degradation and introducing bias between polysaccharides of different liability. Modifications were introduced to reduce these errors [32]. A modified acidolysis process was carried out by Gong et al. [33]. The acetic acid lignin from bamboo shoot shell had a higher yield of lignin (74 wt%) and lower content of associated carbohydrates (2.96 wt%) than MWL (5.16 wt%). Additionally, acetic acid lignin possessed a molecular weight 2789 Da and a narrow polydispersity index (i.e., $M_w/M_n = 1.54$). Higher phenolic hydroxyl group content and S/G ratio were also obtained in this lignin compared to MWL [33]. Enzymatic mild acidolysis lignin (EMAL) is obtained from acidolysis of CEL with dilute acid, such as hydrochloric acid. The remaining carbohydrates linking to lignin can be removed further in the acidolysis producing lignin with higher purity [34].

2.6. Thioacidic Lignin. Modified acidolysis processes were carried out to produce lignin with high yield and purity. Thioacidolysis process, in which ethanethiol is used instead of water, produced more lignin and less complex monomer mixtures. In this process, thioethylated H, G, and S monomers by the cleavage of β-O-4 ether linkages are produced. Traditional thioacidolysis methods require several steps before down streaming analysis or further treatments. Hence, higher-throughput quantitative method is needed for screening various types of LCBMs [35].

2.7. Ionic Liquid Degradation Lignin. IL provides an alternative path for lignin removal to classic organosolv pretreatment for enhancing subsequent enzymatic hydrolysis and isolation. Some ILs, such as 1-ethyl-3-methylimidazolium acetate, can extract lignin from poplar and birch with most structural features retained [36]. Some acidic ILs, such as 1-*H*-3-methylimidazolium chloride, will hydrolyze ether linkages [37] and further degrade lignin. The following are some recent progresses concentrated on the lignin degradation and isolation by ILs.

The ILs containing 1-butyl-3-methylimidazolium (bmim), 1-ethyl-3-methylimidazolium (emim), and 1-allyl-3-methylimidazolium (amim) cations either with acetate or chloride as the anions are commonly used in the lignin dissolution [38]. ILs have the capability to disrupt various linkages between the components in the LCBM by the formation of several types of interactions such as hydrogen bond, dipole-dipole, and van der Waals interactions [39]. Pyridinium formate (PyFor) showed a high capacity for the dissolution of kraft lignin (70 *w/w*%) at a relatively lower temperature (75°C) [40].

Cholinium ILs are novel bio-ILs used in the lignin valorization, in which different chemical reactions take place during the lignin dissolution from imidazolium ILs [41]. In the dissolution of kraft lignin in cholinium ILs, significant changes in the structure and thermal properties of kraft lignin occurred via depolymerization, dehydration, and demethoxylation followed by recondensation. Thermal properties of kraft lignin were altered, that is, increased the maximal decomposition temperature (T_m) and glass transition temperature (T_g); and the molecular weights were reduced after regeneration from cholinium ILs [41].

Other ILs, such as 1-ethyl-3-methylimidazolium xylenesulfonate [emim][ABS] and 1-butyl-3-methylimidazolium methylsulfate [bmim][$MeSO_4$], could promote depolymerization of organosolv lignin and Klason lignin under the oxidative conditions using a Cu/EDTA complex in the presence of a monomeric phenol (4-tert-butyl-2,6-dimethylphenol) [42].

An acidic IL, called 1-(4-sulfobutyl)-3-methyl imidazolium hydrosulfate ([$C_4H_8SO_3$Hmim]HSO_4), was proven to be an efficient catalyst for direct liquefaction of bagasse lignin, where more than 65% degree of liquefaction and 13.5% yield of phenolic monomer without any char formation [43].

A switchable ionic liquid (SIL), synthesized from 1,8-diazabicyclo[5.4.0]undec-7-ene (DBU), monoethanol amine (MEA), and CO_2, named CO_2-switched [DBU][MEASIL], was demonstrated to have high ability to extract the interlinked polysaccharide impurities from the sodium lignosulfonate while the linkages and aromatic subunits remain unaffected during the dissolution-recovery cycle. This SIL can be used as an affordable solvent medium to obtain carbohydrate-free lignin from an impure lignin source [44].

Future developments on the IL degradation of lignin will focus on selective lignin extraction/degradation and functionalization as well as minimization of process costs for recovery and recycling of ILs.

2.8. Multistep Processes. Multistep processes were used to enhance the removal of lignin. A two-step process was carried out in which anhydrous ammonia pretreatment was followed by mild NaOH extraction on corn stover to solubilize and fractionate lignin [45]. Lignin removal of more than 65% with over 84% carbohydrate retention was achieved. Furthermore, a significant reduction in the weight-average molecular weight (M_w) of extracted lignin was also achieved. Synergistic effects were found in the combination of pretreatments to enhance the isolation or conversion of lignins [28, 46]. In the sequential fractionation of *Tamarix spp.*, MWL, organosolv lignin, and alkaline lignin were conducted with dioxane, alkaline organosolv, and alkaline solutions, respectively. The results indicated that the alkaline organosolv extraction released a higher yield of lignin (17.7%) than dioxane and alkaline solution extractions. Small amounts of carbohydrates (0.79%) were detected in the organosolv lignin fraction, suggesting a significant cleavage of α-ether bonds between lignin and carbohydrates in the alkaline organosolv fractionation process [47].

2.9. Comparison of the Processes. Alkaline lignins were found to have higher carbohydrate content (up to 30 wt%) with higher molecular weights around 3000 Da; on the other hand, organosolv lignins had considerable high purity (better than

93 wt%) with molecular weights in the range of 600–1600 Da [48]. The structure and composition of alkali lignin, CEL, and MWL from valonea of *Quercus variabilis* Blume were compared by Yang et al. [49]. The isolation processes of alkali lignin and CEL caused some damages to the structure of lignin. The β-O-4 linkages were largely cleaved during the CEL process since the relative content of β-O-4 linkages in CEL was much lower than those in alkali lignin and MWL. High S/G ratio for alkali lignin was observed, indicating that the S-units were easily released under the alkali conditions.

Yang et al. [50] compared four lignins produced from valonea of *Quercus variabilis*, namely, ethanol lignin, alkali lignin, MWL, and enzyme hydrolysis lignin (EHL). The results showed that the four lignins contained GSH-type with little differences. The MWL contained the least functional groups with the poorest thermostability and the highest antioxidant activity. The EHL had the highest molecular weight (i.e., $M_w = 1429$ g/mol; $M_n = 746.18$ g/mol). In a comparison of pretreatments on hardwood (red oak), softwood (loblolly pine), and herbaceous biomass (corn stover) for lignin valorization through pyrolysis, organosolv lignins contained fewer volatiles in comparison to the corresponding MWLs for all the tested samples [51]. Red oak lignin was affected mostly by the organosolv process, since the greatest decrease in volatile content and increase in carbon content were observed. Corn stover lignin had the highest potential for volatilization because it retained highly branched polymer structure enriched in tricin, ferulate, and coumarate groups.

Clearly, different degradation processes or pretreatments have significant influence on the compositional and structural features of lignin. The selectivity and efficiency of these processes are the main consideration. To elucidate the original structure of lignin, relatively undestroyed and effective degradation methods are feasible, such as IL extraction and organosolv process. To produce value-added chemical from lignin, more aggressive methods aiming at the cleavage of the weak linkages in lignin (i.e., β-O-4 linkages) and the interunits between lignin and polysaccharides can be used in the degradation process. Of course, biological conversion with suitable selectivity might be another orientation for degradation of lignin [52].

3. Spectroscopic Methods

Structural investigation of lignin with spectroscopic techniques has been considered to be promising high-throughput and routine methods, which can provide detailed qualitative and quantitative information on structural features including functional groups, types of chemical bonds, and states of atoms.

3.1. UV Spectroscopy. The content of acid-soluble lignin, the purity, and the components of isolated lignin, can be determined by using UV spectroscopy [53]. National Renewable Energy Laboratory (NREL) proposed an accurate method for the determination of lignin, by which the absorbance of lignin was recorded at the recommended wavelength [54]. According to the intrinsic structure of lignin, several absorption maxima attributed to different functional groups were

TABLE 1: UV spectroscopic absorptions of typical structures in lignin [25, 30, 50, 53–56].

Absorption maxima/nm	Electronic transition style	Chromophores and structures
200	π-π^*	Conjugated bonds/aromatic ring
240	n-π^*	Free -OH
282	π-π^*	Conjugated bonds/aromatic ring
320	π-π^*	Aromatic ring conjugated bond with C = C
320	n-π^*	C = O groups conjugated to aromatic ring
325	n-π^*	Etherified ferulic acid

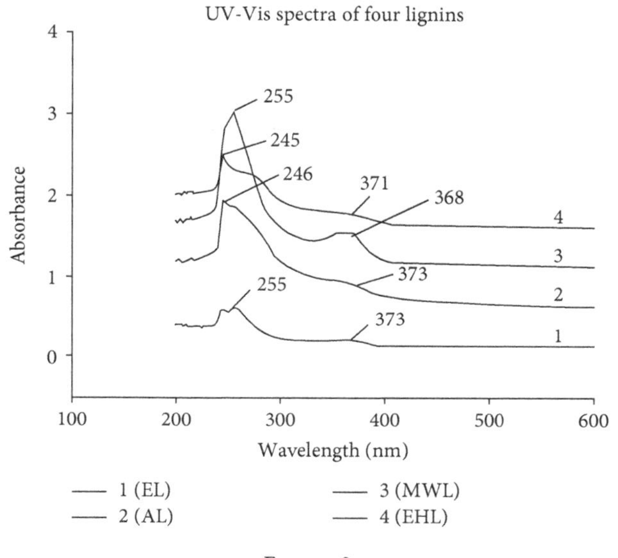

FIGURE 3

observed, as shown in Table 1 [25, 30, 50, 53–56]. The determination of phenolic hydroxyl groups can be achieved basing on the difference in absorption at 292 and 370 nm between phenolic units in neutral and alkaline solutions [30, 50, 55, 56]. Attributed to the symmetrical syringyl unit, the maximum absorbance of lignins produced from different processes exhibited a blue shift. Furthermore, an additional absorbance at approximately 370 nm due to the presence of conjugated phenolic hydroxyl groups was also observed [50] (see Figure 3).

Basing on the Lambert-Beer's Law, UV spectroscopy can be used for the semiquantitative determination of the purity of lignin and its degradation products by using extinction coefficient (EC) [57, 58]. Because of the cross-linking structures of lignin with carbohydrates, cellulose, and hemicelluloses, the isolation of pure lignin is extremely difficult. The low value of EC represents the high nonlignin substance content in the isolated lignin.

3.2. FTIR Spectroscopy. FTIR spectroscopy is the most widely used technique in the functional group determination basing

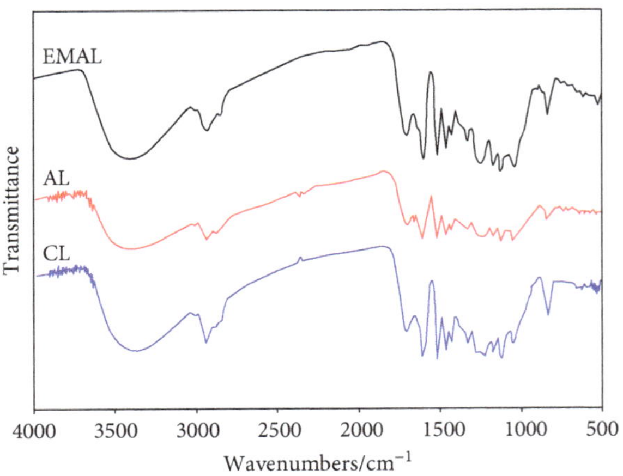

FIGURE 4

on the substances with chromophores. It can be treated as a nondestructive, noninvasive, highly sensitive, and rapid technique. Typical functional groups contained in lignin, such as hydroxyl, carbonyl, methoxyl, carboxyl, and aromatic and aliphatic C-H, can be assigned well in the FTIR spectrum. Figure 4 shows FTIR spectra for detection of different lignins, namely, EMAL, autocooking lignin (AL), and citric acid-catalyzed cooking lignin (CL) [31]. For the assignments of signals in FTIR spectra, Table 2 lists the typical wavelength assigned for possible functional groups and structures in lignin [25, 29, 38].

Attenuated total reflectance- (ATR-) FTIR could be used for the evaluation of kraft lignin in acylation with different acyl chlorides [59] and lignin structural changes during the cooking process with solid alkali and active oxygen [60]. FTIR spectroscopy could also characterize changes in the chemical structure of wood polymers in relation to the tree growth location and conditions [61]. Untreated solid samples (Norway spruce, *P. abies L. Karst.*) from three provenances in Europe were selected. Principal component analysis (PCA) and cluster analysis (CA) were used for evaluation of spectral data obtained by FTIR spectroscopy. The results showed that the samples belonging to the same wood species differ due to the origin. FTIR analysis was able to correctly discriminate samples originating from three different provenances in Europe.

It is known that functional properties of oxyethylated lignins (OELs) and the resulting substances are strongly affected by the degree of oxyethylation (DOE) of phenolic hydroxyl groups (OH_{phen}). Passauer et al. [62] found the strong linear correlations between OH_{phen} contents of lignin/OEL and FTIR vibrations attributed to phenolic and aliphatic acetoxy groups. With appropriate calibration, FTIR spectroscopy combined with sample preacetylation is considered to be a promising tool for rapid and accurate determination of the DOE of OELs with qualitative and quantitative results.

3.3. Raman Spectroscopy. Raman spectroscopy, as the sister spectroscopic technique of FTIR, can provide complementary information on the structural features even for the samples containing water. Furthermore, more absorption bands were detected with Raman spectroscopy than FTIR [63]. Generally, the assignments of the absorption bands in Raman spectra are similar with FTIR spectra.

Raman spectroscopy is suitable for the investigation of the chemical structure of lignin, because it can provide in situ determination on the cell wall of plants even with no sample preparation. However, when analyzing a lignin sample in solutions with various solvents, one should consider the environmental effects of the solvents [64]. Confocal Raman microscopy was used to investigate the structural changes of lignocellulosic cell walls during the dilute acid pretreatments. According to the intensity of the Raman images, the ratio of lignin/cellulose [I(1600 cm^{-1})/I(900 cm^{-1})] was low for oxalic acid-pretreated biomass compared to sulfuric acid-pretreated biomass [65].

3.4. NMR Spectroscopy. NMR spectroscopy provides more precise and comprehensive information on qualitative and quantitative assays for the frequencies of linkages and the composition of H/G/S units in the lignin analysis. The first discovery of dibenzodioxocine and spirodienone structures in lignin was carried out by Ralph et al. [66] and Zhang et al. [67], respectively. ^{1}H, ^{13}C, ^{19}F, and ^{31}P as well as various 2D NMR spectroscopic techniques can be used in the structural and compositional analyses of lignin. Among them, ^{1}H and ^{13}C NMR tend to be the regular tools for the analysis of lignin; and solid-state ^{13}C NMR and 2D heteronuclear single-quantum coherence (HSQC) NMR can provide accurate quantitative results on the functional groups and side chain moieties.

Compared with the spectroscopic methods mentioned above, NMR spectroscopic methods possess much higher resolution and enable a larger amount of information to be obtained. One-dimensional (1D) NMR methods, including ^{1}H, ^{13}C, ^{19}F, and ^{31}P NMR, and two-dimensional (2D) NMR methods, such as 2D HSQC NMR, were applied for the analysis of lignin samples with both solid and liquid states. The distribution of functional groups and amount of linkages and H/G/S units as well as other components in lignin can be qualitatively and quantitatively determined. The chemical shifts of functional groups in the spectra have been established.

^{1}H NMR is the method routinely used in the structural investigation of lignin, because of the simple preparation of samples and fast scanning speed. Almost all the compositional investigations of lignins use ^{1}H NMR for the detection of the chemical environment of proton. In the spectra, the signal observed around 7.5 ppm can be assigned to aromatic protons of H units and the other two chemical shifts around 7.0 ppm and 6.5 ppm are attributed to aromatic protons in G and S units, respectively [68, 69]. The chemical shifts in the range of 6.3–4.0 ppm are assigned to aliphatic protons in the linkages of β-O-4, β-β, and β-5. The signals in the range of 4.0–3.5 ppm are attributed to protons in methoxyl groups. The chemical shifts around 3.10 ppm may be attributed to the protons in anhydroxylose units [31, 70]. Typical peeks are assigned to functional groups in lignin, as shown in Table 3 [25, 31].

TABLE 2: Assignments of signals in FTIR spectrum to functional groups in lignin [25, 29, 38].

Wavenumbers/cm^{-1}	Assignments	Functional groups and structures in lignin
3400–3600	ν (O-H)	Free -OH
3100–3400	ν (O-H)	Associated -OH
2820–2960	ν (C-H)	$-CH_2$, $-CH_3$
2920	ν (C-H)	Carboxylic -OH
2650–2890	ν (C-H)	Methyl group in methoxyl
1771	ν (C=O)	Aromatic
1700–1750	ν (C=O)	Unconjugated ketones, carbonyls, and ester groups
1722	ν (C=O)	Aliphatic
1650–1680	ν (C=O)	Conjugated *p*-substituent carbonyl and carboxyl
1500–1600, 1420–1430	ν (aromatic skeletal)	Benzene ring
1450–1470, 1360–1370	ν (C-H)	$-CH_2$, $-CH_3$
1325–1330, 1230–1235	ν (C-O)	Syringyl ring
1270–1275	ν (C-O)	Guaiacyl ring
1215	ν (C-O)	Ether
1140–1145	ν (C-H)	Guaiacyl
1130	ν (C-H)	Syringyl
1085–1090	ν (C-O)	Secondary alcohol and aliphatic ether
1025–1035	ν (C-O, C-H)	Aromatic ring and primary alcohol
750–860	ν (C-H)	Aromatic ring

TABLE 3: Assignments of signals in ^1H NMR spectrum to typical functional groups in lignin (in CD_3Cl) [25, 31].

Chemical shift/ppm	Assignments
9.7–9.9	Cinamaldehydes and benzaldehydes
6.7–7.1	Aromatic-H in guaiacyl
6.2–6.7	Aromatic-H in syringyl
5.8–6.2	Benzylic OH in β-O-4 and β-1
4.9–5.1	Carbohydrates
3.3-4.0	Methoxyl
3.0–3.1	H_β in β-1
2.2–2.4	Phenolic OH
1.6–2.2	Aliphatic OH

^{13}C NMR can be carried out to overcome the overlapping resonances of some structures in ^1H NMR spectra, providing qualitative and quantitative results with nondestructive detection of solid or solution samples. Although with a higher resolution, it is recommended that relative pure lignin sample is necessary in the ^{13}C NMR analysis, since the unexpected overlapping of spectra was due to the complexity of sample. Typical ^{13}C NMR spectra are shown in Figure 5 [31], and the assignments of signals are presented in Table 4 [31, 38, 49, 71]. By using the data from quantitative ^{13}C NMR, basic parameters which summarizes the main structural characteristics of lignins can be obtained, such as content of β-O-4 structures, degree of condensation, and unit ratio of S/G/H. Radar plots include these parameters and allow a direct classification of different lignins by comparison of the key descriptors [71]. Solid-state ^{13}C NMR analysis is a nondestructive method and not limited by sample insolubility. The cross-polarization/magic angle spinning (CP/MAS) method extensively used NMR technique for elucidating the structure of lignin. The detections take a very short time with high resolution; however, the quantitative analysis of CP/MAS is not sufficient enough [72]. Solid-state ^{13}C NMR is considered to be an advanced method for structural investigation of LCBM at atomic level; however, by using this technique, the structure remains largely unexplored due to the complexity of lignin and the severe spectral crowding of the responding signals [73]. A sensitive hyperpolarization solid-state NMR technique by combining high-field dynamic nuclear polarization (DNP) and MAS was used to improve the resolution of the determination [74]. Furthermore, this technique can provide 2D homonuclear ^{13}C-^{13}C correlation solid-state NMR spectra at natural isotopic abundance, yielding, and an atomic level structural investigation [75, 76]. Most of current lignin content analytical techniques require solo or sequential degradation or dissociation steps, which are time-consuming. By using the solid-state ^{13}C CP/MAS NMR technique with an internal standard (sodium-3-trimethylsilylpropionate, TMSP), a simple yet reliable method was established to analyze content of lignin in various LCBMs without destroying their native structures [77].

Constant et al. [78] carried out the quantification and classification of carbonyls in industrial humins and lignins by ^{19}F NMR. The carbonyl groups were transformed to corresponding hydrazone with 4-(trifluoromethyl)phenylhydrazine before quantification by ^{19}F NMR. By using model compound library, the carbonyl functional groups in Indulin Kraft and Alcell lignins were quantified and classified for the first time.

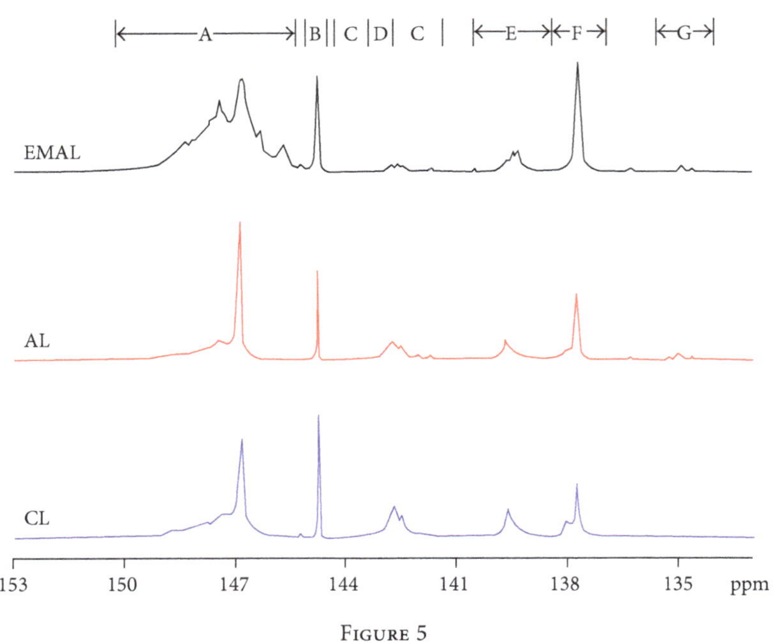

Figure 5

Table 4: Assignments of signals in ^{13}C NMR spectrum to functional groups in lignin [31, 38, 49, 71].

Chemical shift/ppm	Assignments
167–178	Unconjugated -COOH
162–168	Conjugated -COOH
140–155	C_3, C_4 aromatic ether or hydroxyl
127–140	C_1, aromatic C-C
123–127	C_5, aromatic C-C
117–123	C_6, aromatic C-H
114–117	C_5, aromatic C-H
106–114	C_2, aromatic C-H
78–90	Aliphatic C_β-O
67–78	Aliphatic C_α-O
54–57.5	Methoxyl

^{31}P NMR has also been widely used to quantitatively determine the amount of aliphatic and phenolic hydroxyl groups as well as carboxyl groups in lignin after phosphitylation with 2-chloro-4,4,5,5-tetramethyl-1,3,2-dioxaphospholane (TMDP) [29, 79, 80]. The high phenolic OH content reflecting the presence of condensed aromatic units, such as 5–5 units, was found by ^{31}P NMR in a biolignin produced by acetic acid/formic acid/water hydrolysis from wheat straw [81]. The ^{31}P NMR analysis of the insoluble fraction of kraft lignin provided an accurate and quantitative way to illustrate the effects of the laccase-HBT (1-hydroxybenzotriazole) system on lignin chemical bond cleavage [82]. Typical ^{31}P NMR spectra and signal assignments are shown in Figure 6 and Table 5, respectively [29, 82].

Solid-/solution-state ^{13}C NMR spectroscopes are powerful in lignin structural elucidation either in their solid or solution state. However, solid-state ^{13}C NMR spectroscopy is only suitable for the analysis of lignin samples that have restricted solubility and can observe some structural features of lignin due to its low resolution; and lignin is subjected to acetylation by anhydride/pyridine solution before the solution-state ^{13}C NMR spectrum collection [83] since dissolving lignin is difficult.

Various 2D NMR methods were carried out to overcome the overlapping of resonances in 1D NMR with higher resolution and providing more reliability to the assignments of the signals, especially in the determination of lignin [44, 84–93]. 2D NMR methods, such as heteronuclear multiple-quantum coherence (HMQC) spectroscopy, heteronuclear correlation (HETCOR) spectroscopy, homonuclear Hartmann-Hahn (HOHAHA) spectroscopy, total correlation spectroscopy (TOCSY), rotating frame Overhauser experiment spectroscopy (ROESY), heteronuclear single-quantum coherence (HSQC) spectroscopy, and heteronuclear multiple bond coherence (HMBC) spectroscopy, have been employed in lignin structure characterization [44, 84–86]. Among these, 2D HSQC NMR is the most extensively used due to its versatility in illustrating structural features and structural transformations of isolated lignin fractions. Figure 7 presents a typical investigation of lignin with accurate assignments of different structures [71]. 2D HSQC NMR is able to clearly characterize the structures of lignin and polysaccharides in cell walls and the linkages among the lignin without isolating each component [87]. Structural changes of lignin and the other components in LCBM in chemical reaction can be easily monitored by this method. The relationship between the degree of acetylation and the introduction positions of acetyl groups during the acetylation of ground pulp was investigated with 2D HSQC NMR [88]. Acetylation was found to occur firstly on the primary hydroxyl groups of polysaccharides and lignin, followed by the secondary hydroxyl groups of polysaccharides, and finally the hydroxyl groups at the α-position in lignin [88].

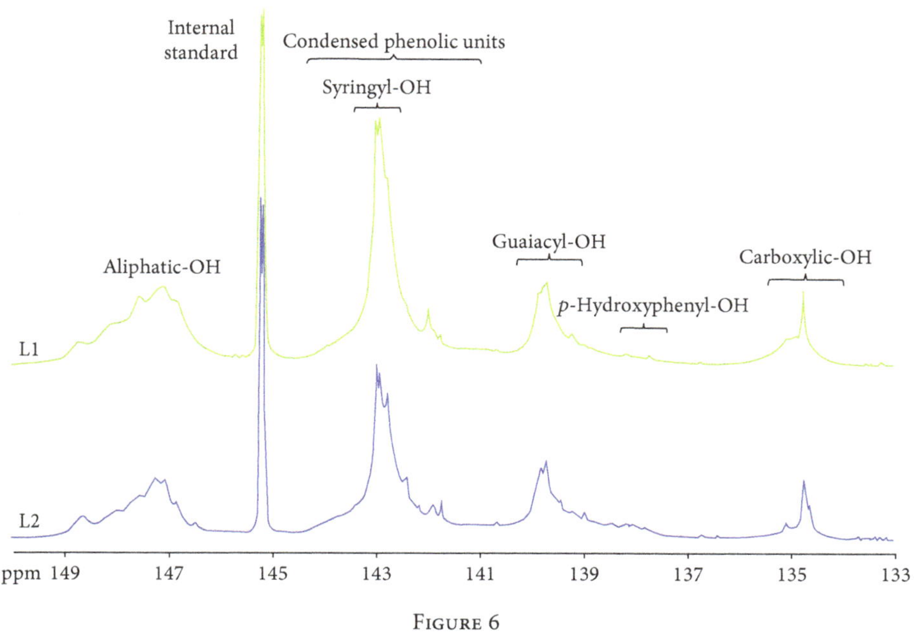

Figure 6

Table 5: Assignments of signals in ^{31}P NMR spectrum to hydroxyl groups in lignin [29, 82].

Chemical shift/ppm	Structural assignments
145.5–150.0	Aliphatic -OH
136.5–144.7	Phenols
140.0–144.5	C_5 substituted
143.5	β-5
142.7	Syringyl
142.3	4-O-5
141.2	5–5
139.0–140.0	Guaiacyl
138.2–139.0	Catechol
137.3–138.2	*p*-Hydroxyphenyl
133.6–136.6	Carboxylic acid -OH

Thioacidolysis was usually used as pretreatment before 2D HSQC NMR analysis in the characterization of the structures of the lignin monomers and oligomers [89]. Changes in the interunit linkage types during solvolysis were investigated. Lignin oligomers ranging from monomers to tetramers were released through considerable cleavage of the β-O-4 linkages [89]. In a study of various lignins derived from brewer's spent grain, 2D HSQC NMR revealed the substructures including β-O-4′ alkyl-aryl ethers (77–79%), β-5′ phenylcoumarans (11–13%), β-β′ resinols (5-6%), and 5–5′ dibenzodioxocins (3–5%); while 2D HMBC NMR and derivatization followed by reductive cleavage analyses showed that *p*-coumarates were acylating at the γ-position of lignin side chains and were mostly occurred in condensed structures [90]. By using high-resolution 2D HSQC NMR, the chemical structures both on low and high molecular weight fractions of bio-oil derived from kraft lignin were determined. In the degradation of kraft lignin to bio-oil, cleavages of both aliphatic carbon-oxygen (C-O) and to some extent carbon-carbon (C-C) bonds as well as repolymerization were observed simultaneously [91].

The combination of quantitative ^{13}C NMR and 2D HSQC NMR has been proven to be a powerful way in structural elucidation of complex samples since it takes advantage of the spectral dispersion afforded by the 2D spectrum to serve as an internal standard to measure the integral values obtained from the quantitative ^{13}C spectrum [92]. This method can overcome the severe overlap of signals and reduce errors in signal quantification due to differential line widths, quantitative abundance of S/G/H units, hydroxycinnamates, and tricin units, as well as various types of side chain substructures by selecting the proper internal standard reference signals [93]. Other combinations of NMR techniques were also reported; for instance, the existence of low energy dipole-dipole interactions and the absence of covalent bond between lignin and chitosan could be revealed clearly by solid-state ^{1}H-^{13}C CP/MAS NMR [94].

4. Conclusions

The comprehensive understanding of the lignin structure relies greatly on the developments of analytical strategies used, which is extremely important for the value-added utilization of biomass. Although significant progresses have been made in the degradation and isolation of the lignin from other components in LCBM, only a fraction of lignin can be identified and analyzed. Structure and composition of lignins from different LCBMs vary significantly according to both issue and age. Furthermore, the analytical results are strongly dependent on the degradation processes and instrumental equipment used.

For the structural investigation of lignin, undestroyed, selective, and efficient isolation methods should be built to

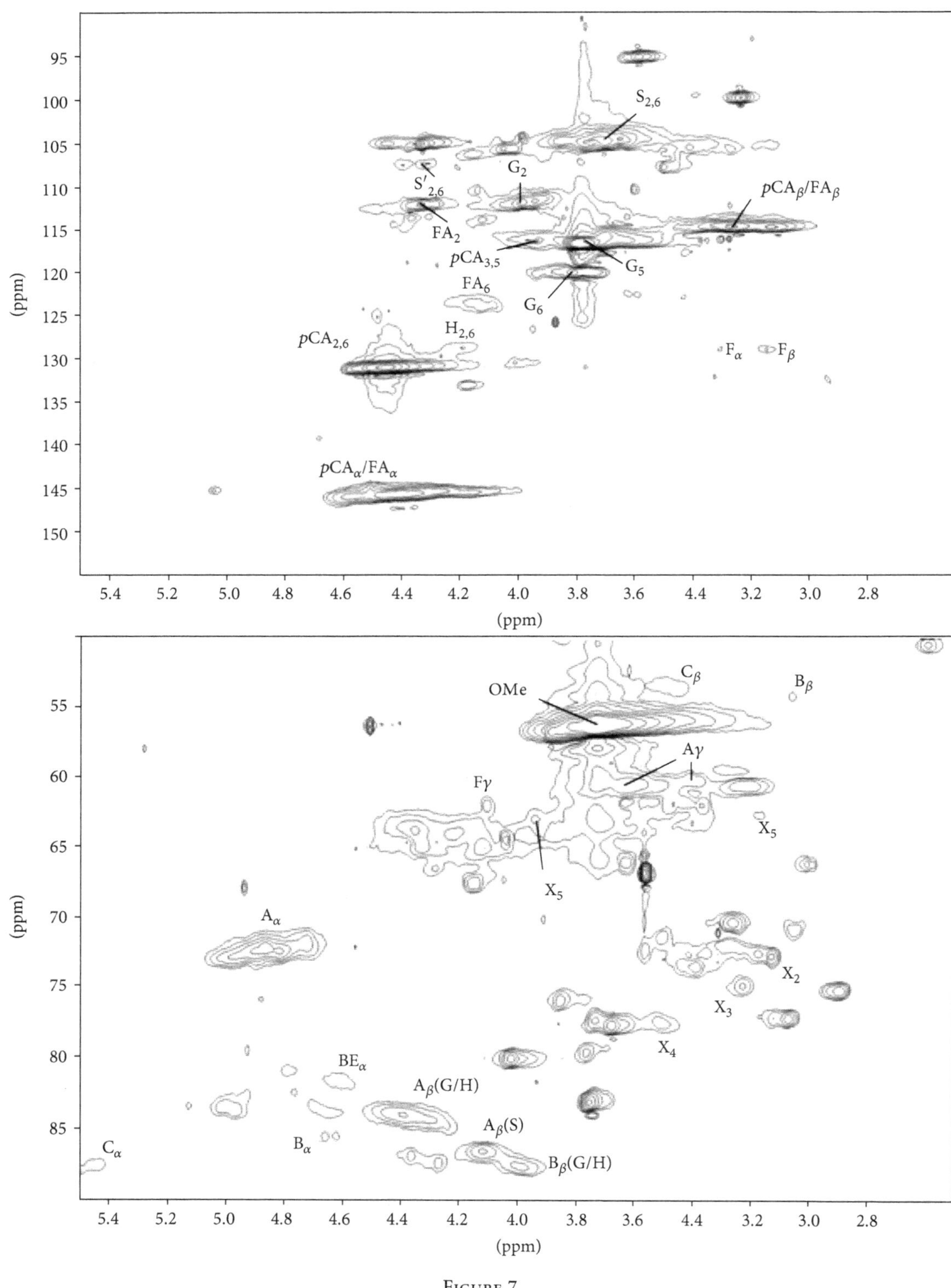

FIGURE 7

preserve the initial structure of lignin and obtain as much sample to be analyzed. Among the wet-chemistry techniques used, IL extraction and organosolv process are the promising methods. They are treated as environmentally friendly methods since relatively mild conditions used and the reagents can be recycled. Biological degradation might be another possible pathway for the oriented isolation of lignin since the outstanding selectivity and rate of conversion.

Various spectroscopic methods are routinely used for the investigation of lignin structures. These methods can provide both qualitative and quantitative information on functional groups and linkages in lignin as well as degradation products of lignin. Among these spectroscopic techniques, UV spectroscopy is less likely to be used since it can provide relatively less information on the structural features of lignin. Generally, FTIR spectroscopy is much more frequently used than Raman spectroscopy. FTIR, ^{1}H NMR, and ^{13}C NMR are commonly used in most of the investigations for the characterization of structure of lignins. Recently, ^{31}P NMR is more adopted in this area. Significant progresses for structural elucidation of lignin rely on the application of quantitative ^{13}C NMR and various 2D NMRs. They are robust techniques by providing detailed qualitative and quantitative results with high resolution and precision and can be treated as ideal methods. Rapid, accurate, and nondestructive spectroscopic techniques can be combined to overcome their individual intrinsic limitations for better elucidation of lignin structure. The data collected from these methods contributes to the understanding of LCBM structure and facilitates the design of effective processes to obtain lignin-based value-added chemicals.

Acknowledgments

This work was supported by the Fundamental Research Funds for the Central Universities (Grant 2015XKMS100), the National Natural Science Foundation of China (Grant nos. 21506250 and 21676293), and the Qing Lan Project of Jiangsu Province (awarded in 2017).

References

[1] J. S. Luterbacher, D. M. Alonso, and J. A. Dumesic, "Targeted chemical upgrading of lignocellulosic biomass to platform molecules," *Green Chemistry*, vol. 16, pp. 4816–4838, 2014.

[2] B. M. Upton and A. M. Kasko, "Strategies for the conversion of lignin to high-value polymeric materials: review and perspective," *Chemical Reviews*, vol. 116, no. 4, pp. 2275–2306, 2016.

[3] K. M. N. Satheesh, A. K. Mohanty, L. Erickson, and M. Misra, "Lignin and its applications with polymers," *Journal of Biobased Materials and Bioenergy*, vol. 3, pp. 1–24, 2009.

[4] D. Stewart, "Lignin as a base material for materials applications: chemistry, applications and economics," *Industrial Crops and Products*, vol. 27, no. 2, pp. 202–207, 2008.

[5] C. R. Poovaiah, M. Nageswara-Rao, J. R. Soneji, H. L. Baxter, and C. N. Stewart Jr, "Altered lignin biosynthesis using biotechnology to improve lignocellulosic biofuel feedstocks," *Plant Biotechnology Journal*, vol. 12, pp. 1163–1173, 2014.

[6] H. B. C. Molinari, T. K. Pellny, J. Freeman, P. R. Shewry, and R. A. C. Mitchell, "Grass cell wall feruloylation: distribution of bound ferulate and candidate gene expression in *Brachypodium distachyon*," *Frontiers in Plant Science*, vol. 4, p. 50, 2013.

[7] J. Shigeto, Y. Ueda, S. Sasaki, K. Fujita, and Y. Tsutsumi, "Enzymatic activities for lignin monomer intermediates highlight the biosynthetic pathway of syringyl monomers in *Robinia pseudoacacia*," *Journal of Plant Research*, vol. 130, no. 1, pp. 203–210, 2017.

[8] A. L. Healey, J. S. Lupoi, D. J. Lee et al., "Effect of aging on lignin content, composition and enzymatic saccharification in *Corymbia* hybrids and parental taxa between years 9 and 12," *Biomass and Bioenergy*, vol. 93, pp. 50–59, 2016.

[9] D. Kai, M. J. Tan, P. L. Chee, Y. K. Chua, Y. L. Yap, and X. J. Loh, "Towards lignin-based functional materials in a sustainable world," *Green Chemistry*, vol. 18, no. 5, pp. 1175–1200, 2016.

[10] Y. Zeng, M. E. Himmel, and S. Y. Ding, "Coherent Raman microscopy analysis of plant cell walls," *Methods in Molecular Biology*, vol. 908, pp. 49–60, 2012.

[11] C. Zhou, Q. Li, V. L. Chiang, L. A. Lucia, and D. P. Griffis, "Chemical and spatial differentiation of syringyl and guaiacyl lignins in poplar wood via time-of-flight secondary ion mass apectrometry," *Analytical Chemistry*, vol. 83, no. 18, pp. 7020–7026, 2011.

[12] S. Jung, M. Foston, U. C. Kalluri, G. A. Tuskan, and A. J. Ragauskas, "3D chemical image using TOF-SIMS revealing the biopolymer component spatial and lateral distributions in biomass," *Angewandte Chemie International Edition*, vol. 51, no. 48, pp. 12005–12008, 2012.

[13] A. J. Ragauskas, G. T. Beckham, M. J. Biddy et al., "Lignin valorization: improving lignin processing in the biorefinery," *Science*, vol. 344, no. 6185, article 1246843, 2014.

[14] A. P. Dodd, J. F. Kadla, and S. K. Straus, "Characterization of fractions obtained from two industrial softwood kraft lignins," *ACS Sustainable Chemistry & Engineering*, vol. 3, no. 1, pp. 103–110, 2015.

[15] N. Giummarella, L. M. Zhang, G. Henriksson, and M. Lawoko, "Structural features of mildly fractionated lignin carbohydrate complexes (LCC) from spruce," *RSC Advances*, vol. 6, no. 48, pp. 42120–42131, 2016.

[16] H. L. Wang, H. X. Ben, H. Ruan et al., "Effects of lignin structure on hydrodeoxygenation reactivity of pine wood lignin to valuable chemicals," *ACS Sustainable Chemistry & Engineering*, vol. 5, no. 2, pp. 1824–1830, 2017.

[17] A. Berlin and M. Balakshin, "Industrial lignins: analysis, properties, and applications," in *Bioenergy Research: Advances and Applications*, V. K. Gupta, M. G. T. P. Kubicek and J. S. Xu, Eds., pp. 315–336, Elsevier, Amsterdam, 2014.

[18] R. Prado, A. Brandt, X. Erdocia, J. Hallet, T. Welton, and J. Labidi, "Lignin oxidation and depolymerisation in ionic liquids," *Green Chemistry*, vol. 18, no. 3, pp. 834–841, 2016.

[19] J. Shi, S. Pattathil, R. Parthasarathi et al., "Impact of engineered lignin composition on biomass recalcitrance and ionic liquid pretreatment efficiency," *Green Chemistry*, vol. 18, no. 18, pp. 4884–4895, 2016.

[20] Y. Lu, X. Y. Wei, J. P. Cao et al., "Characterization of a bio-oil from pyrolysis of rice husk by detailed compositional analysis and structural investigation of lignin," *Bioresource Technology*, vol. 116, pp. 114–119, 2012.

[21] S. Z. Wang, X. Fan, A. L. Zheng et al., "Evaluation of the oxidation of rice husks with sodium hypochlorite using gas chromatography-mass spectrometry and direct analysis in real time-mass spectrometry," *Analytical Letters*, vol. 47, no. 1, pp. 77–90, 2014.

[22] X. Fan, J. L. Zhu, A. L. Zheng et al., "Rapid characterization of heteroatomic molecules in a bio-oil from pyrolysis of rice husk using atmospheric solid analysis probe mass spectrometry," *Journal of Analytical and Applied Pyrolysis*, vol. 115, pp. 16–23, 2015.

[23] V. V. Lobodin, L. Nyadong, B. M. Ruddy et al., "Fourier transform ion cyclotron resonance mass spectrometry for analysis of complex organic mixtures," *International Journal of Mass Spectrometry*, vol. 378, pp. 186–192, 2015.

[24] C. F. Wang, X. Fan, F. Zhang et al., "Characterization of humic acids extracted from a lignite and interpretation for the mass spectra," *RSC Advances*, vol. 7, no. 33, pp. 20677–20684, 2017.

[25] T. T. You and F. Xu, "Applications of molecular spectroscopic methods to the elucidation of lignin structure," *Applications of Molecular Spectroscopy to Current Research in the Chemical and Biological Sciences*, vol. 2016, pp. 235–260.

[26] E. A. Capanema, M. Y. Balakshin, and J. F. Kadla, "A comprehensive approach for quantitative lignin characterization by NMR spectroscopy," *Journal of Agricultural and Food Chemistry*, vol. 52, no. 7, pp. 1850–1860, 2004.

[27] B. Jiang, T. Y. Cao, F. Gu, W. J. Wu, and Y. C. Jin, "Comparison of the structural characteristics of cellulolytic enzyme lignin preparations isolated from wheat straw stem and leaf," *ACS Sustainable Chemistry & Engineering*, vol. 5, no. 1, pp. 342–349, 2017.

[28] C. Zhao, S. Xie, Y. Pu et al., "Synergistic enzymatic and microbial conversion of lignin for lipid," *Green Chemistry*, vol. 18, no. 5, pp. 1306–1312, 2016.

[29] B. Ahvazi, É. Cloutier, O. Wojciechowicz, and T. D. Ngo, "Lignin profiling: a guide for selecting appropriate lignins as precursors in biomaterials development," *ACS Sustainable Chemistry & Engineering*, vol. 4, no. 10, pp. 5090–5105, 2016.

[30] M. Yáñez-S, B. Matsuhiro, C. Nuñez et al., "Physicochemical characterization of ethanol organosolv lignin (EOL) from *Eucalyptus globulus*: effect of extraction conditions on the molecular structure," *Polymer Degradation and Stability*, vol. 110, pp. 184–194, 2014.

[31] Q. Wang, S. S. Liu, G. L. Yang, and J. C. Chen, "Characterization of high-boiling-solvent lignin from hot-water extracted bagasse," *Energy & Fuels*, vol. 28, no. 5, pp. 3167–3171, 2014.

[32] M. B. Whitfield, M. S. Chinn, and M. W. Veal, "Improvement of acid hydrolysis procedures for the composition analysis of herbaceous biomass," *Energy & Fuels*, vol. 30, no. 10, pp. 8260–8269, 2016.

[33] W. H. Gong, Z. Y. Xiang, F. Y. Ye, and G. H. Zhao, "Composition and structure of an antioxidant acetic acid lignin isolated from shoot shell of bamboo (*Dendrocalamus latiforus*)," *Industrial Crops and Products*, vol. 91, pp. 340–349, 2016.

[34] W. Zhang, N. Sathitsuksanoh, B. A. Simmons, C. E. Frazier, J. R. Barone, and S. Renneckar, "Revealing the thermal sensitivity of lignin during glycerol thermal processing through structural analysis," *RSC Advances*, vol. 6, no. 36, pp. 30234–30246, 2016.

[35] A. E. Harman-Ware, C. Foster, R. M. Happs et al., "A Thioacidolysis method tailored for higher-throughput quantitative analysis of lignin monomers," *Journal of Biotechnology*, vol. 11, no. 10, pp. 1268–1273, 2016.

[36] J. L. Wen, S. L. Sun, B. L. Xue, and R. C. Sun, "Quantitative structures and thermal properties of birch lignins after ionic liquid pretreatment," *Journal of Agricultural and Food Chemistry*, vol. 61, no. 3, pp. 635–645, 2013.

[37] B. J. Cox and J. G. Ekerdt, "Pretreatment of yellow pine in an acidic ionic liquid: extraction of hemicelluloses and lignin to facilitate enzymatic digestion," *Bioresource Technology*, vol. 134, pp. 59–65, 2013.

[38] A. Casas, M. Oliet, M. V. Alonso, and F. Rodrígues, "Dissolution of *Pinus radiata* and *Eucalyptus globulus* woods in ionic liquids under microwave radiation: lignin regeneration and characterization," *Separation and Purification Technology*, vol. 97, pp. 115–122, 2012.

[39] S. S. Mohtar, T. N. Z. Tengku Malim Busu, A. M. Md Noor, N. Shaari, and H. Mat, "An ionic liquid treatment and fractionation of cellulose, hemicelluloses and lignin from oil palm empty fruit bunch," *Carbohydrate Polymers*, vol. 166, pp. 291–299, 2017.

[40] T. Rashid, C. F. Kait, I. Regupathi, and T. Murugesan, "Dissolution of kraft lignin using protic ionic liquids and characterization," *Industrial Crops and Products*, vol. 84, pp. 284–293, 2016.

[41] Y. X. An, N. Li, H. Wu, W. Y. Lou, and M. H. Zong, "Changes in the structure and the thermal properties of kraft lignin during its dissolution in cholinium ionic liquids," *ACS Sustainable Chemistry & Engineering*, vol. 3, no. 11, pp. 2951–2958, 2015.

[42] S. Nanayakkara, A. F. Patti, and K. Saito, "Lignin depolymerization with phenol via redistribution mechanism in ionic liquids," *ACS Sustainable Chemistry & Engineering*, vol. 2, no. 9, pp. 2159–2164, 2014.

[43] J. X. Long, W. Y. Lou, L. F. Wang, B. L. Yin, and X. H. Li, "$[C_4H_8SO_3Hmim]HSO_4$ as an efficient catalyst for direct liquefaction of bagasse lignin: decomposition properties of the inner structural units," *Chemical Engineering Science*, vol. 122, pp. 24–33, 2015.

[44] S. G. Khokarale, T. Le-That, and J. P. Mikkola, "Carbohydrate free lignin: a dissolution-recovery cycle of sodium lignosulfonate in a switchable ionic liquid system," *ACS Sustainable Chemistry & Engineering*, vol. 4, no. 12, pp. 7032–7040, 2016.

[45] A. Mittal, R. Katahira, B. S. Donohoe et al., "*Ammonia* pretreatment of corn stover enables facile lignin extraction," *ACS Sustainable Chemistry & Engineering*, vol. 5, no. 3, pp. 2544–2561, 2017.

[46] T. Q. Yuan, T. T. You, W. Wang, F. Xu, and R. C. Sun, "Synergistic benefits of ionic liquid and alkaline pretreatments of poplar wood. Part 2: characterization of lignin and hemicelluloses," *Bioresource Technology*, vol. 136, pp. 345–350, 2013.

[47] Y. C. Sun, M. Wang, and R. C. Sun, "Toward an understanding of inhomogeneities in structure of lignin in green solvents biorefinery. Part 1: fractionation and characterization of cignin," *ACS Sustainable Chemistry & Engineering*, vol. 3, no. 10, pp. 2443–2451, 2015.

[48] C. Nitsos, R. Stoklosa, A. Karnaouri et al., "Isolation and characterization of organosolv and alkaline lignins from hardwood and softwood biomass," *ACS Sustainable Chemistry & Engineering*, vol. 4, no. 10, pp. 5181–5193, 2016.

[49] L. N. Yang, D. M. Wang, D. Zhou, Y. W. Zhang, and T. T. Yang, "Isolation and further structural characterization of lignins from the valonea of *Quercus variabilis*," *International Journal of Biological Macromolecules*, vol. 97, pp. 164–172, 2017.

[50] L. N. Yang, D. M. Wang, D. Zhou, and Y. Zhang, "Effect of different isolation methods on structure and properties of lignin from valonea of *Quercus variabilis*," *International Journal of Biological Macromolecules*, vol. 85, pp. 417–424, 2016.

[51] S. Zhou, Y. Xue, A. Sharma, and X. L. Bai, "Lignin valorization through thermochemical conversion: comparison of hardwood, softwood and herbaceous lignin," *ACS*

Sustainable Chemistry & Engineering, vol. 4, no. 12, pp. 6608–6617, 2016.

[52] G. T. Beckham, C. W. Johnson, E. M. Karp, D. Salvachúa, and D. R. Vardon, "Opportunities and challenges in biological lignin valorization," *Current Opinion in Biotechnology*, vol. 42, pp. 40–53, 2016.

[53] F. Xu, R. C. Sun, M. Z. Zhai, J. X. Sun, J. X. Jiang, and G. J. Zhao, "Comparative study of three lignin fractions isolated from mild ball-milled *Tamarix austromogoliac* and *Caragana sepium*," *Journal of Applied Polymer Science*, vol. 8, no. 2, pp. 1158–1168, 2008.

[54] A. Sluiter, B. Hames, R. Ruiz et al., *Determination of Structural Carbohydrates and Lignin in Biomass*, National Renewable Energy Laboratory (NREL) Laboratory Analytical Procedures (LAP) for Standard Biomass Analysis, National Renewable Energy Laboratory, Golden, USA, 2007, NREL/TP-510-42618.

[55] F. Q. Xiong, Y. M. Han, S. Q. Wang et al., "Preparation and formation mechanism of renewable lignin hollow nanospheres with a single hole by self-assembly," *ACS Sustainable Chemistry & Engineering*, vol. 5, no. 3, pp. 2273–2281, 2017.

[56] B. Hansen, P. Kusch, M. Schulze, and B. Kamm, "Qualitative and quantitative analysis of lignin producedfrom beech wood by different conditions of the organosolv process," *Journal of Polymers and the Environment*, vol. 24, no. 2, pp. 85–97, 2016.

[57] S. O. Prozil, D. V. Evtuguin, A. M. S. Silva, and L. P. C. Lopes, "Structural characterization of lignin from grape stalks (*Vitis vinifera* L.)," *Journal of Agricultural and Food Chemistry*, vol. 62, no. 24, pp. 5420–5428, 2014.

[58] M. Azadfar, H. M. GaoA, M. V. Bule, and S. L. Chen, "Structural characterization of lignin: a potential source ofantioxidants *guaiacol* and *4-vinylguaiacol*," *International Journal of Biological Macromolecules*, vol. 75, pp. 58–66, 2015.

[59] K. A. Y. Koivu, H. Sadeghifar, P. A. Nousiainen, D. S. Argyropoulos, and J. Sipilä, "Effect of fatty acid esterification on the thermal properties of softwood kraft lignin," *ACS Sustainable Chemistry & Engineering*, vol. 4, no. 10, pp. 5238–5247, 2016.

[60] Q. L. Yang, J. B. Shi, and L. Lin, "Characterization of structural changes of lignin in the process of cooking of bagasse with solid alkali and active oxygen as a pretreatment for lignin conversion," *Energy & Fuels*, vol. 26, pp. 6999–7004, 2012.

[61] I. Santoni, E. Callone, A. Sandak, J. Sandak, and S. Dirè, "Solid state NMR and IR characterization of wood polymer structure in relation to tree provenance," *Carbohydrate Polymers*, vol. 117, pp. 710–721, 2015.

[62] L. Passauer, K. Salzwedel, M. Struch, N. Herold, and J. Appelt, "Quantitative analysis of the etherification degree of phenolic hydroxyl groups in oxyethylated lignins: correlation of selective aminolysis with FTIR spectroscopy," *ACS Sustainable Chemistry & Engineering*, vol. 4, no. 12, pp. 6629–6637, 2016.

[63] U. P. Agarwal, R. S. Reiner, A. K. Pandey, S. A. Ralph, K. C. Hirth, and R. H. Atalla, Eds., "Raman spectra of lignin model compounds," in *59th Appita Annual Conference and Exhibition: Incorporating the 13th ISWFPC (International Symposium on Wood, Fiber and Pulping Chemistry), Auckland, New Zealand, 16–19 May 2005: Proceedings*, Appita Inc., 2005.

[64] K. L. Larsen and S. Barsberg, "Environmental effects on the lignin model monomer, vanillyl alcohol, studied by Raman spectroscopy," *Journal of Physical Chemistry B*, vol. 115, no. 39, pp. 11470–11480, 2011.

[65] B. I. Na, S. J. Chang, K. H. Lee, G. Lee, and J. W. Lee, "Characterization of cell wall structure in dilute acid-pretreated biomass by confocal Raman microscopy and enzymatic hydrolysis," *Biomass and Bioenergy*, vol. 93, pp. 33–37, 2016.

[66] J. Ralph, C. Lapierre, F. C. Lu et al., "NMR evidence for benzodioxane structures resulting from incorporation of 5-hydroxyconiferyl alcohol into lignins of O-methyltransferase-deficient poplars," *Journal of Agricultural and Food Chemistry*, vol. 49, no. 7, pp. 86–91, 2001.

[67] L. M. Zhang, G. Gellerstedt, J. Ralph, and L. F. C. , "NMR studies on the occurrence of spirodienone structures in lignins," *Journal of Wood Chemistry and Technology*, vol. 26, no. 1, pp. 65–79, 2006.

[68] G. Wang and H. Chen, "Fractionation and characterization of lignin from steam-exploded corn stalk by sequential dissolution in ethanol-water solvent," *Separation and Purification Technology*, vol. 120, pp. 402–409, 2013.

[69] C. Fernández-Costas, S. Gouveia, M. A. Sanromán, and D. Moldes, "Structural characterization of kraft lignins from different spent cooking liquors by 1D and 2D nuclear magnetic resonance spectroscopy," *Biomass and Bioenergy*, vol. 63, pp. 156–166, 2014.

[70] L. L. An, G. H. Wang, H. Y. Jia, C. Y. Liu, W. J. Sui, and C. L. Si, "Fractionation of enzymatic hydrolysis lignin by sequential extraction for enhancing antioxidant performance," *International Journal of Biological Macromolecules*, vol. 99, pp. 674–681, 2017.

[71] C. A. Esteves Costa, W. Coleman, M. Dube, A. E. Rodrigues, and P. C. Rodrigues Pinto, "Assessment of key features of lignin from lignocellulosic crops: stalks and roots of corn, cotton, sugarcane, and tobacco," *Industrial Crops and Products*, vol. 92, pp. 136–148, 2016.

[72] H. Ben and A. J. Ragauskas, "Torrefaction of loblolly pine," *Green Chemistry*, vol. 14, no. 1, pp. 72–76, 2012.

[73] T. Wang, P. Phyo, and M. Hong, "Multidimensional solid-state NMR spectroscopy of plant cell walls," *Solid State Nuclear Magnetic Resonance*, vol. 78, pp. 56–63, 2016.

[74] D. Lee, S. Hediger, and G. De Paepe, "Is solid-state NMR enhanced by dynamic nuclear polarization?," *Solid State Nuclear Magnetic Resonance*, vol. 66-67, pp. 6–20, 2015.

[75] G. Mollica, M. Dekhil, F. Ziarelli, P. Thureau, and S. Viel, "Quantitative structural constraints for organic powders at natural isotopic abundance using dynamic nuclear polarization solid-state NMR spectroscopy," *Angewandte Chemie International Edition*, vol. 54, no. 20, pp. 6028–6031, 2015.

[76] F. A. Perras, H. Luo, X. M. Zhang, N. S. Mosier, M. Pruski, and M. M. Abu-Omar, "Atomic-level structure characterization of biomass pre- and post-lignin treatment by dynamic nuclear polarization-enhanced solid-state NMR," *Journal of Physical Chemistry A*, vol. 121, no. 3, pp. 623–630, 2017.

[77] X. Gao, D. D. Laskar, J. J. Zeng, G. L. Helms, and S. L. Chen, "A ^{13}C CP/MAS-based nondegradative method for lignin content analysis," *ACS Sustainable Chemistry & Engineering*, vol. 3, no. 1, pp. 153–162, 2015.

[78] S. Constant, C. S. Lancefield, B. M. Weckhuysen, and P. C. Bruijnincx, "Quantification and classification of carbonyls in industrial humins and lignins by ^{19}F NMR," *ACS Sustainable Chemistry & Engineering*, vol. 5, no. 1, pp. 965–972, 2017.

[79] Y. Pu, S. Cao, and A. J. Ragauskas, "Application of quantitative ^{31}P NMR in biomass lignin and biofuel precursors

characterization," *Energy & Environmental Science*, vol. 4, no. 9, pp. 3154–3166, 2011.

[80] S. D. Springer, J. He, M. Chui, R. D. Little, M. Foston, and A. Butler, "Peroxidative oxidation of lignin and a lignin model compound by a manganese SALEN derivative," *ACS Sustainable Chemistry & Engineering*, vol. 4, no. 6, pp. 3212–3219, 2016.

[81] L. Mbotchak, C. L. Morvan, K. L. Duong, B. Roussear, M. Tessier, and A. Fradet, "Purification, structural characterization, and modification of organosolv wheat straw lignin," *Journal of Agricultural and Food Chemistry*, vol. 63, no. 21, pp. 5178–5188, 2015.

[82] S. X. Xie, Q. N. Sun, Y. Q. Pu et al., "Advanced chemical design for efficient lignin bioconversion," *ACS Sustainable Chemistry & Engineering*, vol. 5, no. 3, pp. 2215–2223, 2017.

[83] F. C. Lu and J. Ralph, "Non-degradative dissolution and acetylation of ball-milled plant cell walls: high-resolution solution-state NMR," *The Plant Journal*, vol. 35, no. 4, pp. 535–544, 2003.

[84] J. J. Bozell, C. J. O'Lenick, and S. Warwick, "Biomass fractionation for the biorefinery: heteronuclear multiple quantum coherence-nuclear magnetic resonance investigation of lignin isolated from solvent fractionation of switchgrass," *Journal of Agricultural and Food Chemistry*, vol. 59, no. 17, pp. 9232–9242, 2011.

[85] Y. Le Brech, L. Delmotte, J. Raya, N. Brosse, R. Gadiou, and A. Dufour, "High resolution solid state 2D NMR analysis of biomass and biochar," *Analytical Chemistry*, vol. 87, no. 2, pp. 843–847, 2015.

[86] Y. Le Brech, J. Raya, L. Delmotte, N. Brosse, R. Gadiou, and A. Dufour, "Characterization of biomass char formation investigated by advanced solid state NMR," *Carbon*, vol. 108, pp. 165–177, 2016.

[87] D. Ando, F. Nakatsubo, T. Takano, and H. Yano, "Elucidation of LCC bonding sites via γ-TTSA lignin degradation: crude milled wood lignin (MWL) from *Eucalyptus globulus* for enrichment of lignin xylan linkages and their HSQC-NMR characterization," *Holzforschung*, vol. 70, no. 6, pp. 489–494, 2016.

[88] D. Ando, F. Nakatsubo, and H. Yano, "Acetylation of ground pulp: monitoring acetylation via HSQC-NMR spectroscopy," *ACS Sustainable Chemistry & Engineering*, vol. 5, no. 2, pp. 1755–1762, 2017.

[89] K. Saito, A. Kaiho, R. Sakai, H. Nishimura, H. Okada, and T. Watanabe, "Characterization of the interunit bonds of lignin oligomers released by acid-catalyzed selective solvolysis of *Cryptomeria japonica* and *Eucalyptus globulus* woods via thioacidolysis and 2D-NMR," *Journal of Agricultural and Food Chemistry*, vol. 64, no. 48, pp. 9152–9160, 2016.

[90] J. Rencoret, P. Prinsen, A. Gutiérrez, Á. T. Martínez, and J. C. del Río, "Isolation and structural characterization of the milled wood lignin, dioxane lignin, and cellulolytic lignin preparations from brewer's spent grain," *Journal of Agricultural and Food Chemistry*, vol. 63, no. 2, pp. 603–613, 2015.

[91] C. Mattsson, S. I. Andersson, T. Belkheiri et al., "Using 2D NMR to characterize the structure of the low and high molecular weight fractions of bio-oil obtained from lignoBoost™ kraft lignin depolymerized in subcritical water," *Biomass and Bioenergy*, vol. 95, pp. 364–377, 2016.

[92] L. Zhang and G. Gellerstedt, "Quantitative 2D HSQC NMR determination of polymer structures by selecting suitable internal standard references," *Magnetic Resonance in Chemistry*, vol. 45, no. 1, pp. 37–45, 2007.

[93] J. J. Zeng, G. L. Helms, X. Gao, and S. L. Chen, "Quantification of wheat straw lignin structure by comprehensive NMR analysis," *Journal of Agricultural and Food Chemistry*, vol. 61, no. 46, pp. 10848–10857, 2013.

[94] K. Crouvisier-Urion, P. R. Bodart, P. Winckler et al., "Biobased composite films from chitosan and lignin: antioxidant activity related to structure and moisture," *ACS Sustainable Chemistry & Engineering*, vol. 4, no. 12, pp. 6371–6381, 2016.

16

Application of Permutation Entropy in Feature Extraction for Near-Infrared Spectroscopy Noninvasive Blood Glucose Detection

Xiaoli Li and Chengwei Li

School of Electrical Engineering and Automation, Harbin Institute of Technology, Harbin 150001, China

Correspondence should be addressed to Chengwei Li; lcw@hit.edu.cn

Academic Editor: Feride Severcan

Diabetes has been one of the four major diseases threatening human life. Accurate blood glucose detection became an important part in controlling the state of diabetes patients. Excellent linear correlation existed between blood glucose concentration and near-infrared spectral absorption. A new feature extraction method based on permutation entropy is proposed to solve the noise and information redundancy in near-infrared spectral noninvasive blood glucose measurement, which affects the accuracy of the calibration model. With the near-infrared spectral data of glucose solution as the research object, the concepts of approximate entropy, sample entropy, fuzzy entropy, and permutation entropy are introduced. The spectra are then segmented, and the characteristic wave bands with abundant glucose information are selected in terms of permutation entropy, fractal dimension, and mutual information. Finally, the support vector regression and partial least square regression are used to establish the mathematical model between the characteristic spectral data and glucose concentration, and the results are compared with conventional feature extraction methods. Results show that the proposed new method can extract useful information from near-infrared spectra, effectively solve the problem of characteristic wave band extraction, and improve the analytical accuracy of spectral and model stability.

1. Introduction

Diabetes is one of the major diseases threatening human health, and the number of people with diabetes is growing at an alarming rate. More than one hundred million people suffer from diabetes; furthermore, the number is expected to increase to 592 million by 2035 [1]. Although proper diet and insulin injection can be used to regulate blood glucose levels, serious complications are caused in the later stage of diabetes, such as heart failure and blindness [2]. Therefore, the treatment of diabetes is very important, and the concentration of blood glucose detection is the foundation of diabetes treatment. The noninvasive blood glucose detection technology that measures the glucose concentration in the blood under the condition of no skin damage includes near-infrared spectroscopy, photo acoustic spectroscopy, polarization method, fluorescence method, and dielectric spectroscopy method [3–5]. Compared with the near-infrared spectra method, other noninvasive blood glucose detection methods are not perfectly suitable for real-time detection. The signals are hard to be detected and easy to be interfered by other components. At present, noninvasive blood glucose detection based on near-infrared spectra has become the research focus at home and abroad. Near-infrared spectroscopy (NIR), which is generated from molecular vibrations and reflects the chemical bond information, such as C-H, O-H, N-H, and S-H, can measure most kinds of compounds and their mixtures. Compared with the traditional analytical techniques, NIR has been widely applied because it is highly efficient and causes no damage and pollution [6–8]. The main structure and composition of glucose information is contained in the near-infrared spectra. The useful glucose information can be extracted form spectral data; then, the data after pretreatment are used to establish a mathematical model to calculate the glucose concentration. In the field of biomedicine, NIR combined with chemometrics is considered one of the most effective methods for noninvasive blood glucose concentration detection [3]. The common chemometrics methods include multiple linear regression (MLR), principal component regression (PCR), partial least squares regression (PLSR), and support vector regression (SVR). The MLR is limited by the noise in spectral data, and the irrelevance between some principal components and the actual content appears in the PCR. Therefore, the PLSR method and SVR method are applied in this paper. However, certain technical difficulties exist in noninvasive blood glucose measurement because near-infrared spectral samples cannot be pretreated, namely, the complex background, overlapped spectral peaks, and less effective information rate. Therefore, extracting effective information from the original spectra is critical for establishing an ideal mathematical model. The effective extraction of glucose characteristic information from nonlinear and nonstationary near-infrared spectral signals can improve the detection efficiency and detection precision.

In 1984, Shannon introduced entropy to the field of information theory and proposed the concept of information entropy to measure the uncertainty of events [9]. Subsequently, the concept of entropy was gradually generalized. In 1991, Pincus proposed the concept of approximate entropy (ApEn), which has the advantages of short required calculating data and excellent antinoise ability [10] and offsets the shortcomings of nonlinear analysis. However, the data has no relevance and the errors are produced in the computational process of ApEn. To observably improve the accuracy and efficiency of the ApEn method, Richman and Moorman proposed an improved ApEn in 2000 called sample entropy (SampleEn) [11]. Compared with the ApEn algorithm, SampleEn has short required data and robust antinoise and anti-interference abilities, well consistent in the range of large parameters such unique advantages. The definition of SampleEn must contain a template match; otherwise, it is meaningless. Therefore, Chen et al. improved SampleEn and first defined a new measure of sequence complexity, named fuzzy entropy [12]. This new measure fuzzifies the similarity measure formula with an exponential function to enable the fuzzy entropy value to transition smoothly with changing parameter. Its definition still has significance when the parameter is small, and it inherits the relative consistency and short data set-processing characteristics of SampleEn. Bandt and Pompe proposed a randomness detection method of a time series, namely, permutation entropy (PE), which can detect the randomness of time series and dynamic mutation behavior [13–18]. Permutation entropy calculates entropy based on permutation patterns by comparing the neighboring values of the time series [19]. PE between 0 and 1 has the advantages of simple concept, fast calculation speed, and robust anti-interference ability, and it is especially suitable for nonlinear data.

The key point of near-infrared spectra noninvasive blood glucose detection is to extract the characteristic information from the spectral signal. The near-infrared spectral signals of glucose solution are nonstationary and noisy, but the calculation of PE has a certain antinoise and anti-interference ability. In this study, the feature extraction of spectral information is investigated with glucose solution as the research object from the perspective of whole information from a signal. This paper is organized as follows. Section 2 describes the principles of ApEn, SampleEn, fuzzy entropy, and PE and then briefly introduces the methods of near-infrared spectral characteristic band extraction of a glucose solution, such as fractal dimension, mutual information, and the modeling methods, such as PLSR and SVR. In Section 3, the application of the proposed method is presented, and PLSR and SVR are used to establish calibration models with the extracted characteristic bands, as well as verify the validity and superiority of the proposed method. Finally, the conclusion is drawn in Section 4.

2. Theory and Methods

2.1. Entropy

2.1.1. Approximate Entropy. In 1991, Pincus defined ApEn as a conditional probability that the similarity vector maintains its similarity when it increases from m dimension to $m+1$ dimension. The physical meaning is the probability of generating a new pattern of time series when the dimension changes. ApEn has the following advantages: (1) short required data, (2) robust antinoise and anti-interference abilities, and (3) applicability for deterministic and stochastic signals and a mixed signal composed of deterministic and stochastic signals. The steps of the ApEn algorithm are as follows [10]:

(1) Given a time series of length N, $\{u(i), i=1,\ldots,N\}$, reconstruct a m-dimensional vector X_i, $i=1,2,\ldots,n$, $n=N-m+1$, according to the formula $X_i=\{u(i), u(i+1), \ldots, u(i+m-1)\}$.

(2) Compute the distance between arbitrary the vector X_i and the vector X_j $(j=1,2,\ldots,N-m+1, j\neq i)$.

$$d_{ij}=\max|u(i+j)-u(j+k)|,\quad k=0,1,\ldots,m-1. \quad (1)$$

The distance between the two vectors is the maximum absolute value of the difference between two corresponding elements in two vectors.

(3) Specify the threshold r, which is typically between 0.2 and 0.3. For each vector X_i, find the number of $d_{ij} \leq r \times \mathrm{SD}$ (SD is the standard deviation of the sequence) and calculate the ratio between this number and the total number of distances $(N-m)$, which is denoted as $C_i^m(i)$.

(4) Take the logarithm of $C_i^m(i)$, average for all i, and denote $\phi^m(r)$.

$$\phi^m(r) = \frac{1}{N-m+1} \sum_{i=1}^{N-m+1} \ln C_i^m(r). \tag{2}$$

(5) Increase m by 1 and repeat steps 1 to 4 to obtain $C_i^{m+1}(i)$ and $\phi^{m+1}(r)$.

(6) Obtain the ApEn from ϕ^{m+1}, ϕ^m.

$$\mathrm{ApEn} = \sum_{N\to\infty} \left|\phi^m - \phi^{m+1}\right|. \tag{3}$$

(7) For a finite time series, ApEn can be estimated by a statistical value.

$$\mathrm{ApEn} = \phi^m - \phi^{m+1}. \tag{4}$$

The parameters N, m, and r in the above steps are the length of time series, length of the comparison window, and margin of similarity, respectively. The bigger the value of m is, the more dynamic the process can be reconstructed.

2.1.2. Sample Entropy. The physical meaning of SampleEn is the same as that of ApEn. The larger SampleEn is, the higher the complexity of the sequence and the greater the probability of generating the new pattern will be. The specific algorithm implementation process is as follows [11]:

(1) Given the time series $x(1), x(2), \ldots, x(N)$, compose a set of dimension vectors according to the serial number order.

$$X_m(i) = [x(i), x(i+1), \ldots, x(i+m-1)], \quad i = 1\sim N-m. \tag{5}$$

(2) Define the distance between vector $X_m(i)$ and vector $X_m(j)$ as the largest difference between their corresponding elements, namely,

$$d[X_m(i), X_m(j)] = \max|x(i+k) - x(j+k)|. \tag{6}$$

(3) Define the threshold r. For the value of each $i \leq N-m$, find the number $N(i)$ of $d[X_m(i), X_m(j)] \leq r$, and calculate the ratio between $N(i)$ and the total number of distances $N-m-1$, which is denoted as $B^m(i) = N(i)/(N-m-1)$. The average for all i is as follows:

$$B^m(r) = (N-m)^{-1} \cdot \sum_{i=1}^{N} B^m(i). \tag{7}$$

(4) Increase the dimension to $m+1$ and repeat the above steps to obtain

$$B^{m+1}(r) = (N-m)^{-1} \cdot \sum_{i=1}^{N} B^{m+1}(i). \tag{8}$$

(5) Theoretically, the SamEn of this sequence is as follows:

$$\mathrm{SampleEn}(m, r) = \lim_{N\to\infty}\left\{-\ln\left[\frac{B^{m+1}(r)}{B^m(r)}\right]\right\}, \tag{9}$$

where $N\to\infty$. In practice, N is not an infinite value. When N is a finite value, SamEn is calculated as follows:

$$\mathrm{SampleEn}(m, r, N) = -\ln\left[\frac{B^{m+1}(r)}{B^m(r)}\right]. \tag{10}$$

2.1.3. Fuzzy Entropy. In the definition of fuzzy entropy, the concept of a fuzzy set is introduced, and the exponential function is chosen as the fuzzy function to measure the similarity of two vectors. The exponential function has the following expectation properties: (1) continuity of the exponential function ensures that its value does not have a mutation and (2) the nature of the exponential function ensures that the self-similarity value of the vector is maximum. The definition of fuzzy entropy is as follows [12]:

(1) The sampling sequence with N points is $\{u(i)\text{: } 1 \ll i \ll N\}$.

(2) Compose a set of m-dimensional vectors according to the serial number order.

$$X_i^m = \{u(i), u(i+1), \ldots, u(i+m-1)\} - u_0(i) \quad (i = 1, \ldots, N-m), \tag{11}$$

where $\{u(i), u(i+1), \ldots, u(i+m-1)\}$ represents the continuous m values of u starting from the ith point. $u_0(i)$ is its mean value.

$$u_0(i) = \frac{1}{m} \sum_{j=0}^{m-1} u(i+j). \tag{12}$$

(3) Define the distance d_{ij}^m between vector X_i^m and vector X_j^m as the largest difference between their corresponding elements, namely,

$$d_{ij}^m = d\left[X_i^m, X_j^m\right] = \max_{k\in(0,m-1)}\{|u(i+k) - u_0(i) - (u(j+k) - u_0(j))|\}, \quad (i,j) = 1\sim N-m, j \neq i. \tag{13}$$

(4) Define the similarity D_{ij}^m between vector X_i^m and vector X_j^m, with a fuzzy function $\_(d_{ij}^m, n, r)$, namely,

$$D_{ij}^m = \_\left(d_{ij}^m, n, r\right) = \exp\left(\frac{-\left(d_{ij}^m\right)^n}{r}\right), \tag{14}$$

where the fuzzy function $\_(d_{ij}^m, n, r)$ is an exponential function, and n and r are the gradient and width of the exponential function boundary, respectively.

(5) Define the function as follows:

$$O^m(n,r) = \frac{1}{N-m}\sum_{i=1}^{N-m}\left[\frac{1}{N-m-1}\sum_{j=1,j\neq i}^{N-m} D_{ij}^m\right]. \tag{15}$$

(6) Similarly, repeat steps 2 to 5, reconstruct a set of $m+1$-dimensional vector according to the serial number order, and define the following function:

$$O^m(n,r) = \frac{1}{N-m}\sum_{i=1}^{N-m}\left[\frac{1}{N-m-1}\sum_{j=1,j\neq 1}^{N-m} D_{ij}^{m+1}\right]. \tag{16}$$

(7) Define the fuzzy entropy as follows:

$$\text{FuzzyEn}(m,n,r) = \lim_{N\to\infty}[\ln O^m(n,r) - \ln O^{m+1}(n,r)]. \tag{17}$$

When N is a finite value, the value obtained by the above steps is the estimated value of the fuzzy entropy of the sequence with length N.

$$\text{FuzzyEn}(m,n,r,N) = \ln O^m(n,r) - \ln O^{m+1}(n,r). \tag{18}$$

2.1.4. Permutation Entropy. According to [13], the definition of PE is setting a time sequence $\{X(i), i = 1, 2, \ldots, n\}$, and reconstruct it in phase space to obtain the matrix

$$\begin{bmatrix} x(1) & x(1+\tau) & \cdots & x(1+(m-1)\tau) \\ \vdots & \vdots & & \vdots \\ x(j) & x(j+\tau) & \cdots & x(j+(m-1)\tau) \\ \vdots & \vdots & & \vdots \\ x(K) & x(K+\tau) & \cdots & x(K(m-1)\tau) \end{bmatrix}, \quad j = 1, 2, \ldots, K, \tag{19}$$

where m and τ are the embedding dimension and delay time, respectively, and $K = n - (m-1)\tau$. Each row in the matrix can be regarded as a reconstructed component, with a total of K reconstruction components. The jth reconstruction component $(x(j), x(j+\tau), \ldots, x[j+(m-1)\tau])$ of the reconstruction matrix $X(i)$ is rearranged according to the values in ascending order. $j_1, j_2, \ldots, j_m$ represents the index of the column in which the individual elements of the reconstructed component are as follows:

$$x[i+(j_1-1)\tau] \leq x[i+(j_2-1)\tau] \leq \cdots \leq x[i+(j_m-1)\tau]. \tag{20}$$

If equal values in the reconstructed component are observed,

$$x[i-(j_1-1)\tau] = x[i-(j_m-1)\tau], \tag{21}$$

the components are arranged according to the size of the value of j_1 and j_2, that is, when $j_1 < j_2$, $x[i-(j_1-1)\tau] \ll x[i-(j_2-1)\tau]$.

Therefore, for an arbitrary time series $X(i)$, a set of symbol sequences can be obtained from each row in the reconstructed matrix

$$S(l) = (j_1, j_2, \ldots, j_m), \tag{22}$$

where $l = 1, 2, \ldots, k$ and $k \ll m!$. $m!$ is observed when the m-dimensional phase space map has a different symbolic sequence $(j_1, j_2, \ldots, j_m)$, and the symbolic sequence $S(l)$ is one kind of arrangement. If the probability of the occurrence of each symbol sequence is $P_1, P_2, \ldots, P_k$, the PE of k kinds of different symbol sequences of time series $X(i)$ in terms of Shannon entropy is as follows:

$$H_p(m) = -\sum_{j=1}^{k} P_j \ln P_j. \tag{23}$$

When $P_j = 1/m!$, $H_p(m)$ reaches the maximum value $\ln(m!)$. For convenience, $H_p(m)$ is typically normalized with $\ln(m!)$, namely,

$$0 \ll H_p = \frac{H_p}{\ln(m!) \ll 1}. \tag{24}$$

The magnitude of H_p represents the randomness degree of the time series $X(i)$. The smaller the value of H_p is, the more inerratic the time series will be; otherwise, the more stochastic the time series will be. The change in H_p reflects and amplifies the minute details of the time series.

2.1.5. Application to Simulation Signal. In order to compare the ApEn, SampleEn, fuzzy entropy, and PE, define a mixed signal composed of deterministic signal and stochastic signal with a different probability,

$$\text{MIX}(N,p) = (1-p)x(t) + py(t), \tag{25}$$

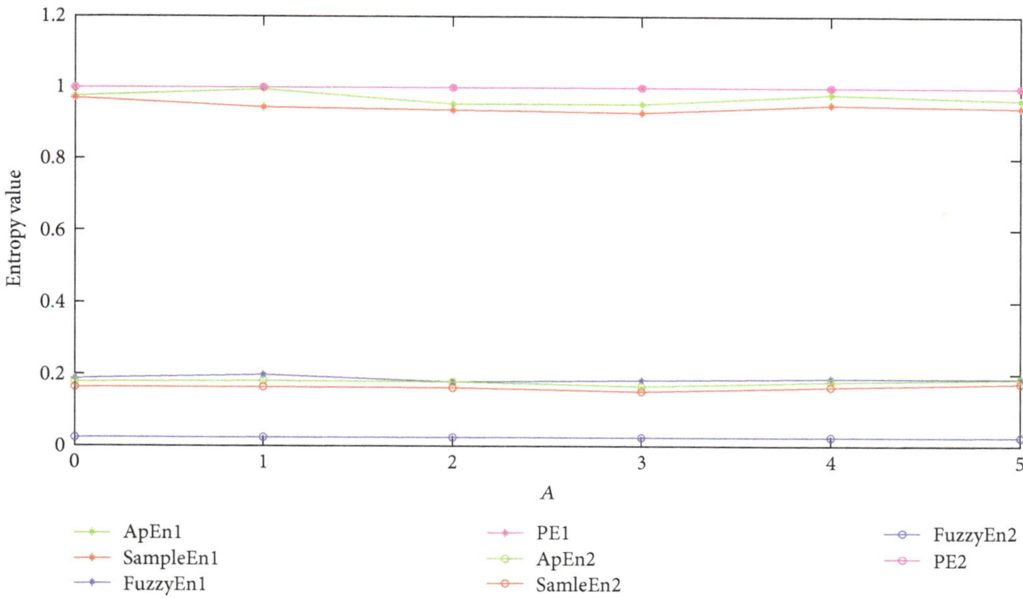

FIGURE 1: Four kinds of entropy values with different amplitudes.

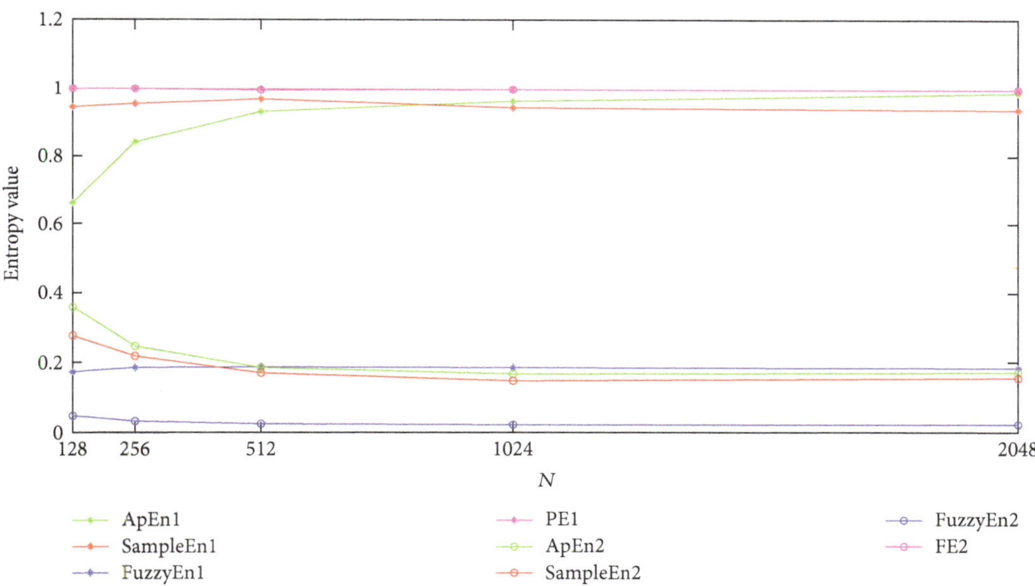

FIGURE 2: Four kinds of entropy values with different data lengths.

where $x(t) = \sqrt{2}\sin(2\pi t)$, $y(t)$ is the stochastic signal in $(-0.5, 0.5)$, N is the data length, and $p \in (0, 1]$. Considering the four kinds of entropy of signal with MIX(1024, 0.4) and MIX(1024, 0.2), note ApEn1, SampleEn1, FuzzyEn1, PE1, ApEn2, SampleEn2, FuzzyEn2, and PE2 for convenience, respectively. Their change relation with signal amplitude A, signal length N, and signal-to-noise ratio of signal are shown in Figures 1–3.

Figure 1 shows that ApEn, SampleEn, and FuzzyEn change slightly with the same probability, when signal amplitude A changes bigger gradually. However, the PE has been the same all the time with a different probability, which illustrates that the PE has excellent stability and consistency.

Figure 2 shows that the ApEn, SampleEn, and FuzzyEn change with the signal length N and remain unchanged after $N = 512$ with the same probability, which illustrates that the values of ApEn, SampleEn, and FuzzyEn are related to the data length. However, the PE has been the same all the time with a different probability, which illustrates that the required time sequence of PE is shorter in the calculation process.

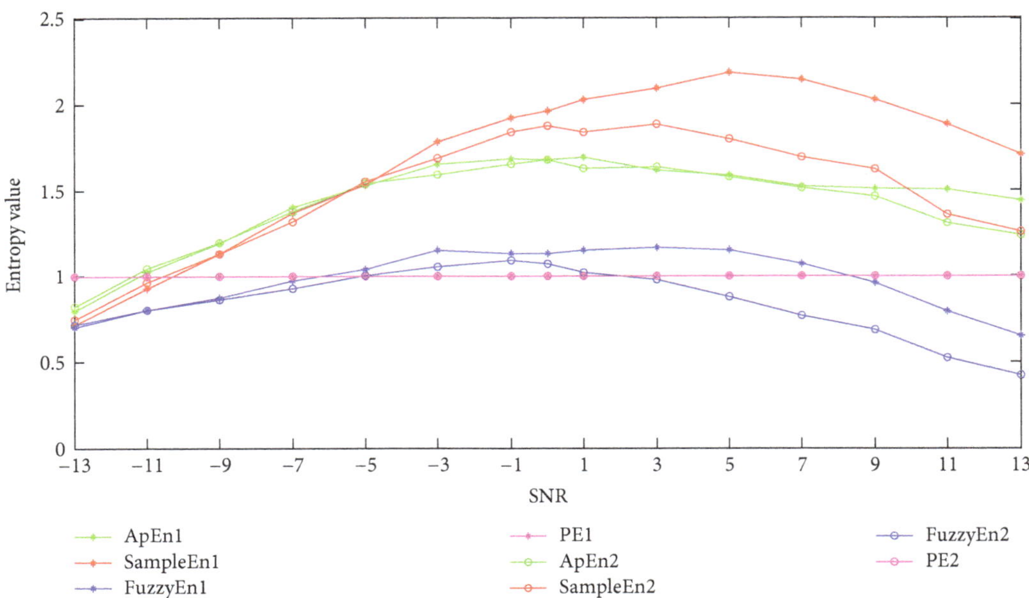

FIGURE 3: Four kinds of entropy values with different signal-to-noise ratios of signal.

Figure 3 shows that the ApEn, SampleEn, and FuzzyEn change with the signal-to-noise ratio of the signal as the same trend with the same probability, which illustrates that the values of ApEn, SampleEn, and FuzzyEn are affected by SNR of the signal. However, the PE has been the same all the time with a different probability, which illustrates that the PE has robust antinoise ability. Therefore, the PE is used for extracting the feature information of near-infrared spectral data of glucose solution in this study.

2.2. Fractal Dimension. Dimension, an important feature of geometry, characterizes the size of the space a shape occupies. Because Euclidean geometry objects are relatively regular shapes, the obtained dimension is an integer. However, Euclidean geometry is not applied for irregular complex shapes. Most natural geometries exhibit similar properties. Therefore, the spatial dimension of an object is not always an integer; it can also be fractional. The noninteger dimension is the real dimension of most geometric shapes; integers are only special cases. The noninteger dimension is a different concept, but it is suitable for all the geometric shapes in nature. In 1919, Hausdorff, who studied the properties of singular sets, first proposed the concept of fractal dimension [20] and defined the Hausdorff measure and dimension theory. Since then, several scholars have developed various dimensions, such as self-similar dimension, box dimension, information dimension, correlation dimension, and Lyapunov dimension. The box dimension is one of the most common fractal dimensions because of its ease of calculation, few parameters, and ease of application. The box dimension is defined as the way that the set X is covered by a hypercube with size ε.

X is a nonempty and bordered subset R^n. If it is covered by $N(\varepsilon)$ hypercube with length ε, then

$$D_B(X) = \lim_{\varepsilon\to 0}\frac{\ln N(\varepsilon)}{\ln(1/\varepsilon)}. \tag{26}$$

The above formula is the definition of the fractal box dimension. The steps are as follows:

(1) Set discrete signal $y(i) \subset Y$, and Y is the closed set in the n-dimensional Euclidean space R^n.

(2) Divide R^n with the ε grids as small as possible, and $N(\varepsilon)$ is the grid counts of set Y. The limit in the above formula cannot be determined by definition; so, an approximate method is used in calculation. The ε grid is used as the reference, and it is enlarged to the $k\varepsilon$ grid, where $k \in Z^+$. In this way, $N_{k\varepsilon}N(k\varepsilon)$ is the grid count of set Y in discrete space, and the following formula can be obtained:

$$P(k\varepsilon) = \sum_{i=1}^{N/k}\Big|\max\{y_{k(i-1)+1}, y_{k(i-1)+2}, \dots, y_{k(i-1)+k+1}\} - \min\{y_{k(i-1)+1}, y_{k(i-1)+2}, \dots, y_{k(i-1)+k+1}\}\Big|, \tag{27}$$

where $i = 1, 2, \dots, N/k$, $k = 1, 2, \dots, M$, $M < N$, and N is the number of sampling points.

The grid count is as follows:

$$N(k\varepsilon) = \frac{P(k\varepsilon)}{k\varepsilon} + 1, \tag{28}$$

where $N(k\varepsilon) > 1$. In the graph of $\lg k\varepsilon - \lg N(k\varepsilon)$, the scale-free region is determined with good linearity. The beginning and end points of the scale-free region are k_1 and k_2; thus,

$$\lg N(k\varepsilon) = a\lg k\varepsilon + b, \quad k_1 \le k \le k_2. \tag{29}$$

Finally, the slope of the line is determined by the least square method,

$$\widehat{a} = \frac{(k_2 - k_1 + 1)\sum \lg k \sum \lg N(k\varepsilon) - \sum \lg k \sum \lg N(k\varepsilon)}{(k_2 - k_1 + 1)\sum \lg^2 k - \left(\sum \lg k\right)^2}. \tag{30}$$

The box dimension D_B is as follows:

$$D_B = \widehat{a}. \tag{31}$$

2.3. Mutual Information. Information is the movement state and the way the movement state changes items that are felt and expressed by the cognitive subject. Two kinds of metric forms are identified for information. One measures how much information the message or message collection itself contains; another one is the measure of how much information is provided between messages or message sets. The former is described by self-information entropy and message entropy, whereas the latter is described by mutual information (MI) and average mutual information. Mutual information measures the degree of interdependence between two variables and represents the amount of shared information between two variables. For two given random variables X and Y, if their respective marginal probability distribution and joint probability distribution are $p(x), p(y)$, and $p(x, y)$, the definition of mutual information between them is as follows:

$$I(X; Y) = \sum_x \sum_y p(x, y) \log \frac{p(x, y)}{p(x)p(y)}. \tag{32}$$

When the variables X and Y are completely unrelated or independent with each other, the mutual information is the minimum 0, which means no information overlaps between the two variables. By contrast, the higher the degree of interdependence between the two variables is, the greater the value of mutual information will be, and the more similar the information it contains.

2.4. Partial Least Squares Regression. In 1984, Wold and Albano first proposed PLSR, which was a new multivariate statistical data analysis method and studied the regression modeling of multiple dependent variables [21].

Given p independent variables $\{x_1, x_2, \dots, x_p\}$ and q dependent variables, the study of the statistical relationship between the independent variable and dependent variable involves the observation of n sample points, which form the data tables X and Y of the independent variable and dependent variable. In PLSR, the components t_1 and u_1 are first extracted form X and Y; namely t_1 is the linear combination of $x_1, x_2, \dots, x_p$ and u_1 is the linear combination of $y_1, y_2, \dots, y_q$. After the first components t_1 and u_1 are extracted, the regression of X on t_1 and Y on u_1 is performed by PLSR. The accuracy of the model is validated; if the regression equation reaches satisfactory accuracy, the algorithm is terminated. Otherwise, the residual information of X interpreted by t_1 and the residual information of Y interpreted by u_1 will be extracted for the second round. This step is repeated until accuracy is satisfactory. Finally, if m components $t_1, t_2, \dots, t_m$ form X, PLSR can be expressed as the regression equation of y_k about the original variables $x_1, x_2, \dots, x_p$ by implementing regression with y_k $(k = 1, 2, \dots, q)$ on $t_1, t_2, \dots, t_m$.

2.5. Support Vector Regression. The support vector machine (SVM) is a new machine learning method proposed by Vapnik et al. based on statistical learning theory. SVM has the characteristics of small sample learning and strong generalization ability, which can avoid the problems of overlearning and local minimum. By introducing the insensitive loss function ε, Vapnik et al. have extended the SVM to the regression estimation of the nonlinear system and established the SVR algorithm. SVR has been widely used in function estimation, nonlinear system modeling, and other fields [22].

For the sample set $G = \{(x_i, y_i)\}_i^n$ (x_i is the input vector, y_i is the corresponding target value, and n is the number of samples), the SVR function is as follows:

$$f(x) = \omega\phi(x) + b, \tag{33}$$

where $\phi(x)$ is the nonlinear map transforming data into a high-dimension feature space and ω and b are coefficients.

2.6. Near-Infrared Spectra Data. In the near-infrared spectra noninvasive blood glucose measurement experiments, the blood glucose solution is temporarily replaced by glucose solution. All the glucose solutions, which concentration ranges from 50 mg/dL to 1000 mg/dL, are continuous and are equally distributed liquid that is uniformly configured under the same conditions. The prepared samples of glucose solution are put into the detection system of spectrometer. All the experimental data are collected by Fourier spectrometer Antaris II FT-NIR, produced by America Thermo Company. Its spectral range is 833–2630 nm, resolution is 4 cm^{-1} across spectral range, wavenumber reproducibility is better than 0.05 cm^{-1}, and wavenumber accuracy is ±0.03 cm^{-1}. The data meet the measurement principle of Lambert-Beer's law. All the collected near-infrared spectral data of glucose solution are measured five times with the same concentration in order to get a small statistical error and shown in Figure 4.

3. Results and Discussion

The principle of selecting optimal wave bands has two key points: (1) information on the selected bands is large and (2) the correlation between bands is small. The widely used extraction methods include the information comparison of each band, information correlation between bands, best

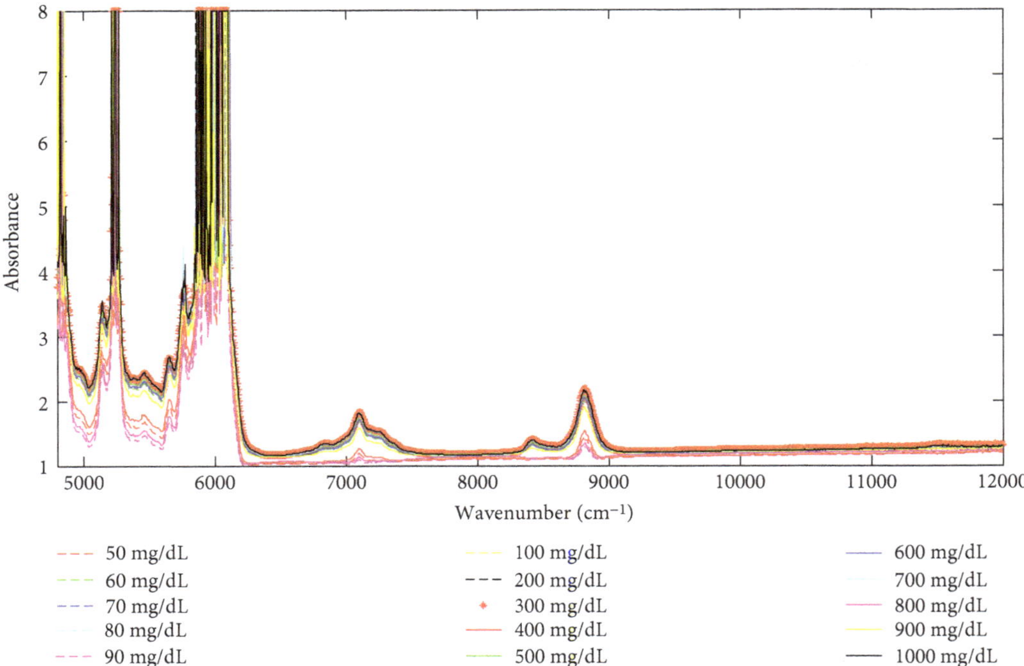

FIGURE 4: Near-infrared spectral data of glucose solution.

index method, entropy, and joint entropy of band data. In feature extraction, the fractal dimension (FD) can be used as a feature because almost all signals have fractal characteristics. The fractal dimension can distinguish different signals on the premise that two signals have different dimensions under the same measure. If two signals come from the same state, they have similarities, and their fractal dimensions are similar. The correlation or correlation coefficient can only reflect the linear correlation between two variables but cannot measure the nonlinear relationship between them. However, from the view of information theory, mutual information can estimate the total information amount between variables, is not limited to a linear relationship, and has greater advantages than correlation comparison. PE analysis can effectively determine the similarity between sequences and show the strong similarity and distinction, which can be further applied to biological sequence analysis. PE of the time series is calculated using the PE algorithm. The ratio value of PE is the basis for analyzing the similarity between sequences. Therefore, the PE, fractal dimension, and mutual information are used in extracting the feature information of near-infrared spectral data of glucose solution in this study.

In the PE calculation, there are three parameters needed to be considered and set, namely, data length N, embedded dimension m, and time delay λ. According to reference [23], λ has a little effect on the PE value and there are very small differences among the PE values with different λ. Therefore, $\lambda = 1$ is chosen in this study. In order to discuss the relationship of N and m with PE value, the PE values of the given signal $\mathrm{MIX}(N, p)$ with data length 128, 256, 512, 1024, and 2048 are calculated in Figure 5. Figure 5 shows that the PE values are almost same in $m = 2, 3, 4$ with different data length N. If m is too small, the algorithms lose significance and effectiveness because of the few states contained in reconstructed sequence. If m is too big, the time series will be homogenized by phase space reconstruction; thus, the computation is time-consuming and the subtle changes in sequence cannot be reacted. Therefor $m = 3$ and $N = 1867$, which is the length of collected spectra data, are chosen in this study.

Because PE cannot be affected by noise ($\lambda = 1$), the characteristic wave bands are extracted from the collected near-infrared spectroscopy data of the glucose solution. Full spectral wavelength data have 1867 points in total, which are divided into wavelength intervals with 50, 100, 150, and 200 points. The ratio of PE of the corresponding wavelength interval between 50 mg/dL, 500 mg/dL, 1000 mg/dL, and pure water solution, the ratio of FD of the corresponding wavelength interval between 50 mg/dL, 500 mg/dL, 1000 mg/dL, and pure water solution and the MI values of the corresponding wavelength interval between 50 mg/dL, 500 mg/dL, 1000 mg/dL, and pure water solution are shown in Figure 6. As shown in Figure 6, the ratios of FD values and MI values of each wavelength interval have no obvious difference. The PE values of some wavelength intervals are substantially consistent, and other wavelength intervals are significantly different. Therefore, the later wavelength intervals are the characteristic wave bands that contained abundant glucose concentration information. However, the PE values of wavelength intervals that are divided with less than 50 points or more than 200 points of different concentration spectra have no obvious distinguished law. In order to improve the precision and accuracy of feature wavelength extraction, the characteristic wavelength intervals are extracted in four uniform ways (50, 100, 150, and 200), and their overlapping

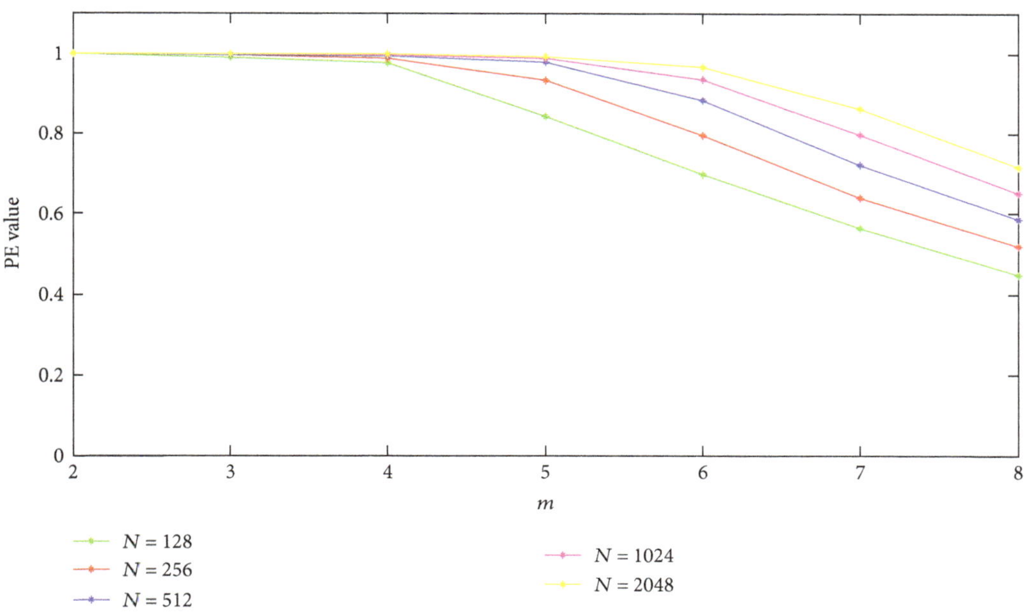

FIGURE 5: The PE value of different lengths N and embedded dimensions m of signal.

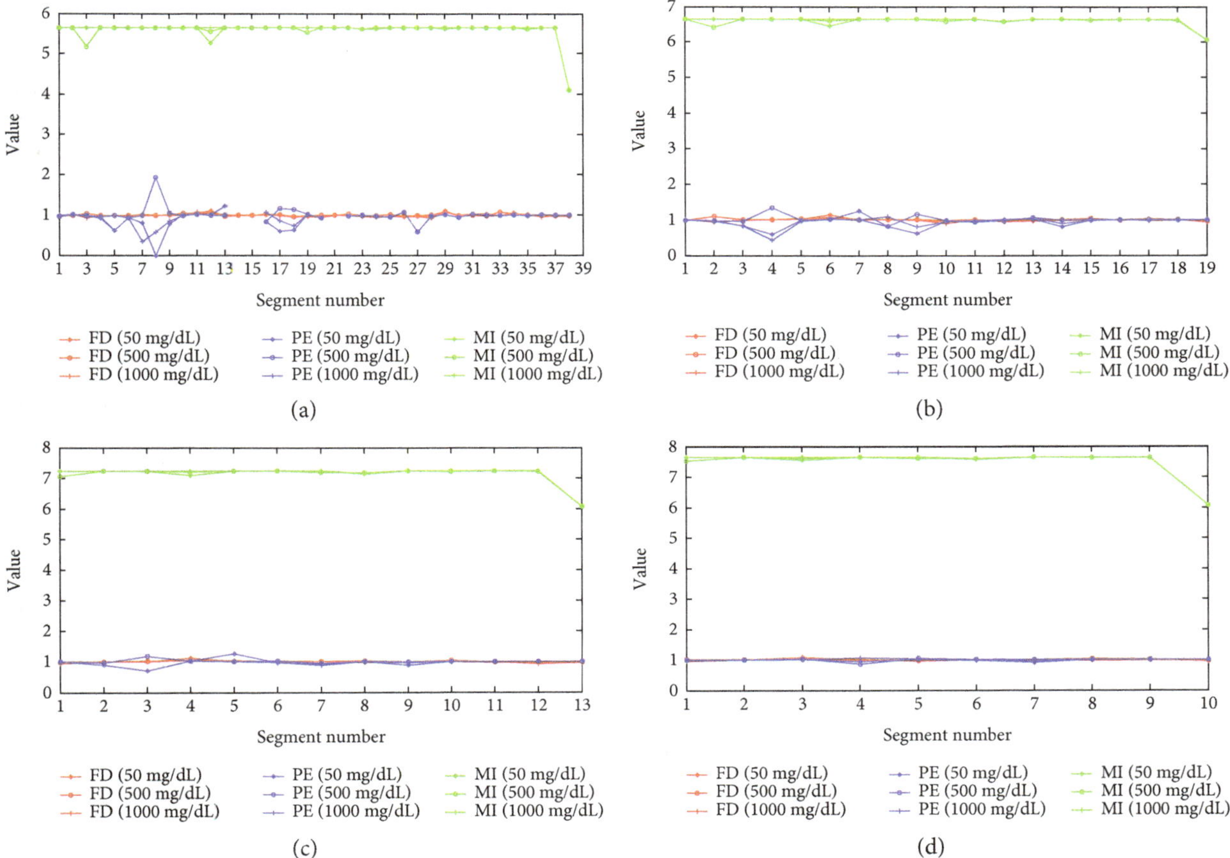

FIGURE 6: PE, FD, and MI of segmented spectral data: (a) 50 mg/dL, (b) 100 mg/dL, (c) 150 mg/dL, and (d) 200 mg/dL.

Table 1: Extraction of feature wave bands by PE.

Number	Segment number	Overlapping wavenumber (cm^{-1})
1	7, 8, 17, 19	6840–7230, 8680–8880
2	4, 9, 14	
3	3, 5	
4	4	

Table 2: R and RMSEP of SVR model.

Method	R	RMSEC (%)
PE	0.9998	0.0346
EMD-SPA	0.9892	0.0670
SPA	0.9542	0.0846
FD	0.9293	0.1050
MI	0.8990	0.1522
Full spectra data	0.8188	0.2499

Table 3: R and RMSEP of PLSR model.

Method	R	RMSEC (%)
PE	0.9897	0.0468
EMD-SPA	0.9688	0.0822
SPA	0.9321	0.1045
FD	0.9089	0.1276
MI	0.8880	0.1601
Full spectra data	0.8076	0.2519

intervals are considered as the final characteristic wavelength intervals (Table 1).

In order to verify the effectiveness of the proposed method, the characteristic wavelength intervals of the collected spectral data of glucose solutions with the FD method, MI method, the proposed method, successive projection algorithm (SPA) method [24], EMD-SPA method [25], and full spectral data are taken into the calibration models that were established by PLSR and SVR. The correlation coefficient and root mean square error correction (RMSEC) of the model are evaluated. The characteristic wavelength points that were extracted based on the PE method are 150, which is much less than the full spectral wavelength points, and the smaller selected characteristic wavelength points are, the shorter the established model time is. The experimental results of SVR and PLSR calibration model (Tables 2 and 3) show that the correlation coefficient (R) and RMSEC of established calibration model by characteristic wavelength intervals that were extracted based on the PE method reach 0.9998/0.9897 and 0.0346%/0.0468%. The results are better than that of the established calibration model by characteristic wavelength intervals that were extracted based on FD method, MI method, SPA method, EMD-SPA method, and by full spectra data. The overall modeling results of SVR are more reliable to that of PLSR modeling method.

4. Conclusions

The feature wave band extraction method that was based on permutation entropy that is proposed in this study for near-infrared spectra noninvasive blood glucose detection. The spectra data do not need denoise because PE has the advantages of robust antinoise and anti-interference abilities, and it is especially suitable for nonlinear data. Taking the near-infrared spectra data of glucose solutions as the object, all of the collected near-infrared spectra are divided with different interval points, and the ratio values of PE of corresponding spectra intervals are calculated. The overlap spectral intervals contained abundant glucose concentration information, which were extracted for reducing the effective range of data. Then, the PLSR and SVR methods are introduced for establishing the calibration models with characteristic spectra that were extracted by the proposed method, FD method, MI method, SPA method, EMD-SPA method, and full spectral data. According to the correlation coefficient and RMSEC of the calibration models, the proposed feature extraction method effectively solves the redundancy problem of near-infrared spectra data, and it also improves the robustness and predictive ability of regression model.

Acknowledgments

This work was supported by the Program for Harbin City Science and Technology Innovation Talents of Special Fund Project (Grant no. 2014RFXXJ065) and the Fundamental Research Funds for the Central Universities (Grant no. HIT. IBRSEM. 201307).

References

[1] L. Guariguata, D. R. Whiting, I. Hambleton, J. Beagley, U. Linnenkamp, and J. E. Shaw, "Global estimates of diabetes prevalence for 2013 and projections for 2035," *Diabetes Research and Clinical Practice*, vol. 103, no. 2, pp. 137–149, 2014.

[2] D. Ding, *Near Infrared Spectroscopy Research and Application in Biomedicine*, Ph.D. thesis, JiLin University, 2004.

[3] C. E. Ferrante do Amaral and B. Wolf, "Current development in non-invasive glucose monitoring," *Medical Engineering & Physics*, vol. 30, no. 5, pp. 541–549, 2008.

[4] A. Duncan, J. Hannigan, S. S. Freeborn et al., "A portable non-invasive blood glucose monitor," in *The International Conference on Solid-State Sensors and Actuators*, pp. 455–458, Stockholm, Sweden, June 1995.

[5] J. R. Blanco, F. J. Ferrero, J. C. Campo et al., "Design of a low-cost portable potentiostat for amperometric biosensors," in *IEEE Instrumentation & Measurement Technology Conference*, pp. 690–694, Sorrento, April 2006.

[6] J. Y. Tang, M. H. Wang, M. K. Chen, and L. S. Jang, "Glucose detection using an electro-optical fluidic device based on pulse width modulation," in *Seventh International Conference on Sensing Technology*, pp. 325–329, Wellington, December 2013.

[7] S. Ramasahayam, K. S. Haindavi, B. Kavala, and S. R. Chowdhury, "Noninvasive estimation of blood glucose using near infrared spectroscopy and double regression analysis," in *Seventh International Conference on Sensing Technology*, pp. 627–631, Wellington, December 2013.

[8] K. A. Unnikrishna Menon, D. Hemachandran, and T. K. Abhishek, "A survey on non-invasive blood glucose monitoring using NIR," in *International Conference on Communications and Signal Processing (ICCSP)*, pp. 1069–1072, Melmaruvathur, April 2013.

[9] C. E. Shannon, "A mathematical theory of communication," *Bell System Technical Journal*, vol. 27, no. 3, pp. 379–423, 1948.

[10] S. M. Pincus, "Approximate entropy as a measure of system complexity," *Proceedings of the National Academy of Sciences of the United States of America*, vol. 88, no. 6, pp. 2297–2301, 1991.

[11] J. S. Richman and J. R. Moorman, "Physiological time-series analysis using approximate entropy and sample entropy," *American Journal of Physiology Heart and Circulatory Physiology*, vol. 278, no. 6, pp. H2039–H2049, 2000.

[12] W. Chen, Z. Wang, H. Xie, and W. Yu, "Characterization of surface EMG signal based on fuzzy entropy," *IEEE Transactions on Neural Systems and Rehabilitation Engineering*, vol. 15, no. 2, pp. 266–272, 2007.

[13] C. Bandt and B. Pompe, "Permutation entropy: a natural complexity measure for time series," *Physical Review Letters*, vol. 88, no. 17, article 174102, 2002.

[14] L. Zunino, M. C. Soriano, I. Fischer, O. A. Rosso, and C. R. Mirasso, "Permutation-information-theory approach to unveil delay dynamics from time-series analysis," *Physical Review E, Statistical, Nonlinear, and Soft Matter Physics*, vol. 82, no. 4, Part 2, pp. 565–590, 2010.

[15] X. Li, D. Ph, S. Cui, and L. J. Voss, "Using permutation entropy to measure the electroencephalographic effects of sevoflurane," *Anesthesiology*, vol. 109, no. 3, pp. 448–456, 2008.

[16] Y. Liu, T.-S. Chon, H. Baek, Y. Do, J. H. Choi, and Y. D. Chung, "Permutation entropy applied to movement behaviors of *Drosophila melanogaster*," *Modern Physics Letters B*, vol. 25, no. 12n13, pp. 1133–1142, 2011.

[17] M. Zanin, L. Zunino, O. A. Rosso, and D. Papo, "Permutation entropy and its main biomedical and econophysics applications: a review," *Entropy*, vol. 14, no. 8, pp. 1553–1577, 2012.

[18] C. Bian, C. Qin, Q. D. Y. Ma, and Q. Shen, "Modified permutation-entropy analysis of heartbeat dynamics," *Physical Review E, Statistical, Nonlinear, and Soft Matter Physics*, vol. 85, no. 2, Part 1, article 021906, 2012.

[19] Z. Shi, W. Song, and S. Taheri, "Improved LMD, permutation entropy and optimized K-means to fault diagnosis for roller bearings," *Entropy*, vol. 18, no. 3, p. 70, 2016.

[20] A. Accardo, M. Affinito, M. Carrozzi, and F. Bouquet, "Use of the fractal dimension for the analysis of electroencephalographic time series," *Biological Cybernetics*, vol. 77, no. 5, pp. 339–350, 1997.

[21] S. Wold, C. Albano, W. J. Dunn et al., "Modelling data tables by principal components and PLS: class patterns and quantitative predictive relations," *Analusis*, vol. 12, no. 10, pp. 477–485, 1984.

[22] A. J. Smola and B. Schölkopf, "A tutorial on support vector regression," *Statistics and Computing*, vol. 14, no. 3, pp. 199–222, 2002.

[23] J. Zheng, J. Cheng, and Y. Yang, "Multiscale permutation entropy based rolling bearing fault diagnosis," *Shock and Vibration*, vol. 1, pp. 1–8, 2014.

[24] G. Y. Hui, L. J. Sun, J. N. Wang, L. K. Wang, and C. J. Dai, "Research on the pre-processing methods of wheat hardness prediction model based on visible-near infrared spectroscopy," *Spectroscopy and Spectral Analysis*, vol. 36, pp. 2111–2116, 2012.

[25] Z. Y. Zhang, G. Li, L. Lin, and B. J. Zhang, "Application of EMD and SPA algorithm in the detection of benzoyl peroxide addition in flour by spectroscopy," *Spectroscopy and Spectral Analysis*, vol. 32, no. 10, pp. 2815–2819, 2012.

17

Shot-Noise Limited Time-Encoded Raman Spectroscopy

Sebastian Karpf,[1] Matthias Eibl,[2] Wolfgang Wieser,[3] Thomas Klein,[3] and Robert Huber[2]

[1]*Department of Electrical Engineering, University of California, Los Angeles, Los Angeles, CA, USA*
[2]*Institut für Biomedizinische Optik, Universität zu Lübeck, Peter-Monnik-Weg 4, 23562 Lübeck, Germany*
[3]*Optores GmbH, Gollierstr. 70, 80339 Munich, Germany*

Correspondence should be addressed to Robert Huber; robert.huber@bmo.uni-luebeck.de

Academic Editor: Christoph Krafft

Raman scattering, an inelastic scattering mechanism, provides information about molecular excitation energies and can be used to identify chemical compounds. Albeit being a powerful analysis tool, especially for label-free biomedical imaging with molecular contrast, it suffers from inherently low signal levels. This practical limitation can be overcome by nonlinear enhancement techniques like stimulated Raman scattering (SRS). In SRS, an additional light source stimulates the Raman scattering process. This can lead to orders of magnitude increase in signal levels and hence faster acquisition in biomedical imaging. However, achieving a broad spectral coverage in SRS is technically challenging and the signal is no longer background-free, as either stimulated Raman gain (SRG) or loss (SRL) is measured, turning a sensitivity limit into a dynamic range limit. Thus, the signal has to be isolated from the laser background light, requiring elaborate methods for minimizing detection noise. Here, we analyze the detection sensitivity of a shot-noise limited broadband stimulated time-encoded Raman (TICO-Raman) system in detail. In time-encoded Raman, a wavelength-swept Fourier domain mode locking (FDML) laser covers a broad range of Raman transition energies while allowing a dual-balanced detection for lowering the detection noise to the fundamental shot-noise limit.

1. Introduction

Optical spectroscopy can yield a wealth of information for biomedical imaging and diagnostics. In fluorescent imaging, for example, a set of diverse fluorescent dyes can help to visualize different functional sites and dynamics at the subcellular level. The high optical resolution and the accessibility of the information play a crucial role in the study of biological research and diagnostics. Raman spectroscopy promises to provide this diverse information by inelastic light scattering, without the need of staining the sample. The big advantage of this label-free operation is that it is noninvasive and does not risk altering the biological function of the sample. However, for in vivo application of Raman spectroscopy, high speed and specificity is needed, so that prominent candidates employ nonlinear Raman techniques for signal level enhancements [1–16].

We previously reported on a new technique for nonlinear, stimulated Raman scattering (SRS), called time-encoded Raman (TICO-Raman) technique [17]. In TICO-Raman, the SRS effect is encoded and detected in time domain rather than in frequency domain. Latest developments in telecommunication-driven, high-speed analog-to-digital converter (ADC) technology promise faster acquisition speeds in Raman spectroscopy [15]. Broadband spectra are recorded by encoding the Raman spectral coverage in time by a wavelength-swept Fourier domain mode locking (FDML) laser [18]. An unambiguous time-to-Raman energy mapping is achieved when the Raman signal height is measured as intensity change of the FDML laser. A unique set of measures were introduced that result in a maximally reduced noise contribution to the SRS spectra, where the ultimate shot-noise limit was reached. In this paper, we show the signal extraction process in detail and visualize the two-stage,

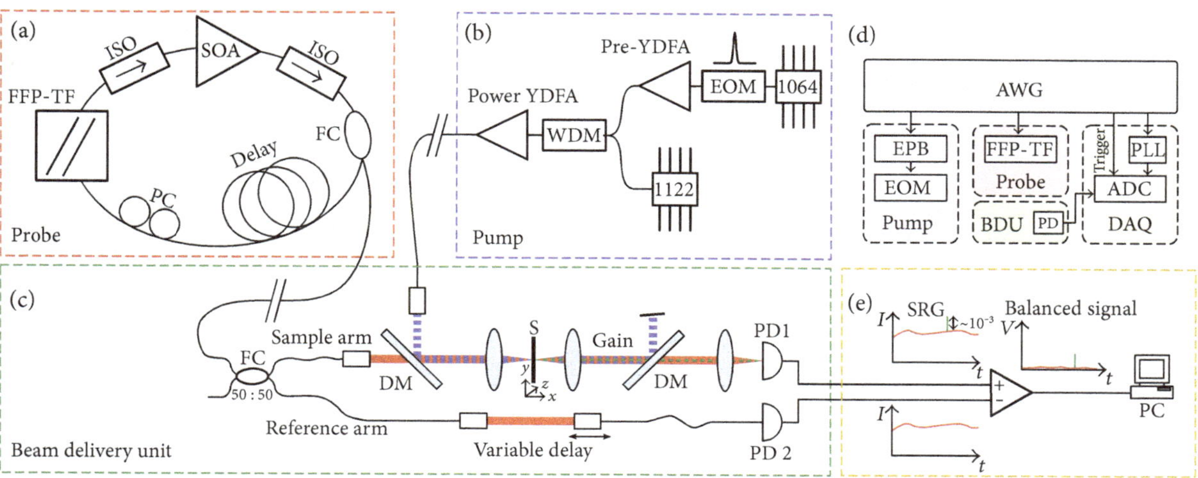

FIGURE 1: The schematic of the shot-noise limited TICO-Raman spectroscopy system is presented. (a) The FDML probe laser comprises a fiber Fabry-Pérot tunable filter (FFP-TF), optical isolators (ISO) for unidirectional lasing, a semiconductor optical amplifier (SOA), a fiber-coupler (FC) output, a resonant time delay, and a polarization controller (PC) paddle. (b) The pump laser is a home-built master oscillator fiber power amplifier (MOFPA) with additional Raman shifter. It consists of a narrow-line laser diode at 1064 nm, followed by a fast electro-optical amplitude modulator (EOM). This EOM is used to digitally modulate the TICO pulse pattern and to adjust the pulse length to the detection bandwidth. The pulses are amplified by ytterbium-doped fiber amplifiers (YDFA). A wavelength-division multiplexer (WDM) can be used to couple in additional 1122 nm seed light for Raman shifting inside the double-clad YDFA power amplifier. (c) The pump and probe lasers are collimated and combined by dichroic mirrors (DM) before being focused onto the Raman active sample (S). After recollimation, the pump light is blocked by additional dichroic filters and detected on a photodiode (PD1). A second photodiode (PD2) is used to detect reference light for differential, balanced detection. (e) The differential transimpedance amplifier subtracts FDML offset light and common mode noise, so only intensity changes induced in the Raman-active sample are digitized by the data acquisition (DAQ). (d) The whole system is synchronized by an arbitrary waveform generator (AWG).

analog and digital balanced detection mechanism for shot-noise limited performance.

2. Materials and Methods

The system comprises two home-built lasers, an FDML Raman probe laser and a master oscillator fiber power amplifier (MOFPA) laser as a programmable Raman pump laser (Figure 1). The home-built MOFPA laser incorporates a fast electro-optical modulator (EOM) which allows for active modulation of the pulses. The system employs a pulse pattern that probes all possible Raman transitions via the TICO-Raman technique [17]. The detection comprises a balanced photodetector and a fast analog-to-digital converter (ADC) card. The pump pulse length is digitally set to match the analog bandwidth of the detection system. As the SRS effect is an instantaneous effect, the stimulated Raman gain (SRG) signal follows the time characteristic of the Raman pump pulse. By matching the pump pulse length to the detection bandwidth, the SRG effect height is fully and optimally digitized.

The process from SRG signal detection to Raman spectra generation is demonstrated using SRG data of neat benzene (Figure 2). The SRG signal is recorded as power change of the probe laser power. We employ a balanced detector with differential amplifier to remove the DC offset and only detect relative power changes. This technique has two advantages: (a) laser noise can be reduced by the common mode rejection of the photoreceiver and (b) the range of the ADC can be utilized optimally to digitize the power change and no bits are lost to digitizing the DC offset. The first analog balancing step consists of the differential amplification of the FDML probe light (rainbow color, left). The probe laser is split in equal parts, one with pump laser interaction and the other without, thus serving as reference light. This analog balancing step additionally leads to common mode noise rejection. For highest possible common mode noise rejection ratio, it is crucial to adjust the two arm lengths to match as closely as possible in order to match the phase of any common electronic noise. Considering the bandwidth of our detection of 400 MHz, the fastest electronic signal may have a period of 2.5 ns. This corresponds to 2Pi phase. For a differential balancing suppression of a factor of 1000, the phase difference of the two arms should be less than one thousandth of Pi/2 (linear approximation), that is, less than 625 fs. In free-space, this corresponds to a distance travelled by light of $3E8*625E-15 \approx 190\,\mu m$. We employed a micrometer screw in the reference arm to adjust for this length. Furthermore, the power was adjusted to carefully match the two arm lengths' powers.

A zoom into the subtracted, amplified signal is presented in the second column of Figure 2. Shown are 512 consecutive sweeps of the FDML laser centered at 0 mV. They only contain intensity changes occurring in the sample arm. First, this shows the very good sweep-to-sweep correlation of the FDML laser. Even though a residual modulation is visible, which stems from chromatic imperfections between the sample arm and the reference arm, it is clear that this spurious

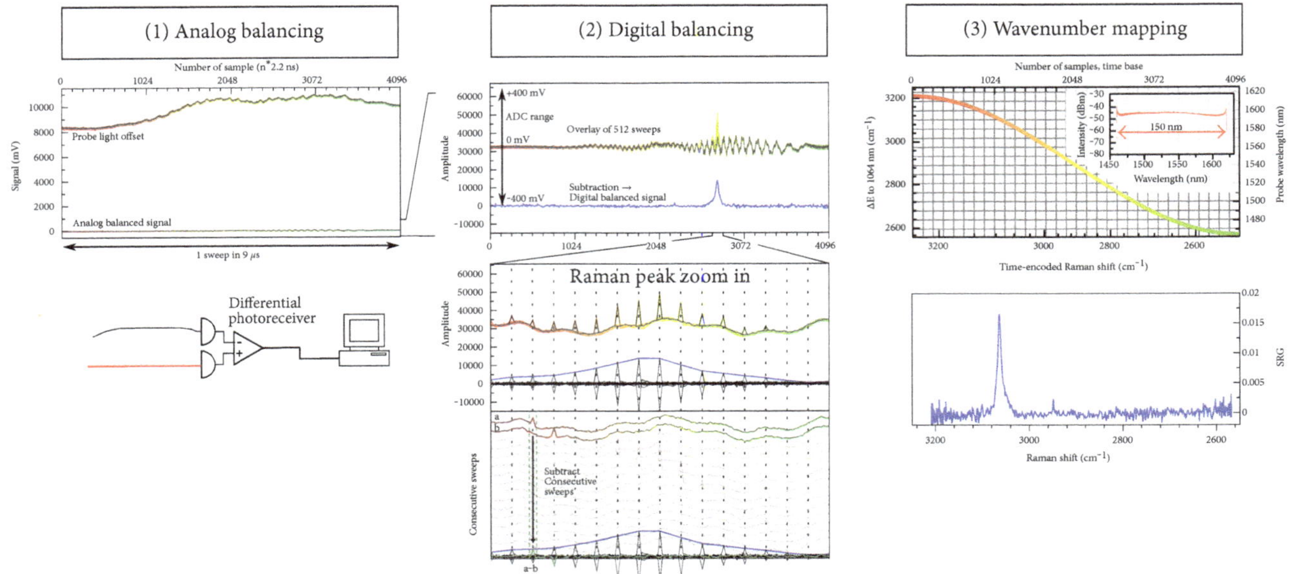

FIGURE 2: Raman spectra extraction by two-stage balanced detection, leading to an SRG spectrum of benzene. Left: The first analog balancing step employs a balanced, differentially amplified photoreceiver to remove the probe light offset. Only power changes in the sample arm remain (cf. Figure 1). Middle: The analog-to-digital converter (ADC) card is used to synchronously sample the FDML probe laser sweeps. The 800 mV range is sampled at 12-bit depth resolution. The lower part shows a zoom-in on a part where SRG signals occur. The vertical lines represent the positions of the pump pulses following the TICO-Raman excitation technique [17]. The SRG signals are clearly visible and fully recorded as ensured by the synchronized detection (see text). The second digital balancing step subtracts consecutive sweeps to further lower the noise level to the shot-noise limit. Only pump-induced intensity changes remain. This step makes use of TICO-Raman pattern, where pulse positions increase from sweep to sweep. Thus, through subtraction, any artifacts due to chromatic imperfections are removed. Right: The last step assigns wavenumber values to the Raman spectrum. First, the spectrum of the FDML laser (inset) is recorded on an optical spectrum analyzer (OSA). Then, the wavelength-to-time mapping of the FDML laser is calculated and from that the energy difference to the pump laser is determined. The result is accurate Raman shift energies for each spectral point. Shown in blue is the final shot-noise limited TICO-Raman spectrum of benzene, acquired in 9 ms.

signal is present on all 512 sweeps and can thus be subtracted digitally. Hence, we call this second step digital balancing. First, the 800 mV range of the ADC (Alazartech ATS9360) samples the balanced intensity changes of the FDML probe laser at 12-bit depth resolution. Now, consecutive sweeps are subtracted (lower graph, middle column) to remove any remaining intensity differences between sample and reference arm. The SRG signals are not subtracted, as the pump pulse position increases stepwise for consecutive sweeps in the TICO-Raman technique (Figure 2, lower middle column). The effect of this digital balancing step is that only pump intensity changes remain. The result is shown in blue and already represents the SRG Raman spectrum. This step ultimately brings the noise level down to the fundamental shot-noise limit, which lies at around 3.6×10^{-4} of the probe laser power of 2 mW. To assess the overall advantage of the dual balancing method, we will compare the dynamic range increase. First, the DC-offset of the FDML probe laser light lies around 10 V (cf. Figure 2). Calculating the effective number of bits (ENOB) for the 12-bit ADC using $\text{ENOB} = (\text{SNR} - 1.76\,\text{dB})/6.02$ and 57 dB SNR (manufacturer datasheet) gives ~9-bit ENOB resolution for the used Alazartech ATS9360. Thus, the full 800 mV scale is divided into 512 units. Thus, with DC-offset, a measurement sensitivity at the shot-noise level of 10^{-4} is not possible. Consequently, our first analog balancing step reduces the required voltage range from 10 V to 800 mV, that is, a factor of 12.5. Hence, the effective voltage scale division after this step is $512 * 12.5 = 6400$, which increases the dynamic range such that digitization can distinguish relative changes of $1/6400 = 1.56 \times 10^{-4}$, that is, over a factor of 2 smaller than the relative shot-noise. In the second balancing step, following the digitization step, the aim is to reduce the contribution of any digitized noise or residual modulation. By subtracting consecutive sweeps, the digital balancing steps act as a filter which increases the dynamic range. Before this digital balancing step, the residual modulation height is about 1/7th of the 800 mV range, that is, ~120 mV. After digital balancing, all of the additional noise and modulation contributions are cancelled out and the spectrum is recorded at its shot-noise limit at 3.6×10^{-4} relative signal height. In the last step of the TICO-Raman technique, Raman transition energies are assigned to the recorded SRG values. This is shown in the right column of Figure 2. The well-defined wavelength-to-time characteristic of the FDML allows mapping energy differences between pump laser and FDML sweep by recording a spectrum of the FDML (see also supplementary information of [17]). In combination, these steps enable the recording of Raman spectra of SRG values with accurate Raman transition energy values.

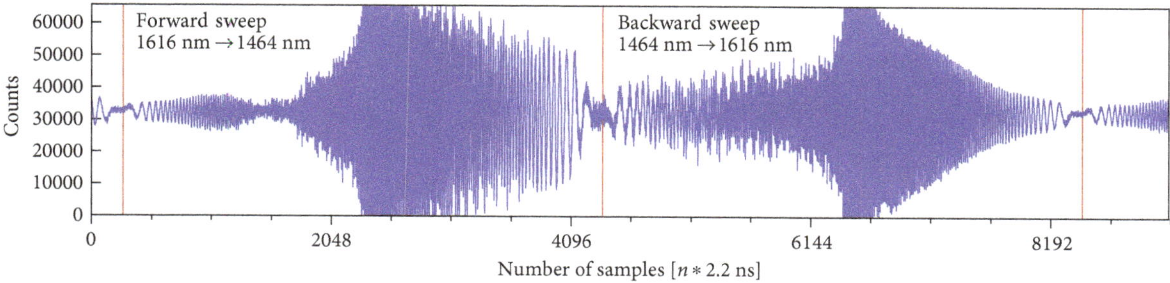

FIGURE 3: Strategy for time synchronization of the FDML sweep to the TICO-Raman technique. Interference fringes from the FDML probe light are generated by inserting an objective glass slide (~1 mm thick) into the beam. These fringes are used to determine the phase of the FDML sweep. This phase can then be adjusted on the arbitrary waveform generator (AWG), such that the first pulse (vertical line) of the TICO-Raman pattern [17] coincides with the lowest wavelength of the sweep. This is crucial for a correct time-to-wavenumber mapping in the TICO-Raman technique.

Precise timing and synchronization is crucial for the TICO-Raman technique. The whole system is driven by an arbitrary waveform generator (AWG), which electronically drives the Fabry-Pérot filter in the FDML laser. A second channel of the AWG is used to generate the TICO-Raman pump pulse pattern [17]. A third channel is used to trigger the acquisition and to provide a synchronized sample clock to the ADC card by means of a phase-locked loop (PLL). This electronic synchronization of the whole system requires an initial setting of the phases according to the relative time delays. In order to correctly time the start of the FDML probe laser sweep, an objective glass slide is inserted in the free-space beam. This ~1 mm thick glass creates interference fringes on the FDML light (see Figure 3). These interference fringes can be used to adjust the phase of the FDML sweep to the start of the TICO-Raman pattern. When the probe and pump laser are synchronized, the glass plate is removed. Now, the detection is synchronized. This step requires more accurate fine-tuning as the 1.8 ns pump pulse length needs to be perfectly timed with respect to the ADC sample clock. To this end, the pulses are accurately timed by employing a home-built electronic pulse board (EPB) by adjusting a delay with respect to the trigger. The delay is set digitally (via USB) from 0 ns to 10 ns with 0.01 ns accuracy. Then, the pump wavelength filters are removed, the FDML laser blocked, and the pump pulses are recorded on the ADC. Low power pump pulses without amplifiers are employed, so a saturation of the photodetector is avoided. The digitized pulse height is set maximal by tuning the time delay on the EPB. This ensures full effect height recordings of the instantaneous SRG effect (following the pump pulse timing). Once the delay on the EPB is set, the pump blocking filters are put back and the FDML laser is unblocked. Finally, a DC-coupled version of the FDML probe sweep is recorded, such that the SRS values can be transformed to SRG values as intensity changes of the probe light power. Therefore, a 99/1 fiber coupler after the FDML laser output (before the beam delivery unit) provides a 1% output, which is plugged into a DC-coupled photodiode of known transimpedance gain. This curve is recorded on the second channel of the ADC and adjusted by the power and transimpedance difference of the sample path (cf. Figure 2, left).

3. Results and Discussion

To assess possible noise limitations, it is important to test experimentally how the noise level decreases upon averaging and how close it lies to the ideal square root behavior. To test this, we acquired a series of spectra with different averaging values. Figure 4 shows the TICO-Raman spectrum of benzene, from single acquisition up to 10000 times averaging. The recording time for a single spectrum was 9 ms. Upon averaging, two smaller Raman transitions around 3180 cm^{-1} become clearly visible. For quantitative analysis of the improvement, we analyzed the signal-to-noise ratio (SNR) by considering the SRG signal height of 1.69×10^{-3} for the 3063 cm^{-1} peak of benzene and dividing by the standard deviation in regions where no Raman transitions occur. We found, that the SNR increases from 34 to 508 when considering the spectral region from 2800 cm^{-1} to 2900 cm^{-1} (region I, 52 samples). However, when comparing to region II (30 samples), the SNR increases up to 1594. This 3-fold difference can be explained by an underlying baseline in the spectrum, which supposedly stems from cross-phase modulation (XPM). The effect shows a chromatic dependency, increasing with decreasing wavenumber values. This linearly increasing baseline adds to the standard deviation in region I. However, in region II, the FDML sweep is close to its turning point, where the sweep direction changes. Therefore, in region II, almost no change in wavelength occurs. Hence, in region II, the effect is nearly constant and has no effect on the standard deviation, so the noise decreases close to the expected value from the ideal square root behavior (Figure 4, right).

The effectiveness of the signal detection is best shown by considering the signal-to-noise ratio (SNR). The SNR depends on both the noise level, which is shot-noise limited here as explained, and the signal level. In order to fully record the intensity change, it is crucial to fully resolve the time response of the probe laser. We used pump pulses which directly correspond to the analog bandwidth of the detection system, such that the whole signal due to SRG interaction is digitized. This direct detection makes the recovery of absolute transmission changes straightforward, and it is possible since we employ nanosecond lasers. The fact that the light

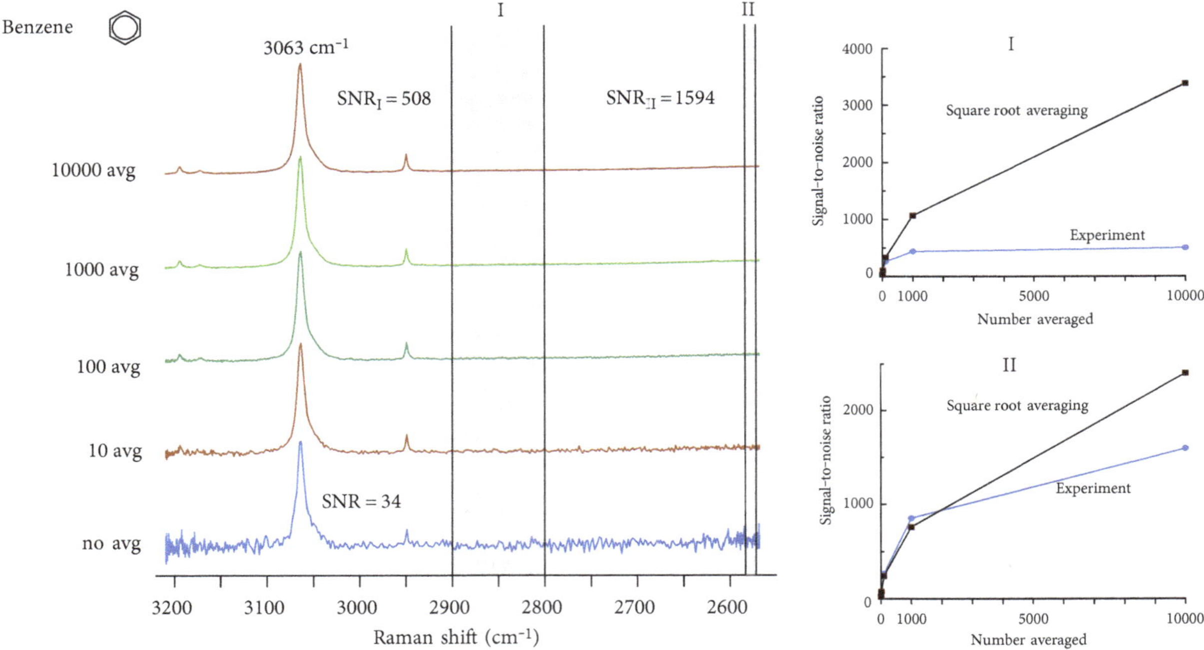

FIGURE 4: The result of averaging on the TICO-Raman spectrum of benzene. The lower, blue spectrum is the nonaveraged spectrum. The signal (3060 cm^{-1} peak) is 34 times higher than the standard deviation of the areas in I or II. Averaging leads to much increased Raman spectra and new spectral features become visible. The different spectra were vertically offset for clarity. The increase in signal-to-noise ratio (SNR) over the amount of averaging is plotted on the right, where two different regions of the spectrum were used to calculate the noise. Region I shows a higher standard deviation due to the contribution of an additional effect—we assume cross-phase modulation (XPM, see text). In region II, this effect is constant and does not enter the standard deviation, leading to an almost optimal square root increase in SNR.

is delivered through a single mode fiber with a defined beam profile additionally helps any quantitative measurements of Raman cross sections.

Our SRG signal levels are in very good agreement with values previously reported in literature [19]. Owyoung et al. report a value of 1.75×10^{-5} for the intense Raman transition of benzene at 992 cm^{-1} using 50 mW of pump power. We can compare this value to our recorded SRG intensity of the 3060 cm^{-1} line of benzene (Figure 2). Taking into consideration the quadratic dependency of the pump and probe wavelengths (factor 4.5), the employed peak pump power of 1.6 kW, and a factor of 6.9 for the lower spectral cross section of the 3060 cm^{-1} line compared to the 992 cm^{-1} line, one calculates an expected SRG signal of 1.81×10^{-2}. Our measured value of 1.69×10^{-2} agrees very well and shows the advantage of using the nanosecond approach, especially if quantitative measurements are desired. One interesting new approach is the application of ultrashort pulses in combination with photonic time stretch, effectively enhancing the detection bandwidth and sampling in time domain [15].

We also found imperfections with the lenses when using wavelengths in the extended near infrared (exNIR). The current study employed standard fused glass lenses (cf. [17]). Chromatic aberrations currently lead to imperfections of the recorded SRG values over the whole spectral range. If quantitative measurements are to be performed, it is important that better engineered exNIR objectives are employed ensuring same spot sizes for pump and probe lasers along the broad spectral coverage.

4. Conclusion

We presented strategies for a shot-noise limited system for broadband SRS spectroscopy by employing a wavelength swept FDML laser and a time domain-based acquisition. Measures were described in detail how to effectively reduce the detection noise down to the fundamental shot-noise limit. These include one analog and one digital balancing step. The nanosecond pulses of our pump laser match the detection bandwidth, allowing a full recovery of the SRG signal height. This capability can lead to precise quantitative measurements while effect heights are nonlinearly enhanced due to SRS. The laser output from a single mode fiber with a well defined mode field diameter further aids quantitative analysis. Future investigations will focus on improving the sensitivity of the system through enhancing the SNR. To this end, the relative shot-noise will be lowered by employing higher, pulsed probe laser powers. Another promising approach is increasing the detection bandwidth for shorter pump pulses, thereby gaining linearly in effect height while the shot-noise contribution only scales with the square root. Overall, this powerful new tool for SRS spectroscopy can propel a biological and medical application in label-free imaging and diagnostics.

Acknowledgments

The authors thank A. Vogel from the University of Lübeck and further acknowledge the funding from the European Union project ENCOMOLE-2i (Horizon 2020, ERC CoG no. 646669) and the German Research Foundation (DFG project HU1006/6).

References

[1] E. Ploetz, S. Laimgruber, S. Berner, W. Zinth, and P. Gilch, "Femtosecond stimulated Raman microscopy," *Applied Physics B: Lasers and Optics*, vol. 87, no. 3, pp. 389–393, 2007.

[2] C. W. Freudiger, W. Min, B. G. Saar et al., "Label-free biomedical imaging with high sensitivity by stimulated Raman scattering microscopy," *Science*, vol. 322, no. 5909, pp. 1857–1861, 2008.

[3] A. Gambetta, V. Kumar, G. Grancini et al., "Fiber-format stimulated-Raman-scattering microscopy from a single laser oscillator," *Optics Letters*, vol. 35, no. 2, pp. 226–228, 2010.

[4] S. Bégin, B. Burgoyne, V. Mercier, A. Villeneuve, R. Vallée, and D. Côté, "Coherent anti-Stokes Raman scattering hyper-spectral tissue imaging with a wavelength-swept system," *Biomedical Optics Express*, vol. 2, no. 5, pp. 1296–1306, 2011.

[5] M. Baumgartl, T. Gottschall, J. Abreu-Afonso et al., "Alignment-free, all-spliced fiber laser source for CARS microscopy based on four-wave-mixing," *Optics Express*, vol. 20, no. 19, pp. 21010–21018, 2012.

[6] Y. Ozeki, W. Umemura, Y. Otsuka et al., "High-speed molecular spectral imaging of tissue with stimulated Raman scattering," *Nature Photonics*, vol. 6, no. 12, pp. 845–851, 2012.

[7] D. Zhang, P. Wang, M. N. Slipchenko, D. Ben-Amotz, A. M. Weiner, and J.-X. Cheng, "Quantitative vibrational imaging by hyperspectral stimulated Raman scattering microscopy and multivariate curve resolution analysis," *Analytical Chemistry*, vol. 85, no. 1, pp. 98–106, 2012.

[8] M. Ji, D. A. Orringer, C. W. Freudiger et al., "Rapid, label-free detection of brain tumors with stimulated Raman scattering microscopy," *Science Translational Medicine*, vol. 5, no. 201, p. 201ra119, 2013.

[9] T. Ideguchi, S. Holzner, B. Bernhardt, G. Guelachvili, N. Picque, and T. W. Hansch, "Coherent Raman spectro-imaging with laser frequency combs," *Nature*, vol. 502, no. 7471, pp. 355–358, 2013.

[10] C.-R. Hu, M. N. Slipchenko, P. Wang et al., "Stimulated Raman scattering imaging by continuous-wave laser excitation," *Optics Letters*, vol. 38, no. 9, pp. 1479–1481, 2013.

[11] Z. Meng, G. I. Petrov, and V. V. Yakovlev, "Microscopic coherent Raman imaging using low-cost continuous wave lasers," *Laser Physics Letters*, vol. 10, no. 6, p. 065701, 2013.

[12] C. H. Camp Jr, Y. J. Lee, J. M. Heddleston et al., "High-speed coherent Raman fingerprint imaging of biological tissues," *Nature Photonics*, vol. 8, no. 8, pp. 627–634, 2014.

[13] P. Berto, E. R. Andresen, and H. Rigneault, "Background-free stimulated Raman spectroscopy and microscopy," *Physical Review Letters*, vol. 112, no. 5, p. 053905, 2014.

[14] M. J. B. Moester, F. Ariese, and J. F. de Boer, *Optimized Signal-to-Noise Ratio with Shot Noise Limited Detection in Stimulated Raman Scattering Microscopy*, 2015.

[15] F. Saltarelli, V. Kumar, D. Viola et al., "Broadband stimulated Raman scattering spectroscopy by a photonic time stretcher," *Optics Express*, vol. 24, no. 19, pp. 21264–21275, 2016.

[16] H. Kerdoncuff, M. R. Pollard, P. G. Westergaard, J. C. Petersen, and M. Lassen, *Probing Molecular Symmetry with Polarization-Sensitive Stimulated Raman Spectroscopy*, 2016, https://arxiv.org/abs/1611.06252.

[17] S. Karpf, M. Eibl, W. Wieser, T. Klein, and R. Huber, "A time-encoded technique for fibre-based hyperspectral broadband stimulated Raman microscopy," *Nature Communications*, vol. 6, p. 6784, 2015.

[18] R. Huber, M. Wojtkowski, and J. G. Fujimoto, "Fourier domain mode locking (FDML): a new laser operating regime and applications for optical coherence tomography," *Optics Express*, vol. 14, no. 8, pp. 3225–3237, 2006.

[19] A. Owyoung and E. D. Jones, "Stimulated Raman spectroscopy using low-power cw lasers," *Optics Letters*, vol. 1, no. 5, pp. 152–154, 1977.

18

Rapid Detection of Pesticide Residues in Chinese Herbal Medicines by Fourier Transform Infrared Spectroscopy Coupled with Partial Least Squares Regression

Tianming Yang, Rong Zhou, Du Jiang, Haiyan Fu, Rui Su, Yangxi Liu, and Hanbo Su

School of Pharmaceutical Sciences, South-Central University for Nationalities, Wuhan 430074, China

Correspondence should be addressed to Haiyan Fu; fuhaiyan@mail.scuec.edu.cn

Academic Editor: Khalique Ahmed

This paper reports a simple, rapid, and effective method for simultaneous detection of cartap (Ca), thiocyclam (Th), and tebufenozide (Te) in Chinese herbal medicines including *Radix Angelicae Dahuricae* and *Liquorices* using Fourier transform infrared spectroscopy (FT-IR) coupled with partial least squares regression (PLSR). The proposed method can handle the intrinsic interferences of herbal samples; satisfactory average recoveries attained from near-infrared (NIR) and mid-infrared (MIR) PLSR models were 99.0 ± 10.8 and $100.2 \pm 1.0\%$ for Ca, 100.2 ± 6.9 and $99.7 \pm 2.5\%$ for Th, and 99.1 ± 6.3 and $99.6 \pm 1.0\%$ for Te, respectively. Furthermore, some statistical parameters and figures of merit are fully investigated to evaluate the performance of the two models. It was found that both models could give accurate results and only the performance of MIR-PLSR was slightly better than that of NIR-PLSR in the cases suffering from herbal matrix interferences. In conclusion, FT-IR spectroscopy in combination with PLSR has been demonstrated for its application in rapid screening and quantitative analysis of multipesticide residues in Chinese herbal medicines without physical or chemical separation pretreatment step and any spectral processing, which also implies other potential applications such as food and drug safety, herbal plants quality, and environmental evaluation, due to its advantages of nontoxic and nondestructive analysis.

1. Introduction

Chinese herbal medicines (CHMs) which were considered to be gentle, nontoxic, and even harmless have been widely used as a means of medication or dietary supplement [1, 2]. In order to prevent, repel, or mitigate the effects of pest, the commercial cultivation of CHMs receives frequent application of diverse pesticides which are highly effective and broad-spectrum but have long half-life, complex degradation, and highly toxic substances. Consequently, the widespread use of pesticides poses high risks to the environment and induces heavy adverse effects on human health [3–6]. Facing such a serious crisis, for safety and health, it is therefore important to establish an effective routine method for quantifying multipesticide residues in Chinese herbal medicines.

Based on the available literature, a number of analytical methods have been proposed for the determination of pesticide residues in different matrices [7–12], including gas chromatography (GC), gas chromatography-mass spectrometry (GC-MS), high performance liquid chromatography (HPLC), and ultrahigh performance liquid chromatography-mass spectrometry (UPLC-MS). Unfortunately, these techniques are time-consuming and reagent-demanding and require highly skilled operators. Therefore, a faster, more accurate, and sensitive identification method is urgent to be developed for its practical application. In recent years, Fourier transform near-infrared (NIR) and mid-infrared (MIR) spectroscopies with advantages such as high efficiency, low cost, simply measuring, little sample preparation, quick data analysis, and nondestructive analytical technique have been widely used in several scientific fields, such as medical and biomedical, food science, pharmaceutical, and petroleum industries [13–17]. However, it should be noticed that only a few papers have described analytical approaches for monitoring multipesticide residues in Chinese herbal medicines by using FT-IR. This is due to the fact that fingerprint information of multipesticide residues in Chinese herbal provided by NIR and MIR spectra may be difficult to directly interpret due to low resolution and overlapping peak bands, so effective and robust chemometrics methods have been extensively concerned to extract and relate the abundant IR spectra information [18–20].

The main role of chemometrics for quantitative analysis of NIR and MIR data is to establish a quantitative model relating the measured NIR signals to certain properties of samples, say the component content. The quantitative model was subsequently applied to predict the same properties of samples in the prediction set. So, a good prediction result relies a good multivariate calibration method. Many methods such as principal component regression (PCR) [21], multiple linear regression (MLR) [22], and partial least squares (PLS) [23] are often used for chemometric calibration. In particular, partial least squares regression (PLSR), as a well-performed multivariate data analysis technique which generalizes and combines features from principal component analysis and multiple linear regression, is especially suitable for modeling from a full spectrum [24]. It is a procedure used to relate a large number of independent variables (predictors) to one (PLSR1) or few (PLSR2) response variables (observations) when a reduced number of cases are available and is useful in predicting a set of dependent variables from a large set of independent collinear variables. Since it reduces a great amount of redundant information, PLSR is ideal for multivariate calibration of spectroscopic data [25, 26]. In recent years, attention was paid to the application of PLSR in various disciplines and a large number of studies reported successful results [27–29].

In the present paper, a rapid and effective strategy for simultaneous determination of cartap (Ca), thiocyclam (Th), and tebufenozide (Te) in CHMs including *Radix Angelicae Dahuricae* and *Liquorices* has been proposed, by combining FI-IR with PLSR algorithm. The results have revealed that the direct determination of pesticide residues in complicated CHMs can be achieved, which adequately exploits the simple, rapid, and accuracy advantage with no requirement of physical or chemical separation and spectral processing, indicating a promising quantitative alternative for CHMs quality control and online monitoring of pesticide residues. Moreover, the accuracy and figures of merit of both NIR-PLSR and MIR-PLSR methods, including average recoveries, root mean square error, and limit of detection (LOD), were investigated and compared to evaluate the performance of the developed methods.

2. Materials and Methods

2.1. Apparatus. A NICOLET 6700 FT-IR, OMNIC 8.2 spectral collecting software (Thermo Fisher Scientific Inc., USA), Antaris II FT-NIR spectrometer, and RESULT 3.0 spectral collecting software (Thermo Electron Co., USA) were used.

2.2. Reagents and Samples. Radix Angelicae Dahuricae and *Liquorices* samples were obtained from Gansu in China. Cartap, thiocyclam, and tebufenozide were collected from Agro-Environmental Protection Institute of the Ministry of Agriculture (Tianjin, China). Chromatography-grade carbinol was purchased from TEDIA (TEDIA, USA). KBr (99.8% purity) was purchased from Aladdin Industrial Corporation (Shanghai, China).

2.3. Sample Preparation. The purchased standard solutions ($100\,\mu g\cdot mL^{-1}$) of individual pesticides were stored in dark glass vials at −20°C. The standard solutions were further diluted with carbinol to prepare the working solutions for calibration and verification in recovery studies. Calibration standards were $8\,\mu g\cdot mL^{-1}$. All solutions were stored in dark glass vials at 4°C.

Radix Angelicae Dahuricae and *Liquorices* samples used in NIR and MIR were crushed with the grinder and sifted into fine powders by 200 mesh sieve and then vacuum-dried at 60°C for 24 hours. Each *Radix Angelicae Dahuricae* or *Liquorices* powdered sample was weighed accurately (about 1.0 g). Different volumes of the working solutions of pesticide were further added into each CHMs sample and completely mixed into the prepared samples. The *Radix Angelicae Dahuricae* and *Liquorices* samples were divided into seven groups (R1–R7) and (L1–L7), respectively. That is, in each of the 7 groups (R1–R7) and (L1–L7), 30 samples with different *Radix Angelicae Dahuricae* and *Liquorices* matrices were prepared, respectively, and then vacuum-dried at 60°C for 24 hours. The sieved powders were stored in a dryer spare. Detailed information about the added content of the working pesticide solutions was listed in Table 1.

2.4. Spectra Acquisition. MIR spectra were recorded using KBr pellets and the wavenumber ranged from 4000 to $400\,cm^{-1}$ with a resolution of $4\,cm^{-1}$ and 30 samples with different herbal matrices in each group was collected, 30 spectra every day. NIR spectra were acquired by the diffuse reflectance mode; the spectral range was from 10000 to $4000\,cm^{-1}$ with a resolution of $8\,cm^{-1}$ and 30 samples with different herbal matrices in each group was collected, 30 spectra every day. The average spectra of three parallel measured spectra for each sample in each group were adopted to construct model. The same operation was repeated for three consecutive days.

2.5. Method of Chemometrics. PLSR programs were written and performed using a Matlab 2010a (Math Works, Natick, MA, USA).

Partial least squares regression (PLSR) was used for developing quantitative NIR and MIR models of cartap (Ca), thiocyclam (Th), and tebufenozide (Te) in CHMs including *Radix Angelicae Dahuricae* and *Liquorices*. In this work, considering $n \times p$ matrix $\mathbf{X}$ including p predictor variables (NIR or MIR spectral variables) for n CHMs samples and $n \times 1$ vector including the corresponding dependent variable for n CHMs samples, for simplicity and without loss of generality, both $\mathbf{X}$ and $\mathbf{y}$ are column centered; the goal of PLS is to find a set of orthogonal latent variables that are the linear combinations of the original predictor variables. The dependent variable is subsequently regressed against the latent variables. Because a calibration model with high

TABLE 1: Concentrations of Ca, Th, and Te in CHMs samples.

CHMs	Sample groups	Added concentration ($\mu g \cdot g^{-1}$)		
		Ca	Th	Te
Radix Angelicae Dahuricae	R1	40.00	0.00	0.00
	R2	32.00	8.00	0.00
	R3	24.00	12.00	4.00
	R4	16.00	16.00	8.00
	R5	8.00	20.00	12.00
	R6	0.00	24.00	16.00
	R7	8.00	12.00	20.00
Liquorices	L1	40.00	0.00	0.00
	L2	32.00	8.00	0.00
	L3	24.00	12.00	4.00
	L4	16.00	16.00	8.00
	L5	8.00	20.00	12.00
	L6	0.00	24.00	16.00
	L7	8.00	12.00	20.00

TABLE 2: A detail calibration and validation CHMs samples of PLSR models based on NIR and MIR.

Sample groups	NIR-PLSR		MIR-PLSR	
	Training number	Prediction number	Training number	Prediction number
R1	22	8	18	12
R2	19	11	14	16
R3	19	11	19	11
R4	19	11	16	14
R5	21	9	22	8
R6	17	13	17	13
R7	18	12	21	9
L1	23	7	18	12
L2	22	8	14	16
L3	20	10	19	11
L4	23	7	16	14
L5	17	13	22	8
L6	21	9	17	13
L7	19	11	21	9

complexity tends to give degraded prediction performance and has a higher risk of overfitting, it is important to make a proper tradeoff between model complexity and accuracy. Therefore, different models with the optimum latent variables (Lvs) were investigated by n-fold cross-validation.

2.6. Method Validation. The 210 NIR spectra data of *Radix Angelicae Dahuricae* or *Liquorices* samples were randomly divided into two sets: 135 spectra were used for calibration and 75 spectra were used for validation. The 210 MIR spectra data of *Radix Angelicae Dahuricae* or *Liquorices* samples were randomly divided into two sets: 127 spectra were used for calibration and 83 spectra were used for validation of constructed PLSR models; detailed sample information was listed in Table 2.

Root mean square error of 8-fold cross-validation (RMSECV) was used as an index to select the number of latent variables. The correlation coefficient (r), regression equation, LOD, LOQ, average recoveries, and intraday and interday precisions were used to validate this method. Precision and repeatability were determined by calculating intraday and interday relative standard deviation of calibration (RSDC) and prediction (RSDP) to evaluate the performance of instrument. The limit of detection (LOD) [30, 31] and the limit of quantification (LOQ) are computed as follows:

$$\begin{aligned} \mathrm{LOD} &= 3.3s(0) \\ \mathrm{LOQ} &= 10s(0), \end{aligned} \quad (1)$$

where $s(0)$ is the standard deviation in the predicted concentration for herbal sample background blank samples.

Performance of the models was estimated in terms of calibration root-mean-squared error (RMSEC), prediction root-mean-squared error (RMSEP) with the optimal latent

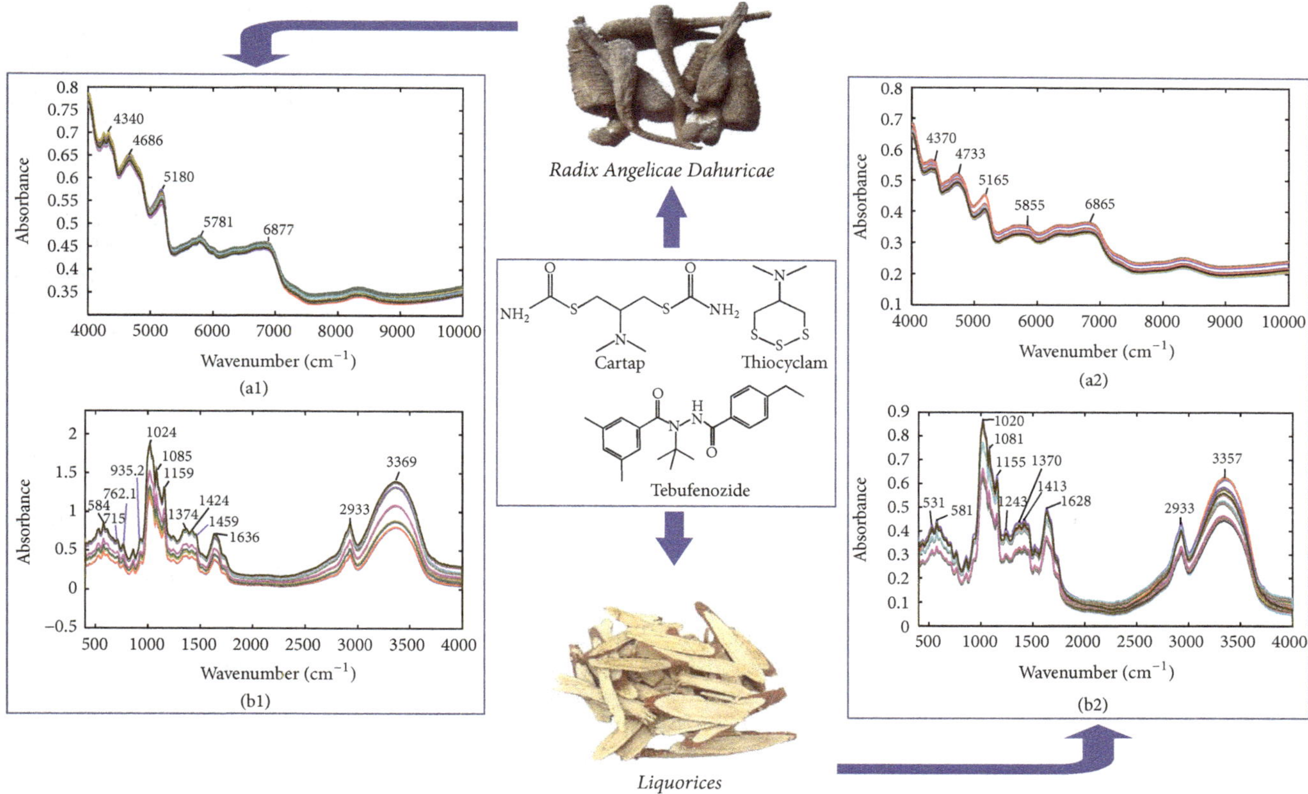

FIGURE 1: The molecular structures of cartap, thiocyclam, and tebufenozide and the original NIR ((a1) and (a2)) or MIR ((b1) and (b2)) spectra of multipesticide residues in *Radix Angelicae Dahuricae* and *Liquorices* samples.

variable, and the correlation coefficient between the reference and the predicted values as the following equations [32]:

$$\begin{aligned} \text{RMSEC} &= \sqrt{\frac{\sum_{i=1}\left(y_{ci}-\widehat{y}_{ci}\right)^2}{Nc}} \\ \text{RMSEP} &= \sqrt{\frac{\sum_{i=1}\left(y_{pi}-\widehat{y}_{pi}\right)^2}{Np}} \qquad (2) \\ R &= \frac{\sum_{i=1}\left(y_{i}-\widehat{y}_{i}\right)^2}{\sqrt{s^2\left(y_i\right)\times s^2\left(\widehat{y}_i\right)}}, \end{aligned}$$

where Nc and Np correspond to the number of calibration samples and validation samples, respectively. y_{ci} and $\widehat{y}_{ci}$ are the references and predicted values of the property y for ith calibration sample, respectively. y_{pi} and $\widehat{y}_{pi}$ represent the reference and predicted values of the property y for ith validation sample. $s^2(y_i)$ and $s^2(\widehat{y}_i)$ denote the variances for the reference and predicted values for the interesting component y_i, respectively.

3. Results and Discussions

3.1. Analysis of NIR and MIR Spectral Fingerprints of Multipesticide Residues in CHMs Samples. The molecular structures of three pesticide residues including Ca, Th, and Te and the NIR and MIR spectra of multipesticide residues in *Radix Angelicae Dahuricae* and *Liquorices* samples are plotted in Figure 1.

The raw NIR or MIR spectra are highly overlapped and have a poor peak resolution, which makes the accurate assignments of specific peaks very difficult. For ease of peak attributions, chemical bonds are denoted as atom-atom, where an atom can be carbon (C), hydrogen (H), oxygen (O), and nitrogen (N). For the NIR spectra of multipesticide residues in *Radix Angelicae Dahuricae* samples as shown in Figure 1(a1), it can be seen that the characteristic absorption peaks can be interpreted as follows [33, 34]: the peak 4251 cm^{-1} is the C-H stretch/C-H deformation in the phenyl or CH_2 bend second overtone and around 4340 cm^{-1} can be attributed to the combination absorbance of C-H anti-symmetric stretching and C-H bending. Around 4686 cm^{-1} peak brand is due to the combination stretching vibration of C=C, =C-H bands and combination of the base bands of N-H stretching and bending; peak 5180 cm^{-1} can be explained as second overtone of C=O stretching bands, stretching first overtone of C-H bands in aromatic rings and combination of the basebands of O-H stretching and bending. Other perk assignments 5781 cm^{-1} as the second overtones of C-H stretching in various groups and 6877 cm^{-1} as the first overtone of O-H stretching. Figure 1(b1) showed MIR spectra of multipesticide residues in *Radix Angelicae Dahuricae* samples. Seen from Figure 1(b1), variation around the peak at 1200–900 cm^{-1} can be associated with C-H group and the peak at 1500–1200 cm^{-1} can be associated with C-O. The wide scope between 3500 and 1700 cm^{-1} mainly consists of the overlapping of -OH stretching (3500–2933 cm^{-1}) and various -NH bending and stretching vibrations of amide compounds (3400–1636 cm^{-1}). Other peaks might be assigned as asymmetric vibrations of CH_2 at 2933 cm^{-1} and characteristic key bands around 715 cm^{-1} and 762.1 cm^{-1} can be regarded as fine features of aromatic substitution. In general, different functional groups of the multipesticide residues or chemical constituents in *Radix Angelicae Dahuricae* such as isoimperatorin, imperatorin, oxypeucedanin, byakangelicol, and byakangelicin could be assigned to different vibration modes. The low spectral resolution can be attributed to the contributions of multicomponents and the shifts and distortions resulted from their interactions. Similarly, NIR and MIR spectral fingerprints of multipesticide residues in *Liquorices* also can reflect characteristic of chemical bonds and still suffered from low spectral resolution and overlapping. Therefore, chemometric methods are required to extract useful information for simultaneous quantitative analysis of Ca, Th, and Te in CHMs.

3.2. Simultaneous Determination of Ca, Th, and Te in CHMs Samples. For simultaneous quantitative analysis of Ca, Th, and Te in CHMs samples, linear PLS models were developed to relate the raw FT-NIR spectra or FT-MIR to multipesticide residues of *Radix Angelicae Dahuricae* and *Liquorices*, respectively. The optimum latent variables numbers of all PLSR models were determined as 6 by 8-fold cross-validation. The corresponding predicted concentration to the analyte (every pesticide residue) as a function with its actual concentration can be found to evaluate relation and deviations through a linear regression equation. Herein, variable y was correlated with the prediction value of pesticide residues, while variable x was correlated with the actual value of pesticide residues. The correlation coefficients of Ca, Th, and Te obtained by using both NIR-PLSR and MIR-PLSR is nearly close to 1, respectively. The prediction results for the different CHMs using the NIR and MIR methods based on PLSR are summarized in Table 3.

For Ca, Th, and Te in *Liquorices*, the average predicted recoveries gained from NIR are 99.0 ± 10.8, 100.2 ± 6.9, and 99.1 ± 6.3% and from MIR are 100.2 ± 1.0, 99.7 ± 2.5, and 99.6 ± 1.0%, respectively. For Ca, Th, and Te in *Radix Angelicae Dahuricae*, the average predicted recoveries gained from NIR are 100.4 ± 6.1, 98.8 ± 6.6, and 100.4 ± 4.6% and from MIR are 100.0 ± 0.1, 99.7 ± 2.8, and 100.0 ± 1.1%, respectively. These results show that both methods can provide a satisfactory prediction capacity to potentially determine Ca, Th, and Te in complicated CHMs matrices. In addition, figures of merit (FOM) such as LOD and LOQ are very important in developing, comparing, and assessing the reliability of analytical methodologies and analytical results. For NIR-PLSR method, the limits of detection (LODs) for Ca, Th, and Te in *Radix Angelicae Dahuricae* are 2.45, 1.20, and 1.15 $\mu g \cdot g^{-1}$, respectively; in *Liquorices* are 2.03, 1.75, and 2.36 $\mu g \cdot g^{-1}$, respectively. For MIR-PLSR method, LODs for Ca, Th, and Te in *Radix Angelicae Dahuricae* are 0.40, 0.27, and 0.14 $ng \cdot mL^{-1}$, respectively; in *Liquorices* are 0.38, 0.32, and 0.30 $\mu g \cdot g^{-1}$. One can find that the proposed NIR and MIR method based on PLSR can yield satisfactory predictive capacity for determination of ATR, AME, and PRO in *Radix Angelicae Dahuricae* and *Liquorices* samples, respectively. The predicted concentrations versus the actual ones for Ca (a), Th (b), and Te (c) in *Radix Angelicae Dahuricae* using NIR-PLSR without spectra preprocessing were also showed in Figure 2. It is observed that prediction values were very close to actual value. Similarly, prediction value of the three pesticide residues in *Radix Angelicae Dahuricae* and *Liquorices* by NIR-PLSR or MIR-PLSR is very slightly deviated from actual value (similar figures were not shown). These results confirmed that the NIR method based on PLSR was fairly effective.

Moreover, for the sake of a further investigation into the accuracy and of the proposed methods, 30 *Radix Angelicae Dahuricae* samples of each group from seven different groups (R1–R7) and 30 *Liquorices* samples of each group from seven different groups (L1–L7) were prepared and analyzed in triplicate in a day; this assay was repeated for 3 days. Statistical parameters including RMSEC, RMSEP, RSDC, and RSDP in intraday and interday testing are demonstrated in Table 4.

It is observed that NIR-PLSR gives intraday RMSEP as 0.59, 0.79, and 0.43 and interday RMSEP as 1.54, 1.14, and 0.67 for Ca, Th, and Te in *Radix Angelicae Dahuricae*, respectively. MIR-PLSR gives intraday RMSEP as 0.11, 0.39, and 0.08 and interday RMSEP as 0.57, 0.53, and 0.69 for Ca, Th, and Te in *Radix Angelicae Dahuricae*, respectively. It can be also seen that all the relative standard deviations (RSDPs) were less than 10% by NIR-PLSR and 7% by MIR-PLSR. These results further verify that the proposed two methods could give accurate results as an alternative to each other; only the performance of MIR-PLSR was slightly better than that of NIR-PLSR under the circumstances suffering from matrix effects of CHMs. This might be due to the fact that the MIR shows the strongest absorption in the infrared active vibration.

4. Conclusion

In the present study, a simple, rapid, and effective method has been successfully developed for quantitative analysis of Ca, Th, and Te in *Radix Angelicae Dahuricae* and *Liquorices* based on the PLSR using both FT-NIR and FT-MIR methods. The proposed quantitative method proved to be capable of performing the simultaneous determination of multi-pesticide residues in complex Chinese herbal medicines without physical or chemical separation pretreatment step and spectral processing. Furthermore, the figures of merit and statistical parameters indicated that both methods could give accurate and stable results; only the performance of MIR-PLSR was slightly better than that of NIR-PLSR in the cases suffering from matrix effects. It is expected that the potential advantages of this determination of trace pesticide in Chinese herbal medicines, such as accuracy, rapidity, and low cost, can

TABLE 3: Results for multipesticide residues of the different CHMs using the NIR and MIR methods based on PLSR.

Sample groups	Predicted average concentration by NIR-PLSR model			Predicted average concentration by MIR-PLSR model		
	Ca ($\mu g \cdot g^{-1}$)	Th ($\mu g \cdot g^{-1}$)	Te ($\mu g \cdot g^{-1}$)	Ca ($\mu g \cdot g^{-1}$)	Th ($\mu g \cdot g^{-1}$)	Te ($\mu g \cdot g^{-1}$)
R1	39.95 ± 0.31	0.12 ± 0.36	0.19 ± 0.31	40.00 ± 0.05	0.0091 ± 0.08	−0.011 ± 0.05
R2	32.01 ± 0.85	7.69 ± 0.64	0.29 ± 0.38	32.02 ± 0.13	8.01 ± 0.13	−0.019 ± 0.04
R3	24.09 ± 0.62	11.93 ± 0.62	4.03 ± 0.30	24.00 ± 0.05	11.99 ± 0.06	4.01 ± 0.07
R4	16.18 ± 0.40	15.64 ± 0.42	8.19 ± 0.30	15.97 ± 0.08	16.04 ± 0.12	8.00 ± 0.07
R5	8.28 ± 0.80	20.03 ± 0.95	11.68 ± 0.44	8.04 ± 0.14	19.98 ± 0.16	11.96 ± 0.17
R6	0.24 ± 0.83	23.45 ± 0.89	16.29 ± 0.46	−0.027 ± 0.16	24.03 ± 0.16	15.99 ± 0.08
R7	7.83 ± 0.80	12.41 ± 0.95	19.81 ± 0.51	7.99 ± 0.20	12.00 ± 0.22	20.04 ± 0.05
r	0.9999	0.9998	0.9998	1.000	0.9999	1.000
Regress equation	$y = 0.9998x + 0.0035$	$y = 0.9995x + 0.0061$	$y = 0.9997x + 0.0028$	$y = x + 0.00061$	$y = 0.9998x + 0.0026$	$y = x + 0.00039$
Average recovery (%)	100.4 ± 6.1	98.8 ± 6.6	100.4 ± 4.6	100.0 ± 1.1	99.7 ± 2.8	100.0 ± 1.1
LOD	2.45	1.20	1.15	0.40	0.27	0.14
LOQ	7.43	3.64	3.47	1.21	0.81	0.43
L1	39.58 ± 0.78	0.39 ± 0.53	0.93 ± 1.00	40.08 ± 0.18	0.047 ± 0.10	0.040 ± 0.11
L2	31.77 ± 1.07	8.02 ± 0.82	0.29 ± 0.27	31.98 ± 0.12	8.01 ± 0.07	0.010 ± 0.07
L3	23.83 ± 0.63	12.12 ± 0.46	3.91 ± 0.44	24.06 ± 0.15	11.95 ± 0.01	3.98 ± 0.08
L4	16.60 ± 0.89	15.73 ± 0.50	7.58 ± 0.49	16.04 ± 0.10	16.03 ± 0.00	7.94 ± 0.07
L5	7.40 ± 1.08	20.23 ± 0.61	12.28 ± 0.49	8.09 ± 0.16	19.97 ± 0.12	11.95 ± 0.06
L6	0.087 ± 0.66	24.05 ± 0.52	16.03 ± 0.38	0.046 ± 0.12	23.99 ± 0.08	15.96 ± 0.06
L7	8.28 ± 1.37	12.11 ± 1.14	19.70 ± 0.42	7.95 ± 0.12	12.02 ± 0.08	20.02 ± 0.07
r	0.9994	0.9988	0.9993	1.000	1.000	1.000
Regress equation	$y = 0.9989x + 0.0211$	$y = 0.9976x + 0.0302$	$y = 0.9987x + 0.011$	$y = x + 0.00046$	$y = x + 0.00045$	$y = x + 0.00026$
Average recovery (%)	99.0 ± 10.8	100.2 ± 6.9	99.1 ± 6.3	100.2 ± 1.0	99.7 ± 2.5	99.6 ± 1.0
LOD	2.03	1.75	2.36	0.38	0.32	0.30
LOQ	6.14	5.29	7.14	1.16	0.97	0.91

TABLE 4: The results obtained for validation assays of accuracy and precision in CHMs samples.

Methods	CHMs	Pesticides	Intraday				Interday			
			Accuracy		Precision		Accuracy		Precision	
			RMSEC	RMSEP	RSDC%	RSDP%	RMSEC	RMSEP	RSDC%	RSDP%
NIR-PLSR	*Radix Angelicae Dahuricae*	Ca	0.48	0.59	2.09	2.55	1.12	1.54	4.66	7.30
		Th	0.47	0.79	2.83	4.69	0.76	1.14	4.49	6.74
		Te	0.33	0.43	3.18	3.94	0.47	0.67	4.30	6.37
MIR-PLSR		Ca	0.08	0.11	0.36	0.42	0.48	0.57	2.26	2.41
		Th	0.10	0.39	0.61	1.06	0.33	0.53	1.92	3.09
		Te	0.05	0.08	0.43	0.77	0.50	0.69	4.71	6.80
NIR-PLSR	*Liquorices*	Ca	0.45	1.01	1.90	5.00	1.50	1.96	6.40	9.20
		Th	0.36	0.85	2.30	4.20	0.86	1.24	5.30	7.00
		Te	0.26	0.55	2.60	4.60	0.67	0.85	6.40	6.60
MIR-PLSR		Ca	0.067	0.14	0.30	0.60	0.38	0.46	1.60	1.80
		Th	0.042	0.34	0.20	0.80	0.34	0.43	1.90	2.80
		Te	0.038	0.082	0.30	0.80	0.22	0.31	1.90	3.30

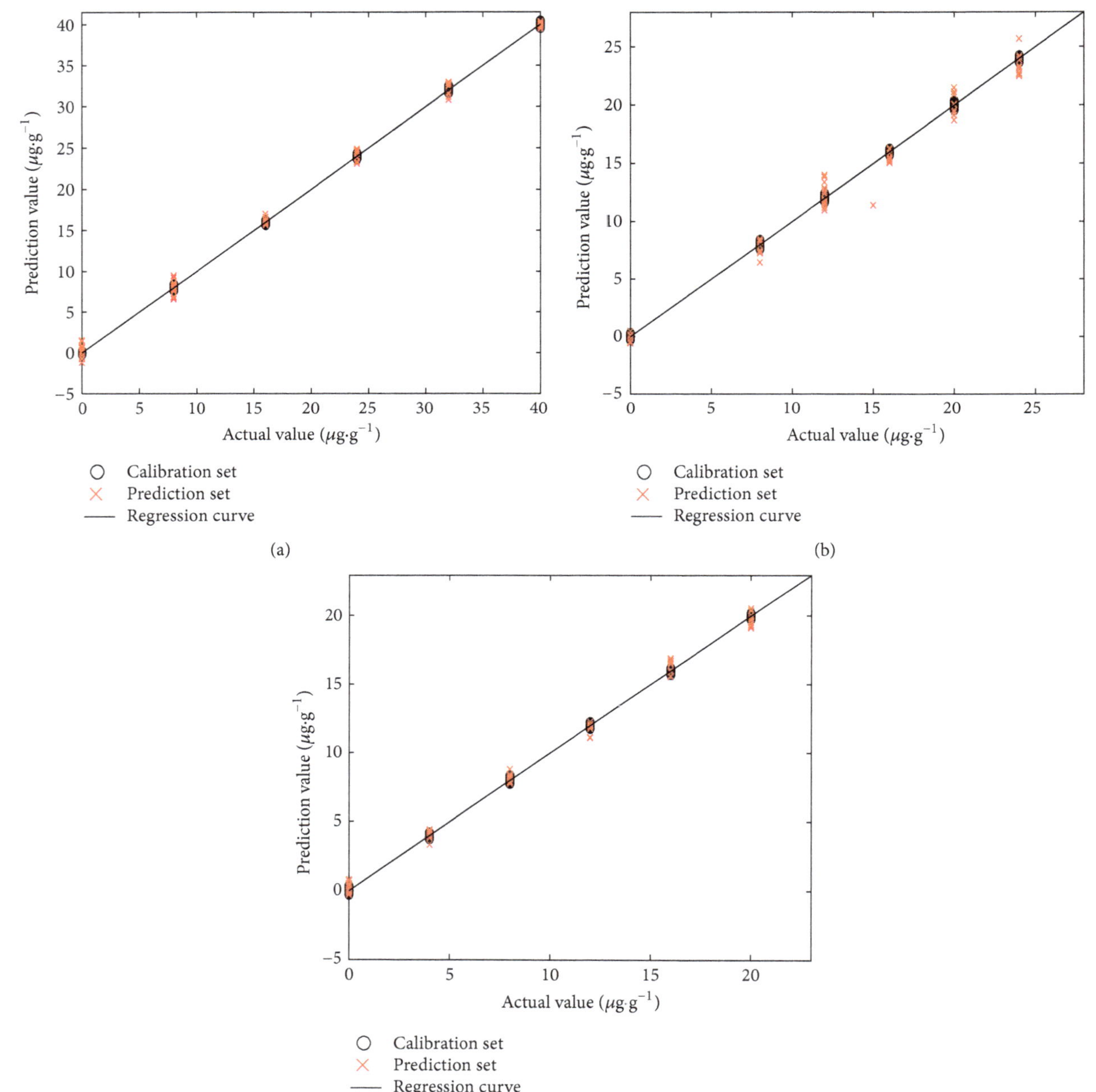

FIGURE 2: Correlation plots of the predicted concentrations versus the actual ones for Ca (a), Th (b), and Te (c) in *Radix Angelicae Dahuricae* using NIR-PLSR.

be even more highlighted by considering the possibility of automating the proposed methods for online detection.

Competing Interests

The authors declare that they have no competing interests.

Acknowledgments

This work was financially supported by the National Natural Science Foundation of China [nos. 21576297 and 21205145]; the Open Funds of State Key Laboratory Breeding Base of Green Chemistry-Synthesis Technology of Zhejiang University of Technology [no. GCTKF2014003]; the Open Research Program [nos. 2015ZD001 and 2015ZD002] from the Modernization Engineering Technology Research Center of Ethnic Minority Medicine of Hubei province (South-Central University for Nationalities).

References

[1] K. Chan, "Some aspects of toxic contaminants in herbal medicines," *Chemosphere*, vol. 52, no. 9, pp. 1361–1371, 2003.

[2] C. W. Huie, "A review of modern sample-preparation techniques for the extraction and analysis of medicinal plants," *Analytical and Bioanalytical Chemistry*, vol. 373, no. 1-2, pp. 23–30, 2002.

[3] K. Banerjee, S. Utture, S. Dasgupta et al., "Multiresidue determination of 375 organic contaminants including pesticides, polychlorinated biphenyls and polyaromatic hydrocarbons in fruits and vegetables by gas chromatography-triple quadrupole mass spectrometry with introduction of semi-quantification approach," *Journal of Chromatography A*, vol. 1270, pp. 283–295, 2012.

[4] Z. H. Rivera, E. Oosterink, L. Rietveld, F. Schoutsen, and L. Stolker, "Influence of natural organic matter on the screening of pharmaceuticals in water by using liquid chromatography with full scan mass spectrometry," *Analytica Chimica Acta*, vol. 700, no. 1-2, pp. 114–125, 2011.

[5] A. Garcia-Ac, P. A. Segura, L. Viglino et al., "On-line solid-phase extraction of large-volume injections coupled to liquid chromatography-tandem mass spectrometry for the quantitation and confirmation of 14 selected trace organic contaminants in drinking and surface water," *Journal of Chromatography A*, vol. 1216, no. 48, pp. 8518–8527, 2009.

[6] F. Hernández, T. Portolés, E. Pitarch, and F. J. López, "Target and nontarget screening of organic micropollutants in water by solid-phase microextraction combined with gas chromatography/high-resolution time-of-flight mass spectrometry," *Analytical Chemistry*, vol. 79, no. 24, pp. 9494–9504, 2007.

[7] Q. Gou, X. Lv, L. Tan, and B.-Y. Yu, "Simultaneous determination of 26 pesticide residues in 5 Chinese medicinal materials using solid-phase extraction and GC-ECD method," *Chinese Journal of Natural Medicines*, vol. 7, no. 3, pp. 210–216, 2009.

[8] R. Xu, J. Wu, Y. Liu et al., "Analysis of pesticide residues using the Quick Easy Cheap Effective Rugged and Safe (QuEChERS) pesticide multiresidue method in traditional Chinese medicine by gas chromatography with electron capture detection," *Chemosphere*, vol. 84, no. 7, pp. 908–912, 2011.

[9] H. Tong, Y. Tong, J. Xue, D. Liu, and X. Wu, "Multi-residual pesticide monitoring in commercial chinese herbal medicines by gas chromatography-triple quadrupole tandem mass spectrometry," *Food Analytical Methods*, vol. 7, no. 1, pp. 135–145, 2014.

[10] X. Mao, Y. Wan, A. Yan, M. Shen, and Y. Wei, "Simultaneous determination of organophosphorus, organochlorine, pyrethriod and carbamate pesticides in *Radix astragali* by microwave-assisted extraction/dispersive-solid phase extraction coupled with GC-MS," *Talanta*, vol. 97, pp. 131–141, 2012.

[11] B. Guo, S. Ji, F. Zhang, B. Yang, J. Gu, and X. Liang, "Preparation of C_{18}-functionalized Fe_3O_4@SiO_2 core-shell magnetic nanoparticles for extraction and determination of phthalic acid esters in Chinese herb preparations," *Journal of Pharmaceutical and Biomedical Analysis*, vol. 100, pp. 365–368, 2014.

[12] L. Chen, F. Song, Z. Liu, Z. Zheng, J. Xing, and S. Liu, "Multi-residue method for fast determination of pesticide residues in plants used in traditional chinese medicine by ultra-high-performance liquid chromatography coupled to tandem mass spectrometry," *Journal of Chromatography A*, vol. 1225, pp. 132–140, 2012.

[13] R. M. Balabin and R. Z. Safieva, "Near-infrared (NIR) spectroscopy for biodiesel analysis: fractional composition, iodine value, and cold filter plugging point from one vibrational spectrum," *Energy & Fuels*, vol. 25, no. 5, pp. 2373–2382, 2011.

[14] A. P. Craig, A. S. Franca, L. S. Oliveira, J. Irudayaraj, and K. Ileleji, "Application of elastic net and infrared spectroscopy in the discrimination between defective and non-defective roasted coffees," *Talanta*, vol. 128, pp. 393–400, 2014.

[15] Y. Feng, D. Lei, and C. Hu, "Rapid identification of illegal synthetic adulterants in herbal anti-diabetic medicines using near infrared spectroscopy," *Spectrochimica Acta—Part A: Molecular and Biomolecular Spectroscopy*, vol. 125, pp. 363–374, 2014.

[16] J. Peerapattana, K. Otsuka, and M. Otsuka, "Application of NIR spectroscopy for the quality control of mangosteen pericarp powder: quantitative analysis of alpha-mangostin in mangosteen pericarp powder and capsule," *Journal of Natural Medicines*, vol. 67, no. 3, pp. 452–459, 2013.

[17] C. Zhang and J. Su, "Application of near infrared spectroscopy to the analysis and fast quality assessment of traditional Chinese medicinal products," *Acta Pharmaceutica Sinica B*, vol. 4, no. 3, pp. 182–192, 2014.

[18] J. J. Liu, H. Xu, W. S. Cai, and X. G. Shao, "Discrimination of industrial products by on-line near infrared spectroscopy with an improved dendrogram," *Chinese Chemical Letters*, vol. 22, no. 10, pp. 1241–1244, 2011.

[19] O. Galtier, O. Abbas, Y. Le Dréau et al., "Comparison of PLS1-DA, PLS2-DA and SIMCA for classification by origin of crude petroleum oils by MIR and virgin olive oils by NIR for different spectral regions," *Vibrational Spectroscopy*, vol. 55, no. 1, pp. 132–140, 2011.

[20] W. Q. Luo, S. Y. Huan, H. Y. Fu et al., "Preliminary study on the application of near infrared spectroscopy and pattern recognition methods to classify different types of apple samples," *Food Chemistry*, vol. 128, no. 2, pp. 555–561, 2011.

[21] P. Geladi and B. R. Kowalski, "Partial least-squares regression: a tutorial," *Analytica Chimica Acta*, vol. 185, pp. 1–17, 1986.

[22] Q. Shen, J.-H. Jiang, C.-X. Jiao, G.-L. Shen, and R.-Q. Yu, "Modified particle swarm optimization algorithm for variable selection in MLR and PLS modeling: QSAR studies of antagonism of angiotensin II antagonists," *European Journal of Pharmaceutical Sciences*, vol. 22, no. 2-3, pp. 145–152, 2004.

[23] H. Kubinyi, "Evolutionary variable selection in regression and PLS analyses," *Journal of Chemometrics*, vol. 10, no. 2, pp. 119–133, 1996.

[24] M. Mirzaie, R. Darvishzadeh, A. Shakiba, A. A. Matkan, C. Atzberger, and A. Skidmore, "Comparative analysis of different uni- and multi-variate methods for estimation of vegetation water content using hyper-spectral measurements," *International Journal of Applied Earth Observation and Geoinformation*, vol. 26, no. 1, pp. 1–11, 2014.

[25] P. Lin, Y. Chen, and Y. He, "Identification of geographical origin of olive oil using visible and near-infrared spectroscopy technique combined with chemometrics," *Food and Bioprocess Technology*, vol. 5, no. 1, pp. 235–242, 2012.

[26] M. M. Pojić and J. S. Mastilović, "Near infrared spectroscopy—advanced analytical tool in wheat breeding, trade, and processing," *Food and Bioprocess Technology*, vol. 6, no. 2, pp. 330–352, 2013.

[27] L. Ragni, A. Berardinelli, C. Cevoli, and E. Valli, "Assessment of the water content in extra virgin olive oils by Time Domain Reflectometry (TDR) and Partial Least Squares (PLS) regression methods," *Journal of Food Engineering*, vol. 111, no. 1, pp. 66–72, 2012.

[28] I. Gouvinhas, J. M. M. M. de Almeida, T. Carvalho, N. Machado, and A. I. R. N. A. Barros, "Discrimination and characterisation of extra virgin olive oils from three cultivars in different maturation stages using Fourier transform infrared spectroscopy in tandem with chemometrics," *Food Chemistry*, vol. 174, no. 4, pp. 226–232, 2015.

[29] S. A. Moreira, J. Sarraguça, D. F. Saraiva, R. Carvalho, and J. A. Lopes, "Optimization of NIR spectroscopy based PLSR models for critical properties of vegetable oils used in biodiesel production," *Fuel*, vol. 150, pp. 697–704, 2015.

[30] A. Lorber, K. Fabert, and B. R. Kowalski, "Net analyte signal calculation in multivariate calibration," *Analytical Chemistry*, vol. 69, no. 8, pp. 1620–1626, 1997.

[31] M. J. Rodríguez-Cuesta, R. Boqué, F. X. Rius, J. L. M. Vidal, and A. G. Frenich, "Development and validation of a method for determining pesticides in groundwater from complex overlapped HPLC signals and multivariate curve resolution," *Chemometrics and Intelligent Laboratory Systems*, vol. 77, no. 1-2, pp. 251–260, 2005.

[32] A. M. K. Pedro and M. M. C. Ferreira, "Nondestructive determination of solids and carotenoids in tomato products by near-infrared spectroscopy and multivariate calibration," *Analytical Chemistry*, vol. 77, no. 8, pp. 2505–2511, 2005.

[33] B. M. Nicolaï, K. Beullens, E. Bobelyn et al., "Nondestructive measurement of fruit and vegetable quality by means of NIR spectroscopy: a review," *Postharvest Biology and Technology*, vol. 46, no. 2, pp. 99–118, 2007.

[34] B. C. Tang, H. Y. Fu, Q. B. Yin et al., "Combining near-infrared spectroscopy and chemometrics for rapid recognition of an Hg-contaminated plant," *Journal of Spectroscopy*, vol. 2016, Article ID 3597451, 7 pages, 2016.

19

Diagnosis of High Voltage Insulators Made of Ceramic Using Spectrophotometry

Paweł Frącz, Ireneusz Urbaniec, Tomasz Turba, and Sławomir Krzewiński

Faculty of Electrical Engineering, Automatic Control and Computer Science, Opole University of Technology, Prószkowska 76, 45-758 Opole, Poland

Correspondence should be addressed to Paweł Frącz; pawelfracz2@gmail.com

Academic Editor: Jau-Wern Chiou

The paper presents results of comparative analysis of optical signals emitted by partial discharges occurring on three types of high voltage insulators made of porcelain. The research work consisted of diagnosis of the following devices: a long rod insulator, a cap insulator, and an insulating cylinder. For optical signal registration a spectrophotometer was applied. All measurements were performed under laboratory conditions by changing the value of partial discharges generation voltage. For the cylindrical insulator also the distance between high voltage and ground electrodes was subjected for investigation as a factor having influence on partial discharges. The main contribution which resulted from the studies is statement that application of spectrophotometer enables faster recognition of partial discharges, as compared to standard methods.

1. Introduction

This research concerns a field of science which is related to the generation and development of electric discharges on surfaces of high voltage (HV) insulation systems. An important problem of insulation systems is the ageing process progressing during their operation and causing deterioration of insulating properties [1–3]. Important ageing factors that occur in operational practice include UV radiation, ozone and nitrogen oxides, temperature fluctuations, rainfall (including acid rain), rime deposition, dirt, and partial discharges (PD). The origin of research and scientific studies conducted by our team is the need to analyze the physical phenomena that occur during electric discharges on the surface on nonorganic (ceramic) dielectrics. The main objective of research conducted within this scope is to determine as accurately as possible how, when, and at what pace a process of deterioration of insulation elements is progressing. Another issue is to determine the conditions under which the complete breakdown occurs. The studies involved measurements of optical emission spectra of electric discharges occurring on the surface of ceramic insulation using spectrophotometer. Based on the gathered data, the initial voltage values of electric discharges U_{10} were estimated with a greater sensitivity and faster than previously possible, based on the measurement of initial corona voltage U_0. The presented study results were aimed at confirming the hypothesis that measurement of emission spectra, especially in the range from 250 nm to 280 nm, generated by electric discharges occurring on surfaces of ceramic dielectrics, can serve as a sensitive and effective indicator of their surface strength.

2. Materials and Methods

2.1. Metrological Parameters of the Measuring System. Nowadays, there are many measurement methods of detecting PD occurring in the insulating systems of electrical power equipment. Diagnostic methods can be divided into invasive methods and noninvasive methods. Invasive methods involve detection and recording PD currents using special measuring probes that express PD value in pC. The disadvantage of such methods is the need to disconnect the tested device from the power supply during connection and disconnection of the measuring apparatus [4–6]. Alternative methods do not require invasion in the operation of the device, making them safer for people performing measurements with measuring equipment, usually hand-held and portable [7, 8].

The acoustic method allows the detection of ultrasound (sound pressure) within the range of up to 500 kHz generated in the early stage of surface PD using the specialized devices [9, 10]. The electromagnetic method, depending on the instrument used, allows the recording of the intensity of spectrum for varying radiation range. The knowledge available today in the literature allowed us to conclude that the emitted spectrum is in the range from ultraviolet to infrared radiation [11, 12] and in X-rays in the range 10 pm to 10 nm [13].

This paper concerns electromagnetic method, and particularly recording of UV and visible radiation. The properties of electromagnetic radiation emitted during the PD depend on the type of insulating material and parameters of the propagation medium and the environment surrounding the area of PD generation, such as pressure, temperature, and humidity [14, 15].

The basic technique for measuring optical radiation is spectrophotometry. Its main advantage is the galvanic separation of the measuring position from the tested device, which is under high voltage [16]. Thanks to that, the obtained results do not affect external electromagnetic fields with high intensity and other types of disturbances that have negative impact on the signals recorded using electric or acoustic method. A condition necessary for the detection of PD with the use of optical method is a direct line of sight. A method for recording optical radiation is the spectrophotometry technique, allowing the recording of not only the radiation intensity but, above all, the occurrence and shape of the spectrum present in the emitted signals [17]. In this technique spectrophotometers are used, which are currently being constructed from sets of photodiodes with variable wavelengths, light-sensitive CCD detectors, and special software for processing [18, 19]. An important parameter affecting the accuracy and efficiency of the obtained results of measurements is the amount of attenuation and dispersion of optical signals depending, among other things, on the distance of the recording equipment from the point of PD generation and the propagation medium itself.

Table 1 presents basic technical parameters of the Ocean Optics HR4000 spectrophotometer used in the measurements. Measurement data archiving and processing were performed using numerical procedures implemented in the MATLAB programming environment.

All the measurements were conducted in a darkened room of the laboratory. The measurement procedure included increase of the PD generation voltage from 0 to 100 kV while recording the intensity of optical radiation. For each of the considered voltage levels, five measurements using a spectrophotometer were made. The power supply system consisted of a testing transformer and a control panel, which included an autotransformer, an overcurrent protection system, and a voltmeter for measurement of the momentary voltage value. The voltage values were controlled by the autotransformer and forwarded to primary winding of the single-phase testing transformer. The output of the secondary winding was connected to a water resistor for limitation of the short-circuit current with which the tested insulator was powered.

Table 1: Characteristics of technical parameters of spectrophotometer applied in the study.

Parameter	Value	Unit
CCD detector	Toshiba TCS1304AP - 3648	Pixel
Integration time (counting time)	from 3.8×10^{-3} to 20	s
Type of optical link	SMA 905	
Width of apertures lit	5, 10, 25, 50, 100, 200	μm
Optical resolution	~0.02–8.4	nm FWHM
Light scattering	<0.05% for 600; <0.1% for 435	nm
Visible light range	190–1100	nm
UV spectral range	200–1100	nm
Dynamic range	1300 : 1	
Corrected linearity	>99.8	%
A/D resolution	14	bits
Maximum pixel digitalization range	1	MHz
Size of pixel, CCD element	8×200	μm
Sensitivity	100 per count for 800 nm	Photons

Measurement cycles were conducted for supply voltage varying in the range from initial voltages U_0 to U_p. The initial voltages U_0 and U_{01} were determined based on voltmeter reading, while the spectrophotometer showed intensity exceeding 300.

The breakdown voltage depended on the type of tested porcelain and the distance between electrodes. An optical transducer was placed in a specially designed holder mounted on a stand, allowing us to adjust height, angle, and distance between the measuring head and the area of PD occurrence on the tested dielectric. For each tested dielectric, a value of initial voltage U_0 and a breakdown voltage U_p was determined five times prior to the measurement of PD. Measurements of emission spectra were conducted for averaged values relative to U_0 and U_p.

The tests involved the preparation of three ceramic insulation systems for PD generation, that is,

(i) LS-type long rod insulator (Figure 1(a)),

(ii) WPK-type porcelain cylinder, filled with quartz (Figure 2(b)),

(iii) LK-type cap insulator (Figure 2(a)).

For the WPK-type insulator, PD measurements were performed at various distances between the HV and ground electrodes in the range from 3 cm to 11 cm. In consequence, the following definitions are used in this paper: WPK3, WPK5, WPK7, WPK9, and WPK11, where the numbers correspond to the distance between electrodes in cm.

2.2. Numerical Methods Applied for Measurement Results Analysis. The obtained results of measurements of emission spectra were analyzed statistically in order to determine differences and similarities between signals and to determine

(a)

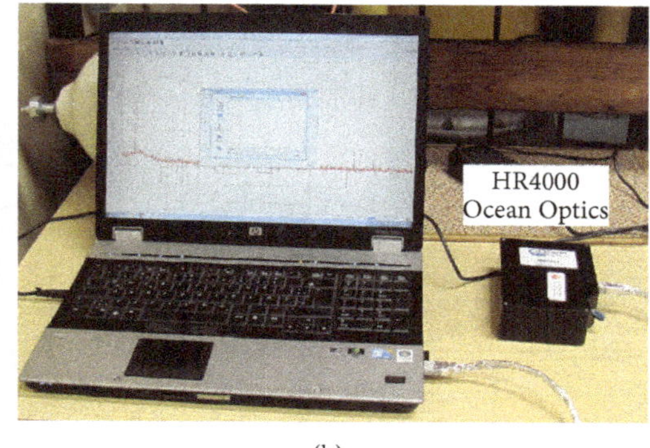

(b)

FIGURE 1: (a) The long rod insulator, type LS, and (b) the spectrophotometer applied in the study.

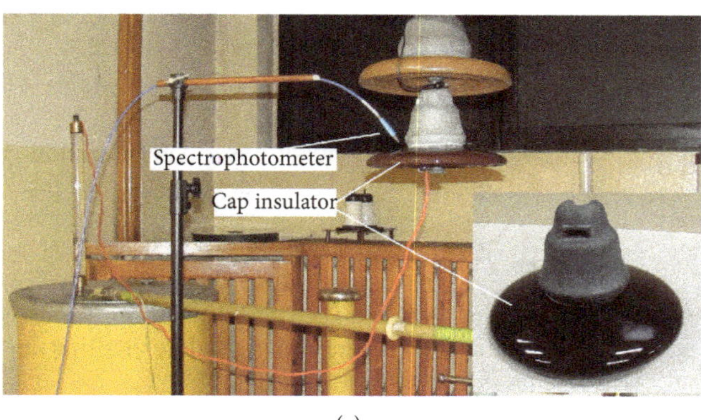

(a)

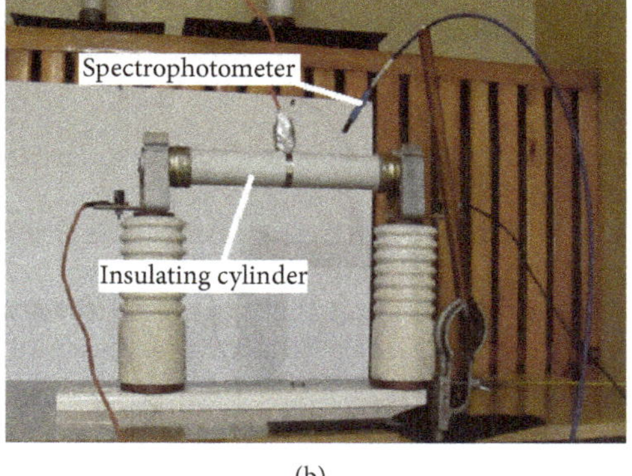

(b)

FIGURE 2: (a) The cap insulator, type LK, and (b) the cylindrical insulator, type WPK.

the effect of PD generation voltage on the obtained emission spectra. Light emission intensity spectra were visualized for various PD generation voltages, as well as using cumulative graph for all tested voltages. Histograms showing the number of variable intensity radiation I were calculated for individual wavelengths. A histogram is a graphical method to present the empirical distribution, which is determined by calculating the frequency distribution. Histograms were presented on cumulative graphs showing the dependence of obtained values on PD generation voltage. Wavelength ranges were determined for the light emission of the highest intensity. These values are dependent on the voltage of PD generation and the type of studied insulation system. Recorded characteristics underwent a process of mathematical regression. For that, components of linear spectrum were determined using a series of Gaussian functions. The developed procedures use Nelder-Mead Simplex method to look for optimal parameters of the function in question, that is, parameters that would provide minimum deviations between theoretical and empirical data (3). A final effect was the sum of spectral components in the form of relation designated as M2

$$\mathrm{M2}(\lambda) = \sum_{i=1}^{N} A_i * e^{-((\lambda - B_i)/C_i)^2}, \quad (1)$$

where λ is independent variable wavelength, A_i is amplitude of component (spectral line) of width C_i, N is varying number of components in the model, and A, B, and C are model parameters.

In order to determine most predominant components of the linear spectrum, a mathematical model M1 was developed that can be described using relation (2). M1 model is a sum of eight Gaussian functions:

$$\mathrm{M1}(\lambda) = \sum_{i=1}^{8} A_i * e^{-((\lambda - B_i)/C_i)^2}, \quad (2)$$

where λ is independent variable wavelength, A_i is amplitude of component (spectral line) of width C_i, and A, B, and C are model parameters:

$$\delta = \sqrt{\sum_{\forall i} (x_i - \tilde{x}_i)^2}, \quad (3)$$

where x_i is ith empirical variable value and $\tilde{x}_i$ is ith theoretical variable value (estimated).

In order to perform analysis of obtained regression results, values of coefficients which are the measures of matching the studied models to the studied dependence were

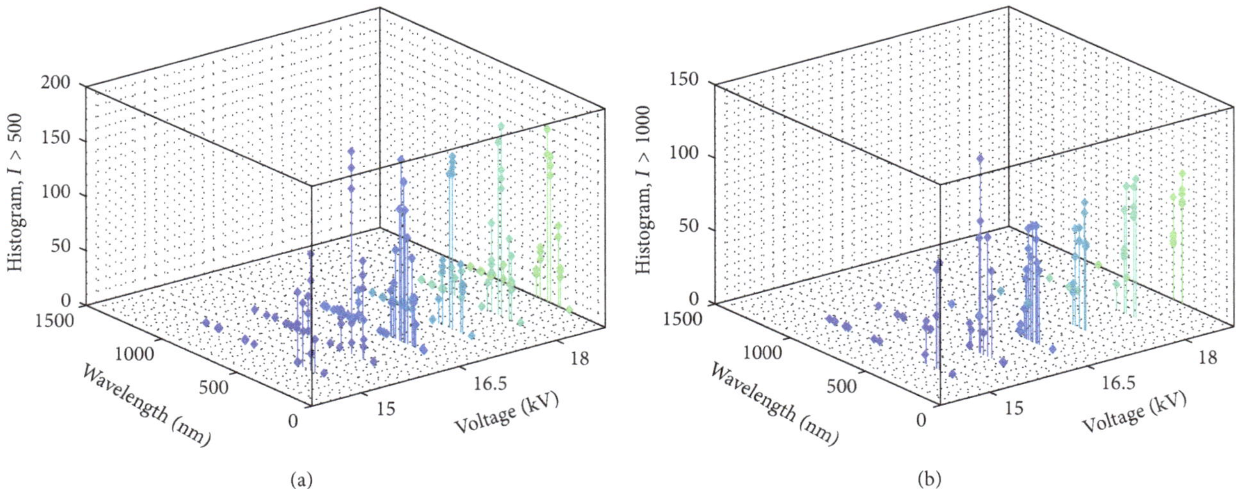

FIGURE 3: Summary of spectrum intensity histograms for different voltages: (a) $I > 500$ and (b) $I > 1000$.

determined, including SSE (4), RMSE (5), R-square (6), and adj-R-square (7).

(a) SSE is the sum of squares of residuals. It determines the total deviation of estimated values from empirical data. A value close to zero indicates that the model has a smaller random error component, and the fitting can be more useful for prediction:

$$\text{SSE} = \sum_{\forall i} (y_i - \tilde{y}_i)^2, \tag{4}$$

where y_i is ith empirical variable value and $\tilde{y}_i$ is ith theoretical variable value (estimated).

(b) RMSE is the standard error of regression constituting the root mean square error. The value close to one means higher utility of the considered model for prediction:

$$\text{RMSE} = \sqrt{\frac{\sum_{\forall i} (y_i - \tilde{y}_i)^2}{n - p}}, \tag{5}$$

where y_i is ith empirical variable value, $\tilde{y}_i$ is ith theoretical variable value (estimated), n is number of samples contained in estimated course/dependence, and p is number of model parameters.

(c) R-square (R^2) determines the variability in the data. This value is the square of the correlations between empirical and estimated data. Values close to one indicate that most of the variance is included in the model. In this study, we apply R^2 as a fitting indicator. Values above 0.6 and 0.8 indicate well and very well fitting results, respectively. Values equal or less than 0.5 indicate poor or no fitting, respectively:

$$R\text{-square} = 1 - \frac{\sum_{\forall i} (y_i - \tilde{y}_i)^2}{\sum_{\forall i} (y_i - \mu)^2}, \tag{6}$$

where y_i is ith empirical variable value, $\tilde{y}_i$ is ith theoretical variable value (estimated), and μ is arithmetic mean of the empirical data.

(d) Adj-R-square (adj-R^2) is a determination coefficient adjusted by the number of degrees of freedom. It is an indicator that allows the comparison of results obtained by models with different numbers of parameters. The values are in the range below one. Values close to one indicate a good matching of the model to empirical data. Negative values indicate that the model contains elements that do not help in predicting model response:

$$\text{adj-}R\text{-square} = 1 - \frac{\text{SSE}(n-1)}{\text{SST}(n-p)}, \tag{7}$$

where n is number of samples contained present in estimated curve/dependence and p is number of model parameters:

$$\text{SST} = \sum_{\forall i} (y_i - \mu)^2, \tag{8}$$

where y_i is ith empirical variable value and μ is arithmetic mean of the empirical data.

M1 and M2 models, despite apparent similarities, are very different in terms of number of parameters and parameter estimation method. The paper presents fit coefficients calculated using regression for both models, that is, SSE, R-square, and RMSE in form of 3D graphs visualizing dependence as a function of PD generation voltage.

3. Example Results and Discussion

3.1. Analysis of Emission Spectra of Discharges Occurring on the Surface of Porcelain Insulation Cylinder Filled with Quartz: Example Distance between Electrodes Is 3 cm. Figure 3 presents a cumulative comparison of histograms calculated for spectra of the highest intensities obtained during measurements as a function of PD generation value. The graph on the left refers to the intensity exceeding the value of 500 ($I > 500$), while the graph on the right indicates intensity exceeding the value of 1000 ($I > 1000$). On the basis of this analysis, it is possible to determine the frequency of light emission generated by PD for a given wavelength.

TABLE 2: Dominant wavelengths present in light emission of value $I > 1000$.

Wavelength interval (nm)	Dominant wavelengths and intervals present in light emission of value $I > 1000$ (nm)
λ = [200–299]	203, 296–298
λ = [300–399]	311–318, 333–340, 351–360, 369–382, 390–396, 397–399
λ = [400–499]	400-401, 404–407, 420, 425–428, 434-435
λ > 500	745, 933, 1073

Based on the obtained histograms for intensities of values exceeding 1000, dominant light wavelengths occurring in the recorded signals were determined for the individual intervals and presented in Table 2.

Figure 4 presents the results of modeling intensity spectra using Gaussian series, model M2 for selected values of PD generation voltage. A red color denotes measurement results, while blue denotes modeling results. The legend contains value of determination coefficient R^2.

Figures 5 and 6 present graphically the values of fitting parameters, R^2, RMSE, and SSE obtained from the regressions using Gaussian series M2 for data recorded during each of five measurements conducted for various PD generation values.

Based on the R^2 coefficient analysis, it was found that the M2 model was performed with an excellent adequacy level in the majority of cases, which was confirmed by the obtained values, close to 1. No fitting was obtained for lower PD generation voltages. This is due to a lack of dominant wavelengths emitted by PD at these voltages.

Based on the analysis of the values of the coefficients RMSE and SSE that reached values above 100, it was concluded that the model should not be used for prediction.

Figure 7 presents the results of modeling intensity spectra using a sum of eight Gaussian functions, model M1 for selected values of PD generation voltage. The red color denotes measurement results and blue corresponds to modeling results. The legend contains the value of determination coefficient R^2.

Figures 8 and 9 present graphical values of fitting parameters, R^2, RMSE, and SSE obtained by the regression of model M1 for all data recorded during each five measurements conducted by different PD generation values.

It was found that the M1 model was performed in a very similar way to the M2 model with an excellent adequacy level in the majority of cases, which was confirmed by the obtained values of R^2, close to one. Only in individual cases for the lowest PD generation voltages, poor fitting or no fitting was observed.

The application of the M1 model allowed precise determination of intensity and wavelength of dominant light waves in recorded spectrum emitted by PD. Figure 9(b) presents intensity of individual wavelengths depending on the value of voltage supplying given system.

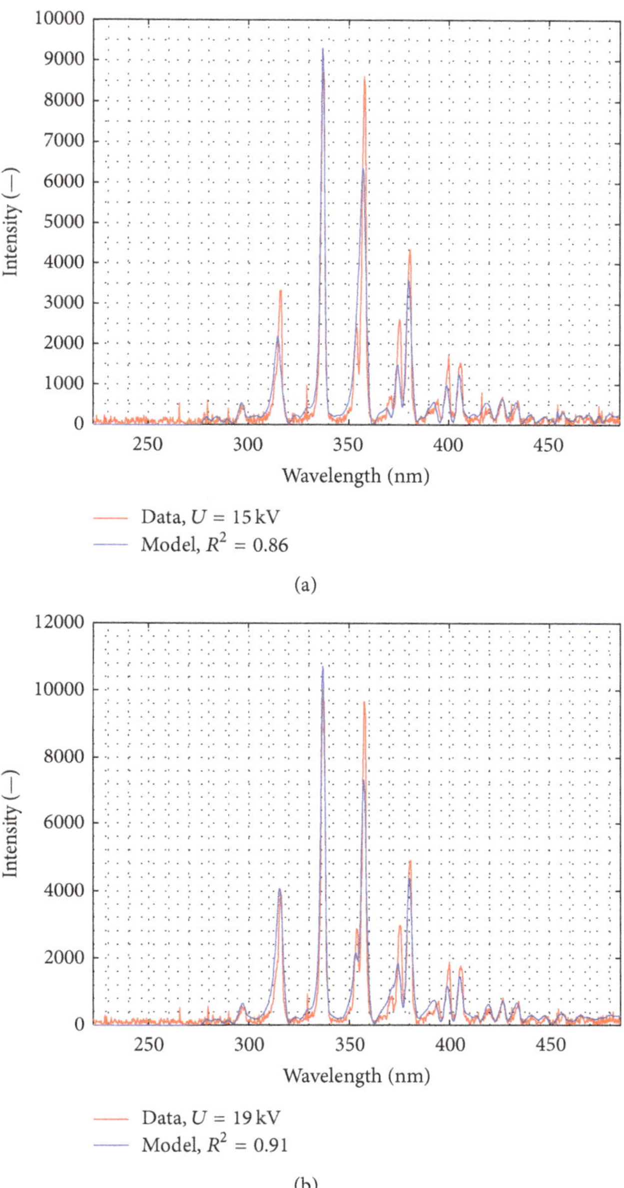

FIGURE 4: Intensity spectra of recorded and modeled light emission for selected values of PD generation voltages: (a) 0.75 U_p = 15 kV and (b) 0.94 U_p = 19 kV.

3.2. Analysis of Emission Spectra of Discharges Occurring on the Surface of LK-Type Cap Insulator. Figure 10 depicts cumulative comparison of histograms calculated for spectra of highest intensities obtained during measurements as a function of PD generation value. Figure 10(a) refers to the intensity exceeding the value of 500 ($I > 500$) and Figure 10(b) intensity exceeding the value of 1000 ($I > 1000$).

For intensities of values exceeding 1000, dominant light wavelengths occurring in the recorded signals were determined for the individual intervals and presented in Table 3.

Figure 11 shows the results of modeling (blue line) intensity spectra using model M2 for selected values of PD

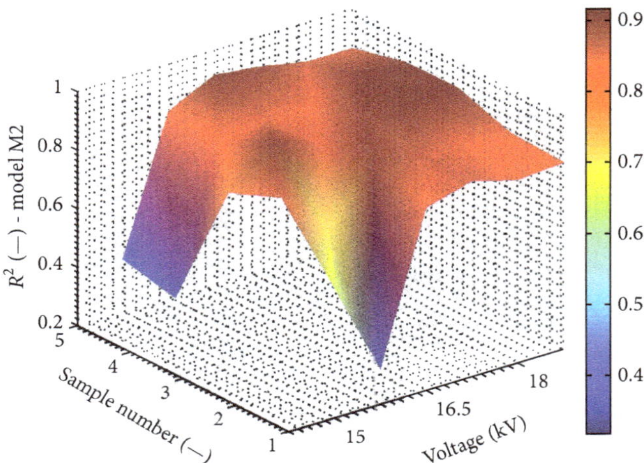

Figure 5: Comparison of values of R^2 coefficient obtained by the regression of the model M2 for all the recorded signals.

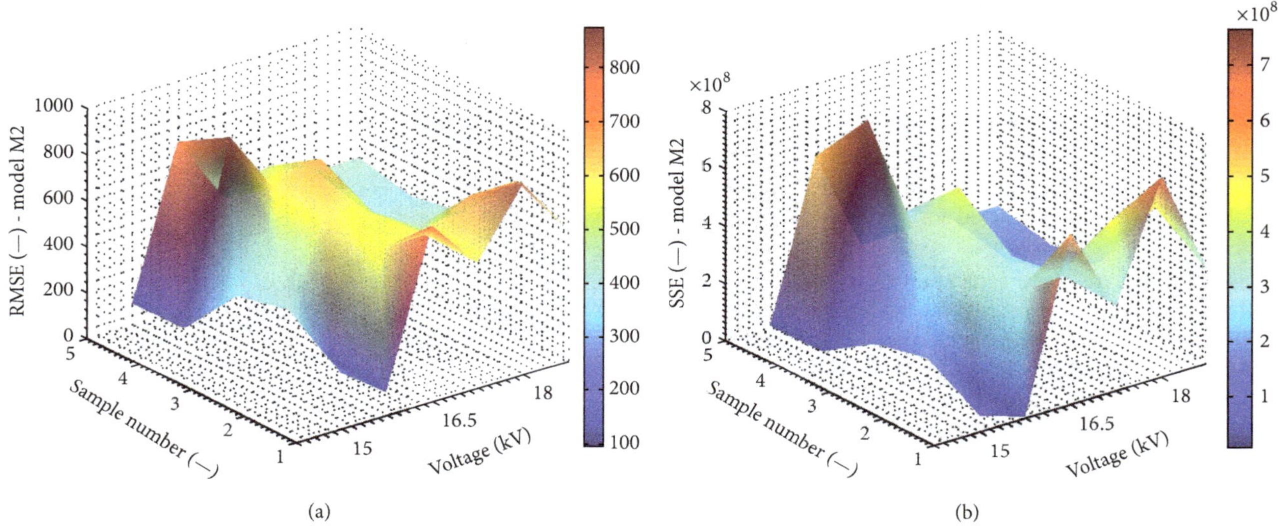

Figure 6: Comparison of values of coefficients RMSE (a) and SSE (b) obtained by the regression of the model M2 for all the recorded signals.

Table 3: Dominant wavelengths present in light emission of value $I > 1000$.

Wavelength interval (nm)	Dominant wavelengths and intervals for $I > 1000$ (nm)
$\lambda = [200–299]$	243
$\lambda = [300–399]$	312–317, 329, 333–339, 352–371, 373–382, 398-399
$\lambda = [400–499]$	400, 404–406, 435, 460
$\lambda > 500$	724, 933

generation voltage compared to measurement results (red line).

Figures 12 and 13 present values of fitting parameters, R^2, RMSE, and SSE obtained from the regressions using model M2 for data recorded during each five measurements conducted by different PD generation values.

It was found that model M2 obtained an excellent adequacy level for PD generation voltage values of 60 kV and 70–80 kV. However, no fitting was obtained for other voltages. The coefficients RMSE and SSE that reached values above 100 indicate that the model cannot be used for prediction.

Figure 14 shows the results of modeling intensity spectra using model M1 for all selected values of PD generation voltage.

Figures 15 and 16(a) depict values of fitting parameters: R^2, RMSE, and SSE obtained by the regression using model M1 for data recorded during each five measurements conducted by different PD generation values.

Model M1 for PD generation voltage values of 65 kV and 70–80 kV obtained an excellent adequacy level, which was confirmed by the obtained values, over 0.6. However, no fitting was obtained for other voltages. The coefficients RMSE and SSE that reached values above 100 indicate that the model cannot be used for prediction.

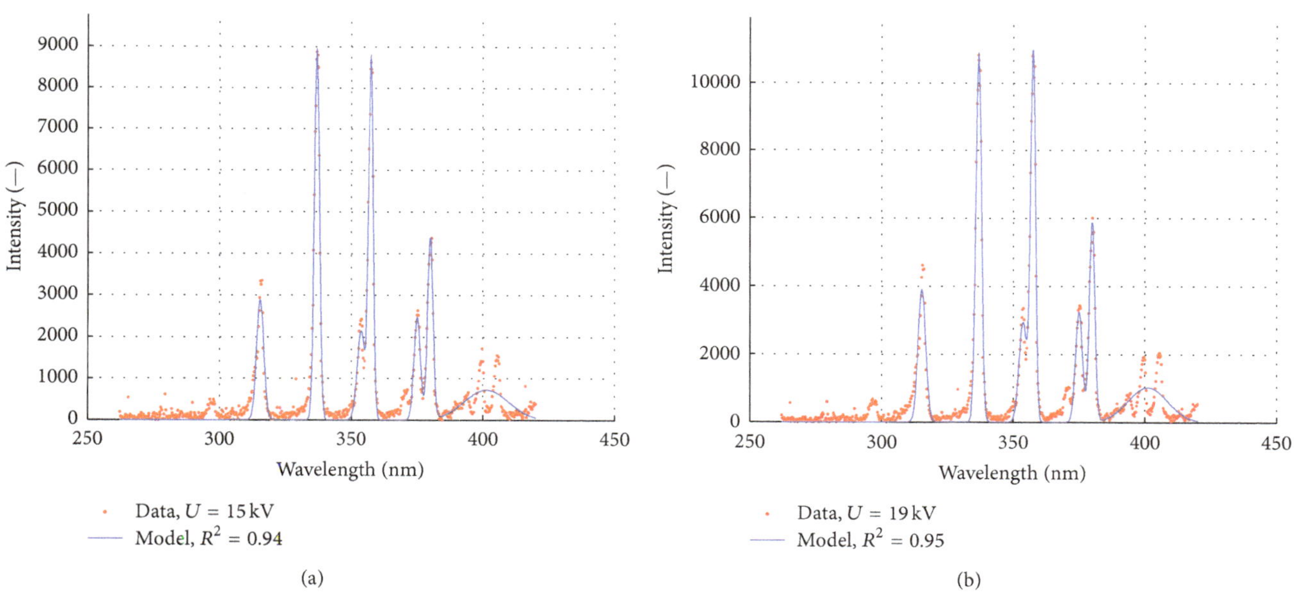

FIGURE 7: Intensity spectra of recorded and modeled light emission for selected values of PD generation voltages: (a) 0.75 U_p = 15 kV and (b) 0.94 U_p = 19 kV.

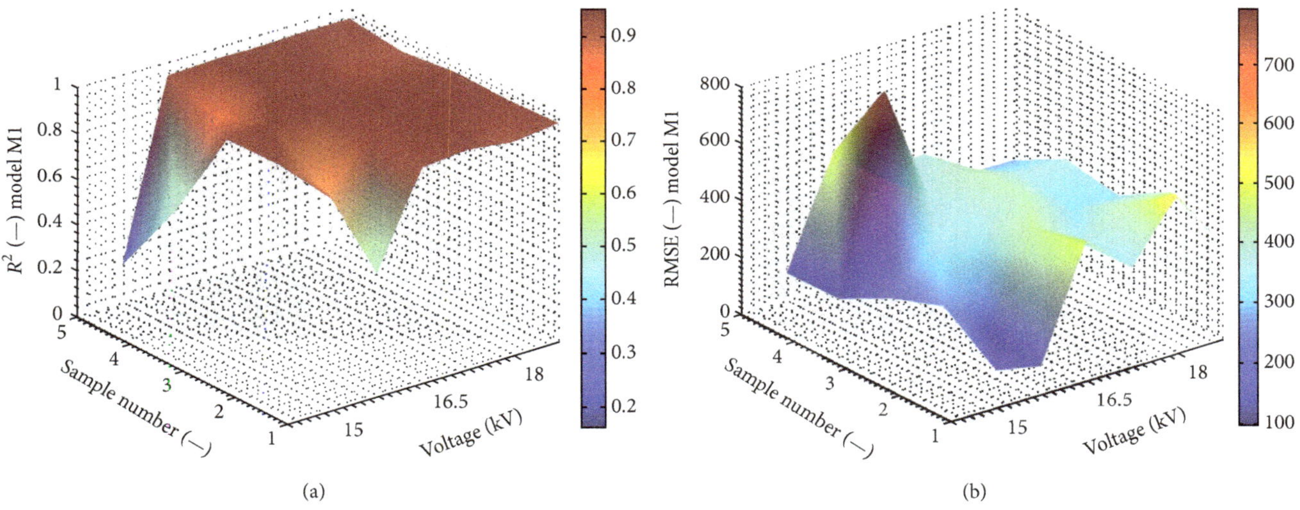

FIGURE 8: Comparison of values of R^2 (a) and RMSE (b) coefficients obtained by the regression of the model M1 for all the recorded signals.

Figure 16(b) presents the intensity of individual wavelengths depending on the value of voltage supplying given system.

3.3. Analysis of Emission Spectra of Discharges Occurring on the Surface of LS-Type Long Rod Insulator. Figure 17 depicts a cumulative comparison of histograms calculated for spectra of highest intensities obtained during measurements as a function of PD generation value. Figure 17(a) refers to the intensity exceeding the value of 500 ($I > 500$) and Figure 17(b) to the intensity exceeding the value of 2000 ($I > 2000$).

For intensities of values exceeding 2000, dominant light wavelengths occurring in the recorded signals were determined for the individual intervals and presented in Table 4.

TABLE 4: Dominant wavelengths present in light emission of value $I > 2000$.

Wavelength interval (nm)	Dominant wavelengths and intervals present in light emission of value $I > 2000$ (nm)
λ = [200–299]	—
λ = [300–399]	315–317, 335–338, 353–358, 379–381
λ = [400–499]	—
$\lambda > 500$	724

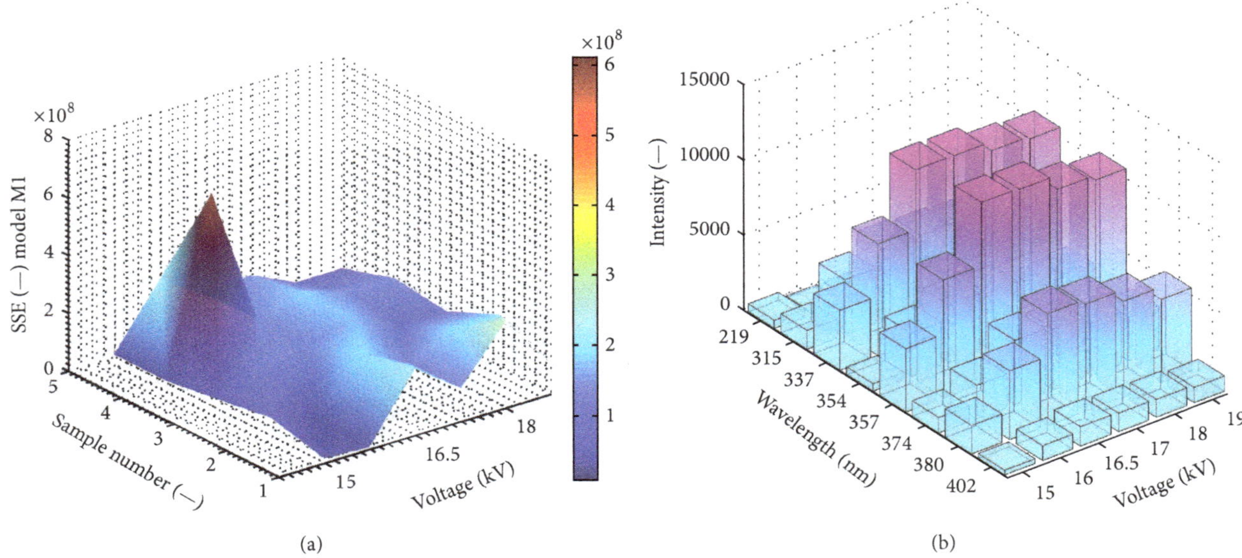

FIGURE 9: (a) Voltage dependency on SSE obtained by model M1 for all data samples. (b) Intensity of wavelength components obtained by model M1 as a function of PD generation voltage.

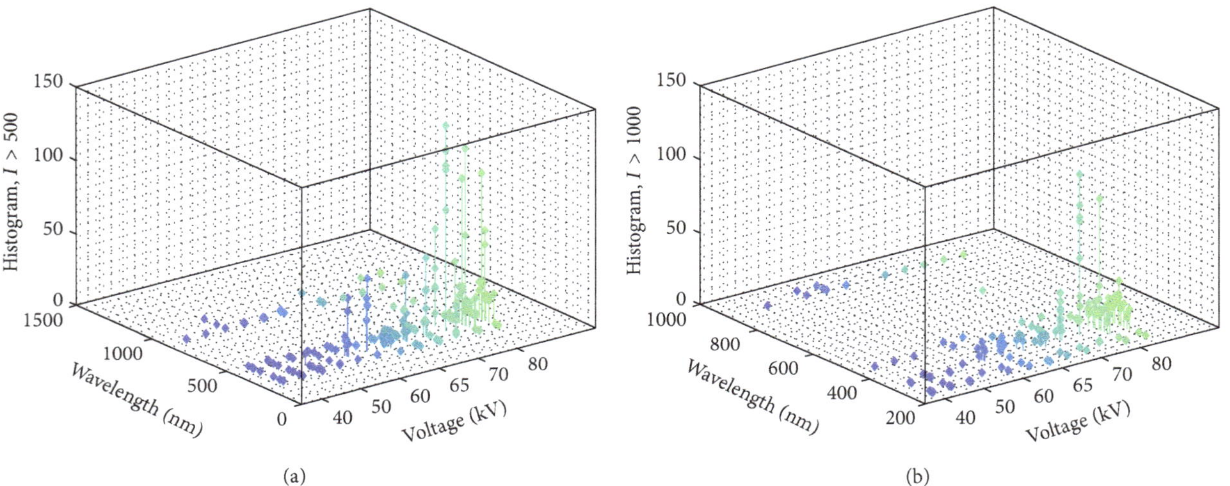

FIGURE 10: Summary of spectrum intensity histograms for different voltages: (a) $I > 500$ and (b) $I > 1000$.

Figure 18 presents the results of modeling (blue line) intensity spectra using model M2 for selected values of PD generation voltage and experimental results (red line).

Figures 19 and 20 show values of fitting parameters: R^2, RMSE, and SSE obtained from the regressions using model M2 for data recorded during each five measurements conducted by various PD generation values. It was found that model M2 obtained an excellent adequacy level in most cases, which was confirmed by the obtained values, R^2 above 0.6. Only for the 37.5 kV no fitting was obtained. From Figure 20 it is to conclude that the model cannot be used for prediction.

Figure 21 presents results of modeling (blue color) intensity spectra using model M1 for selected values of PD generation voltage. Red dots denote experimental results.

Figures 22 and 23 depict values of fitting parameters: R^2, RMSE, and SSE obtained by the regression using model M1 for data recorded during each of five measurements conducted for various PD generation values.

It was found that model M1 gives very good fitting for signals registered in voltage values over 40 kV, which was proven by the R^2 values over 0.8. No fitting was obtained at 37.5 kV in 4 of 5 samples. Figure 23(b) depicts the PD generation voltage dependency on the intensity of individual wavelengths.

4. Conclusions

Conducting tests with spectrophotometer allowed precise determination of lines length and corresponding intensities

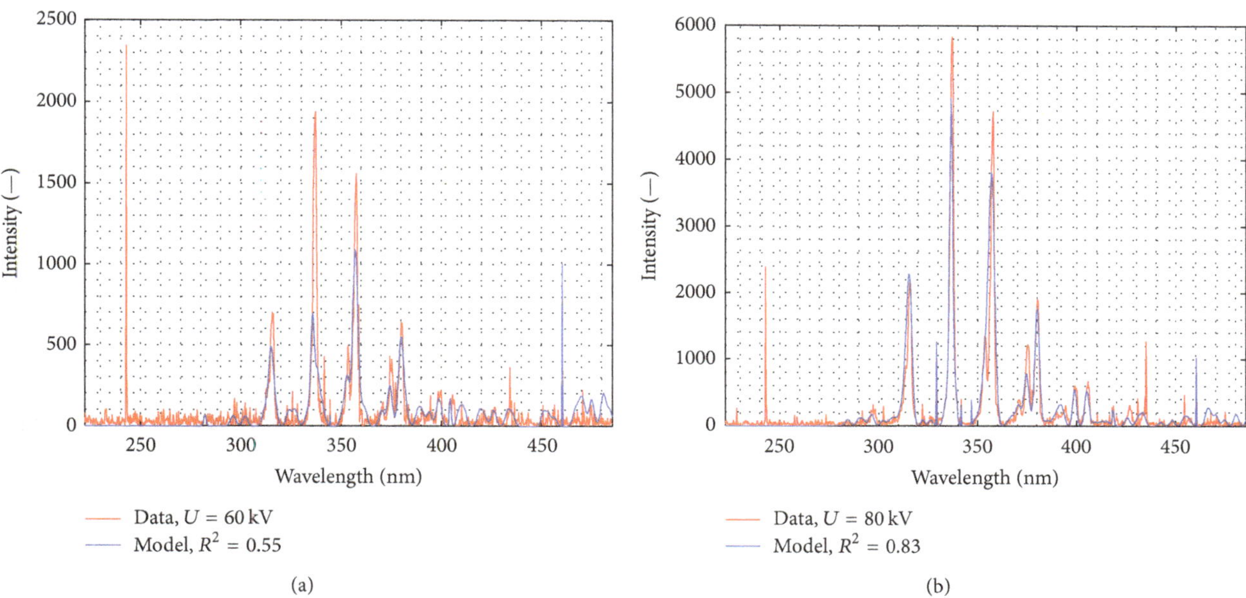

FIGURE 11: Intensity spectra of recorded and modeled light emission for selected values of PD generation voltages: (a) 0.56 U_p = 60 kV and (b) 0.89 U_p = 80 kV.

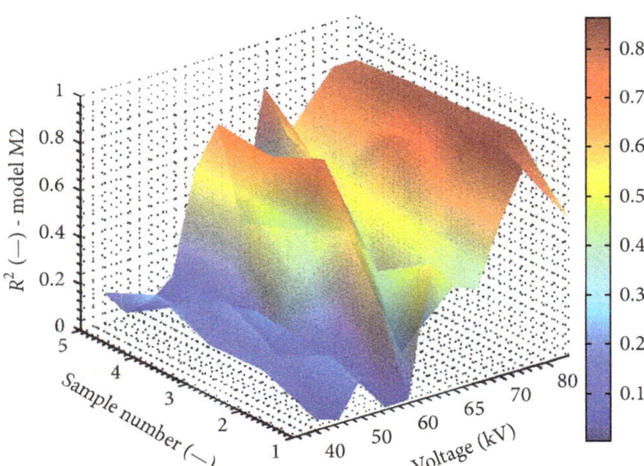

FIGURE 12: Comparison of R^2 obtained by regression of model M2 for all the recorded signals.

in the recorded optical signals. Based on the obtained dependencies of intensities of emission spectrum emitted by PD generated on studied dielectrics and insulation system the following was concluded:

(i) During PD generation on the surface of a porcelain insulating cylinder filled with quartz, while the distance between electrodes was 3 cm, the highest intensities reaching 10,000 were obtained for spectral lines in a wavelength range of 333–340 and 351–360 nm, when the system was supplied with voltage in a range from 16.5 kV to 19 kV.

(ii) During PD generation on the surface of a porcelain insulating cylinder filled with quartz, while the distance between electrodes was 5 cm, the highest intensities reaching 10,000 were obtained for spectral lines in a wavelength range of 331–341 and 349–362 nm, when system was supplied with voltage in a range from 25.5 kV to 27 kV.

(iii) During PD generation on the surface of a porcelain insulating cylinder filled with quartz, while the distance between electrodes was 7 cm, the highest intensities reaching 6000 were obtained for spectral lines in a wavelength range of 335–339 and 352–359 nm, when the system was supplied with voltage in a range from 31.5 kV to 33.5 kV.

(iv) During PD generation on the surface of a porcelain insulating cylinder filled with quartz, while the distance between electrodes was 9 cm, the highest

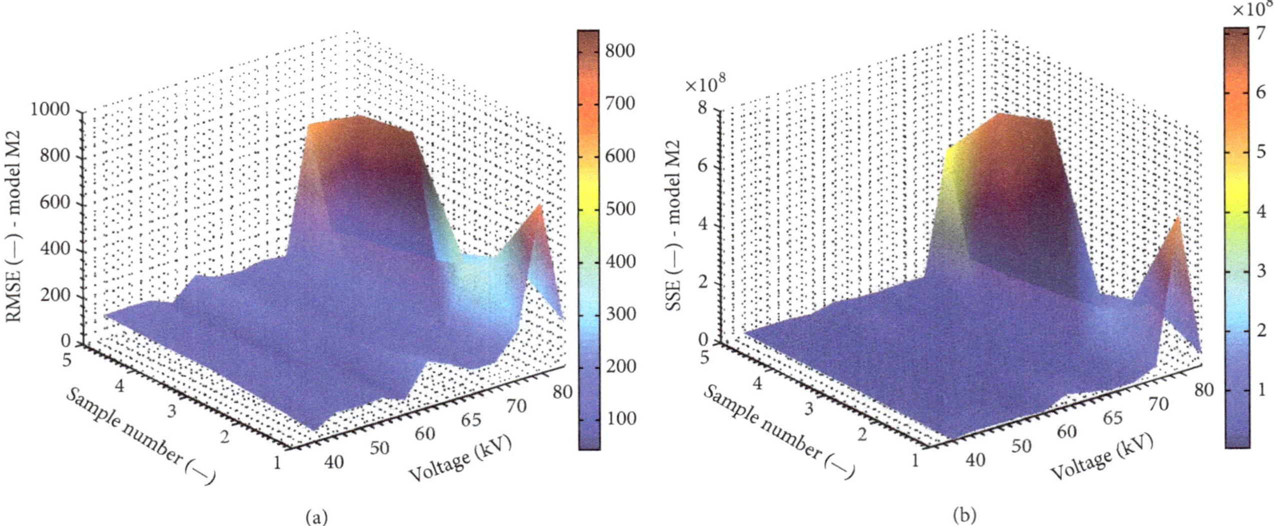

FIGURE 13: Comparison of RMSE (a) and SSE (b) obtained by regression of model M2 for all data samples.

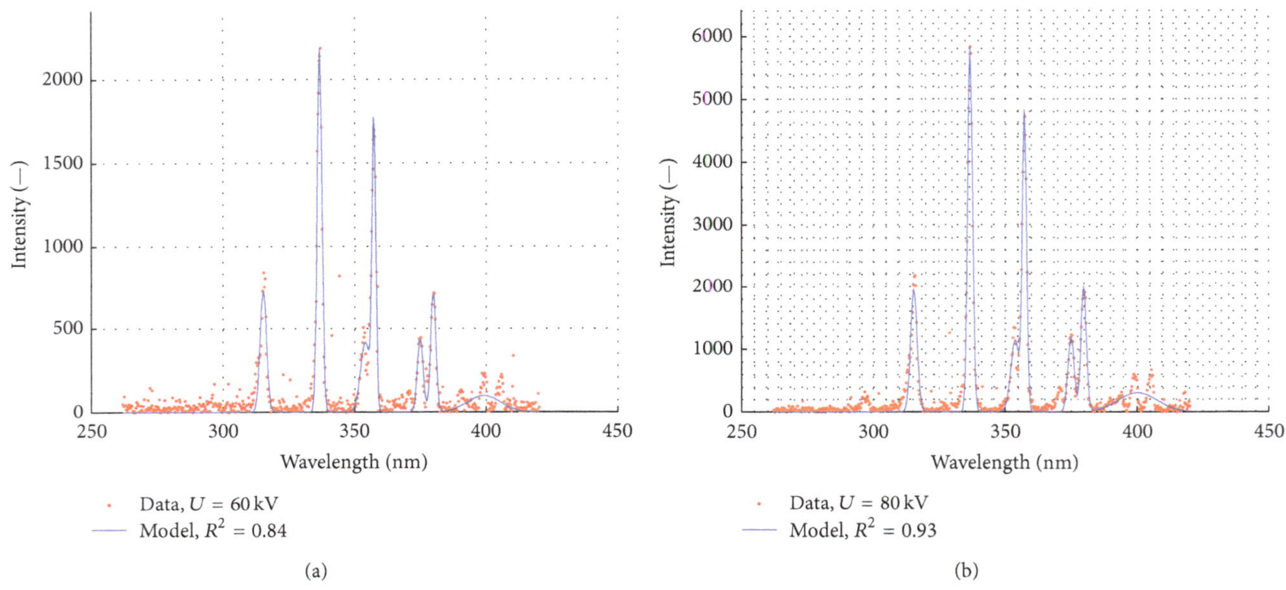

FIGURE 14: Intensity spectra of recorded and modeled light emission for selected values of PD generation voltages: (a) 0.56 U_p = 60 kV and (b) 0.89 U_p = 80 kV.

intensities reaching 2000 were obtained for spectral lines in a wavelength range of 336–338, 354, and 356–359 nm, when the system was supplied with voltage in a range from 39 kV to 41 kV and 43 kV to 46 kV.

(v) During PD generation on the surface of a porcelain insulating cylinder filled with quartz, while the distance between electrodes was 9 cm, the highest intensities reaching 2000 were obtained for spectral lines in a wavelength range of 335–339 and 353–359 nm, when the system was supplied with voltage in a range from 44 kV to 45 kV and for approximately 50 kV.

(vi) By analyzing the influence of distance on the obtained results, one can conclude that the range of components decreases with increasing distance, which is probably a result of the suppression of individual spectral lines in the air.

(vii) During PD generation on a cap insulator made of porcelain, the highest intensities, over 9000, were obtained for spectral lines in ranges 333–339 and 352–371. The insulator was supplied with voltage of values in a range from 72 kV to 76 kV.

(viii) During PD generation on an individual cap of porcelain long rod insulator, the highest intensities, over

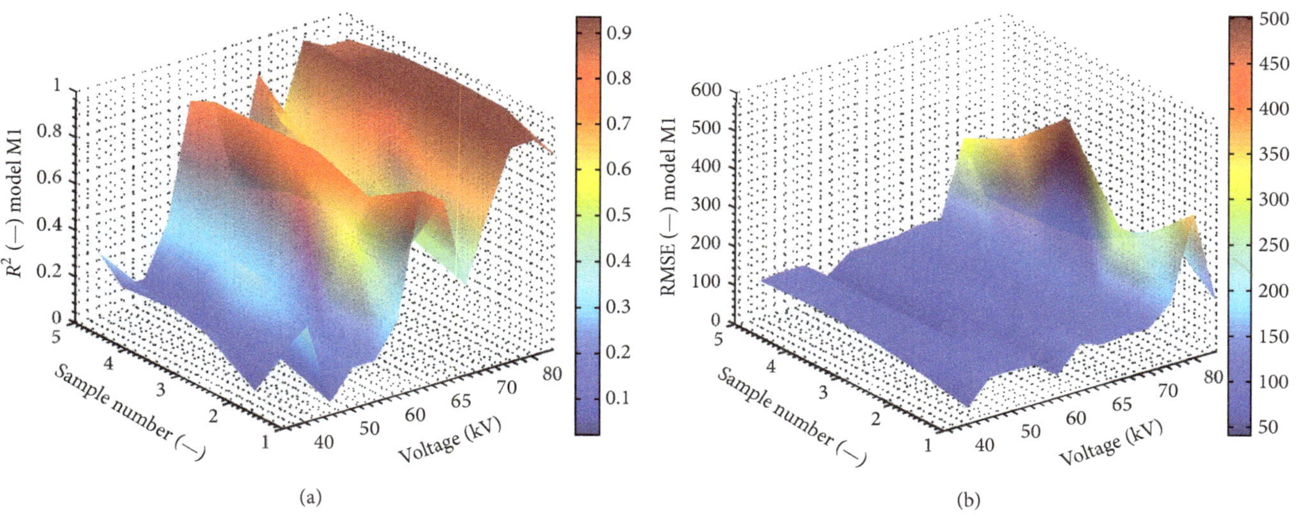

FIGURE 15: Comparison of R^2 (a) and RMSE (b) obtained by regression of model M1 for all data samples.

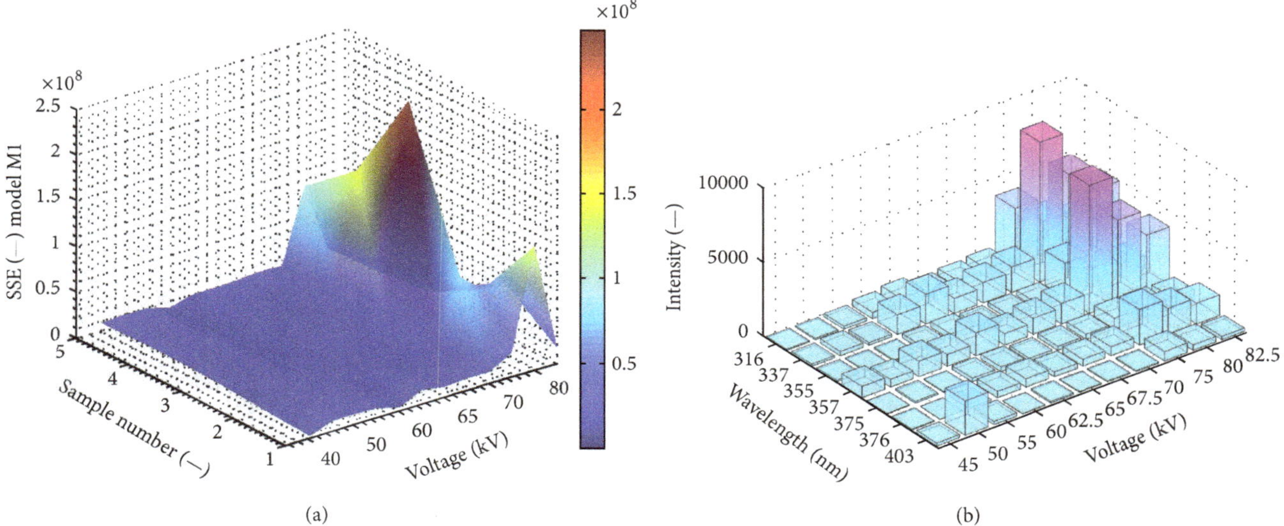

FIGURE 16: (a) Comparison of SSE coefficient obtained by regression of model M1 for all recorded signals. (b) Voltage dependence of intensity of individual wavelengths obtained by model M1.

6000, were obtained for spectral lines in ranges 335–338 and 353–358. The insulator was supplied with voltage of values in a range from 42.5 kV to 46 kV.

Table 5 shows a comparison between PD initial voltage values U_0 and U_{01}, during which the measurement device recorded first light emission waves (U_{01}) or when the recorded spectra included the most components of increased (over 300) intensities (U_0), depending on the test system. The analysis of the values presented in Table 5 allows one to conclude that the application of the spectrophotometer enables earlier recognition of PD generation, which confirms the thesis of the study. On average, the increased performance was estimated as a lower level of voltage, 42%.

Based on the obtained results, Table 6 presents a comparison of dominant wavelengths contained in light emission of intensities exceeding 1000, obtained for individual dielectrics and ceramic insulators.

The comparison of the data presented in Table 6 allows one to formulate the following general conclusions:

(i) PD emitted electromagnetic waves in a range of 300–400 nm regardless of in which dielectric and insulator system they are generated.

(ii) PD emitted waves of 203 nm length on a porcelain cylinder filled with quartz, except when the HV electrode was placed at a distance between electrodes of 11 cm.

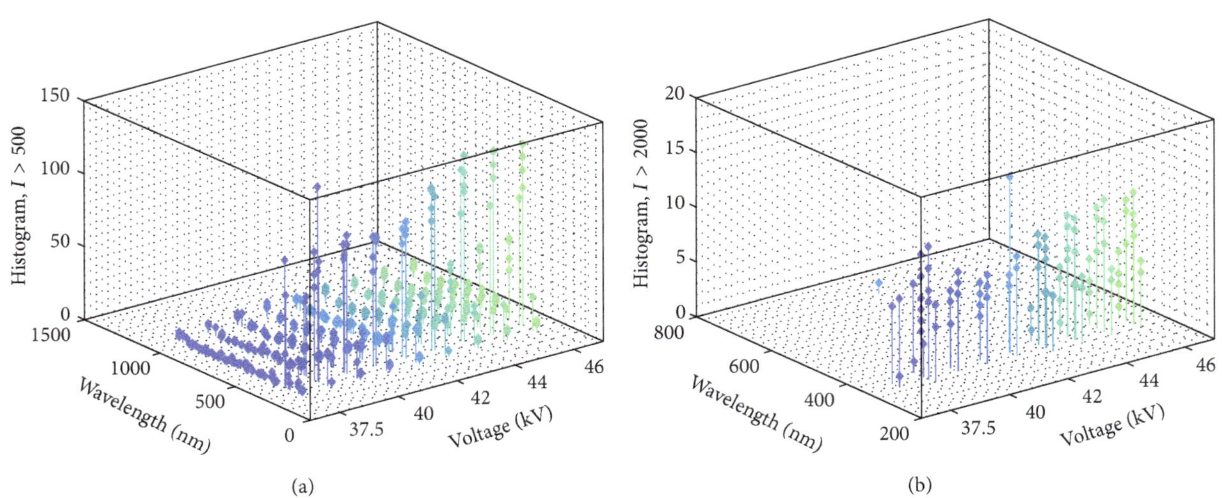

FIGURE 17: Summary of spectrum intensity histograms for different voltages: (a) $I > 500$ and (b) $I > 2000$.

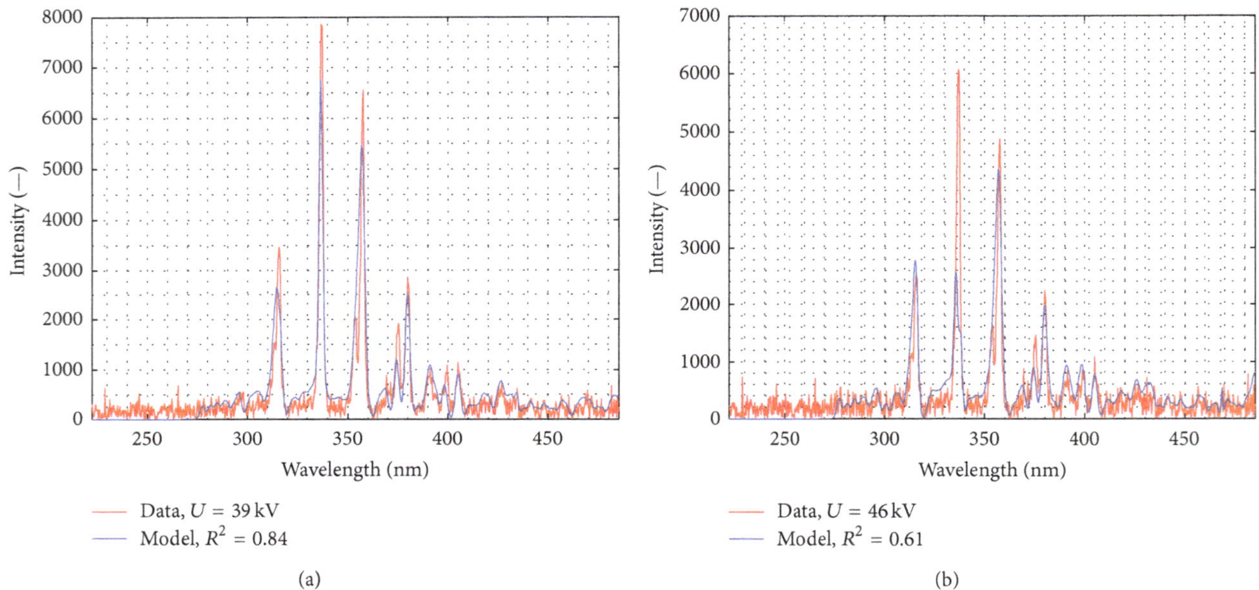

FIGURE 18: Intensity spectra of recorded and modeled light emission for selected values of PD generation voltages: (a) 0.81 U_p = 39 kV and (b) 0.95 U_p = 46 kV.

TABLE 5: Comparison of PD initial voltages for tested insulation systems, during which emission spectra may have increased intensity and the broadest range.

	Values of initial voltages U_0 and U_{01} for tested insulation systems						
	WPK 3 cm	WPK 5 cm	WPK 7 cm	WPK 9 cm	WPK 11 cm	LK —	LS —
U_0	9.9%	18.2%	23.6%	24.3%	25.7%	23.3%	12%
U_{01}	15.5%	25.5%	30.5%	37%	39%	72%	37.5%
Difference	36.2%	28.6%	22.6%	34.3%	34.1%	67.6%	68%

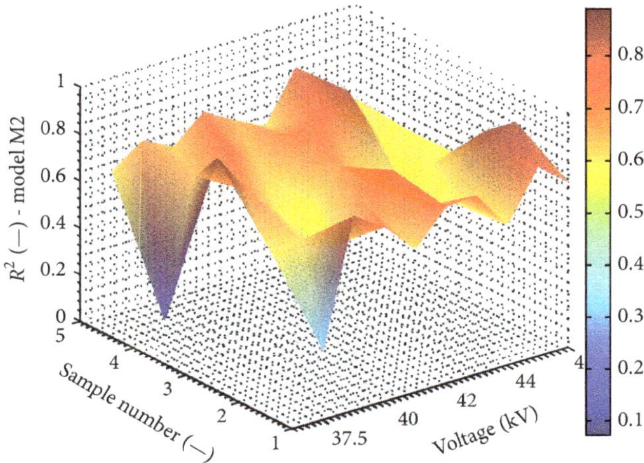

FIGURE 19: Comparison of R^2 coefficient obtained by regression of model M2 for all the recorded signals.

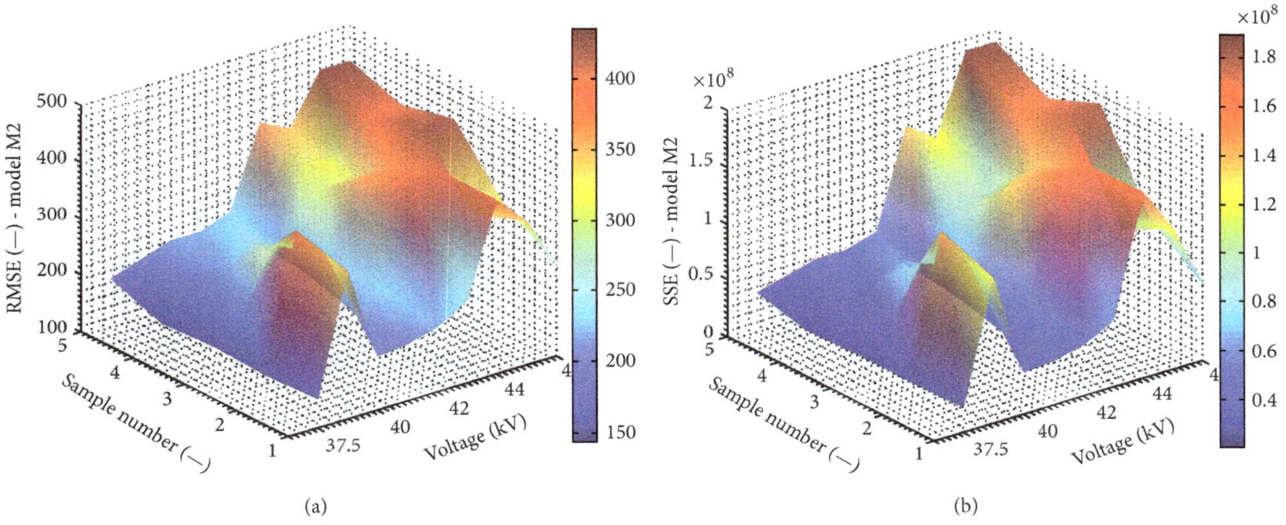

FIGURE 20: Comparison of RMSE (a) and SSE (b) obtained in model M2 for all the recorded signals.

TABLE 6: Comparison of the dominating wavelengths present in light emission of value $I > 1000$ for all tested dielectrics and ceramic insulators.

System symbol	Values and ranges of recorded optical waves λ (nm) exhibiting intensity $I > 1000$			
WPK 3	203, 296–298	311–318, 333–340, 351–360, 369–382, 390–396, 397–399	400-401, 404–407, 420, 425–428, 434-435	745, 933, 1073
WPK 5	203, 295–299	309–318, 328–329, 331–341, 349–362, 366–383, 386, 388–399	400–407, 414, 416, 418–421, 424–428, 431–435, 449, 456–458	607, 653-654, 658–663, 667–671, 676–678, 724, 745, 774, 793, 913, 933, 1062, 1073
WPK 7	203	306–317, 335–339, 352–359, 374–382, 386, 393-394, 399	400, 405-406, 423	587–594, 640, 720, 745, 766–768, 770–771, 793, 933
WPK 9	203	315–317, 336–338, 354, 356–359, 375-376, 379–381	—	660, 724, 813, 933
WPK 11	—	313–317, 335–339, 353–359, 374–377, 379–382, 399	400, 406	660, 724, 739
LS	—	315–317, 335–338, 353–358, 379–381	—	724
LK	243	312–317, 329, 333–339, 352–371, 373–382, 398-399	400, 404–406, 435, 460	724, 933

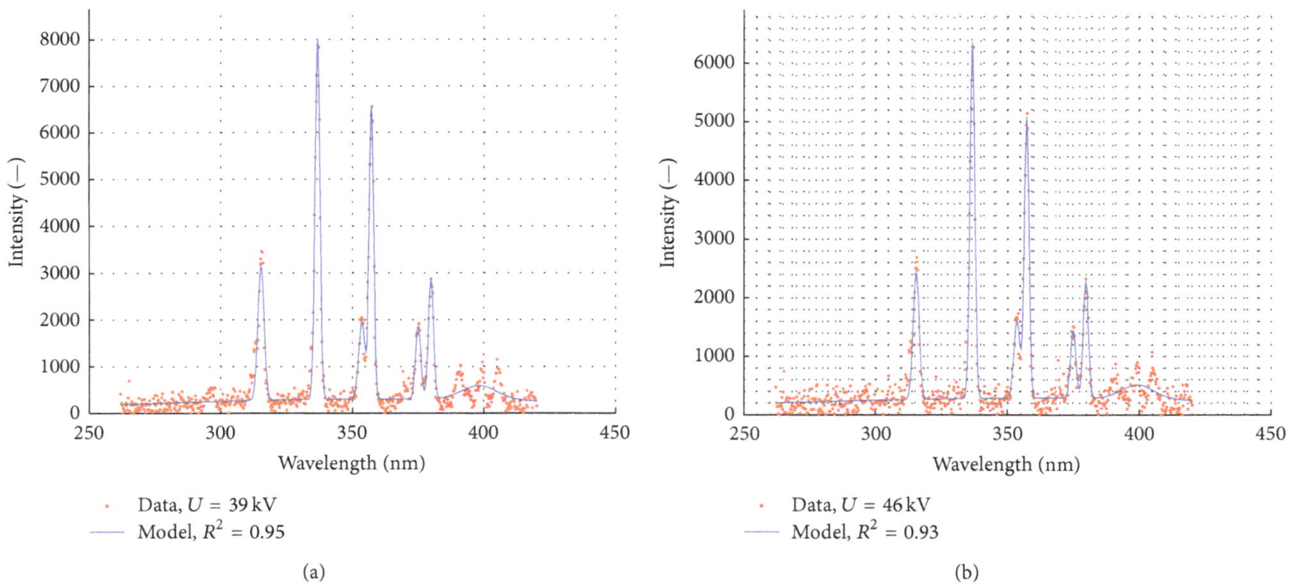

FIGURE 21: Intensity spectra of recorded and modeled light emission for selected values of PD generation voltages: (a) 0.81 U_p = 39 kV and (b) 0.95 U_p = 46 kV.

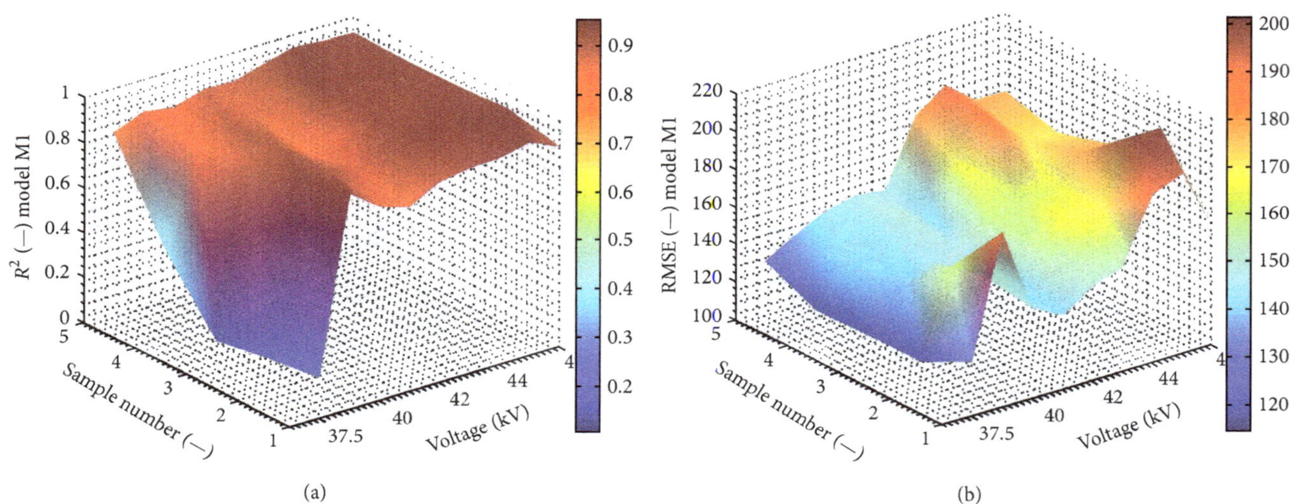

FIGURE 22: Comparison of R^2 (a) and RMSE (b) coefficient obtained in model M1 for all the recorded signals.

(iii) PD occurring on the surface of the tested cap insulator are characterized by the presence of waves of 243 nm length.

(iv) PD occurring on the porcelain cylinder with a HV electrode placed at distances between electrodes of 3 cm and 5 cm emitted waves of 295–299 nm length.

(v) PD emitted many waves in a range of 400–500 when they occurred on a porcelain cylinder with a HV electrode placed at distances between electrodes of 3 cm and 5 cm. Single waves in this range were emitted on cap insulators and other types of insulating cylinders, except for cylinder tested at a distance between electrodes of 9 cm. PD occurring on the porcelain long rod insulator did not emit any optical waves in this range.

(vi) EM waves in visible light range, above 500 nm, were present in spectra in different ways. Most components from this range were recorded on porcelain roll insulators during PD generation at distances of 5 cm and 7 cm.

(vii) The lowest number of spectral components was generated by PD occurring on the long rod insulator made of porcelain.

Figures 24–26 present a comparison showing the averaged values of fitting coefficients of regression models M1 and M2 to empirical data: SSE (4), R^2 (6), and RMSE (5).

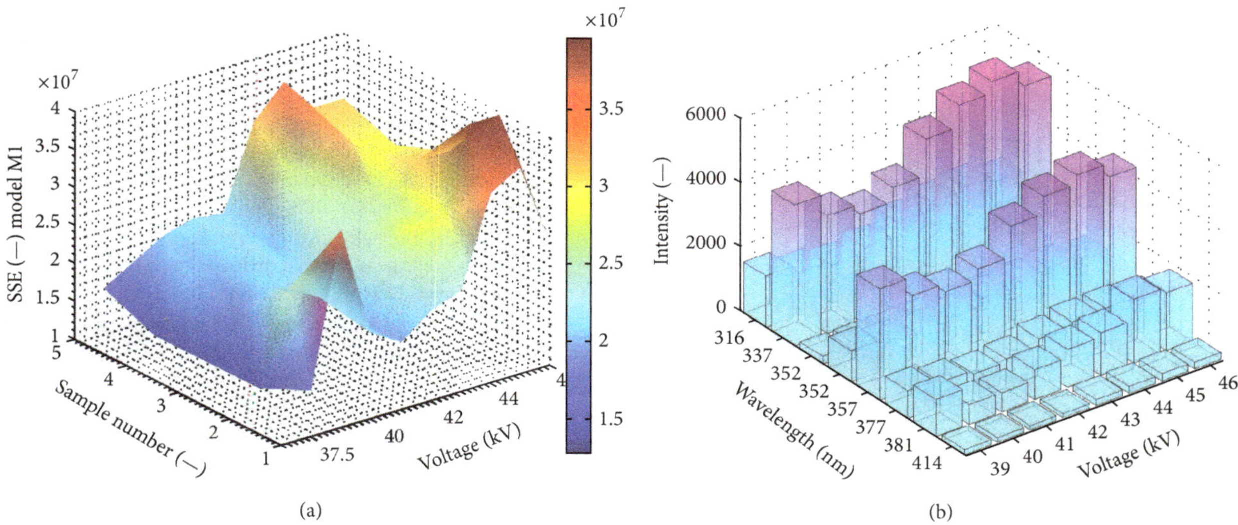

(a)

(b)

FIGURE 23: (a) Comparison of SSE coefficient values obtained in model M1 for all the recorded signals. (b) Dependence of the intensity of individual wavelengths as a function of PD generation voltage.

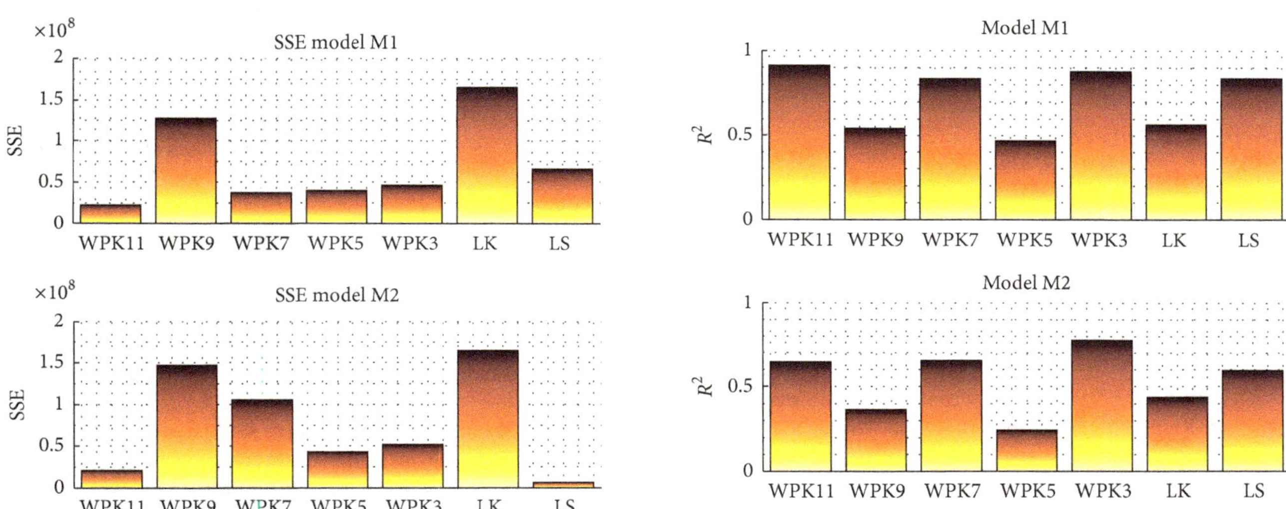

FIGURE 24: Comparison of the averaged values of the SSE coefficient obtained by the regression of models M1 and M2 for all the tested dielectrics and insulation systems.

FIGURE 25: Comparison of the averaged values of the R^2 coefficient obtained by the regression of models M1 and M2 for all the tested dielectrics and insulation systems.

The comparative analysis of models M1 and M2 fitting coefficients, which describe the dependency of intensity of emission spectra as a function of wavelength, may serve as a justification for the following statements:

(i) The values of coefficients SSE and RMSE are similar in both models. The tested insulation system exhibited high values of these parameters, which proves that these models are not optimal for prediction applications.

(ii) Regarding the comparison of determination coefficient values, R^2, obtained for the M1 model, we may conclude that good fitting was observed only in selected insulation models, including porcelain cylinder at distances of 3, 7, and 11 cm and in LS-type insulator. The rest of values indicate rather moderate and poor fitting. In case of model M2, all the fitting coefficient values had results of R^2 below 0.7, except for the WPK3 system, which also indicates poor fitting of the model to empirical data.

(iii) It should be noted that all of the conclusions mentioned above apply to values that have been averaged "twice"; that is, firstly, the results from five measurements were averaged and then a second averaging was conducted for all PD generation voltages. Therefore, it seems more reasonable to take into consideration

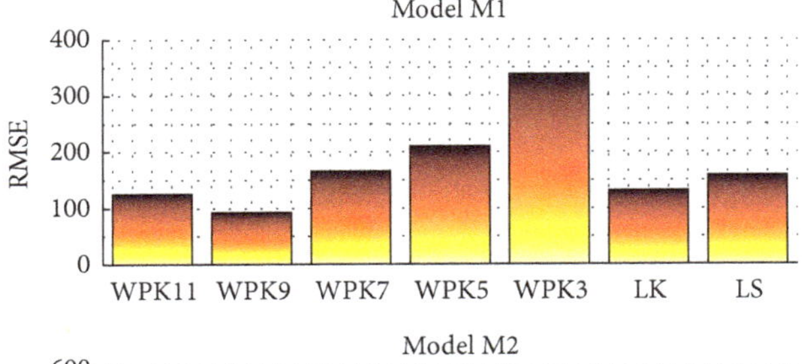

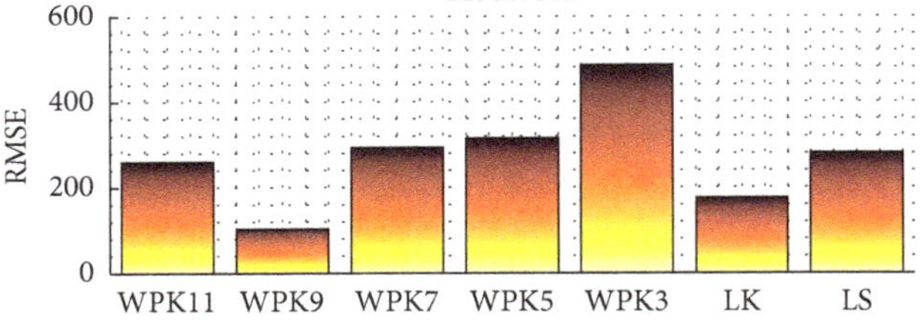

FIGURE 26: Comparison of the averaged values of the RMSE coefficient obtained by the regression of models M1 and M2 for all the tested dielectrics and insulation systems.

the results obtained for values averaged just "once" for individual voltages, which are depicted in Figures 5, 6, 8, 9(a), 12, 13, 15, 16(a), 19, 20, 22, and 23(a).

Competing Interests

The authors declare that there is no conflict of interests regarding the publication of this paper.

Acknowledgments

The work was cofinanced with funds from the National Science Centre (NCS) as part of the OPUS program, Project no. 2013/09/B/ST8/01736. The work was cofinanced by the European Regional Development Fund "Increase of Scientific Research and Innovation for Enterprises in Terms of Sustainable Development through the Creation of a Modern Diagnostics Laboratory of Surge Voltage at the Opole University of Technology" Part I (2010-2011) and Part II (2011-2013), Project nos. RPO.01.03.0101-16-007/10-00 and WND-RPOP.01.03.01-16-007/10.

References

[1] R. J. van Brunt, "Physics and chemistry of partial discharge and corona—recent advances and future challenges," in *Proceedings of the IEEE Conference on Electrical Insulation and Dielectric Phenomena*, pp. 29–70, October 1994.

[2] I. J. Kemp, "Partial discharge plant-monitoring technology: present and future developments," *IEE Proceedings—Science, Measurement and Technology*, vol. 142, no. 1, pp. 4–10, 1995.

[3] A. Aman, M. M. Yaacob, M. A. Alsaedi, and K. A. Ibrahim, "Polymeric composite based on waste material for high voltage outdoor application," *International Journal of Electrical Power and Energy Systems*, vol. 45, no. 1, pp. 346–352, 2013.

[4] R. Lopatkiewicz, Z. Nadolny, and P. Przybylek, "The influence of water content on thermal conductivity of paper used as transformer windings insulation," in *Proceedings of the IEEE 10th International Conference on the Properties and Applications of Dielectric Materials (ICPADM '12)*, pp. 1–4, IEEE, Bangalore, India, July 2012.

[5] F. Witos and Z. Gacek, "Application of the joint electro-acoustic method for partial discharge investigations within a power transformer," *The European Physical Journal Special Topics*, vol. 154, no. 1, pp. 239–247, 2008.

[6] A. Cichoń, S. Borucki, and D. Wotzka, "Modeling of acoustic emission signals generated in on load tap changer," *Acta Physica Polonica A*, vol. 125, no. 6, pp. 1396–1399, 2014.

[7] E. Veldhuizen and W. Rutgers, "Corona discharges: fundamentals and diagnostics," in *Proceedings of the 4th Conference on Frontiers in Low Temperature Plasma Diagnostic*, pp. 40–49, 2001.

[8] I. A. D. Giriantari, "Monitoring the insulator condition by on-line voltage distribution measurement," in *Proceedings of the International Conference on Condition Monitoring and Diagnosis (CMD '08)*, pp. 392–394, Beijing, China, April 2008.

[9] A. Blachowicz, T. Boczar, and D. Wotzka, "Application of a mobile system in diagnostics of power capacitors using the acoustic emission method," *Insight*, vol. 58, no. 2, pp. 94–100, 2016.

[10] A. Cichoń, P. Frącz, and D. Zmarzły, "Characteristic of acoustic signals generated by operation of on load tap changers," *Acta Physica Polonica A*, vol. 120, no. 4, pp. 585–588, 2011.

[11] R. Badent, K. Kist, A. Schwab, and M. Wurster, "Light emission measurements of predischarges in insulation oil," in *Proceedings of IEEE Annual Report Conference on EI and Dielectric Phenomena*, vol. 2, pp. 452–455, Atlanta, Ga, USA, 2008.

[12] H. Kojima, N. Hayakawa, F. Endo, and H. Okubo, "Novel measurement and analysis system for investigation of partial discharge mechanism in SF6 gas," in *Proceedings of the 12th International Middle East Power System Conference (MEPCON '08)*, pp. 75–79, Aswan, Egypt, March 2008.

[13] D. Zmarzly, Ł. Nagi, S. Borucki, and T. Boczar, "Analysis of ionizing radiation generated by partial discharges," *Acta Physica Polonica A*, vol. 125, no. 6, pp. 1377–1379, 2014.

[14] M. Brockschmidt, F. Pohlmann, S. Kempen, and P. Gröppel, "Testing of nano-insulation materials: some ideas, some experiences," in *Proceedings of the 30th Electrical Insulation Conference (EIC '11)*, pp. 506–510, Annapolis, Md, USA, June 2011.

[15] P. Morshuis and E. Gulski, "Diagnostic tools for condition monitoring of insulating materials," in *Proceedings of the Conference on Electrical Insulation and Dielectric Phenomena*, pp. 327–330, Virginia Beach, Va, USA, October 1995.

[16] W. Tan, W. Huang, K. Wang, and Z. Zhang, "The study of ultraviolet pulse for partial discharge of transformers," in *Proceedings of the IEEE World Automation Congress (WAC '08)*, pp. 1–4, Waikoloa, Hawaii, USA, October 2008.

[17] H. Zhang, Q. Pang, and X. Chen, "The characteristics of high-voltage corona and its detection," *Electrical Measurement and Instrumentation*, vol. 43, pp. 6–8, 2006.

[18] P. Frącz, T. Boczar, D. Zmarły, and T. Szczyrba, "Analysis of optical radiation generated by electrical discharges on support insulator," *Acta Physica Polonica A*, vol. 124, pp. 413–416, 2013.

[19] P. Frącz, "Measurement of optical signals emitted by surface discharges on bushing and post insulator," *IEEE Transactions on Dielectrics and Electrical Insulation*, vol. 20, no. 5, pp. 1909–1914, 2013.

20

Kinetic Study of Atmospheric Pressure Nitrogen Plasma Afterglow Using Quantitative Electron Spin Resonance Spectroscopy

A. Tálský, O. Štec, M. Pazderka, and V. Kudrle

Department of Physical Electronics, Masaryk University, Kotlarska 2, 61137 Brno, Czech Republic

Correspondence should be addressed to V. Kudrle; kudrle@sci.muni.cz

Academic Editor: Nikša Krstulović

Quantitative electron spin resonance spectroscopy is used to measure nitrogen atom density in atmospheric pressure dielectric barrier discharge afterglow. The experiment shows that oxygen injection into early afterglow increases the nitrogen dissociation in certain parts of the afterglow while it is decreased in the rest of the afterglow. Numerical kinetic modelling supports and explains the experimental data while the best fit provides some a priori unknown parameters such as initial concentrations and rate constants.

1. Introduction

Nonequilibrium cold plasmas at atmospheric pressure and especially dielectric barrier discharges are getting increased attention of both basic and applied research. They are widely used in many industrial applications from ozone production [1] and lighting [2] to plasma surface modifications [3]. Their main advantages over other competing technologies are the ease-of-use, economy, and environmental considerations [4]. Besides the already established applications, there is rapid expansion of atmospheric pressure plasmas into new areas, such as plasma-chemical synthesis of substances which are difficult to attain by other techniques [5–7], plasma medicine [8], material disinfection and sterilisation [9], and use in cosmetics [10] or in fashion industry [11].

Nitrogen plasmas are often used as a source of high density nitrogen atoms. Molecular nitrogen is rather inert gas, but atomic nitrogen is quite reactive, which is made use of in plasma deposition of nitride films [12] or plasma nitridation [13]. Although the low pressure plasmas are currently dominating this field, there is strong incentive for research and development of nitrogen plasma sources operating at atmospheric pressure [14].

In order to develop new plasma-chemical technologies it is important to understand the elementary processes taking place in both the active plasma and plasma afterglow. Nitrogen, despite being a simple diatomic molecule, has quite complex plasma chemistry and kinetics, especially in mixtures with oxygen [15, 16]. As stated above, the nitrogen atoms play a significant kinetic role due to their high reactivity. Their concentration is then one of the most important parameters to be experimentally determined. Although there is broad range of experimental techniques able to detect N atoms, for example, optical emission spectroscopy, only few of them are suitable for absolute, not relative, measurements. UV absorption spectroscopy [17], NO titration [18], mass spectroscopy (MS) [19], laser induced fluorescence (LIF) [20], and catalytic probes [21], for example, are widely used. Another challenge is an operation of such diagnostic technique in atmospheric pressure, which, for example, greatly increases quenching in LIF, complicates pumping in MS, or totally changes the plasma chemistry (NO titration).

In this paper, the electron spin/paramagnetic resonance (ESR/EPR) [30–32] method is used for N atom density determination. This method is very useful for detection and identification of paramagnetic particles and especially radicals [33–35]. Although the method is routinely used in chemistry, its use in plasma physics is relatively rare. Since the pioneering works [36–38], other, mostly laser based methods appeared which are now considered mainstream.

Electron spin or paramagnetic resonance (ESR/EPR) is essentially microwave absorption spectroscopy [39] on

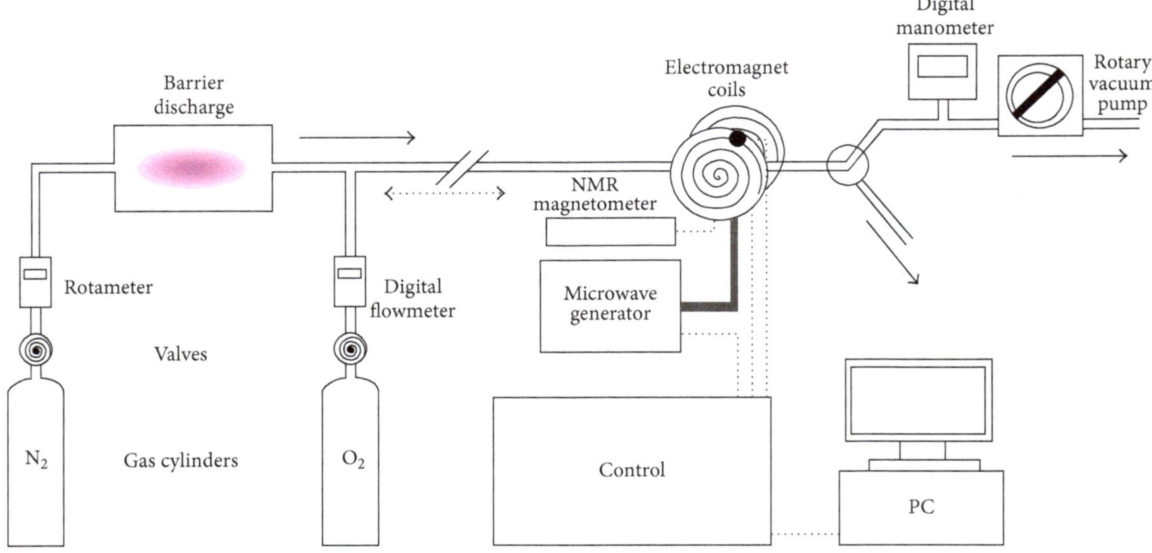

FIGURE 1: Schematic drawing of the experimental set-up.

Zeeman split [40] levels. The transitions with very low energy separation between the upper and the bottom levels (as is the case in microwave spectroscopy) are very easily disturbed by the collisions with other particles. The collisional/pressure broadening of absorption line can be quite pronounced [41, 42], especially for gas phase atomic and molecular lines with very low natural linewidths. This makes most of the species (for extensive list, see [43]), including, for example, atomic oxygen, difficult to detect in atmospheric pressure plasmas. However, the particles in s-state, such as nitrogen or hydrogen atoms in ground state, do not exhibit pressure broadening and can be detected by ESR/EPR even at atmospheric pressure.

There are a number of works, both theoretical and experimental, dealing with low pressure plasma diagnostics by ESR/EPR method [44–50]. However, due to above-mentioned difficulties of pressure broadening, the diagnostics of atmospheric pressure plasmas by EPR/ESR is rather unresearched topic, as our previous paper [51] is, to authors' best knowledge, still the only one published.

Although the nitrogen and oxygen plasmas are studied for very long time, there is still intensive research [52–54] going on. The discharges in N_2-O_2 mixtures have very complicated plasma kinetics and some effects, especially in mixtures with low O_2/N_2 ratio, are not fully explained yet. In our previous work [55] it was reported that adding of small amount of oxygen into a low pressure nitrogen plasma afterglow causes an increase of nitrogen atom density. In this paper we extend the study of the influence of oxygen admixture on the nitrogen afterglow to atmospheric pressure.

2. Experimental Apparatus and Methods

2.1. Experimental Set-Up. The experimental apparatus is depicted in Figure 1. Nitrogen plasma at atmospheric pressure was produced using industrial ozoniser LifeTech 50. It is based on coaxial dielectric barrier discharge, excited by 25–35 W power at 15 kHz frequency. The stainless steel inner electrode has approx. 2 cm diameter and is separated by 0.7 mm plasma gap and 2.5 mm thick corundum (Al_2O_3) ceramics from the outer electrode which is realised by an aluminium foil tightly wrapped around the ceramic tube. The discharge tube is approx. 15 cm long. Although the metal electrode in direct contact with plasma can significantly reduce N atom density in the effluent due to increased surface recombination/reassociation, this design [56] is well established in industrial ozonisers.

The discharge was operated in flowing (12 standard litres per minute) nitrogen, coming from pressurised cylinder (Messer-Griesheim, purity 99.995%) via pressure reduction valve. The volumetric flow rate of nitrogen was measured by flowmeter with floating element (rotameter UPLS-R3). Downstream from the plasma source, the oxygen is optionally introduced. The oxygen comes from pressurised cylinder (Messer-Griesheim, purity 99.995%) via Hastings mass flow-controller maintaining the oxygen flow at 7.5 sccm (standard cubic centimetre per minute). The tubing is made from polyethylene and stainless steel.

The oxygen is introduced into the afterglow tube just downstream from the plasma generator using an inlet three-way valve. This configuration permits presetting the oxygen flow on the flow-controller and rapidly switches on and off the oxygen admixture. The distance between the output port of the plasma generator and the inlet of oxygen is 6 cm. The afterglow tube is made of fused silica with 10 mm outer diameter and 8 mm inner diameter. This tube passes through ESR/EPR spectrometer resonator and ends in another outlet three-way valve. In normal operation the plasma afterglow vents into open atmosphere and the afterglow tube length (approx. 2.5 metres) together with high gas flow prevents any back-diffusion of air into the afterglow or plasma. In second position of this outlet valve the whole apparatus can be pumped down by the rotary vane oil vacuum pump. This is used for calibration by molecular oxygen at reduced pressure and for degassing/cleaning as before every measurement series the whole plasma system is evacuated to pressure about 10 Pa to remove residual impurities. After that, the nitrogen flow is switched on and the outlet three-way valve is turned to normal position. Then the discharge is run 1 hour at maximum power to burn-in the plasma generator and remove the possible contamination off the electrode. Only after such procedure are the experiments started.

The distance from the inlet of oxygen admixture to the measuring ESR/EPR resonator can be changed by shifting the whole plasma system. To ease this, the plasma system including afterglow tube is placed on nonmagnetic rails parallel to ESR/EPR resonator axis. The furthest possible position is limited by the rail length to 90 cm. However, in the present study the maximum extension was not used as the ESR/EPR signal in this extreme position was already difficult to measure. The minimum distance in which the measure could be carried out is about 20 cm. Although it is mechanically possible to place the discharge closer to the resonator, at shorter distances the magnetic field of the ESR/EPR spectrometer electromagnet affected the plasma generator.

The ESR/EPR spectrometer JEOL JES-PE is specially adapted for use in plasma physics. It is classical continuous wave spectrometer [34] using klystron source operating at X band and with standard sensitivity around 10^{10} paramagnetic particles in the interaction volume of TE_{011} resonator. The output voltage of the analogue spectrometer is measured by digital voltmeter (Metra M1T390, 5 digits) and transmitted to a personal computer via GPIB. The ESR/EPR signal is analysed and postprocessed using custom software. It automatically identifies the spectral lines and calculates their peak-to-peak height, peak-to-peak width w, and area under the absorption line I_N (as ESR/EPR lines are typically [34] recorded in the form of derivation, double integration is needed).

2.2. The Principle of Measurement and Calibration Procedure. The ESR/EPR phenomenon is based on resonant absorption of microwave photons by transition between Zeeman split [40] energy levels. Typically, the resonance is achieved by variable magnetic field (which sets the energy level splitting) while the microwave frequency (i.e., photon energy) is fixed. The area under absorption spectral line is proportional [38] to the concentration of absorbing paramagnetic particles. The proportionality constant depends on spectrometer settings and the transition probability (Einstein coefficient). After a calibration of the spectrometer by a known sample it is possible to measure the concentration of paramagnetic particles absolutely [57].

In this work, the spectrometer calibration [44] is carried out using molecular oxygen O_2. The measuring resonator is filled with gaseous molecular oxygen at known concentration (calculated from pressure and temperature), its intensive spectral line C (the traditional naming, found in many papers, e.g., [38, 41, 43, 44]) is recorded, and corresponding double integral I_C is calculated. Using the same settings of the spectrometer, the ESR/EPR line of atomic nitrogen is

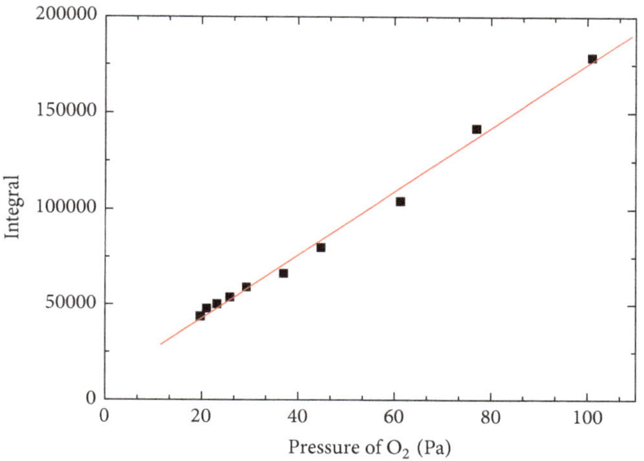

FIGURE 2: Calibration of EPR spectrometer using the line C of molecular oxygen. Linear dependency of absorption line integral I_C (i.e., double integral of measured signal) on oxygen pressure validates the calibration process and its slope provides a constant $[O_2]/I_C$ needed for the calibration.

recorded and its double integral I_N is calculated. Knowing the Einstein coefficients of both species the concentration of atomic nitrogen is given by simple relation

$$[N] = 5.88 \cdot 10^{-3} I_N / I_C \times [O_2].$$

The numeric constant in this formula is the ratio of Einstein coefficient of molecular oxygen line C and that of one component of atomic nitrogen triplet [43].

As the oxygen calibration is carried out in the flow-regime, due to Hagen–Poiseuille law [58] there exist pressure gradients along the tube, making the pressure in the ESR/EPR resonator slightly different from the one indicated by pressure meter. The vacuum conductivity of the tube depends [59] on pressure, too. Moreover, as at low pressures the partial pressure of O_2 could be smaller than the total pressure indicated due to small vacuum leaks, finite base pressure of the pump used, or degassing from walls, it is better to record O_2 line for several indicated pressures than to rely on single value only. The result of such calibration measurement is shown in Figure 2.

The ESR/EPR lines of atomic nitrogen are extremely narrow and therefore sensitive to any broadening [41]. The ground state of nitrogen $N(^4S_{3/2})$ has zero orbital magnetic momentum and its paramagnetism is given only by the electron spin. In that case there should be no collision induced broadening; its spectral linewidth should be independent of pressure [60]. In work [61] this was studied in the pressure range from 50 to 600 Pa and in temperature range from 80 to 300 K. It was found that with decreasing temperature the apparent linewidth $w = 6\,\mu T$ remained constant, while the integral and thus the concentration [N] decreased. The linewidth was thus independent of N atom density, which suggested that the role of spin-spin relaxation at these N concentrations (around 10^{13}–10^{14} cm^{-3}) is negligible.

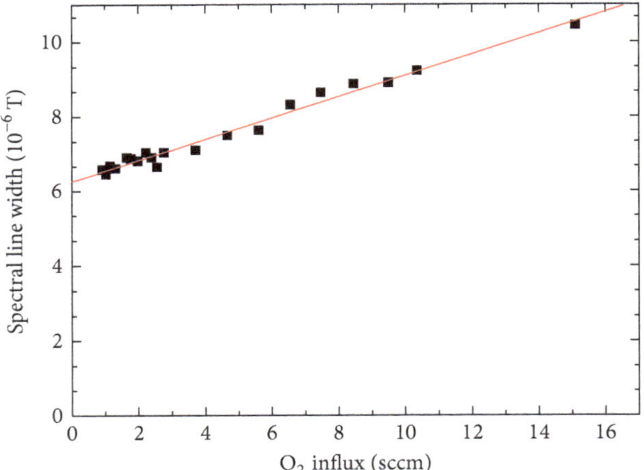

FIGURE 3: Linewidth of N atom ground state ($^4S_{3/2}$) EPR line as a function of oxygen admixture.

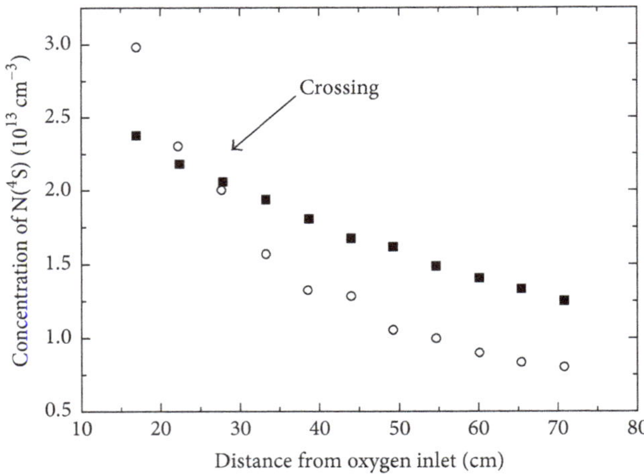

○ N_2 with O_2 admixture
■ Pure N_2

FIGURE 4: Spatial distribution of the nitrogen atom concentration (in ground state) along the afterglow tube for two cases: with and without 625 ppm oxygen admixture.

In present work, a similar measurement of $N(^4S_{3/2})$ linewidth was carried out in pure nitrogen afterglow but at atmospheric pressure. Practically the same value $w = 6.4\,\mu T$ was observed. The linewidth (and spectral line shape) being constant, the concentration of N atoms should be directly proportional to the absorption line height and by consequence also to the peak-to-peak height of the measured (i.e., absorption derivative) line. It can be advantageous to use this peak-to-peak height over the spectral line integral as the latter exhibits much higher statistical variations due to an increased influence of a noise overlaying the line shoulders.

However, a presence of other paramagnetic particles can affect even the s-states. The experimental result, nitrogen spectral linewidth as a function of molecular oxygen admixture, is shown in Figure 3. The changing nitrogen linewidth prevents the use of simple line height as concentration indicator and the integral must be used instead. This effect is consistent with [60].

3. Results and Discussions

3.1. Experimental Results. Electron paramagnetic/spin resonance was used to measure the concentration of atomic nitrogen $N(^4S_{3/2})$ in atmospheric pressure discharge afterglow. This density [N] was measured along fused silica tube in pure nitrogen afterglow and with 625 ppm of oxygen injected into the early afterglow; see Figure 4.

The most important result is the fact that the two curves apparently intersect. It means that a small admixture of oxygen can cause both an increase and a decrease in concentration of nitrogen atoms, depending on which part of the afterglow is observed. This nontrivial behaviour is a typical demonstration of complexities inherent in N_2-O_2 kinetics.

In flowing kinetic studies, if the gas velocity is known, the spatial distribution of measured species can be transferred to their temporal evolution. Sometimes the same gas velocity in the whole cross-section (i.e., plug flow) is assumed without any further consideration. However, there can be a significant radial gradient of gas velocity due to a shear flow and a boundary layer. While a laminar flow in circular tube produces well-known parabolic velocity profile, in a well-developed turbulent flow the shear zone is thin compared to the tube diameter, so the plug flow is appropriate. The character of flow in a tube can be deduced from the Reynolds number [62]

$$R = ud/\nu,$$

where u is mean flow velocity, d is the inner tube diameter, and ν is cinematic viscosity. For nitrogen at standard pressure and temperature $\nu = 1.50 \cdot 10^{-5}\,\mathrm{m^2\,s^{-1}}$ according to [63], which gives $R = 2100$ for our experimental conditions. This value falls in critical Reynolds number range $R = 1800$–2300 where a transition between the laminar and the turbulent flow happens [64]. One can expect that the inhomogeneity caused by the lateral oxygen inlet together with other imperfections introduces some additional turbulence. Based on this reasoning, there is enough turbulence to consider the plug flow as a valid approximation in this paper, too.

In that case, the radially uniform flow velocity is approx. 4 m/s. Using this value, the measured dependence of the nitrogen atom density on position in the afterglow (Figure 4) was recalculated to the dependence on time; see Figure 5.

It is experimentally impossible to measure the concentration near the oxygen inlet (i.e., at times close to $t = 0$ s) as the T-piece (needed for the inlet) cannot pass through the ESR/EPR measuring resonator opening and the close distance between the plasma generator and the ESR/EPR electromagnet would magnetically influence the plasma generator and the plasma itself. However, if one excludes a back-diffusion, it is evident that with or without oxygen admixture the value at $t = 0$ s must be the same; that is, both curves must start at the same initial value depicted in Figure 5 by red circle. While this common initial value is a priori unknown

TABLE 1: Main plasma-chemical reactions in the pure nitrogen afterglow.

Reaction number	Reaction	Rate coefficient	Reference
(1)	$N(^4S) + N(^4S) + N_2 \rightarrow N_2(A) + N_2$	$k_1 = 1.05 \cdot 10^{-13}\ cm^3\ s^{-1}$	[22]
(2)	$N(^4S) + wall \rightarrow N_2 + wall$	k_2 = see text	
(3)	$N_2(A) + N(^4S) \rightarrow N_2(X) + N(^2P)$	$k_3 = 5 \cdot 10^{-11}\ cm^3\ s^{-1}$	[23]
(4)	$N(^2P) + N_2 \rightarrow N(^2D) + N_2$	$k_4 = 2 \cdot 10^{-18}\ cm^3\ s^{-1}$	[24]
(5)	$N(^2P) + N \rightarrow N(^2D) + N$	$k_5 = 1.8 \cdot 10^{-12}\ cm^3\ s^{-1}$	[25]
(6)	$N(^2D) + N_2 \rightarrow N(^4S) + N_2$	$k_6 = 6 \cdot 10^{-15}\ cm^3\ s^{-1}$	[26]
(7)	$N_2(A) + N_2(A) \rightarrow N_2(C) + N_2$	$k_7 = 2 \cdot 10^{-12}\ cm^3\ s^{-1}$	[27]

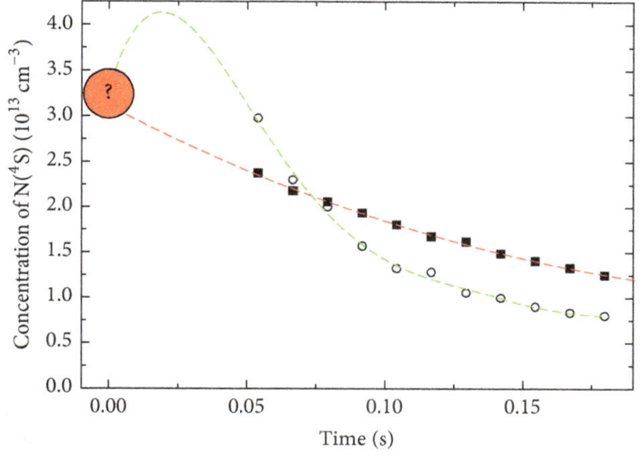

FIGURE 5: Data from the previous figure recalculated from distance to time. The rough estimation of initial [N] value at $t = 0$ s is shown, too. The sketch of a reasonable time evolution is shown by dashed lines.

it effectively sets the probable shapes of both curves (with and without oxygen admixture) in the unmeasurable region. These are shown in Figure 5 using the dashed lines. One can see that the case without oxygen admixture should be simply governed by losses while the case with oxygen admixture initially exhibits some process producing the nitrogen atoms and only later the losses dominate. Higher losses of N with O_2 present cause the steeper descent of this curve and so both curves intersect.

3.2. Kinetic Modelling of Plasma Afterglow in Pure Nitrogen. The model is based on collisional processes between the principal species in the nitrogen afterglow: N_2 molecules, electronically excited molecules $N_2(A)$, atoms N(S), N(P), and N(D), and it takes into account also the wall processes. The model is focused on $N(^4S)$ and so the species $N_2(X,v)$ despite being another important energy carrier is not included as it cannot directly cause N_2 dissociation and it does not exhibit very strong quenching by nitrogen atoms (which $N_2(A)$ does). An overview of the main plasma-chemical processes based on [28] is presented in Table 1.

The kinetic equations describing the processes in Table 1 were transformed to a set of differential equations and solved numerically using the Euler method. As this method might be unstable for rapidly changing functions, it is necessary to use sufficiently small time-step. This convergence was verified in the presented model. Some parameters, such as k_2 and initial concentrations of $N(^4S)$ and $N_2(A)$, are a priori unknown and must be determined by fitting to the experimental data. The concentration of molecular nitrogen $[N_2] = 2.77{\cdot}10^{19}\ cm^{-3}$ at standard pressure and temperature was used as one of initial values.

The reaction (1) describes the volume recombination and the reaction (2) the wall recombination of atomic nitrogen in ground state $N(^4S)$. The model includes the reactions of the $N(^2P)$, $N(^2D)$, and $N_2(A)$ metastables, too. However, significant simplifications are possible. In the experiment, only the $N(^4S)$ concentration is actually measured. Moreover, the metastable N atoms generally end in ground state (reactions (4)–(6)). So the reaction (3) effectively produces one $N(^4S)$ atom and is equivalent to $N_2(A)$ deexcitation.

Further simplification stems from the mutual ratio of $[N_2(A)]$ and $[N(^4S)]$. Although these values are a priori unknown, they can be roughly estimated. A rough estimation of the initial concentration $[N_2(A)]_0$ of $10^{11}\ cm^{-3}$ can be taken from papers [65–67], which used sufficiently similar conditions to the present ones (atmospheric pressure nitrogen afterglow operated in similar tube diameter and similar input power). As the visual extrapolation of pure nitrogen curve (red hollow diamonds) in Figure 5 gives the initial value of $[N(^4S)]_0$ around $10^{13}\ cm^{-3}$, one may safely assume that $[N(^4S)]_0 > [N_2(A)]_0$.

Using these simplifications, the "full" model of Table 1 is effectively reduced to the set of kinetic equations (1), (2), (3), and (7). The relative difference $\delta = ([N(^4S)]|_{reduced\ model} - [N(^4S)]|_{full\ model})/[N(^4S)]|_{reduced\ model}$ between the reduced and full models is shown in Figure 6 for different initial ratios of $[N(^4S)]$ and $[N_2(A)]$.

In the first few milliseconds there is a big difference between the full and the reduced models due to the reaction (3). But in later reaction times, in which the real experiment is carried out (as discussed above, due to experimental constraints it is not possible to measure at afterglow positions before 0.06 s), there is nearly negligible difference between the full and the reduced models. Taking into account the experimental error around 10% and an assumption about initial values $10^{11}\ cm^{-3} < [N_2(A)]_0 < [N(^4S)]_0$, one can safely use the reduced model only.

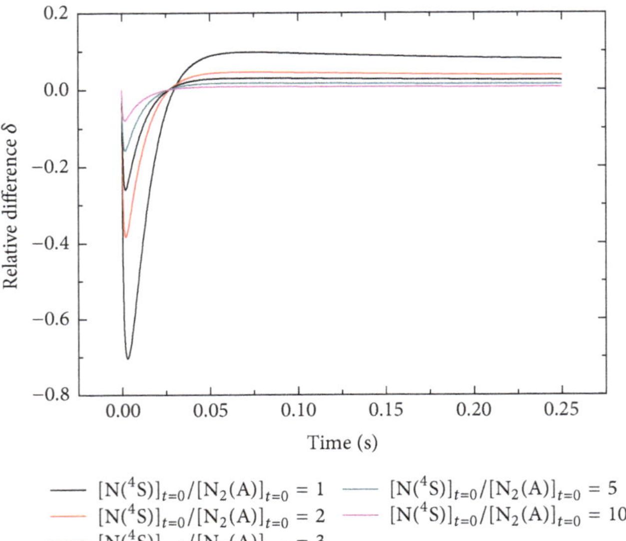

FIGURE 6: Relative difference δ between full (see (1)–(7)) and reduced (see (1), (2), (3), and (7)) models for several initial ratios between $[N(^4S)]$ and $[N_2(A)]$.

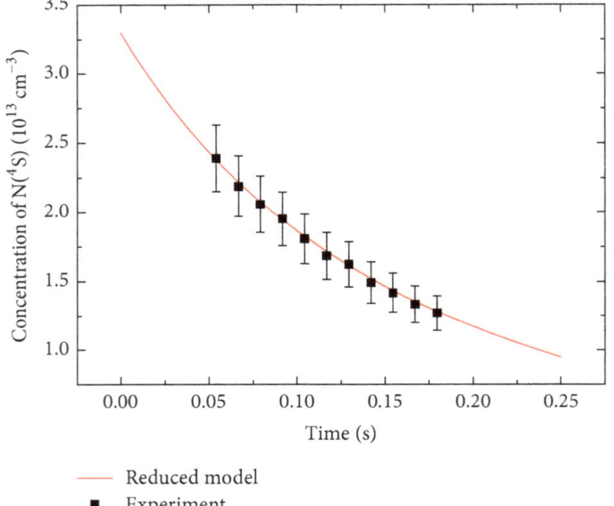

FIGURE 7: Fit of the kinetic model of the pure nitrogen afterglow given by reduced reactions from Table 1 to the experimental data, giving the previously unknown initial value $[N(^4S)]_0 = 3.3 \cdot 10^{13}\,cm^{-3}$ and the value of k_2.

The least squares fit of experimental data by this reduced model is shown in Figure 7 and gives the values of previously unknown parameters: $[N(^4S)]_0 = 3.3 \cdot 10^{13}\,cm^{-3}$ and $k_2 = 3.1\,s^{-1}$. To the rate constant k_2 exists corresponding coefficient (probability) of surface recombination/reassociation $\gamma = 1.24 \cdot 10^{-6}$ which is in good agreement with [68] where they get $\gamma_{lit} = 1.35 \cdot 10^{-6}$. Following the reasoning of the previous paragraph, one may assume that these values do not strongly depend (see Figure 6) on the initial concentration of $N_2(A)$ (for sufficiently small $[N_2(A)]_0$).

3.3. Kinetic Modelling of Nitrogen Afterglow with Oxygen Admixture. There are many published sets of kinetic processes for N_2-O_2 mixtures with varying degree of detail [16, 28, 53–56, 69–72]. Taking into account the time scales and typical values of concentrations, these extensive sets can be substantially reduced and simplified. Essentially, it is possible to extend the model of pure nitrogen afterglow (see reactions (1)–(7)) by including $N_2(X,v)$, O_2, O, NO, NO_2, and N_2O (see Table 2).

The inclusion of vibrationally excited nitrogen molecules is necessary as these have even greater importance [73] in $N_2 + O_2$ plasma afterglow. In a nitrogen pink afterglow the vibrationally excited molecules carry a significant part of energy in the afterglow [74, 75]. This energy is then responsible for the formation of $N_2(A)$, $N_2(a')$ and finally for the ionisation of N_2 molecules. Although the pink afterglow was not observed [76] in atmospheric pressure plasma, one may assume [73, 77] the concentration of $N_2(v = 12)$ greater than $5 \cdot 10^{13}\,cm^{-3}$ at afterglow position of 15 ms.

The concentration of molecular oxygen $[O_2] = 1.56 \cdot 10^{16}\,cm^{-3}$ is calculated according to the ideal gas law from the experimental conditions (625 ppm of O_2 in N_2 at atmospheric pressure and room temperature).

The reaction set of Table 2 can be further reduced as follows. Kinetic reaction (22) is negligible to the reaction (12) because k_{22} is about two orders of magnitude smaller than k_{12}. For the same reason, reaction (15) can be neglected with respect to reaction (14). Furthermore, (26), according to the numerical calculations, has not significant influence on the concentration of $N(^4S)$, so it can be omitted, too.

Most of the equations describe the loss of ground state metastable atom $N(^4S)$. Equation (13) describes production of $N(^4S)$ but its contribution is not significant in comparison with the losses due to reactions (8), (9), and (10). This is in contrast with the experiment, where an increase in $N(^4S)$ concentration is observed when oxygen is added. Using the larger equation set of [28] does not help either. To explain the $[N(^4S)]$ increase, (23) is needed, where vibrationally excited molecule in the electronic ground state $N_2(X,v)$ produces N atom and nitric oxide molecule by reacting with oxygen atom [69, 77].

The heterogeneous reactions in the model include the wall deexcitation of $N_2(X,v)$, see (24), and $N_2(A)$, see (25).

As usual, the kinetic model must be supplemented by the initial concentrations. The initial value of $[N(^4S)]_0$ was taken from the pure nitrogen model. Due to the adsorption of oxygen on the walls of the afterglow tube [78–81], a smaller value of the wall recombination coefficient k_2 (change from $3.1\,s^{-1}$ to $0.5\,s^{-1}$) was assumed. Atomic oxygen wall recombination coefficient k_{16} was considered to be same as k_2 for simplicity. Rate constants for wall deexcitation (24) and (25) were taken to be the same for both species, that is, $k_{24} = k_{25}$, with estimated value in range of 1–$10\,s^{-1}$ to correspond to [82]. The precise value of k_{24} and k_{25} is calculated by fitting the model to the experimental data.

The final simplified set of equations considered in the reduced model of N_2-O_2 afterglow consists of (1)–(3), (7)–(14), (16)–(21), and (23)–(25). The remaining unknowns

TABLE 2: Additional set of kinetic reactions for nitrogen afterglow with oxygen admixture, mostly based on [28].

Reaction number	Reaction	Rate coefficient	Reference
(8)	$N + O_2 \rightarrow NO + O$	$k_8 = 1.03 \cdot 10^{-16}$ cm^3 s^{-1}	[22]
(9)	$N + O + N_2 \rightarrow NO + N_2$	$k_9 = 2.74 \cdot 10^{-13}$ cm^3 s^{-1}	[22]
(10)	$N + NO \rightarrow N_2 + O$	$k_{10} = 1.82 \cdot 10^{-13}$ cm^3 s^{-1}	[22]
(11)	$O + NO + N_2 \rightarrow NO_2 + N_2$	$k_{11} = 9.54 \cdot 10^{-13}$ cm^3 s^{-1}	[22]
(12)	$N_2(A) + O_2 \rightarrow N_2(X) + O + O$	$k_{12} = 2.54 \cdot 10^{-12}$ cm^3 s^{-1}	[23]
(13)	$N_2(A) + O \rightarrow NO + N(^2D) \rightarrow NO + N(^4S)$	$k_{13} = 7 \cdot 10^{-12}$ cm^3 s^{-1}	[29]
(14)	$O + O + N_2 \rightarrow O_2 + N_2$	$k_{14} = 7.91 \cdot 10^{-14}$ cm^3 s^{-1}	[28]
(15)	$O + O + O_2 \rightarrow 2O_2$	$k_{15} = 1{,}05 \cdot 10^{-16}$ cm^3 s^{-1}	[28]
(16)	O + wall $\rightarrow O_2$ + wall	k_{16} = see text	
(17)	$N + NO_2 \rightarrow N_2 + O_2$	$k_{17} = 7 \cdot 10^{-13}$ cm^3 s^{-1}	[28]
(18)	$N + NO_2 \rightarrow N_2 + O + O$	$k_{18} = 9.1 \cdot 10^{-13}$ cm^3 s^{-1}	[28]
(19)	$N + NO_2 \rightarrow N_2O + O$	$k_{19} = 3 \cdot 10^{-12}$ cm^3 s^{-1}	[28]
(20)	$N + NO_2 \rightarrow 2NO$	$k_{20} = 2.3 \cdot 10^{-12}$ cm^3 s^{-1}	[28]
(21)	$O + NO_2 \rightarrow NO + O_2$	$k_{21} = 2.54 \cdot 10^{-12}$ cm^3 s^{-1}	[28]
(22)	$N_2(A) + O_2 \rightarrow N_2O + O$	$k_{22} = 7.8 \cdot 10^{-14}$ cm^3 s^{-1}	[23]
(23)	$N_2(X,v) + O \rightarrow NO + N$	k_{23} = see text	
(24)	$N_2(X,v)$ + wall $\rightarrow N_2$ + wall	k_{24} = see text	
(25)	$N_2(A)$ + wall $\rightarrow N_2$ + wall	k_{25} = see text	
(26)	$NO_2 + NO_2 + N_2 \rightarrow N_2O_4 + N_2$	$k_{26} = 5.64 \cdot 10^{-13}$ cm^3 s^{-1}	[22]

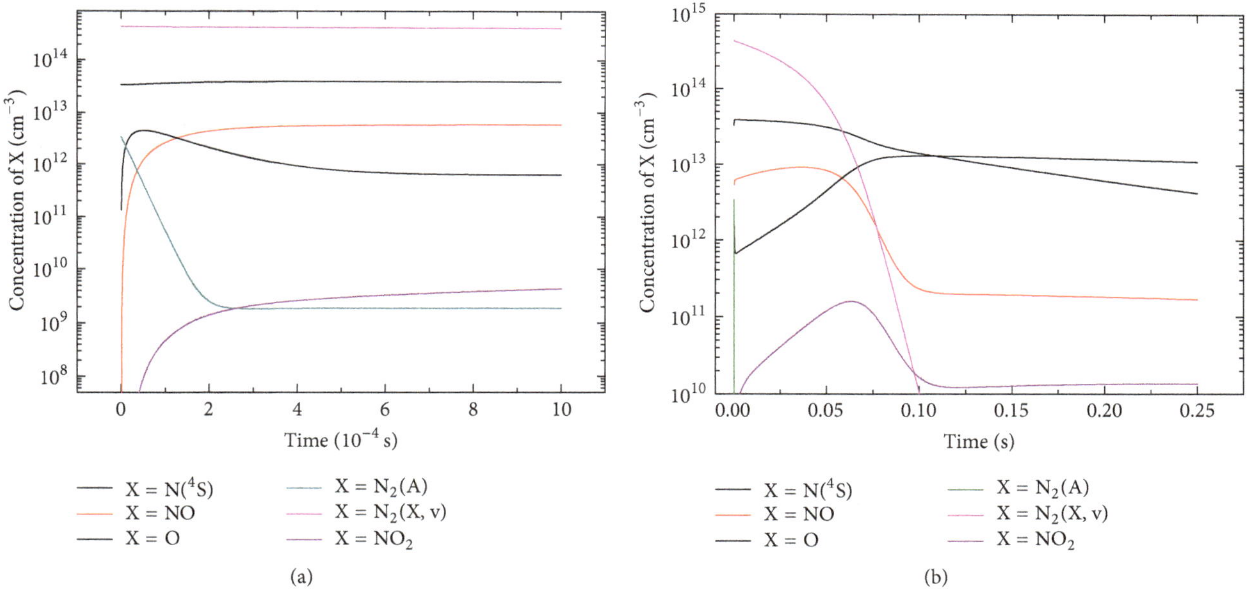

FIGURE 8: The temporal evolution of main species considered in the reduced kinetic model of N_2-O_2 afterglow. For clarity, the data are shown in two time scales.

are the initial concentrations of $N_2(A)$, $N_2(X,v)$ and the rate constants k_{23}, k_{24}, and k_{25}. These parameters were determined from the least squares fit of model to the experimental data.

Figure 8 shows the calculated temporal evolution of concentrations of main species considered in the present model in short (a) and long (b) time scales. The dominant species (besides the parent N_2 and O_2) are the nitrogen and oxygen atoms. Interestingly, despite the very unfavourable ratio of $[O_2]/[N_2]$ = 625 ppm, there are more oxygen atoms than nitrogen atoms in the late afterglow. Electronically and vibrationally excited nitrogen molecules disappear quickly. Both oxides of nitrogen, nitrogen dioxide NO_2 and nitric oxide NO, remain after 0.1 s constant, [NO] being approx. 10 times higher than $[NO_2]$.

Full comparison of the model and the experiment is shown in Figure 9 for both cases, with and without the oxygen admixture. Values of a priori unknown parameters obtained

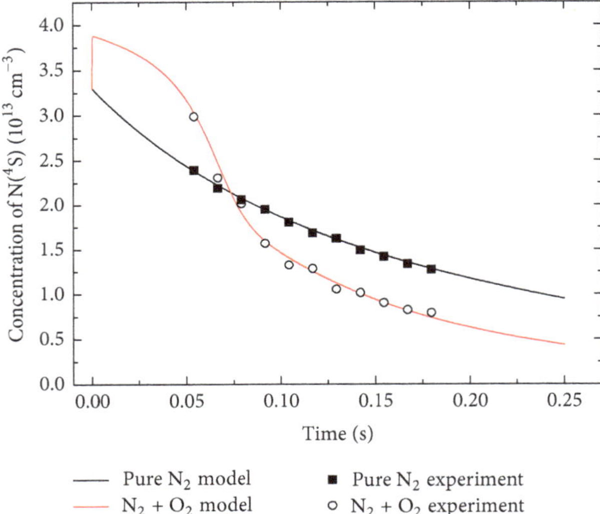

FIGURE 9: The parameters from the fit are used to calculate the extrapolation to shorter afterglow times of kinetic models of atmospheric pressure nitrogen afterglow with and without oxygen admixture.

by the fit of N_2-O_2 model are $[N_2(A)]_0 = 3.4 \cdot 10^{12}\,cm^{-3}$, $[N_2(X,v)]_0 = 4.5 \cdot 10^{14}\,cm^{-3}$, $k_{23} = 1.6 \cdot 10^{-11}\,cm^3\,s^{-1}$, and $k_{24} = k_{25} = 8.7\,s^{-1}$. The value of k_{23} obtained from the fit is in very good agreement with [16] where this constant was estimated to be at least $10^{-11}\,cm^3\,s$.

The agreement between the models and the experimental data in Figure 9 is very good. Moreover, the model covers also the times shorter than 50 ms which are inaccessible by the experiment. The effect of initially increased [N] just after oxygen admixture, which was predicted in Figure 5, is therefore verified and explained by reaction of vibrationally excited $N_2(X,v)$ with atomic oxygen.

4. Conclusions

Quantitative electron spin/paramagnetic resonance spectroscopy was used to measure the concentration of nitrogen atoms in flowing atmospheric pressure dielectric barrier discharge afterglow with typical [N] values around $2 \cdot 10^{13}\,cm^{-3}$. The evolution of this concentration along the afterglow tube was shown to be significantly affected by the relatively small amount (625 ppm) of oxygen added into the early afterglow. The nitrogen dissociation is increased just after the oxygen inlet and it is decreased in later parts of the afterglow.

Numerical kinetic model explains this behaviour by a balance of production and loss terms, both of which are affected by the presence of oxygen. Main reaction producing N atoms is the collision of $N_2(X,v)$ with oxygen atoms. By fitting the model to the experimental data, it was possible to estimate several a priori unknown and not directly measurable quantities, such as the initial concentrations of $N(^4S)$, $N_2(A)$, and $N_2(X,v)$, wall recombination coefficient of N atoms, and rate coefficient of some reactions.

Competing Interests

The authors declare that they have no competing interests.

Acknowledgments

This research has been supported by the Project LO1411 (NPU I) funded by Ministry of Education, Youth and Sports of Czech Republic.

References

[1] B. Eliasson, M. Hirth, and U. Kogelschatz, “Ozone synthesis from oxygen in dielectric barrier discharges,” *Journal of Physics D: Applied Physics*, vol. 20, no. 11, pp. 1421–1437, 1987.

[2] J. P. Boeuf, “Plasma display panels: physics, recent developments and key issues,” *Journal of Physics D: Applied Physics*, vol. 36, no. 6, pp. R53–R79, 2003.

[3] M. Strobel, C. S. Lyons, and K. Mittal, Eds., *Plasma Surface Modification of Polymers: Relevance to Adhesion*, VSP, 1994.

[4] U. Kogelschatz, “Dielectric-barrier discharges: their history, discharge physics, and industrial applications,” *Plasma Chemistry and Plasma Processing*, vol. 23, no. 1, pp. 1–46, 2003.

[5] L. M. Zhou, B. Xue, U. Kogelschatz, and B. Eliasson, “Nonequilibrium plasma reforming of greenhouse gases to synthesis gas,” *Energy & Fuels*, vol. 12, no. 6, pp. 1191–1199, 1998.

[6] L. Zajíčková, M. Eliáš, O. Jašek et al., “Atmospheric pressure microwave torch for synthesis of carbon nanotubes,” *Plasma Physics and Controlled Fusion*, vol. 47, no. 12, pp. B655–B666, 2005.

[7] P. Synek, O. Jašek, L. Zajíčková, B. David, V. Kudrle, and N. Pizúrová, “Plasmachemical synthesis of maghemite nanoparticles in atmospheric pressure microwave torch,” *Materials Letters*, vol. 65, no. 6, pp. 982–984, 2011.

[8] G. Fridman, G. Friedman, A. Gutsol, A. B. Shekhter, V. N. Vasilets, and A. Fridman, “Applied plasma medicine,” *Plasma Processes and Polymers*, vol. 5, no. 6, pp. 503–533, 2008.

[9] G. Fridman, A. D. Brooks, M. Balasubramanian et al., “Comparison of direct and indirect effects of non-thermal atmospheric-pressure plasma on bacteria,” *Plasma Processes and Polymers*, vol. 4, no. 4, pp. 370–375, 2007.

[10] J. Heinlin, G. Morfill, M. Landthaler et al., “Plasma medicine: possible applications in dermatology,” *JDDG: Journal der Deutschen Dermatologischen Gesellschaft*, vol. 8, no. 12, pp. 968–976, 2010.

[11] V. Stěpánová, P. Slavíček, M. Stupavská, J. Jurmanová, and M. Černák, “Surface chemical changes of atmospheric pressure plasma treated rabbit fibres important for felting process,” *Applied Surface Science*, vol. 355, pp. 1037–1043, 2015.

[12] W. A. P. Claassen, W. G. J. N. Valkenburg, M. F. C. Willemsen, and W. M. Wijgert, “Influence of deposition temperature, gas pressure, gas phase composition, and RF frequency on composition and mechanical stress of plasma silicon nitride layers,” *Journal of the Electrochemical Society*, vol. 132, no. 4, pp. 893–898, 1985.

[13] G. G. Tibbetts, “Role of nitrogen atoms in ‘ion–nitriding,’” *Journal of Applied Physics*, vol. 45, no. 11, pp. 5072–5073, 1974.

[14] A. Fridman, *Plasma Chemistry*, Cambridge University Press, 2008.

[15] I. Stefanovic, N. K. Bibinov, A. A. Deryugin, I. P. Vinogradov, A. P. Napartovich, and K. Wiesemann, “Kinetics of ozone and

nitric oxides in dielectric barrier discharges in O_2/NO_x and $N_2/O_2/NO_x$ mixtures," *Plasma Sources Science and Technology*, vol. 10, no. 3, p. 406, 2001.

[16] V. Guerra and J. Loureiro, "Non-equilibrium coupled kinetics in stationary N_2-O_2 discharges," *Journal of Physics D: Applied Physics*, vol. 28, no. 9, pp. 1903–1918, 1995.

[17] S. Tada, S. Takashima, M. Ito, M. Hori, T. Goto, and Y. Sakamoto, "Measurement and control of absolute nitrogen atom density in an electron-beam-excited plasma using vacuum ultraviolet absorption spectroscopy," *Journal of Applied Physics*, vol. 88, no. 4, pp. 1756–1759, 2000.

[18] P. Vašina, V. Kudrle, A. Tálský, P. Botoš, M. Mrázková, and M. Meško, "Simultaneous measurement of N and O densities in plasma afterglow by means of NO titration," *Plasma Sources Science and Technology*, vol. 13, no. 4, pp. 668–674, 2004.

[19] S. Agarwal, B. Hoex, M. C. M. Van de Sanden, D. Maroudas, and E. S. Aydil, "Absolute densities of N and excited N_2 in a N_2 plasma," *Applied Physics Letters*, vol. 83, no. 24, pp. 4918–4920, 2003.

[20] J. Amorim, G. Baravian, and J. Jolly, "Laser-induced resonance fluorescence as a diagnostic technique in non-thermal equilibrium plasmas," *Journal of Physics D: Applied Physics*, vol. 33, no. 9, pp. R51–R65, 2000.

[21] F. Gaboriau, U. Cvelbar, M. Mozetic, A. Erradi, and B. Rouffet, "Comparison of TALIF and catalytic probes for the determination of nitrogen atom density in a nitrogen plasma afterglow," *Journal of Physics D: Applied Physics*, vol. 42, no. 5, Article ID 055204, 2009.

[22] O. E. Krivonosova, S. A. Losev, V. P. Nalivaiko, Y. K. Mukoseev, and O. P. Shatalov, *Plasma Chemistry*, vol. 14 of *edited by B. M. Smirnov*, Energoatomizdat, Moscow, Russia, 1987.

[23] M. P. Iannuzi, J. B. Jeffries, and F. Kaufman, "Product Channels of the N_2 $(A^3\Sigma_u^+) + O_2$ interaction," *Chemical Physics Letters*, vol. 87, no. 6, pp. 570–574, 1982.

[24] R. J. Donovan and D. Husain, "Recent advances in the chemistry of electronically excited atoms," *Chemical Reviews*, vol. 70, no. 4, pp. 489–516, 1970.

[25] D. I. Slovetskii, *Mechanisms of Chemical Reactions in Nonequilibrium Plasma*, Mir, Moscow, Russia, 1980 (Russian).

[26] M. H. Bortner and T. Bauer, *Defense Nuclear Agency Reaction Rate Handbook*, DNA 1948H, US GPO, Washington, DC, USA, 1971.

[27] V. P. Silakov, *Mechanism of Supporting the Long-Lived Plasma in Molecular Nitrogen at High Pressure*, 010-90M, Moscow Engineering Physical Institute, 1990.

[28] I. A. Kossyi, A. Y. Kostinsky, A. A. Matveyev, and V. P. Silakov, "Kinetic scheme of the non-equilibrium discharge in nitrogen-oxygen mixtures," *Plasma Sources Science and Technology*, vol. 1, no. 3, 1992.

[29] L. G. Piper, "The excitation of O(1S) in the electronic energy transfer between N2(A3S+ u) and O," *The Journal of Chemical Physics*, vol. 77, pp. 2373–2377, 1982.

[30] E. K. Zavoisky, "Paramagnetic relaxation of liquid solutions for perpendicular fields," *Journal of Physics-USSR*, vol. 9, pp. 211–216, 1945.

[31] E. K. Zavoisky, "Spin-magnetic resonance in paramagnetics," *Journal of Physics-USSR*, vol. 9, pp. 211–245, 1945.

[32] E. K. Zavoisky, "Spin magnetic resonance in the decimeter-wave region," *Journal of Physics-USSR*, vol. 10, pp. 197–198, 1946.

[33] A. Carrington and A. D. McLachlan, *Introduction to Magnetic Resonance with Applications to Chemistry and Chemical Physics*, Harper and Row, New York, NY, USA, 1967.

[34] C. P. Poole, *Electron Spin Resonance*, John Wiley & Sons, New York, NY, USA, 2nd edition, 1983.

[35] J. A. Well and J. R. Bolton, *Electron Paramagnetic Resonance*, John Wiley & Sons, New York, NY, USA, 2007.

[36] A. Abragam and J. H. Van Vleck, "Theory of the microwave Zeeman effect in atomic oxygen," *Physical Review*, vol. 92, no. 6, pp. 1448–1455, 1953.

[37] M. Tinkham and M. W. P. Strandberg, "Theory of the fine structure of the molecular oxygen ground state," *Physical Review*, vol. 97, no. 4, pp. 937–950, 1955.

[38] S. Krongelb and M. W. P. Strandberg, "Use of paramagnetic-resonance techniques in the study of atomic oxygen recombinations," *The Journal of Chemical Physics*, vol. 31, no. 5, pp. 1196–1210, 1959.

[39] T. K. Ishii, *Handbook of Microwave Technology: Applications*, Academic Press, New York, NY, USA, 1995.

[40] P. Zeeman, "Over den invloed fleener magnetisatie op den aard van het door een stof uitgezonden licht," *Verslagen Der Koninklijke Akademie van Wetenschappen te Amsterdam*, vol. 5, p. 181, 1896.

[41] A. A. Westenberg and N. Dehaas, "Observations on ESR linewidths and concentration measurements of gas–phase radicals," *The Journal of Chemical Physics*, vol. 51, no. 12, pp. 5215–5225, 1969.

[42] T. J. Cook and T. A. Miller, "Gas-phase EPR linewidths and intermolecular potentials. I. Theory," *The Journal of Chemical Physics*, vol. 59, no. 3, pp. 1342–1351, 1973.

[43] A. A. Westenberg, "Use of ESR for the quantitative determination of gas phase atom and radical concentrations," *Progress in Reaction Kinetics*, vol. 7, pp. 23–82, 1973.

[44] A. A. Westenberg and N. De Haas, "Quantitative measurements of gas phase O and N Atom concentrations by ESR," *The Journal of Chemical Physics*, vol. 40, no. 10, pp. 3087–3098, 1964.

[45] T. J. Cook, B. R. Zegarski, W. H. Breckenridge, and T. A. Miller, "Gas phase EPR of vibrationally excited O_2," *The Journal of Chemical Physics*, vol. 58, no. 4, pp. 1548–1552, 1973.

[46] W. H. Breckenridge and T. A. Miller, "Detection of metastable 3P argon atoms by gas phase EPR," *Chemical Physics Letters*, vol. 12, no. 3, pp. 437–442, 1972.

[47] V. Doležal, M. Mrázková, P. Dvořák, A. Tálský, and V. Kudrle, "Hydrogen line broadening in afterglow observed by means of EPR," *Acta Physica Slovaca*, vol. 55, no. 5, pp. 435–439, 2005.

[48] V. Kudrle, A. Tálský, A. Kudláč, V. Křápek, and J. Janča, "Influence of admixtures on production rate of atomic nitrogen," *Czechoslovak Journal of Physics*, vol. 50, no. 3, pp. 305–308, 2000.

[49] P. Dvořák, V. Doležal, M. Mrázková, V. Kudrle, A. Tálský, and J. Janča, "EPR measurements in hydrogen post-discharge," *Czechoslovak Journal of Physics*, vol. 54, no. 3, pp. C539–C543, 2004.

[50] V. Kudrle, P. Vašina, A. Tálský, and J. Janča, "Measurement of concentration of N atoms in afterglow," *Czechoslovak Journal of Physics*, vol. 52, pp. 589–595, 2002.

[51] V. Kudrle, P. Vašina, A. Tálský, M. Mrázková, O. Štec, and J. Janča, "Plasma diagnostics using electron paramagnetic resonance," *Journal of Physics D: Applied Physics*, vol. 43, no. 12, Article ID 124020, 2010.

[52] C. D. Pintassilgo, V. Guerra, O. Guaitella, and A. Rousseau, "Study of gas heating mechanisms in millisecond pulsed discharges and afterglows in air at low pressures," *Plasma Sources Science and Technology*, vol. 23, no. 2, Article ID 025006, 2014.

[53] C. D. Pintassilgo, O. Guaitella, and A. Rousseau, "Heavy species kinetics in low-pressure dc pulsed discharges in air," *Plasma Sources Science and Technology*, vol. 18, no. 2, Article ID 025005, 2009.

[54] A. Ricard, S. G. Oh, and V. Guerra, "Line-ratio determination of atomic oxygen and metastable absolute densities in an RF nitrogen late afterglow," *Plasma Sources Science and Technology*, vol. 22, no. 3, Article ID 035009, 2013.

[55] M. Mrázková, P. Vašina, V. Kudrle, A. Tálský, C. D. Pintassilgo, and V. Guerra, "On the oxygen addition into nitrogen post-discharges," *Journal of Physics D: Applied Physics*, vol. 42, no. 7, Article ID 075202, 2009.

[56] A. A. Westenberg, "Intensity relations for determining gas–phase OH, Cl, Br, I, and free–electron concentrations by quantitative ESR," *The Journal of Chemical Physics*, vol. 43, no. 5, pp. 1544–1549, 1965.

[57] W. Siemens, "Ueber die elektrostatische induction und die verzögerung des stroms in flaschendrähten," *Annalen der Physik*, vol. 178, no. 9, pp. 66–122, 1857.

[58] S. P. Sutera and R. Skalak, "The history of Poiseuille's law," *Annual Review of Fluid Mechanics*, vol. 25, no. 1, pp. 1–20, 1993.

[59] A. Roth, *Vacuum Technology*, Elsevier, 2012.

[60] J. M. Gershenzon, A. B. Nalbandyan, and V. B. Rozenshtein, *Magnetic Resonance in Gases*, Publishing House of Academy of Sciences of Armenian Soviet Socialist Republic, 1987.

[61] A. Tálský and M. Kunovský, "Die dissoziation des stickstoffen in der mikrowellen entladung," *Scripta Facultatis Scientiarum Naturalium, Universitatis Purkynianae Brunensis Physica*, vol. 18, no. 6, article 229, 1988.

[62] O. Reynolds, "An experimental investigation of the circumstances which determine whether the motion of water shall be direct or sinuous, and of the law of resistance in parallel channels," *Philosophical Transactions of the Royal Society of London*, vol. 174, no. 0, pp. 935–982, 1883.

[63] J. Kestin and W. Leidenfrost, "An absolute determination of the viscosity of eleven gases over a range of pressures," *Physica*, vol. 25, no. 7–12, pp. 1033–1062, 1959.

[64] M. Ohmi and M. Iguchi, "Critical Reynolds number in an oscillating pipe flow," *Bulletin of JSME*, vol. 25, no. 200, pp. 165–172, 1982.

[65] A.-M. Pointu, A. Ricard, E. Odic, and M. Ganciu, "Nitrogen atmospheric pressure post discharges for surface biological decontamination inside small diameter tubes," *Plasma Processes and Polymers*, vol. 5, no. 6, pp. 559–568, 2008.

[66] A. M. Pointu, E. Mintusov, and P. Fromy, "Study of an atmospheric pressure flowing afterglow in N_2/NO mixture and its application to the measurement of N_2(A) concentration," *Plasma Sources Science and Technology*, vol. 19, no. 1, Article ID 015018, 2009.

[67] A. M. Pointu and G. D. Stancu, "N_2(A) as the source of excited species of N_2, N and O in a flowing afterglow of N_2/NO mixture at atmospheric pressure," *Plasma Sources Science and Technology*, vol. 20, no. 2, Article ID 025005, 2011.

[68] V. Mazánková, D. Trunec, and F. Krčma, "Study of nitrogen flowing afterglow with mercury vapor injection," *Journal of Chemical Physics*, vol. 141, no. 15, Article ID 154307, 2014.

[69] B. F. Gordiets, C. M. Ferreira, V. L. Guerra et al., "Kinetic model of a low-pressure N_2-O_2 flowing glow discharge," *IEEE Transactions on Plasma Science*, vol. 23, no. 4, pp. 750–768, 1995.

[70] G. Cartry, L. Magne, and G. Cernogora, "Experimental study and modelling of a low-pressure N_2-O_2 time afterglow," *Journal of Physics D: Applied Physics*, vol. 32, no. 15, pp. 1894–1907, 1999.

[71] C. D. Pintassilgo, J. Loureiro, and V. Guerra, "Modelling of a N_2–O_2 flowing afterglow for plasma sterilization," *Journal of Physics D: Applied Physics*, vol. 38, no. 3, 2005.

[72] W. Van Gaens and A. Bogaerts, "Kinetic modelling for an atmospheric pressure argon plasma jet in humid air," *Journal of Physics D: Applied Physics*, vol. 46, no. 27, Article ID 275201, 2013.

[73] D. Blois, P. Supiot, M. Barj et al., "The microwave source's influence on the vibrational energy carried by in a nitrogen afterglow," *Journal of Physics D: Applied Physics*, vol. 31, no. 19, article 2521, 1998.

[74] J. Loureiro, P. A. Sá, and V. Guerra, "Role of long-lived N2($X^1\Sigma$g+,v) molecules and N2(A3Σu+) and N2(a'1Σu-) states in the light emissions of an N2 afterglow," *Journal of Physics D: Applied Physics*, vol. 34, no. 12, pp. 1769–1778, 2001.

[75] N. Sadeghi, C. Foissac, and P. Supiot, "Kinetics of $N_2(A^3\sum_u^+)$ molecules and ionization mechanisms in the afterglow of a flowing N_2 microwave discharge," *Journal of Physics D: Applied Physics*, vol. 34, no. 12, pp. 1779–1788, 2001.

[76] A.-M. Pointu, A. Ricard, B. Dodet, E. Odic, J. Larbre, and M. Ganciu, "Production of active species in N_2-O_2 flowing post-discharges at atmospheric pressure for sterilization," *Journal of Physics D: Applied Physics*, vol. 38, no. 12, pp. 1905–1909, 2005.

[77] M. Capitelli, C. M. Ferreira, B. F. Gordiets, and A. I. Osipov, *Plasma Kinetics in Atmospheric Gases*, vol. 31, Springer Science & Business Media, 2013.

[78] V. Zvonicek, V. Guerra, J. Loureiro, A. Talsky, and M. Touzeau, "Surface and volume kinetics of O(3P) atoms in a low-pressure O2-N2 microwave discharge," in *Proceedings of the 23rd International Conference on Phenomena in Ionized Gases (ICPIG '97)*, vol. 4, p. 160, Toulouse, France, July 1997.

[79] G. Cartry, L. Magne, G. Cernogora, M. Touzeau, and M. Vialle, "Effect of wall treatment on oxygen atoms recombination," in *Proceedings of the 23rd International Conference on Phenomena in Ionized Gases (ICPIG '97)*, vol. 2, pp. 70–71, Toulouse, France, July 1997.

[80] G. Cartry, L. Magne, and G. Cernogora, "Atomic oxygen recombination on fused silica: modelling and comparison to low-temperature experiments (300 K)," *Journal of Physics D: Applied Physics*, vol. 33, no. 11, 2000.

[81] K. M. Evenson and D. S. Burch, "Atomic–nitrogen recombination," *The Journal of Chemical Physics*, vol. 45, no. 7, pp. 2450–2460, 1966.

[82] G. Black, H. Wise, S. Schechter, and R. L. Sharpless, "Measurements of vibrationally excited molecules by Raman scattering. II. Surface deactivation of vibrationally excited N_2," *The Journal of Chemical Physics*, vol. 60, no. 9, pp. 3526–3536, 1974.

21

Application of Near-Infrared Spectroscopy to Quantitatively Determine Relative Content of *Puccnia striiformis* f. sp. *tritici* DNA in Wheat Leaves in Incubation Period

Yaqiong Zhao,[1] Yilin Gu,[1] Feng Qin,[1] Xiaolong Li,[1] Zhanhong Ma,[1] Longlian Zhao,[2] Junhui Li,[2] Pei Cheng,[1] Yang Pan,[1] and Haiguang Wang[1]

[1]*College of Plant Protection, China Agricultural University, Beijing 100193, China*
[2]*College of Information and Electrical Engineering, China Agricultural University, Beijing 100083, China*

Correspondence should be addressed to Haiguang Wang; wanghaiguang@cau.edu.cn

Academic Editor: Wee Chew

Stripe rust caused by *Puccinia striiformis* f. sp. *tritici* (*Pst*) is a devastating wheat disease worldwide. Potential application of near-infrared spectroscopy (NIRS) in detection of pathogen amounts in latently *Pst*-infected wheat leaves was investigated for disease prediction and control. A total of 300 near-infrared spectra were acquired from the *Pst*-infected leaf samples in an incubation period, and relative contents of *Pst* DNA in the samples were obtained using duplex TaqMan real-time PCR arrays. Determination models of the relative contents of *Pst* DNA in the samples were built using quantitative partial least squares (QPLS), support vector regression (SVR), and a method integrated with QPLS and SVR. The results showed that the *k*QPLS-SVR model built with a ratio of training set to testing set equal to 3 : 1 based on the original spectra, when the number of the randomly selected wavelength points was 700, the number of principal components was 8, and the number of the built QPLS models was 5, was the best. The results indicated that quantitative detection of *Pst* DNA in leaves in the incubation period could be implemented using NIRS. A novel method for determination of latent infection levels of *Pst* and early detection of stripe rust was provided.

1. Introduction

Wheat stripe rust, caused by the biotrophic pathogen *Puccinia striiformis* f. sp. *tritici* (*Pst*), is an important wheat disease worldwide [1–5]. This disease can cause severe yield losses. Once an epidemic of this disease occurs, at least 10%–30% of wheat yield can be reduced [1]. In China, wheat stripe rust occurs in almost all wheat-growing regions, and it is the most devastating wheat disease [1, 4, 5]. This disease is always a great potential threat to the national wheat production in China. Since 1950, many severe epidemics of wheat stripe rust have occurred in China [1, 5], among which the four most destructive epidemics occurred in 1950, 1964, 1990, and 2002, resulting in yield losses of 6.0, 3.2, 1.8, and 1.3 million tons, respectively [4, 6].

Generally, an infection process of wheat stripe rust in the field can be divided into four stages, that is, the contact period, the penetration period, the incubation period, and the diseased period [1]. Under favorable environmental conditions, *Pst* urediospores landing on the surface of wheat leaves germinate, produce germ tubes and appressoria, and then penetrate into wheat leaf tissues [1, 5]. During the incubation period, a large quantity of *Pst* hyphae accumulate gradually in the infected leaf tissues [1, 5], but it is very difficult to observe disease symptoms on the surface of the infected wheat leaves with naked eyes, causing lots of difficulties for early monitoring and prediction of this disease. Once uredinia appear on the surface of wheat leaves, a large number of urediospores may be liberated after rupture of the uredinia and then be spread by air, which may result in secondary infections. So it is of great significance to conduct the early detection of *Pst* infections and the quantitative detection of pathogen amounts in the infected plants for early precise site-specific control of the disease, disease prediction, and control strategy making. In particular, it is very important to realize the early detection of pathogen amounts in the infected plants in the overwintering regions and the oversummering regions of *Pst* for the macro control of wheat stripe rust. Traditionally, the monitoring of wheat stripe rust is carried out via field investigation. This method is time-consuming and laborious. Moreover, using this conventional method, only the diseased wheat fields with disease symptoms can be surveyed, and the infected wheat leaves without symptom appearance in the incubation period cannot be identified accurately and rapidly. At present, some techniques and methods, including molecular biology techniques [7–10], hyperspectral remote sensing technology [11–14], thermal infrared imaging technology [15], and near-infrared spectroscopy (NIRS) [16, 17], have been applied to the monitoring and early detection of wheat stripe rust.

There are many reports on early qualitative detection of wheat stripe rust. The *Pst* infection of a wheat leaf in the incubation period of stripe rust can be detected accurately and qualitatively by detecting the presence of wheat stripe rust pathogen using molecular biology techniques such as polymerase chain reaction (PCR) assay and loop-mediated isothermal amplification (LAMP) assay [7–9]. Hyperspectral remote sensing technology has been used to identify healthy wheat plants and *Pst*-infected wheat plants without symptoms [18]. It was reported that *Pst* latent infections in wheat leaves could be detected using thermal infrared imaging technology [15] and NIRS technology [16, 17] before symptoms appear. A study on the detection of *Pst* latent infections in wheat leaves incubated in an artificial climate chamber using NIRS technology was conducted by Li et al. [17], and the results obtained by analyzing the spectral data demonstrated that the latently infected leaves could be distinguished from healthy wheat leaves as early as one day after artificial inoculation with the suspension of *Pst* urediospores. However, there are relatively few studies on the early quantitative detection of *Pst* in the incubation period of wheat stripe rust. The quantitative detection of *Pst* in the latently infected wheat leaves was implemented mainly relying on a real-time PCR method [10, 19].

The detection process using a molecular biological method is complex, and the method has high requirements on technologies and instruments [20, 21]. With the development of science and technology, portable fluorescence quantitative PCR instruments have been applied gradually in recent years and the instruments for quantitative analysis in the field have been available [20]. However, it is still necessary for users to master the relevant molecular biology skills, and a large number of samples to be tested are required to be obtained. Only using molecular biology technology, it is very difficult to quickly get enough data in a large scale to provide information for disease forecasting and disease warning. The instruments used in hyperspectral remote sensing are very expensive. Thermal infrared imaging technology has high requirements on imaging resolution and temperature resolution of the used instrument. At the present time, detection methods of latent infection of pathogens based on molecular biology technology, hyperspectral remote sensing technology, and thermal infrared imaging technology are very difficult to be popularized in practice. Therefore, it is critical to explore a simple, convenient, accurate, and rapid method for early detection of wheat stripe rust.

As a kind of nondestructive analytical technology, NIRS can be used to carry out both qualitative analysis and quantitative analysis of samples [22, 23]. The analysis process of NIRS is simple and rapid, and it is a fast detection technology with low cost and is suitable for online analysis [22, 23]. It has been widely used in many fields such as agriculture, food industry, chemical industry, pharmaceutical industry, and petroleum industry [21–27]. NIRS has been applied to implement early qualitative identification of wheat stripe rust and wheat leaf rust (caused by *Puccinia triticina*) before disease symptoms appeared [16] and assess disease severity of wheat stripe rust [28]. It has also been used to qualitatively identify urediospores of *Pst* and *P. triticina* and quantitatively determine the content of each pathogen in a mixture of two kinds of pathogens including *Pst* and *P. triticina* [29]. Moreover, using NIRS, wheat leaves infected with *Pst* could be identified before symptoms appear [17]. However, to the best of our knowledge, there are no reports on quantitative detection of *Pst* in wheat leaves in the incubation period of wheat stripe rust using NIRS.

In this study, a method based on NIRS was explored for quantitative determination of the relative content of *Pst* DNA in wheat leaves in the incubation period. Based on the acquired near-infrared spectral data of the latently *Pst*-infected wheat leaf samples and the data on the relative content of *Pst* DNA in the corresponding samples obtained using the duplex TaqMan real-time PCR arrays, dynamic changes of the relative contents of *Pst* DNA in the latently *Pst*-infected wheat leaves during the incubation period were investigated and quantitative determination models were built to realize the quantitative and rapid detection of the pathogen amounts in the latently *Pst*-infected wheat leaves. The aim of this study was to provide a method for rapid, nondestructive, and quantitative determination of latent infection levels of *Pst* in wheat leaves and early detection of wheat stripe rust. Furthermore, support information can be provided for prediction and control of this disease. Some methodological references can also be provided for the early quantitative and nondestructive detection of other diseases.

2. Materials and Methods

2.1. Materials. CYR 33, a dominant physiological race of *Pst* in China, was used in this study. A highly susceptible wheat cultivar Mingxian 169 was used to multiply the pathogen in an artificial climate chamber in the Laboratory of Plant Disease Epidemiology, Department of Plant Pathology, China Agricultural University. Artificial inoculation by spraying urediospores of CYR 33 onto the surface of seedling leaves of Mingxian 169 was conducted to obtain the latently *Pst*-infected wheat leaves.

2.2. Multiplication of Wheat Stripe Rust Pathogen. Urediospores of the *Pst* physiological race CYR33 were multiplied using a similar method as described by Cheng et al. [30].

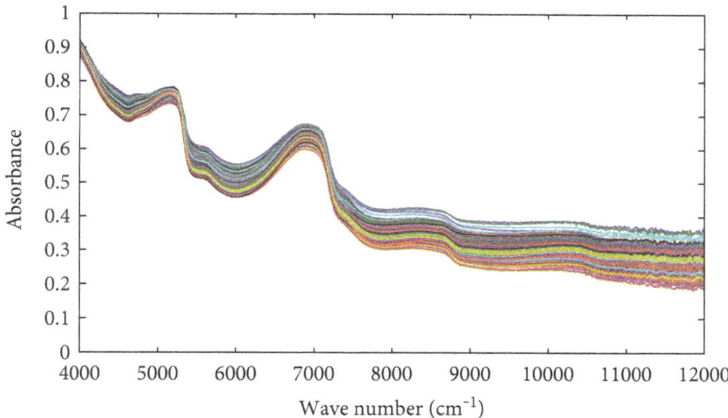

FIGURE 1: Near-infrared spectra of wheat leaves during the incubation period of wheat stripe rust.

After seeds of Mingxian 169 were soaked for 24 h in sterile water, the plump and well-germinated seeds were selected and sown in pots (10 cm in diameter) with approximately 25 seeds per pot. Then the pots were incubated in an artificial climate chamber at 11–13°C with 60%–70% relative humidity (RH) and 14 h of light per day (10,000 lux) and were watered when necessary. When the first leaves of wheat seedlings fully expanded, urediospores of CYR 33 stored in a liquid nitrogen container were taken out, reactivated in warm water of 40–45°C for 5 min, and then hydrated at 4°C for 12 h. A spore suspension was made with 0.2% Tween 80 solution for artificial spray inoculation. Immediately after inoculation, each pot with the inoculated wheat seedlings was covered with a clear glass cylinder that was covered with two layers of sterile cotton gauze on the top, and all pots with the inoculated wheat seedlings were transferred into a moist chamber under dark conditions at 11–13°C for 24 h. Subsequently, the inoculated wheat seedlings were placed into the artificial climate chamber under the conditions described above and incubated until a large number of urediospores were produced on the surface of the seedling leaves. The fresh urediospores were harvested for the following experiments or stored in the liquid nitrogen container for later use.

2.3. Collection of Latently Pst-Infected Wheat Leaves. Using the artificial spray inoculation method described above, healthy wheat seedlings with the fully expanded first leaves were inoculated with a 0.15 mg/mL spore suspension prepared with 3 mg harvested fresh urediospores of CYR33 and 0.2% Tween 80 solution. A total of 50 pots of wheat seedlings were inoculated for further experiments. Sixty wheat leaves with uniform size and same growth vigor were collected every 24 h after inoculation until disease symptoms appeared and the incubation period of wheat stripe rust ended. The incubation period was defined as the number of days from inoculation to rupture of the first uredinium [31], and it was 10 days in this study. Two leaves were treated as a sample for acquisition of near-infrared spectral data, and 30 samples per day were collected. A total of 300 samples were collected in this study.

2.4. Acquisition of Near-Infrared Spectral Data. The near-infrared spectra of the latently *Pst*-infected wheat leaf samples were acquired by using FT-NIR MPA spectrometer (Bruker, Germany). Wheat leaves of each sample were cut into small square fragments and then were placed into a sample cup (20 mm in diameter) for spectral measurement. Using integrating sphere diffuse reflectance method, the spectra in a range of 4000–12,000 cm^{-1} were measured with the spectral resolution of 8 cm^{-1} and the number of scan processes of 32. Each acquired spectrum contained 2100 wavelength points. After spectral acquisition, each sample was put into a 2 mL grinding tube. Each tube was numbered and immediately kept at −80°C for subsequent DNA extraction. On each day during the incubation period, the near-infrared spectra of 30 collected samples were acquired. In this study, a total of 300 spectra were obtained as shown in Figure 1.

2.5. DNA Extraction and Quantitative Measurements of DNA in Wheat Leaf Samples. Healthy wheat leaves, 3 mg of urediospores of *Pst*, and the latently *Pst*-infected wheat leaves after acquisition of near-infrared spectral data were used as samples for DNA extraction. The extracted wheat DNA from the healthy leaves and the extracted *Pst* DNA from the *Pst* urediospores were used to generate standard curves. DNA was extracted using a modification of the procedure described by Justesen et al. [32]. Each sample in a 2 mL grinding tube was mixed with 0.4 g of quartz sands, one glass bead, and 600 μL of 2% cetyl trimethyl ammonium bromide (CTAB) buffer. The CTAB buffer was prepared with 2 g of CTAB and 1 g of polyvinyl pyrrolidone by adding 10 μL of 1 M Tris-HCl (pH = 8.0), 2 μL of 0.5 M ethylene diamine tetraacetic acid (pH = 8.0), and 28 mL of 5 M NaCl into a final volume of 100 mL adjusted with deionized water, then was sterilized under high temperature and high pressure, and finally was added with 100 μL β-mercapto ethanol after cooling. The tube was shaken in a FastPrep-24 instrument (MP Biomedicals, Santa Ana, CA, USA) at 6.0 m/s for two periods of 40 s, with 5 min cooling on ice between them. After incubation in water bath at 65°C for 1 h with gentle shakes every

Table 1: Duplex TaqMan real-time PCR primers and probes used in this study.

Primer/probe	NCBI accession number	Sequence of primers and corresponding probes (5′-3′)	Amplified fragment length (bp)
TAG2315F	AF280605.1	CAGAAAGCGAGTGGAAAGATGAAAG	181
TAG2473R		GCAAGGAGGACAAAGATGAGGAA	
TAG-Pr1		HEX-CAAGCATCAAAGGCAAGCAAGCAGTAGT-BHQ1	
Pst-F	GU382673.1	AACCCTCTCATTAAATAATTTTG	102
Pst-R		CCAACTTATAGAAAAGTGACTTA	
Pst-P		FAM-ATTACAGCAGCACTCAACATCCATT-BHQ1	

15 min, the tube was added with 600 μL of chloroform/isoamyl alcohol ($v:v=24:1$) with mixing and then was centrifuged at the speed of 12,000 rpm for 10 min. The supernatant was transferred to a new clean 1.5 mL centrifuge tube, and 0.6 volumes of cold isopropanol at −20°C was added. The mixture in the tube was shaken gently and then was incubated at −20°C for 1 h. The tube was centrifuged at the speed of 12,000 rpm for 10 min, and then the liquid supernatant was abandoned. The resultant was added with 500 μL of 70% ethanol for rinse, shaken gently, and then centrifuged at 12000 rpm for 10 min. After discarding the supernatant and drying the resultant in the air, the DNA was dissolved in 50 μL of sterile double distilled water and kept at −20°C for later use.

Primers and probes used for the duplex TaqMan real-time PCR assays in this study were listed in Table 1. When the quantity of wheat DNA was determined using duplex TaqMan real-time PCR, TAG2315F and TAG2473R reported by Sandberg et al. [33] that were designed based on the DNA sequence of a prolamin gene of wheat were used as the primers, and the probe TAG-Pr1 was designed based on this DNA sequence. When the quantity of *Pst* DNA was determined using duplex TaqMan real-time PCR, the primers (*Pst*-F and *Pst*-R) and the probe *Pst*-P reported by Li et al. [34] that were designed based on the internal transcribed spacer (ITS) region sequence of *Pst* were used.

Each real-time PCR assay was performed in a volume of 20 μL containing 3.20 μL of $MgCl_2$ (25 μM), 2.00 μL of dNTP (2500 μM), 2.00 μL of Taq buffer (10x), 0.60 μL of Taq (5 U/μL), 0.30 μL of each primer (*Pst*-F, *Pst*-R, TAG2315F, and TAG2473R, 10 μM each), 0.25 μL of each probe (*Pst*-P and TAG-Pr1, 10 μM each), 2.00 μL of template DNA, and double distilled water added to a final volume of 20 μL.

The real-time PCR amplification conditions were as follows: 1 cycle of initial denaturation at 95°C for 3 min and 40 cycles consisting of denaturation at 94°C for 20 s, annealing at 56°C for 30 s and extension at 72°C for 30 s (fluorescence signal detection was conducted under this condition).

To generate standard curves, tenfold serial dilutions of *Pst* DNA (10, 1, 10^{-1}, 10^{-2}, 10^{-3}, and 10^{-4} ng/μL) and wheat DNA (100, 10, 1, 10^{-1}, and 10^{-2} ng/μL) were made. The real-time PCR amplifications were conducted as described above, and the corresponding cycle threshold (Ct) values were recorded. Then two standard curves to quantify *Pst* DNA and wheat DNA from the samples consisting of the latently *Pst*-infected wheat leaves after acquisition of near-infrared spectral data were generated.

Using the extracted DNA from the latently *Pst*-infected samples as template DNA, the real-time PCR amplifications were conducted as described above, and the corresponding Ct value of each sample was obtained. According to the standard curves, the contents of *Pst* DNA and wheat DNA were quantitatively determined. And the relative content of *Pst* DNA could be calculated using the following formula: $RCP = CP \times 100\%/(CP + CW)$, where RCP is the relative content of *Pst* DNA, CP is the content of *Pst* DNA (ng), and CW is the content of wheat DNA (ng).

2.6. Establishment of Determination Models for Quantification of the Relative Content of Pst DNA in Wheat Leaves in the Incubation Period. The obtained near-infrared spectra of the latently *Pst*-infected wheat leaf samples were preprocessed by using three methods including multiplication scatter correction (MSC), standard normalized variate (SNV), and vector normalization (VN). The 300 obtained spectra of the samples were divided into training set and testing set based on a ratio of the training set to the testing set equal to 3 : 1. By the combined use of quantitative partial least squares (QPLS) and support vector regression (SVR), the *k*QPLS-SVR models to quantify the relative content of *Pst* DNA in wheat leaves in the incubation period were built based on the original near-infrared spectra and the spectral data obtained by using the three preprocessing methods.

As shown in Figure 2, to build a *k*QPLS-SVR model, firstly, *m* features were randomly selected from the spectral features (spectral attributes) of 2100 wavelength points, the number of principal components was set as *n*, and a QPLS model was built. Using this method, *k* QPLS models were built; secondly, a SVR model was built by using the predicted values of the *k* QPLS models as variable. Thus, a *k*QPLS-SVR model was obtained. The SVR model was built with radial basis function as the kernel function. Using the grid search algorithm, the penalty parameter *C* and the kernel function parameter *g* for the SVR model were optimized in a range of 2^{-8}–2^{8} with the searching step of 0.8. As the minimum mean squared error of the training set was achieved at a point within the grid, the corresponding values of *C* and *g* were regarded as the optimal parameters for building the SVR model. In this study, the number of the randomly selected spectral features (*m*) was set as 700 or 1400, the number of principal components (*n*) while building a QPLS model was set as 4, 8, or 12, and the number of the built QPLS models (*k*) was set as 5, 10, or 15. All the calculations and modeling processes described above were implemented using the

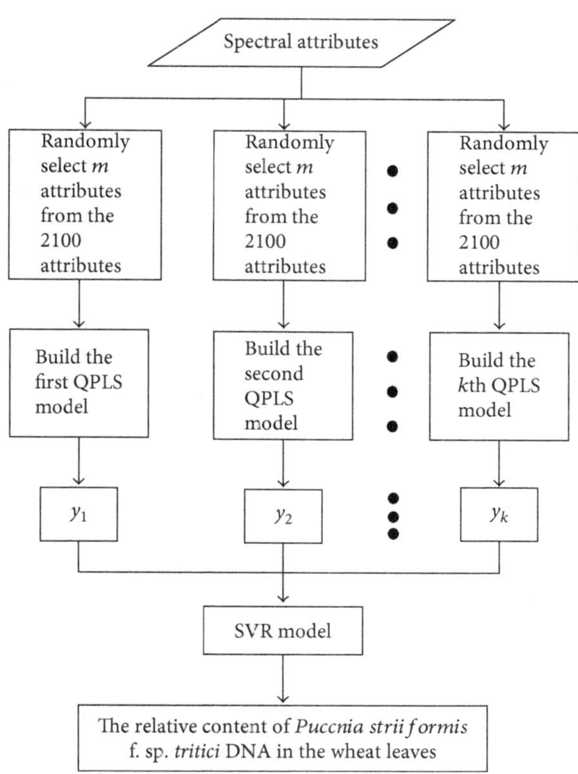

Figure 2: The algorithm flowchart for building the kQPLS-SVR models to quantify the relative content of *Pst* DNA in wheat leaves in the incubation period.

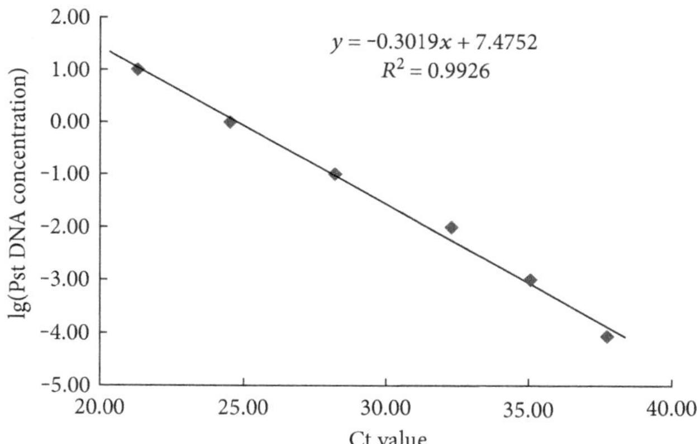

Figure 3: The standard curve of duplex TaqMan real-time PCR for quantification of *Pst* DNA and the corresponding linear regression equation.

software MATLAB 7.8.0 (R2009a) (MathWorks, Natick, MA, USA).

The coefficient of determination (R^2), standard error of calibration (SEC), average absolute relative deviation (AARD) and relative prediction deviation (RPD) of the training set and R^2, standard error of prediction (SEP), AARD, and RPD of the testing set were used to evaluate the kQPLS-SVR models built for quantification of the relative content of *Pst* DNA in wheat leaves in the incubation period. A value of R^2 more than 0.5 denotes that the corresponding model can be used for rough screening and actual application [35]. The closer the value of R^2 is to 1, the higher the accuracy of the model is. The accuracy of a model is also related to the value of RPD, and a higher value of RPD denotes that the model has greater prediction ability [35]. The less value of SEC, SEP, or AARD denotes that higher accuracy can be obtained using the model and that the model has greater prediction ability. According to these evaluation indicators described above, the selection of the optimal kQPLS-SVR model was conducted.

Determination models of the relative content of *Pst* DNA in wheat leaves in the incubation period were also built using individual methods including QPLS and SVR based on the same training set and testing set as used for building the optimal kQPLS-SVR model. A comparison of the effects of the three models was conducted according to R^2, SEC, AARD, and RPD of the training set and R^2, SEP, AARD, and RPD of the testing set. The SVR model was built and optimized as described above. The number of principal components used during building the QPLS model was determined by evaluating the prediction residual error sum of square (PRESS) [36]. Generally, when the minimum PRESS value is obtained, the value of the corresponding number of principal components may be the optimum, but it may lead to overfitting in this case. The value of PRESS can be calculated using the following formula [36]:

$$\text{PRESS} = \sum_{i=1}^{n}\sum_{j=1}^{d}\left(y_{p,ij} - y_{ij}\right)^2, \quad (1)$$

where n is the number of the samples in the training set, d is the number of principal components used during modeling, $y_{p,ij}$ is the predicted value of the ith sample of the training set, and y_{ij} is the actual value of the ith sample of the training set. In this study, the optimal number of principal components was determined using F statistical method described as the following formula: $F(f) = \text{PRESS}(f)/\text{PRESS}(f^*)$, where f^* is the number of principal components corresponding to the minimum PRESS value; the optimal number of principal components (f) is less than f^*, and it should be as small as possible and satisfy the following condition: $F(f) < F_{\alpha,m,m}$ ($\alpha = 0.25$, m is the number of degrees of freedom) [36].

3. Results

3.1. Quantitative Measurements of Pst DNA in Wheat Leaves in the Incubation Period Using Duplex TaqMan Real-Time PCR. After the real-time PCR amplifications were conducted with tenfold serial dilutions of *Pst* DNA, the standard curve for quantification of *Pst* DNA (Figure 3) was generated using common logarithmic value of the concentration of *Pst* DNA (lg(*Pst* DNA concentration)) as the ordinate and Ct value as the abscissa, and the corresponding linear regression equation was obtained. As shown in Figure 3, the equation of the standard curve for quantification of *Pst* DNA was as follows: $y = -0.3019x + 7.4752$ ($R^2 = 0.9926$), where y is lg(*Pst* DNA concentration) and x is the Ct value. After the real-time PCR amplifications were conducted with tenfold serial

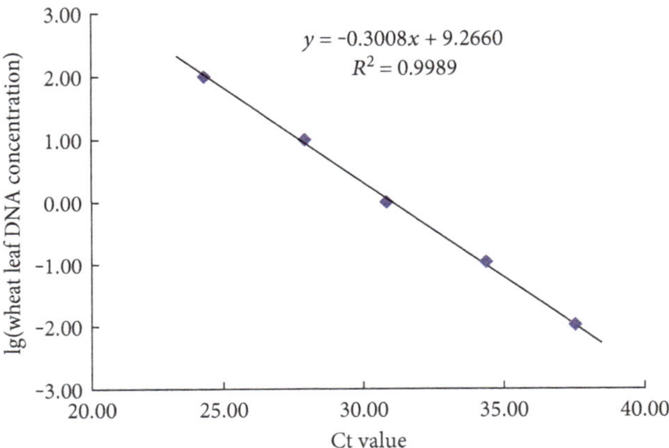

FIGURE 4: The standard curve of duplex TaqMan real-time PCR for quantification of wheat DNA and the corresponding linear regression equation.

dilutions of wheat DNA from healthy leaves, the standard curve for quantification of wheat DNA (Figure 4) was generated using common logarithmic value of the concentration of wheat DNA from leaves (lg(wheat leaf DNA concentration)) as the ordinate and Ct value as the abscissa, and the corresponding linear regression equation was also obtained. As shown in Figure 4, the equation of the standard curve for quantification of wheat DNA was as follows: $y = -0.3008x + 9.2660$ ($R^2 = 0.9989$), where y is lg(wheat leaf DNA concentration) and x is the Ct value. Moreover, satisfactory amplification efficiencies of the real-time PCR arrays were obtained for both *Pst* DNA and wheat DNA, and they were 99.7% and 100.1%, respectively.

After the real-time PCR arrays were conducted with the extracted DNA from the latently *Pst*-infected wheat leaf samples, the contents of *Pst* DNA and wheat DNA in each sample were figured out based on the linear regression equations of the generated standard curves. The results showed that the relative content of *Pst* DNA increased exponentially with time (Figure 5). An equation to fit the daily changes of the relative contents of *Pst* DNA in wheat leaves after inoculation was built: $y = 0.0096e^{0.9287x}$ ($R^2 = 0.8784$), where y is the relative content of *Pst* DNA (%) in a wheat leaf sample after inoculation and x is the days post inoculation. In this study, attempts were made to build a linear regression model by using partial least squares method. The relative contents of *Pst* DNA in wheat leaves after inoculation changed in a range of 0.00385%–90.09%. Especially, all relative contents of *Pst* DNA in the latently *Pst*-infected wheat leaves in the first six days were no more than 0.2%. To make the data follow a normal distribution, a logarithmic transformation of the relative contents of *Pst* DNA in wheat leaves after inoculation was conducted in this study. Meanwhile, to ensure that each value after transformation was positive, each value of the relative contents of *Pst* DNA in wheat leaves after inoculation was multiplied by 10^5 before the logarithmic transformation. The linear regression model ($y = 0.4036x + 0.9861$, where y is lg(relative content *Pst* DNA $\times 10^5$) and x is the days post inoculation) built based on the common logarithmic values

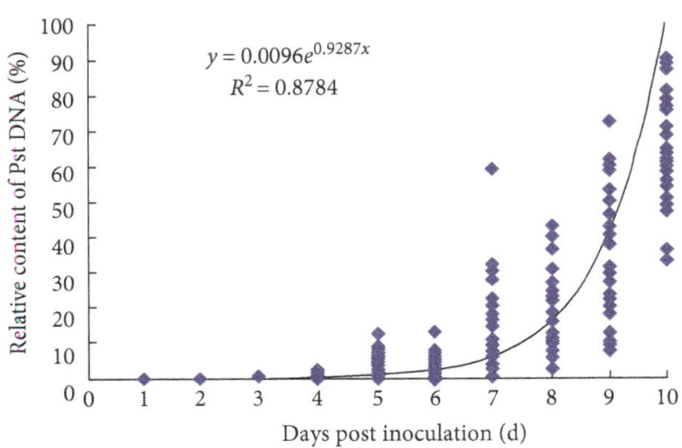

FIGURE 5: The daily changes of the relative contents of *Pst* DNA in wheat leaves after inoculation.

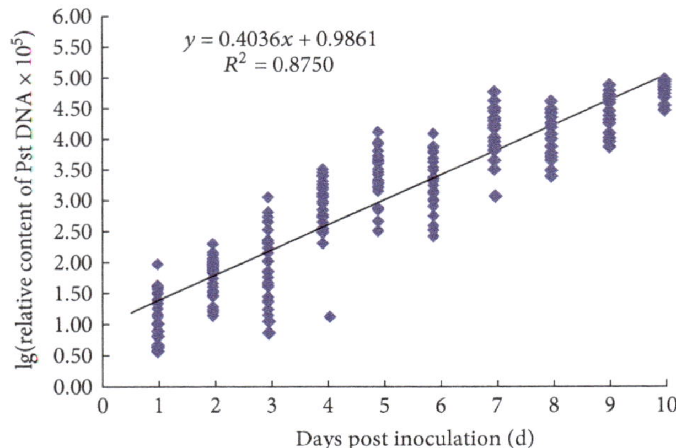

FIGURE 6: The daily changes of the common logarithmic values after transformation of the relative contents of *Pst* DNA in wheat leaves after inoculation.

was shown in Figure 6. As shown in Figures 5 and 6, the coefficients of determination (R^2) of the equations built with the days post inoculation as independent variable before and after logarithmic transformation were 0.8784 and 0.8750, respectively. It was demonstrated that there was no great difference between the values of R^2 of the two equations. Therefore, in combination with the near-infrared spectral data, lg(CP $\times$ 100%/(CP + CW) $\times 10^5$) in which CP was the content of *Pst* DNA in ng and CW was the content of wheat DNA in ng, that is, lg(relative content *Pst* DNA $\times 10^5$) was treated as chemical value for further modeling to develop a determination model for quantification of the relative content of *Pst* DNA in wheat leaves in the incubation period.

3.2. Results of the kQPLS-SVR Models for Quantification of the Relative Content of Pst DNA in Wheat Leaves in the Incubation Period. To quantitatively determine relative content of *Pst* DNA in wheat leaves in the incubation period using NIRS technology, the results of the *k*QPLS-SVR models built with original near-infrared spectra were shown in Table 2. The results demonstrated that the best effects were achieved for the *k*QPLS-SVR model built when the number

TABLE 2: The results of the *k*QPLS-SVR models for quantification of the relative content of *Pst* DNA in wheat leaves in the incubation period built with the original near-infrared spectra.

The number of attributes	The number of principal components	The number of QPLS models	Training set R^2	Training set SEC	Training set AARD	Training set RPD	Testing set R^2	Testing set SEP	Testing set AARD	Testing set RPD
700	4	5	0.7179	0.6559	0.2283	1.8826	0.6773	0.7190	0.2624	1.7604
700	4	10	0.7664	0.5968	0.1847	2.0692	0.7147	0.6760	0.2391	1.8723
700	4	15	0.7729	0.5884	0.1836	2.0986	0.6732	0.7236	0.2645	1.7492
700	8	5	0.9027	0.3852	0.1257	3.2057	0.8820	0.4349	0.1274	2.9107
700	8	10	0.9029	0.3849	0.1282	3.2085	0.8788	0.4406	0.1400	2.8727
700	8	15	0.9026	0.3854	0.1263	3.2042	0.8802	0.4382	0.1353	2.8888
700	12	5	0.9564	0.2579	0.0872	4.7879	0.8514	0.4880	0.1458	2.5938
700	12	10	0.9580	0.2531	0.0816	4.8797	0.8604	0.4729	0.1416	2.6767
700	12	15	0.9590	0.2501	0.0808	4.9374	0.8535	0.4844	0.1414	2.6128
1400	4	5	0.7193	0.6543	0.2158	1.8874	0.6309	0.7689	0.2636	1.6461
1400	4	10	0.7047	0.6711	0.2173	1.8401	0.6422	0.7571	0.2598	1.6718
1400	4	15	0.7729	0.5884	0.1836	2.0986	0.6732	0.7236	0.2645	1.7492
1400	8	5	0.9072	0.3762	0.1241	3.2826	0.8806	0.4373	0.1342	2.8944
1400	8	10	0.9039	0.3827	0.1273	3.2266	0.8771	0.4437	0.1385	2.8530
1400	8	15	0.9036	0.3834	0.1284	3.2209	0.8785	0.4411	0.1356	2.8692
1400	12	5	0.9668	0.2249	0.0739	5.4900	0.8180	0.5400	0.1634	2.3440
1400	12	10	0.9666	0.2256	0.0751	5.4742	0.8215	0.5347	0.1572	2.3672
1400	12	15	0.9659	0.2280	0.0739	5.4164	0.8223	0.5335	0.1555	2.3723

TABLE 3: The results of the *k*QPLS-SVR models for quantification of the relative content of *Pst* DNA in wheat leaves in the incubation period built when MSC was used as the preprocessing method of the original near-infrared spectra.

The number of attributes	The number of principal components	The number of QPLS models	Training set R^2	Training set SEC	Training set AARD	Training set RPD	Testing set R^2	Testing set SEP	Testing set AARD	Testing set RPD
700	4	5	0.8674	0.4497	0.1427	2.7460	0.8640	0.4668	0.1485	2.7117
700	4	10	0.8792	0.4292	0.1325	2.8769	0.8689	0.4582	0.1464	2.7623
700	4	15	0.9081	0.3744	0.1070	3.2980	0.8694	0.4574	0.1420	2.7674
700	8	5	0.9215	0.3460	0.1148	3.5691	0.8712	0.4542	0.1367	2.7869
700	8	10	0.9238	0.3408	0.1105	3.6234	0.8710	0.4547	0.1372	2.7838
700	8	15	0.9227	0.3434	0.1127	3.5961	0.8730	0.4510	0.1361	2.8062
700	12	5	0.9649	0.2312	0.0780	5.3402	0.8338	0.5160	0.1552	2.4530
700	12	10	0.9645	0.2326	0.0737	5.3083	0.8534	0.4846	0.1444	2.6117
700	12	15	0.9650	0.2311	0.0753	5.3430	0.8395	0.5071	0.1462	2.4962
1400	4	5	0.8726	0.4408	0.1320	2.8013	0.8606	0.4725	0.1467	2.6786
1400	4	10	0.8750	0.4367	0.1304	2.8280	0.8586	0.4760	0.1416	2.6590
1400	4	15	0.9019	0.3869	0.1111	3.1922	0.8652	0.4648	0.1452	2.7232
1400	8	5	0.9272	0.3332	0.1100	3.7056	0.8592	0.4750	0.1442	2.6648
1400	8	10	0.9257	0.3366	0.1159	3.6688	0.8613	0.4713	0.1403	2.6855
1400	8	15	0.9257	0.3366	0.1156	3.6683	0.8634	0.4677	0.1407	2.7060
1400	12	5	0.9709	0.2108	0.0686	5.8573	0.7932	0.5756	0.1653	2.1990
1400	12	10	0.9722	0.2058	0.0636	6.0017	0.7911	0.5785	0.1675	2.1881
1400	12	15	0.9718	0.2073	0.0645	5.9559	0.7929	0.5760	0.1673	2.1973

of the spectral attributes of the randomly selected wavelength points was 700, the number of principal components was 8, and the number of the built QPLS models was 5. For this model, the R^2 value, SEC, AARD, and RPD of the training set were 0.9027, 0.3852, 0.1257, and 3.2057, respectively, and the R^2 value, SEP, AARD, and RPD of the testing set were 0.8820, 0.4349, 0.1274, and 2.9107, respectively.

The results of the *k*QPLS-SVR models built based on the data obtained after preprocessing of original near-infrared

Table 4: The results of the *k*QPLS-SVR models for quantification of the relative content of *Pst* DNA in wheat leaves in the incubation period built when SNV was used as the preprocessing method of the original near-infrared spectra.

The number of attributes	The number of principal components	The number of QPLS models	Training set R^2	Training set SEC	Training set AARD	Training set RPD	Testing set R^2	Testing set SEP	Testing set AARD	Testing set RPD
700	4	5	0.8800	0.4278	0.1337	2.8863	0.8721	0.4527	0.1418	2.7959
700	4	10	0.9009	0.3747	0.1069	3.2959	0.8643	0.4663	0.1501	2.7144
700	4	15	0.9080	0.3747	0.1051	3.2961	0.8683	0.4593	0.1446	2.7560
700	8	5	0.9255	0.3370	0.1096	3.6641	0.8725	0.4519	0.1320	2.8009
700	8	10	0.9186	0.3522	0.1161	3.5058	0.8725	0.4520	0.1389	2.8005
700	8	15	0.9225	0.3439	0.1151	3.5912	0.8712	0.4543	0.1379	2.7862
700	12	5	0.9636	0.2356	0.0757	5.2422	0.8393	0.5074	0.1504	2.4947
700	12	10	0.9660	0.2275	0.0723	5.4270	0.8456	0.4973	0.1486	2.5450
700	12	15	0.9660	0.2276	0.0719	5.4257	0.8462	0.4963	0.1476	2.5503
1400	4	5	0.8744	0.4376	0.1376	2.8217	0.8544	0.4831	0.1461	2.6203
1400	4	10	0.8821	0.4241	0.1263	2.9120	0.8632	0.4681	0.1439	2.7039
1400	4	15	0.8963	0.3976	0.1117	3.1057	0.8870	0.4255	0.1319	2.9748
1400	8	5	0.9257	0.3367	0.1125	3.6676	0.8614	0.4712	0.1437	2.6861
1400	8	10	0.9283	0.3307	0.1083	3.7346	0.8604	0.4729	0.1423	2.6763
1400	8	15	0.9273	0.3330	0.1102	3.7087	0.8626	0.4693	0.1394	2.6973
1400	12	5	0.9705	0.2122	0.0695	5.8193	0.7924	0.5768	0.1739	2.1945
1400	12	10	0.9718	0.2074	0.0654	5.9552	0.7958	0.5720	0.1626	2.2128
1400	12	15	0.9722	0.2060	0.0646	5.9937	0.7930	0.5759	0.1658	2.1978

spectra by using MSC were shown in Table 3. The model built when the number of the spectral attributes randomly selected was 700, the number of principal components was 8, and the number of the built QPLS models was 15 was better than others among the *k*QPLS-SVR models as shown in Table 3. For this model, the R^2 value, SEC, AARD, and RPD of the training set were 0.9227, 0.3434, 0.1127, and 3.5961, respectively, and the R^2 value, SEP, AARD, and RPD of the testing set were 0.8730, 0.4510, 0.1361, and 2.8062, respectively.

When the original near-infrared spectra were preprocessed by using the method SNV, the results of the built *k*QPLS-SVR models were shown in Table 4. Among the *k*QPLS-SVR models as shown in Table 4, the better effects were obtained for the model built when the number of the spectral attributes randomly selected was 1400, the number of principal components was 4, and the number of the built QPLS models was 15. For this model, R^2, SEC, AARD, and RPD of the training set were 0.8963, 0.3976, 0.1117, and 3.1057, respectively, and R^2, SEP, AARD, and RPD of the testing set were 0.8870, 0.4255, 0.1319, and 2.9748, respectively.

The results of the built *k*QPLS-SVR models based on the data obtained when VN was used as the preprocessing method were shown in Table 5. As demonstrated in Table 5, the effects of the model built when the number of the spectral attributes randomly selected was 700, the number of principal components was 8, and the number of the built QPLS models was 15 was better than others. For this *k*QPLS-SVR model, the value of R^2, SEC, AARD, and RPD of the training set were 0.8232, 0.5193, 0.1814, and 2.3782, respectively, and the value of R^2, SEP, AARD, and RPD of the testing set were 0.7964, 0.5711, 0.1833, and 2.2164, respectively.

The results shown in Tables 2–5 indicated that satisfactory effects could be obtained using the *k*QPLS-SVR models built based on the original near-infrared spectra and the spectral data obtained by using MSC and SNV. In contrast, for the built *k*QPLS-SVR models when the original near-infrared spectra were preprocessed by using the method VN, the values of both R^2 and RPD of the training set and the testing set were less. For the optimal *k*QPLS-SVR model built based on the data obtained when SNV was used as the preprocessing method, the number of principal components was 4. It was less than the number of principal components used for building the optimal *k*QPLS-SVR model based on the original near-infrared spectra, the spectral data obtained by using MSC, or the spectral data obtained by using VN. In this case, there may be underfitting problem resulting in the reduction of the prediction ability of the optimal *k*QPLS-SVR model built when SNV was used as the preprocessing method. For the optimal *k*QPLS-SVR model built based on the data obtained when MSC was used as the preprocessing method, the value of R^2 of the testing set was 0.8730. And it was less than the value of R^2 of the testing set resulting from the optimal *k*QPLS-SVR model built based on the original near-infrared spectra. Therefore, the optimal *k*QPLS-SVR model built based on the original near-infrared spectra (the number of the spectral attributes of the randomly selected wavelength points was 700, the number of principal components was 8, and the number of the built QPLS models was 5) was regarded as the optimal *k*QPLS-SVR model to quantitatively determine the relative content of *Pst* DNA in wheat leaves in the incubation period.

TABLE 5: The results of the *k*QPLS-SVR models for quantification of the relative content of *Pst* DNA in wheat leaves in the incubation period built when VN was used as the preprocessing method of the original near-infrared spectra.

The number of attributes	The number of principal components	The number of QPLS models	Training set R^2	Training set SEC	Training set AARD	Training set RPD	Testing set R^2	Testing set SEP	Testing set AARD	Testing set RPD
700	4	5	0.7040	0.6718	0.2424	1.8381	0.5965	0.8040	0.2667	1.5743
700	4	10	0.6885	0.6893	0.2360	1.7916	0.6004	0.8001	0.2704	1.5820
700	4	15	0.7313	0.6401	0.2186	1.9291	0.5987	0.8018	0.2584	1.5786
700	8	5	0.8071	0.5424	0.1939	2.2766	0.7907	0.5790	0.1851	2.1860
700	8	10	0.8247	0.5171	0.1826	2.3881	0.7784	0.5959	0.1967	2.1241
700	8	15	0.8232	0.5193	0.1814	2.3782	0.7964	0.5711	0.1833	2.2164
700	12	5	0.8522	0.4748	0.1718	2.6009	0.7556	0.6257	0.2033	2.0229
700	12	10	0.8513	0.4762	0.1731	2.5934	0.7333	0.6540	0.2163	1.9353
700	12	15	0.8697	0.4458	0.1531	2.7703	0.7475	0.6361	0.2063	1.9899
1400	4	5	0.6789	0.6998	0.2459	1.7647	0.5489	0.8501	0.2706	1.4889
1400	4	10	0.6974	0.6793	0.2350	1.8178	0.5382	0.8601	0.2803	1.4716
1400	4	15	0.6804	0.6982	0.2488	1.7688	0.5324	0.8655	0.2836	1.4624
1400	8	5	0.8220	0.5211	0.1827	2.3699	0.7578	0.6229	0.2001	2.0321
1400	8	10	0.8136	0.5331	0.1902	2.3165	0.7668	0.6112	0.1987	2.0707
1400	8	15	0.8095	0.5389	0.1949	2.2914	0.7625	0.6168	0.2050	2.0520
1400	12	5	0.8565	0.4678	0.1629	2.6399	0.7078	0.6842	0.2295	1.8500
1400	12	10	0.8695	0.4462	0.1526	2.7678	0.7479	0.6355	0.1947	1.9916
1400	12	15	0.8717	0.4423	0.1498	2.7917	0.7234	0.6657	0.2118	1.9015

TABLE 6: The results of the optimal models for quantification of the relative content of *Pst* DNA in wheat leaves in the incubation period built using individual methods including QPLS and SVR.

Modeling methods	Optimal parameters C	Optimal parameters g	Training set R^2	Training set SEC	Training set AARD	Training set RPD	Testing set R^2	Testing set SEP	Testing set AARD	Testing set RPD
QPLS	—	—	0.8534	0.4728	0.1689	2.6117	0.8684	0.4592	0.1467	2.7565
SVR	256	0.1895	0.8887	0.4119	0.1277	2.9981	0.7971	0.5702	0.2110	2.2200

3.3. Results of the QPLS Models and the SVR Model for Quantification of the Relative Content of Pst *DNA in Wheat Leaves in the Incubation Period.* Since the selected optimal *k*QPLS-SVR model to quantitatively determine the relative content of *Pst* DNA in wheat leaves in the incubation period was built based on the original near-infrared spectra, the QPLS models and the SVR model were built with the original near-infrared spectra based on the same training set and testing set as used for building the optimal *k*QPLS-SVR model. When the determination models of the relative content of *Pst* DNA in wheat leaves in the incubation period were built using QPLS, the number of principal components corresponding to the minimum PRESS value was calculated, and it was 8. The value of $F_{\alpha,m,m}$ was calculated using the software IBM SPSS Statistics 19.0 (IBM Corporation, Somers, NY, USA), and it was 1.09. When the number of principal components was 6, the value of $F(f)$ was 1.07. And the value of $F(f)$ was 1.02 when the number of principal components was 7. The two values of $F(f)$ were less than the value of $F_{\alpha,m,m}$. So the optimal number of principal components was set as 6 for building the QPLS model to quantitatively determine the relative content of *Pst* DNA in wheat leaves in the incubation period. As shown in Table 6, for the optimal QPLS model, the value of R^2, SEC, AARD, and RPD of the training set were 0.8534, 0.4728, 0.1689, and 2.6117, respectively, and the value of R^2, SEP, AARD, and RPD of the testing set were 0.8684, 0.4592, 0.1467, and 2.7565, respectively. When $C = 256$ and $g = 0.1895$, the optimal SVR model for quantification of the relative content of *Pst* DNA in wheat leaves in the incubation period was obtained. As shown in Table 6, for this SVR model, the value of R^2, SEC, AARD, and RPD of the training set were 0.8887, 0.4119, 0.1277, and 2.9981, respectively, and the value of R^2, SEP, AARD, and RPD of the testing set were 0.7971, 0.5702, 0.2110, and 2.2200, respectively.

A comparison of the effects of the three models, including the optimal *k*QPLS-SVR model, the optimal QPLS model, and the optimal SVR model, was conducted according to R^2, SEC, AARD, and RPD of the training set and R^2, SEP, AARD, and RPD of the testing set. The results showed that the effects of the optimal *k*QPLS-SVR model built based on the original near-infrared spectra (the number of the spectral attributes of the randomly selected wavelength points was 700, the number of principal components was 8, and the number of the built QPLS models was 5) were the best. Therefore, this optimal *k*QPLS-SVR model was selected as the optimal model to quantitatively determine the relative content of *Pst* DNA in wheat leaves in the incubation period.

4. Discussion

For the *k*QPLS-SVR models, the optimal QPLS model, and the optimal SVR model as shown in Tables 2–6, all the values of R^2 of the training set and the testing set were more than 0.5. The results indicated that the correlation between the features of near-infrared spectra (absorbances) and the relative content of *Pst* DNA in wheat leaves in the incubation period was relatively high and that the built models could be used for rough screening. It was indicated that quantitative determination of the relative content of *Pst* DNA in wheat leaves in the incubation period using NIRS technology is feasible. After a comprehensive comparison of the determination results of the models as shown in Tables 2–6, the *k*QPLS-SVR model built with a ratio of the training set to the testing set equal to 3 : 1 based on the original near-infrared spectra in a range of 4000–12,000 cm^{-1} when the number of the spectral attributes of the randomly selected wavelength points was 700, the number of principal components was 8, and the number of the built QPLS models was 5 was regarded as the optimal model to quantitatively determine the relative content of *Pst* DNA in wheat leaves in the incubation period. The satisfactory effects were obtained using this optimal model. For this model, the value of R^2, SEC, AARD, and RPD of the training set were 0.9027, 0.3852, 0.1257, and 3.2057, respectively, and the value of R^2, SEP, AARD, and RPD of the testing set were 0.8820, 0.4349, 0.1274, and 2.9107, respectively. The results indicated that the proposed method based on NIRS could be used as a new method for rapid, nondestructive, and quantitative detection of *Pst* DNA content in wheat leaves. Meanwhile, a reference was provided for nondestructive determination of the DNA contents of other kinds of pathogens in plant hosts.

Changes of the near-infrared spectra of wheat leaves infected with *Pst* during the incubation period may be induced by many factors. The infection of *Pst* and subsequent expansion in plant hosts are dynamic change processes [1, 5]. After germination of a *Pst* urediospore, a growing germ tube emerges. Then an appressorium is formed at the tip of the germ tube. An infection peg grows from the appressorium and then enters the wheat leaf. During the incubation period, with the continuous spread of hyphae in the infected leaf tissues, changes of the photosynthesis, respiration, and transpiration of wheat leaves occur, affecting the synthesis and decomposition of organic substances [1, 37–39]. Thus, the near-infrared spectra of the infected wheat leaf are influenced. Especially in the middle and later stage of the incubation period, a large quantity of *Pst* hyphae accumulate in the infected leaf tissues, the latent lesions on the infected leaves become more and more obvious, and the effects on various physiological indexes of wheat are increasing [1], more directly affecting the near-infrared spectra of the infected wheat leaves. In this study, according to the changes of the near-infrared spectra of the infected wheat leaves during the incubation period, the detection of the relative content of *Pst* DNA in the latently *Pst*-infected wheat leaves was performed using NIRS technology in combination with the duplex TaqMan real-time PCR arrays, providing a basis for modeling the quantitative relationship between the relative content of *Pst* DNA in the latently *Pst*-infected wheat leaves and the corresponding near-infrared spectral data.

The quantitative detection results obtained by using the duplex TaqMan real-time PCR arrays in this study showed that the quantity of *Pst* DNA exponentially increased as *Pst* extended in wheat leaf during the incubation period of wheat stripe rust, and it was consistent with the growth trend of *Pst* DNA reported by Pan et al. [10]. The results in this study showed that the relative contents of *Pst* DNA in wheat leaves after inoculation continuously increased. The relative contents of *Pst* DNA in the latently *Pst*-infected wheat leaves in the incubation period were more than 20% on 7 days post inoculation (dpi) and reached to approximately 90% on 10 dpi. It was demonstrated that there was a particularly obvious increase of the relative contents of *Pst* DNA in the latently *Pst*-infected wheat leaves from the seventh day to the tenth day during the incubation period. This indicated that it would have great significance to perform early detection and control of wheat stripe rust.

Wheat stripe rust is a kind of air-borne disease. As disease symptoms appear on wheat leaves, a large number of urediospores will be produced and serve as inocula for the further spread and epidemic of the disease [1, 3–5]. Accurate quantitative determination of the latent infection levels of wheat leaves by *Pst* can provide important information for estimating potential inoculum level. Early detection of *Pst* infection, early disease warning, and early disease control measure-making can effectively reduce the amount of inoculum sources and the dosage of pesticides. The results of this study demonstrated that latent infection of wheat seedlings caused by *Pst* could be detected as early as 24 h after inoculation based on NIRS. For the first time, dynamic changes of the amounts of wheat stripe rust pathogen in the infected leaves during the incubation period were investigated by using NIRS technology in combination with a real-time PCR method. A modeling method integrated with QPLS and SVR was used to establish the dynamic quantitative detection model of wheat stripe rust pathogen during the incubation period. A reference was provided for prediction of disease epidemic trend and early control of wheat stripe rust in this study.

5. Conclusions

It is critical to implement early quantitative detection of *Pst* and disease prediction accurately for the prevention and control of wheat stripe rust. In this study, a method based on NIRS to implement the quantitative determination of the relative content of *Pst* DNA in wheat leaves in the incubation period was investigated. Based on the data on the relative content of *Pst* DNA in wheat leaves in the incubation period obtained using the duplex TaqMan real-time PCR arrays and the corresponding near-infrared spectral data, the optimal model integrated with QPLS and SVR was obtained to quantify the relative content of *Pst* DNA in wheat leaves in the incubation period. Using this model, satisfactory results were achieved. The results indicated that rapid, nondestructive, and quantitative detection of the amount of *Pst* in the infected leaves during the incubation period could be realized using NIRS technology. A novel method for quantitatively estimating the latent infection levels of *Pst* in wheat leaves and early detection of wheat stripe rust was provided in this study. Furthermore, this proposed method can be used as a reference to establish models for determining the latent amounts of other kinds of pathogens in plant hosts and implement early nondestructive detection of the diseases caused by these agents. It is very helpful in getting the information about disease prevalence and inoculum amount in the field as early as possible, inhibiting pathogen accumulation, reducing the quantity of propagules, performing early precise site-specific control, disease forecasting, and making macro strategy for disease management. And it is also conducive to reducing the dosage of pesticides and increasing efficiency of control measures in plant disease management. Moreover, in this study, some basis was provided for the development of portable near-infrared spectrometers and sensors for quantitative detection of plant diseases.

Conflicts of Interest

The authors declare that there is no conflict of interests regarding the publication of this paper.

Authors' Contributions

Yaqiong Zhao, Yilin Gu, and Feng Qin contributed equally to this paper.

Acknowledgments

This study was supported by the National Key Basic Research Program of China (2013CB127700), National Natural Science Foundation of China (31471726), and National Key Technologies Research and Development Program of China (2012BAD19BA04).

References

[1] Z. Q. Li and S. M. Zeng, *Wheat Rust in China*, China Agriculture Press, Beijing, China, 2002.

[2] R. F. Line, "Stripe rust of wheat and barley in North America: a retrospective historical review," *Annual Review of Phytopathology*, vol. 40, pp. 75–118, 2002.

[3] X. M. Chen, "Epidemiology and control of stripe rust [*Puccinia striiformis* f. sp. *tritici*] on wheat," *Canadian Journal of Plant Pathology*, vol. 27, no. 3, pp. 314–337, 2005.

[4] A. M. Wan, X. M. Chen, and Z. H. He, "Wheat stripe rust in China," *Australian Journal of Agricultural Research*, vol. 58, no. 6, pp. 605–619, 2007.

[5] W. Chen, C. Wellings, X. Chen, Z. Kang, and T. Liu, "Wheat stripe (yellow) rust caused by *Puccinia striiformis* f. sp. *tritici*," *Molecular Plant Pathology*, vol. 15, no. 5, pp. 433–446, 2014.

[6] A. M. Wan, Z. H. Zhao, X. M. Chen et al., "Wheat stripe rust epidemic and virulence of *Puccinia striiformis* f. sp. *tritici* in China in 2002," *Plant Disease*, vol. 88, no. 8, pp. 896–904, 2004.

[7] J. Zhao, X. J. Wang, C. Q. Chen, L. L. Huang, and Z. S. Kang, "A PCR-based assay for detection of *Puccinia striiformis* f. sp. *tritici* in wheat," *Plant Disease*, vol. 91, no. 12, pp. 1669–1674, 2007.

[8] X. J. Wang, C. L. Tang, J. L. Chen et al., "Detection of *Puccinia striiformis* in latently infected wheat leaves by nested polymerase chain reaction," *Journal of Phytopathology*, vol. 157, no. 7-8, pp. 490–493, 2009.

[9] C. Huang, Z. Y. Sun, J. H. Yan, Y. Luo, H. G. Wang, and Z. H. Ma, "Rapid and precise detection of latent infections of wheat stripe rust in wheat leaves using loop-mediated isothermal amplification," *Journal of Phytopathology*, vol. 159, no. 7-8, pp. 582–584, 2011.

[10] J. J. Pan, Y. Luo, C. Huang et al., "Quantification of latent infections of wheat stripe rust by using real-time PCR," *Acta Phytopathologica Sinica*, vol. 40, no. 4, pp. 504–510, 2010.

[11] D. Moshou, C. Bravo, J. West, S. Wahlen, A. McCartney, and H. Ramon, "Automatic detection of 'yellow rust' in wheat using reflectance measurements and neural networks," *Computers and Electronics in Agriculture*, vol. 44, no. 3, pp. 173–188, 2004.

[12] R. Devadas, D. W. Lamb, S. Simpfendorfer, and D. Backhouse, "Evaluating ten spectral vegetation indices for identifying rust infection in individual wheat leaves," *Precision Agriculture*, vol. 10, no. 6, pp. 459–470, 2009.

[13] L. Yuan, Y. B. Huang, R. W. Loraamm, C. W. Nie, J. H. Wang, and J. C. Zhang, "Spectral analysis of winter wheat leaves for detection and differentiation of diseases and insects," *Field Crops Research*, vol. 156, pp. 199–207, 2014.

[14] J. L. Zhao, L. S. Huang, W. J. Huang et al., "Hyperspectral measurements of severity of stripe rust on individual wheat leaves," *European Journal of Plant Pathology*, vol. 139, no. 2, pp. 407–417, 2014.

[15] X. L. Li, K. Wang, Z. H. Ma, and H. G. Wang, "Early detection of wheat disease based on thermal infrared imaging," *Transactions of the Chinese Society of Agricultural Engineering*, vol. 30, no. 18, pp. 183–189, 2014.

[16] X. L. Li, Z. H. Ma, L. L. Zhao, J. H. Li, and H. G. Wang, "Early diagnosis of wheat stripe rust and wheat leaf rust using near infrared spectroscopy," *Spectroscopy and Spectral Analysis*, vol. 33, no. 10, pp. 2661–2665, 2013.

[17] X. L. Li, F. Qin, L. L. Zhao, J. H. Li, Z. H. Ma, and H. G. Wang, "Detection of *Puccinia striiformis* f. sp. *tritici* latent infections in wheat leaves using near infrared spectroscopy technology," *Spectroscopy and Spectral Analysis*, vol. 34, no. 7, pp. 1853–1858, 2014.

[18] J. Li, Y. H. Chen, J. B. Jiang, and H. C. Cai, "Using hyperspectral derivative index to identify winter wheat stripe rust disease," *Science & Technology Review*, vol. 25, no. 6, pp. 23–26, 2007.

[19] Y. Pan, Y. L. Gu, Y. Luo et al., "Study on relationship between the quantity of *Puccinia striiformis* f. sp. *tritici* in latent during overwinter and the disease index in Xiangyang area," *Acta Phytopathologica Sinica*, vol. 46, no. 5, pp. 679–685, 2016.

[20] N. W. Schaad and R. D. Frederick, "Real-time PCR and its application for rapid plant disease diagnostics," *Canadian Journal of Plant Pathology*, vol. 24, no. 3, pp. 250–258, 2002.

[21] S. Sankaran, A. Mishra, R. Ehsani, and C. Davis, "A review of advanced techniques for detecting plant diseases," *Computers and Electronics in Agriculture*, vol. 72, no. 1, pp. 1–13, 2010.

[22] G. T. Xu, H. F. Yuan, and W. Z. Lu, "Development of modern near infrared spectroscopic techniques and its applications," *Spectroscopy and Spectral Analysis*, vol. 20, no. 2, pp. 134–142, 2000.

[23] X. L. Chu and W. Z. Lu, "Research and application progress of near infrared spectroscopy analytical technology in China in the past five years," *Spectroscopy and Spectral Analysis*, vol. 34, no. 10, pp. 2595–2605, 2014.

[24] Q. S. Chen, J. W. Zhao, M. H. Liu, J. R. Cai, and J. H. Liu, "Determination of total polyphenols content in green tea using FT-NIR spectroscopy and different PLS algorithms," *Journal of Pharmaceutical and Biomedical Analysis*, vol. 46, no. 3, pp. 568–573, 2008.

[25] M. K. Ahmed and J. Levenson, "Application of near-infrared spectroscopy to the quality assurance of ethanol and butanol bended gasoline," *Petroleum Science and Technology*, vol. 30, no. 2, pp. 115–121, 2012.

[26] E. Tamburini, G. Ferrari, M. G. Marchetti, P. Pedrini, and S. Ferro, "Development of FT-NIR models for the simultaneous estimation of chlorophyll and nitrogen content in fresh apple (*Malus domestica*) leaves," *Sensors*, vol. 15, no. 2, pp. 2662–2679, 2015.

[27] H. Chen, Z. Lin, H. G. Wu, L. Wang, T. Wu, and C. Tan, "Diagnosis of colorectal cancer by near-infrared optical fiber spectroscopy and random forest," *Spectrochimica Acta Part A: Molecular and Biomolecular Spectroscopy*, vol. 135, pp. 185–191, 2015.

[28] X. L. Li, F. Qin, L. L. Zhao, J. H. Li, Z. H. Ma, and H. G. Wang, "Identification and classification of disease severity of wheat stripe rust using near infrared spectroscopy technology," *Spectroscopy and Spectral Analysis*, vol. 35, no. 2, pp. 367–371, 2015.

[29] X. L. Li, Z. H. Ma, L. L. Zhao, J. H. Li, and H. G. Wang, "Application of near infrared spectroscopy to qualitative identification and quantitative determination of *Puccinia striiformis* f. sp. *tritici* and *P. recondita* f. sp. *tritici*," *Spectroscopy and Spectral Analysis*, vol. 34, no. 3, pp. 643–647, 2014.

[30] P. Cheng, Z. H. Ma, X. J. Wang et al., "Impact of UV-B radiation on aspects of germination and epidemiological components of three major physiological races of *Puccinia striiformis* f. sp. *tritici*," *Crop Protection*, vol. 65, pp. 6–14, 2014.

[31] Y. Y. Xiao and S. M. Zeng, "Comparisons among three equations to predict apparent infection rates of wheat stripe rust," *Scientia Sinica (Series B)*, vol. 2, pp. 151–157, 1985.

[32] A. F. Justesen, C. J. Ridout, and M. S. Hovmoller, "The recent history of *Puccinia striiformis* f. sp. *tritici* in Denmark as revealed by disease incidence and AFLP markers," *Plant Pathology*, vol. 51, no. 1, pp. 13–23, 2002.

[33] M. Sandberg, L. Lundberg, M. Ferm, and I. M. Yman, "Real time PCR for the detection and discrimination of cereal contamination in gluten free foods," *European Food Research and Technology*, vol. 217, no. 4, pp. 344–349, 2003.

[34] Y. Li, Y. L. Gu, B. M. Wu et al., "Establishment of a duplex *Taq*Man real-time PCR method for quantifying *Puccinia striiformis* f. sp. *tritici* and *Blumeria graminis* f. sp. *tritici*," *Acta Phytopathologica Sinica*, vol. 45, no. 2, pp. 205–210, 2015.

[35] A. De Girolamo, S. Cervellieri, A. Visconti, and M. Pascale, "Rapid analysis of deoxynivalenol in durum wheat by FT-NIR spectroscopy," *Toxins*, vol. 6, no. 11, pp. 3129–3143, 2014.

[36] W. Z. Lu, H. F. Yuan, G. T. Xu, and D. M. Qiang, *Modern Near Infrared Spectroscopy Analytical Technology*, China Petrochemical Press, Beijing, China, 2000.

[37] R. C. G. Smith, A. D. Heritage, M. Stapper, and H. D. Barrs, "Effect of stripe rust (*Puccinia striiformis* West.) and irrigation on the yield and foliage temperature of wheat," *Field Crops Research*, vol. 14, no. 1, pp. 39–51, 1986.

[38] Y. R. Li and H. S. Shang, "Effect of stripe rust infection on photosynthesis and transpiration of wheat," *Journal of Triticeae Crops*, vol. 21, no. 2, pp. 51–56, 2001.

[39] H. Zhao, R. Y. Wang, P. L. Ma, and J. L. Zhu, "Effect of stripe rust infection on photosynthesis and transpiration of wheat in the semi-arid region," *Arid Meteorology*, vol. 22, no. 4, pp. 56–59, 2004.

22

Spatial Spectroscopy Approach for Detection of Internal Defect of Component without Zero-Position Sensors

Qizhou Wu, Yong Jin, Zhaoba Wang, and Zhaoqian Xiao

National Key Lab for Electronic Measurement Technology, North University of China, Taiyuan, Shanxi 030051, China

Correspondence should be addressed to Qizhou Wu; wqzpaper@126.com

Academic Editor: Yu Shang

Conventional approach to detect the internal defect of a component needs sensors to mark the "zero" positions, which is time-consuming and lowers down the detecting efficiency. In this study, we proposed a novelty approach that uses spatial spectroscopy to detect internal defect of objects without zero-position sensors. Specifically, the spatial variation wave of distance between the detecting source and object surface is analyzed, from which a periodical cycle is determined with the correlative approaches. Additionally, a wavelet method is adopted to reduce the noise of the periodic distance signal. This approach is validated by the ultrasound detection of a component with round cross section and elliptical shape in axis. The experimental results demonstrate that this approach greatly saves the time spent on the judgment of a complete cycle and improves the detecting efficiency of internal defect in the component. The approach can be expanded to other physical methods for noninvasive detection of internal defect, such as optical spectroscopy or X-ray scanning, and it can be used for hybrid medium, such as biological tissues.

1. Introduction

The components with large change in the surface shape are often found in both industrial and biological fields. Ultrasonic detection of the internal defects in such components is of importance and becoming a hot topic among researchers [1]. In the process of detection, the ultrasonic probe is required to be kept in an invariable and certain distance from the surface of the component. Moreover, the incident angle of the probe should be kept perpendicular to the surface of the component. Traditionally, the component needs a whole scanning in a manner of constant speed and across multisections. In each scanning cycle, the triggering signals of "zero" position are detected by a sensor and taken as the beginning and ending signals of scanning. Then, the scanning on the whole component is made with fixed intervals. Once the scanning in each cycle is completed, the probe is moved at a fixed distance along the axial direction and waits again to receive the triggering signals from the "zero-" position sensor. The time spent for waiting generally is half of that spent in the scanning in one cycle. Therefore, the efficiency of the detection is very low.

In this paper, we proposed a novelty approach with correlative theory, called spatial spectroscopy, to continuously scan the component without waiting for triggering signals from "zero-" position sensor. Here the "spatial spectroscopy" indicates that the periodical signals of ultrasound backscatter, with the spatial change of the component, are used to detect the internal detects.

Additionally, we combined this approach with self-adaptive filter technology for improvement of the detection efficiency.

The self-adaptive filter is a signal processing technology that was originated since 1960s, and many researchers have contributed to this technology. For example, Xu et al. applied self-adaptive filter to the field of Internet engineering [2]. Chen et al. combined self-adaptive filter with neural network for use in image processing [3]. Feng and Lin utilized the self-adaptive filter to enhance the signal intensity and elevate the signal-to-noise ratio for A/D conversion [4]. Zhong et al. utilized self-adaptive filter to improve the searching engine in GPS (Global Positioning System) [5]. Recently, self-adaptive filter was introduced into ultrasonic detection of internal defects in components, with purpose of improving

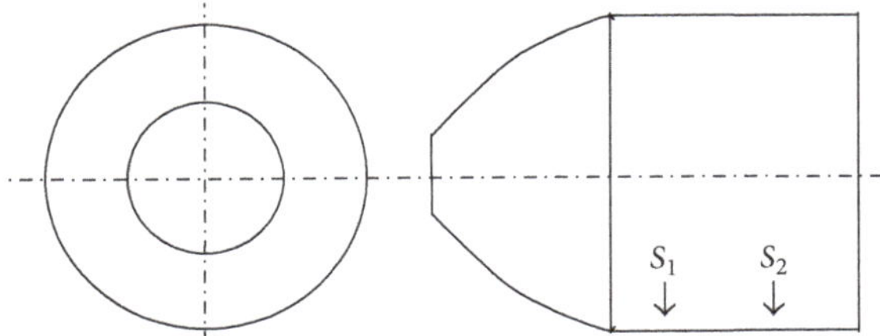

FIGURE 1: Structure of the component with round cross section and elliptical shape in axis.

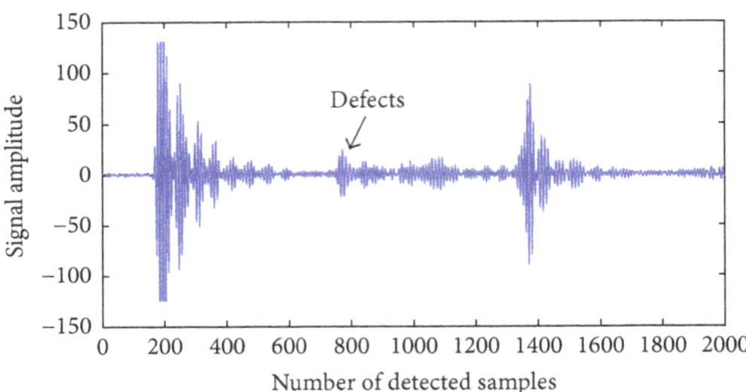

FIGURE 2: The ultrasound backscatter signals received from the surface of component.

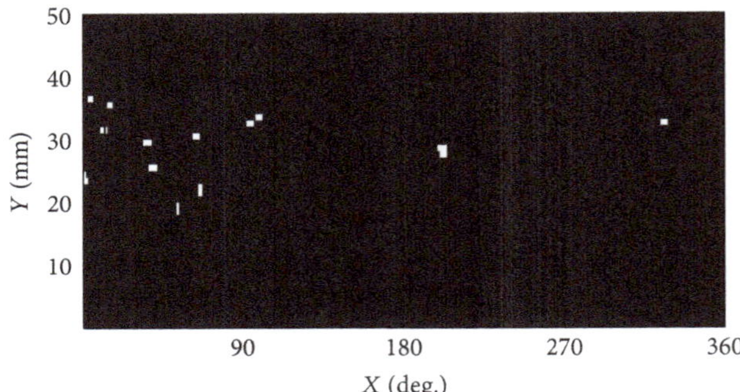

FIGURE 3: Results of the whole scanning on the component.

the signal-to-noise ratio. The aim of using self-adaptive filter in this study is to smooth the ultrasonic signals before using spatial spectroscopy approach.

The correlative theory is more often utilized for signal matching. For example, Ji et al. and Ge et al. proposed methods that utilized correlative theory to extract the true signals from the data with weak signal-to-noise ratio [6, 7]. Wang and Ruan proposed an image-matching algorithm based on correlative theory [8]. Moreover, they proposed a pyramid-layered model to reduce the calculation loads due to correlation of images and improve the precision of image-matching. Lei et al. proposed a recognition method for internal defects based on correlative theory, which has become a standard for online or offline detection of internal defects during structure fluctuation [9].

To overcome the low detection efficiency of "zero-" position sensor method, we realized the fast localization of "zero" position by using the feature of high correlation in periodic signals. Through preprocessing by the self-adaptive filter, the ultrasonic signals exhibit excellent correlations, thus allowing for fast scanning of the component without "zero" sensor.

2. Methods

2.1. Analysis on the Inspection and Measurement on the Component. The component structure and defect positions (S_1 and S_2) are shown in Figure 1. It has round cross section and elliptical shape in axis, 50 mm length and 101 mm diameter. The length and width of the defect are 18 mm and 1 mm, respectively. The distance between the two defect positions is 29 mm. The ultrasound backscatter signals received from the surface of component are depicted in Figure 2.

The approach for scanning without zero-position sensors is illustrated as follows. First, the scanning on the component with constant speed (3.14 rad/s in counterclockwise) started immediately after the ultrasonic probe is moved to the cross sections of the component, and the position of the zero point in each detection cross section can be any point in the cross section. When scanning on a certain component with an elongated defect without zero sensors the same defect can be detected in different angles on different cross sections, which directly affects the judgment on the defect, as depicted in Figure 3.

Because the detection target is a blank component, the time for each cycle is fixed when the component is being rotated with constant speed, and within this cycle the changes of the distance between the probe and the surface of the component can be detected, as shown in Figure 4.

After the scanning (rotate), the periodic signals in each detection cross section are extracted, covering the range of distance between the probes over the cross sections and the surface of the component. The cycles covering any cross sections are the same, if taking the periodic zero point in the first cross section as the circumference zero point. Based on this characteristic, a correlative judgment on the periodic zero points in all the cross sections and the zero point in the first detection cross section is made through the method of signal process, and matching zeros is implemented to make a rectification on the detection result.

2.2. Analyses on the Periodic Signal. The curve in Figure 4 represents the distance signal of 3 rotating cycles in a certain detected cross section, which is obtained with measurement of the distance between the probe and the surface of a certain component.

Affected by the on-engine high-frequency vibration resulting from the motor drive and other factors, there exists a kind of high-frequency noise in the periodic signal. Moreover, there may be some miniature saddle-backing and subsiding because the detected object is the blank component; in order to assure the similarity of the periodic signal in each cross section, the periodic signal obtained at the position of those saddle-backing and subsiding should be processed

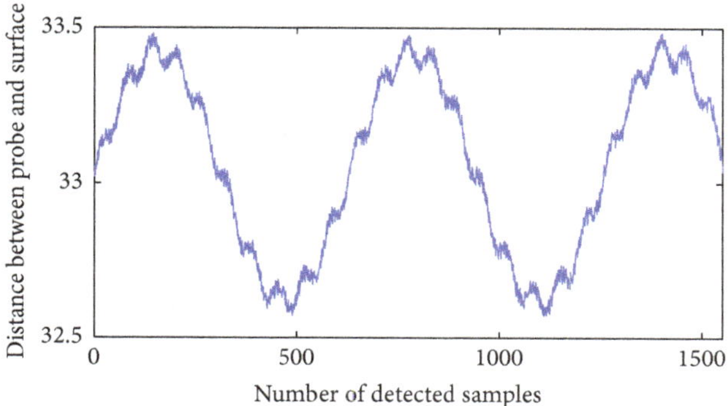

FIGURE 4: The relationship between the distance (from probe to surface, mm) and number of the detected samples.

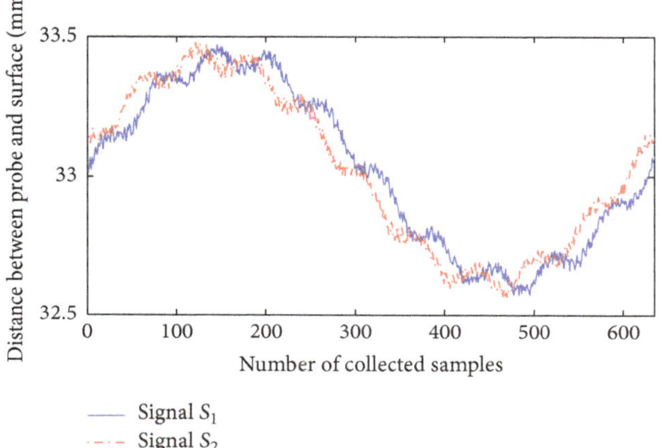

FIGURE 5: Distance (mm) signals S_1, S_2 on the scanning cross sections.

as the frequency noise. Therefore, the noise involved in the signals should be reduced before matching zeros, so as to extract the periodic signals. Figure 5 is the sketch map for the distance signals S_1 and S_2 of two different cross sections obtained in the scanning on a certain component.

2.3. Reduction of the Noise Involved in Signals. During the scanning, the signal of distance between the probe and the surface of the component is treated as stable low-frequency signal, and the high-frequency part in the signal is the noise. As such, the influence of the high-frequency part in the signal should be eliminated before signal analysis is performed. Because the noise with different frequency components needs to be filtered out, a self-adapting noise reduction method with multilevel and multiresolution should be used [10–14]. Here wavelet analysis is used for this purpose [15]. The following are steps of the method for reducing the noise involved in signals.

(1) Select a wavelet and analyze the number of decomposition. Wavelet decomposition is made on the signals involving noise to obtain the low-frequency and high-frequency coefficients.

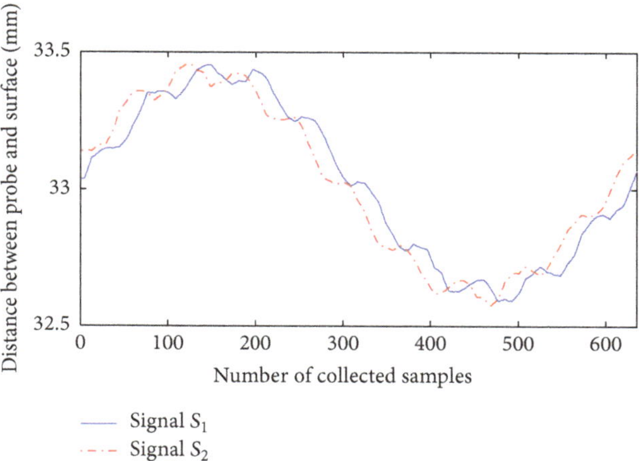

FIGURE 6: Distance signals S_1 and S_2 after the noise reduction.

(2) A suitable method is adopted to make an analysis on all the high-frequency coefficients, so as to remove the noise with frequency components.

(3) According to the levels number of decomposition on the wavelet, a reconstruction of the discrete wavelet is made on the processed high-frequency and low-frequency coefficients. The acquired signals are those with noise reduction.

In addition, according to the selective analysis on the wavelet base and the levels number of decomposition on wavelet described in the reference, the wavelet db5 is used to make decomposition with 5 levels on the distance signals S_1 and S_2.

Furthermore, the method of HeurSure threshold value is adopted to make a soft threshold selection on the high-frequency coefficients in the different levels and filter out the noise with high-frequency, and a wavelet reconstruction is made on the filtered constituents to get the distance signals S_1 and S_2 after noise reduction with the method of multiresolution analysis, as indicated in Figure 6.

2.4. Matching Zeros. In the process of scanning, the signals obtained after the noise reduction on the collected distance signals are the distance signals in one cycle. In other words, the detected values are the measured distance in a single-cycle within one cross section of the component. Therefore, the inertial point in the cycle needs to be found so as to perform the correlative judgment on the zero points in the detected component and realize the matching zeros.

The mutually correlative function of the stochastic signals is used to describe the interdependent relationship between the two sample functions at the different time [16–19]. This function is used to describe the close mutual interdependent relationship between the waveforms of the two stochastic signals with the movement of the time coordinates.

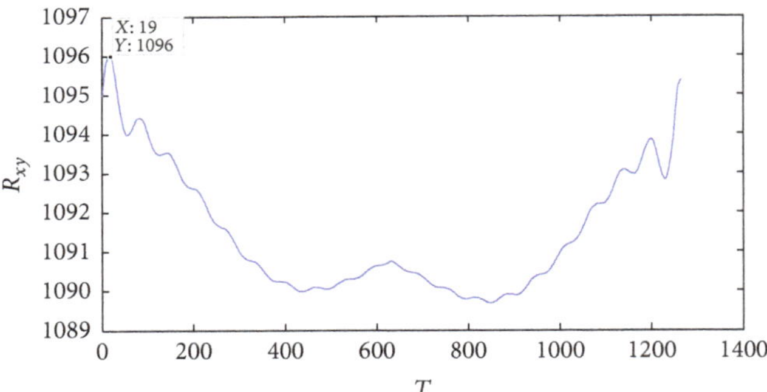

FIGURE 7: Cross correlative function R_{xy} of S_1 and S_2.

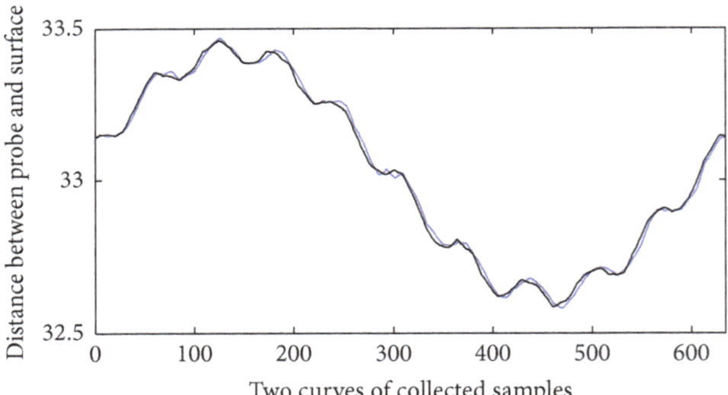

FIGURE 8: Comparison between S_1 (blue) and S_2 (red) in component distance.

The mutual correlative function for the discrete stochastic signals is expressed as follows:

$$R_{xy}(\tau) = \frac{1}{N} \sum_{n=0}^{N-1} x(n)\, y(n+\tau). \qquad (1)$$

Here, $x(n)$ and $y(n)$ are the two signals to be analyzed. $R_{xy}(\tau)$ is the cross-correlation function of the two signals.

Generally, the objects studied in the mutual correlative functions $x(t)$ and $y(t)$ are two different signals; the cross correlative function is neither an even function nor an odd function. As shown in Figure 6, the curves S_1 and S_2 are the two discrete stochastic signals. Therefore, S_1 and S_2 can be defined as $x(t)$ and $y(t)$, respectively, and the method of cross correlative function can be used to localize the corresponding position of the referential inertial point (the detected inertial point of the curve for the distance in the first cycle) to eliminate the errors.

As shown in Figure 7, when R_{xy} is the largest one, the correlation between S_1 and S_2 is ideal; in other words, the two curves are most close. Consequently, $R_{xy}(19, 1096)$ in Figure 7 is the maximal, and the cycle ending point can be achieved with S_2 being moved with detection points $(19, 1096)$ along the time domain. The comparison between S_1 and S_2 in component distance is shown in Figure 8.

This method was applied to the matching zeros of the curve for the distance of the detected cross sections shown in Figure 3. Then, a spatial spectroscopy scanning can be acquired, as shown in Figure 9. Consequently, the internal defects were detected within much shorter time.

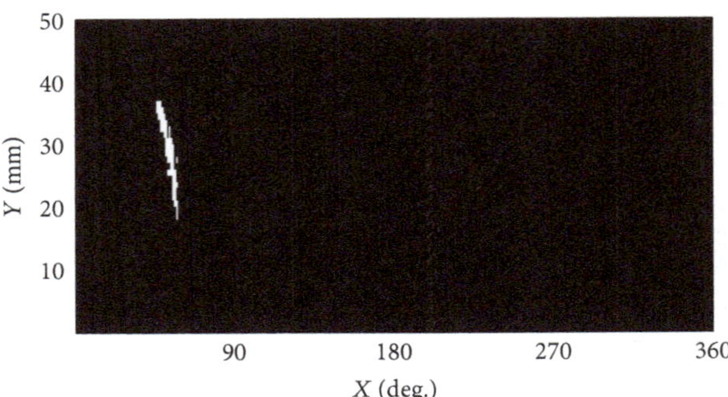

FIGURE 9: Spatial spectroscopy scanning of the component with internal defects.

3. Discussion and Conclusions

In order to solve the problem of low efficiency in the process of the scanning on the component with complex configuration, we proposed in this paper an approach of matching zeros to detect cross sections, through using the fixed cycle caused by the ellipticity of the component in the process of the scanning with constant speed. Then, a noise reduction method is applied to the periodic signals. Additionally, the method of mutual correlation is adopted to adjust the whole scanning detected without zero sensors. Experimental results show that the detection efficiency of this method is improved by 35% compared with that of the conventional approach with zero-position sensor. The approach can be expanded to other physical methods for noninvasive detection of internal defect, such as optical spectroscopy or X-ray scanning, and for hybrid medium, such as biological tissues.

This approach has been utilized to detect the defects in components with various shapes. To emphasize the approach rather than the applications, this paper only presents a representative component. The results from other components are similar to the representative one.

The ultrasound backscatter signals are more easily distinguished on the coarse surface than the smooth one when the component is scanned; thus the proposed method is limited for use to detect defects in the components with smooth surface. More advanced methods of pattern recognition will be adopted in the future for improvement of the proposed spatial spectroscopy approach.

To conclude, we proposed a combination of self-adaptive filer and spatial spectroscopy in this study, so as to speed up the scanning component with large change in the surface shape. With this combination, we realized fast detection of internal defect in components.

Competing Interests

The authors declare that they have no competing interests.

References

[1] L. Qi, Z. Wang, Y. Jin, and Z. Ding, "The study of ultrasonic testing method on claviform feedstock of rocket shell body," *Journal of Projectiles, Rockets, Missiles and Guidance*, vol. 29, no. 4, pp. 269–272, 2009.

[2] B. Xu, X. Yi et al., "Self-adaptive learning based immune algorithm," *Journal of Central South University*, vol. 19, no. 4, pp. 1021–1031, 2012.

[3] X. Chen, H. Hu, J. Zhang, and Q. Zhou, "An ECT system based on improved RBF network and adaptive wavelet image enhancement for solid/gas two-phase flow," *Chinese Journal of Chemical Engineering*, vol. 20, no. 2, pp. 359–367, 2012.

[4] H. Feng and Z. Lin, "Design of adaptive sigma-delta A/D converter," *High Technology Letters*, vol. 11, no. 4, pp. 367–370, 2005.

[5] J. Zhong, Y.-J. Shen, M.-J. Zhao, and L.-Y. Li, "Design of acquisition algorithm based on delay-finding detector for indoor GNSS signals," *Journal of Harbin Institute of Technology*, vol. 19, no. 5, pp. 6–16, 2012.

[6] X. Ge, D. Luo, and Y. Cao, "Application of correlation function in digial signal processing," *Electronics Optics & Control*, vol. 13, no. 6, pp. 78–80, 2006.

[7] L. Ji and H. Zhang, "Study on signal detections based on cross-correlation theory," *Journal of Jilin Teachers Institute of Engineering and Technology*, no. 6, pp. 39–41, 2004.

[8] Y.-S. Wang and Q.-Q. Ruan, "Application of image orientation and matching algorithm based on correlative matching method," *Journal of Northern Jiaotong University*, vol. 26, no. 2, pp. 20–24, 2002.

[9] J. Lei, Q. Yao, Y. Lei, and Z. Liu, "Structural damage detection method based on correlation function analysis of vibration measurement data," *Journal of Vibration and Shock*, vol. 30, no. 8, pp. 221–224, 2011.

[10] A. A. Cavalini, F. S. Lobato, E. H. Koroishi, and V. Steffen, "Model updating of a rotating machine using the self-adaptive differential evolution algorithm," *Inverse Problems in Science and Engineering*, vol. 24, no. 3, pp. 504–523, 2015.

[11] S. Ghosh, N. Senroy, S. Mishra, and S. Kamalasadan, "Fast power system stabilizer tuning in large power systems," in *Proceedings of the IEEE Power & Energy Society General Meeting*, pp. 1-5, Denver, Colo, USA, July 2015.

[12] S. K. Podilchak, J. Shaker, R. Chaharmir, and Y. Antar, "Microwave waveguide transitions using planar anisotropic image guides," in *Proceedings of the International Conference on Electromagnetics in Advanced Applications (ICEAA '15)*, pp. 1416–1418, IEEE, Torino, Italy, September 2015.

[13] N. Bai and B. Lü, "A 200 mV low leakage current subthreshold SRAM bitcell in a 130 nm CMOS process," *Journal of Semiconductors*, vol. 33, no. 6, Article ID 065008, pp. 95–100, 2012.

[14] J. Zhong, Y.-J. Shen, M.-J. Zhao, and L.-Y. Li, "Design of acquisition algorithm based on delay-finding detector for indoor GNSS signals," *Journal of Harbin Institute of Technology*, vol. 19, no. 5, pp. 7–16, 2012.

[15] Y. Deng, *Study of threshold de-noising algorithm of speech signal based on wavelet transformation [M.S. dissertation]*, Chongqing University, 2009.

[16] J. Zhou, J. Wang, H. Yan, S. Li, and G. Gui, "Multiple-response optimization for melting process of aluminum melting furnace based on response surface methodology with desirability function," *Journal of Central South University*, vol. 19, no. 10, pp. 2875–2885, 2012.

[17] H. Azizpour, R. Sotudeh-Gharebagh, R. Zarghami, and N. Mostoufi, "Vibration time series analysis of bubbling and turbulent fluidization," *Particuology*, vol. 10, no. 3, pp. 292–297, 2012.

[18] G.-G. Meng, T. Takemoto, and H. Nishikawa, "Correlations between IMC thickness and three factors in Sn-3Ag-0.5Cu alloy system," *Transactions of Nonferrous Metals Society of China*, vol. 17, no. 4, pp. 686–690, 2007.

[19] X. Li, J. Song, H. Yuan et al., "Forms and functions of inorganic carbon in the Jiaozhou Bay sediments," *Acta Oceanologica Sinica*, vol. 28, no. 6, pp. 30–41, 2009.

23

Solid-State FTIR Spectroscopic Study of Two Binary Mixtures: Cefepime-Metronidazole and Cefoperazone-Sulbactam

Hassan Refat H. Ali,[1] Ramadan Ali,[2] Hany A. Batakoushy,[2] and Sayed M. Derayea[3]

[1]*Department of Pharmaceutical Analytical Chemistry, Faculty of Pharmacy, Assiut University, Assiut 71526, Egypt*
[2]*Department of Analytical Chemistry, Faculty of Pharmacy, Al-Azhar University, Assiut 71524, Egypt*
[3]*Department of Analytical Chemistry, Faculty of Pharmacy, Minia University, Minia, Egypt*

Correspondence should be addressed to Hassan Refat H. Ali; hareha11374@gmail.com

Academic Editor: Feride Severcan

The structural information of the pharmaceuticals and insights on the modes of molecular interactions are very important aspects in drug development. In this work, two cephalosporins and antimicrobial combinations, cefepime-metronidazole and cefoperazone-sulbactam, were studied in the solid state using FTIR spectroscopy for the first time. Quantitation of the studied drugs and their binary mixtures was performed by integrating the peak areas of the characteristic well-resolved bands: υ (C=O) band at 1773 cm^{-1} for cefepime and ring torsion band at 826 cm^{-1} for metronidazole and υ (C=O) band at 1715 cm^{-1} for cefoperazone and ring torsion band at 1124 cm^{-1} for sulbactam. The results of this work were compared with the relevant spectrophotometric reported methods. This study provides data that can be used for the preparative process monitoring of the studied drugs in various dosage forms.

1. Introduction

Cefepime hydrochloride (CPM) is a fourth-generation, semisynthetic cephalosporin antibiotic for parenteral administration. It is 1-[[(6R, 7R)-7-[2-(2-amino-4-thiazolyl)-glyoxylamido]-2-carboxy-8-oxo-5-thia-1-azabicyclo [4.2.0] oct 2-en-3-yl] methyl]-1-methylpyrrolidinium chloride, 72-(Z)-(O-methyloxime), monohydrochloride, and monohydrate (Figure 1). CPM is commonly used in the treatment of moderate-to-severe infections such as pneumonia, intra-abdominal infections, and febrile neutropenia. Metronidazole (MTZ) is (=[1-(2-hydroxyethyl)-2-methyl-5-nitro-1H-imidazole). Metronidazole is the therapeutic agent of choice for amoebiasis and also used in combination with other antimicrobial drugs against yeast infections [1]. Cephalosporin and MTZ combination regimens have been previously studied for this reason [2]. The efficacy of MTZ combined with ceftriaxone [3], cefuroxime [4], and cefepime [5] was well documented. Cefepime and MTZ combination is the optimum choice for mixing into a single bag because both agents may be administered every 12 hours in patients with normal kidney functions and once daily in patients with impaired kidney [5].

Sulbactam sodium (SBT) is 4-thia-1-azabicyclo [3.2.0] heptane 2-carboxylic acid, 3,3-dimethyl-7-oxo-4,4 dioxo sodium salt, and it is official in the British Pharmacopoeia [6]. Cefoperazone (CFZ), (6R, 7R)-7-[[(2R)-[[(4-ethyl-2, 3-dioxo-1-piperazinyl) carbonyl] amino] (4 hydroxy phenyl) acetyl]amino]-3-[[(1-methyl-1H-tetrazol-5-yl)thio]methyl]-8-oxo-5-thia-1-azabicyclo [4.2.0] oct-2-ene-2-carboxylic acid (Figure 1), has been combined with SBT in a dosage form (Sulperazone® or Peractam®) for intra-abdominal infections [7].

FTIR spectroscopy is a prime vibrational spectroscopic technique classified within category I of analytical methods according to the United States Pharmacopeia (USP) [8]. It is considered as a primary and simple tool in providing specific information on the identification and characterization of materials at the molecular level. It was successfully applied for the determination of many pharmaceuticals [9–16].

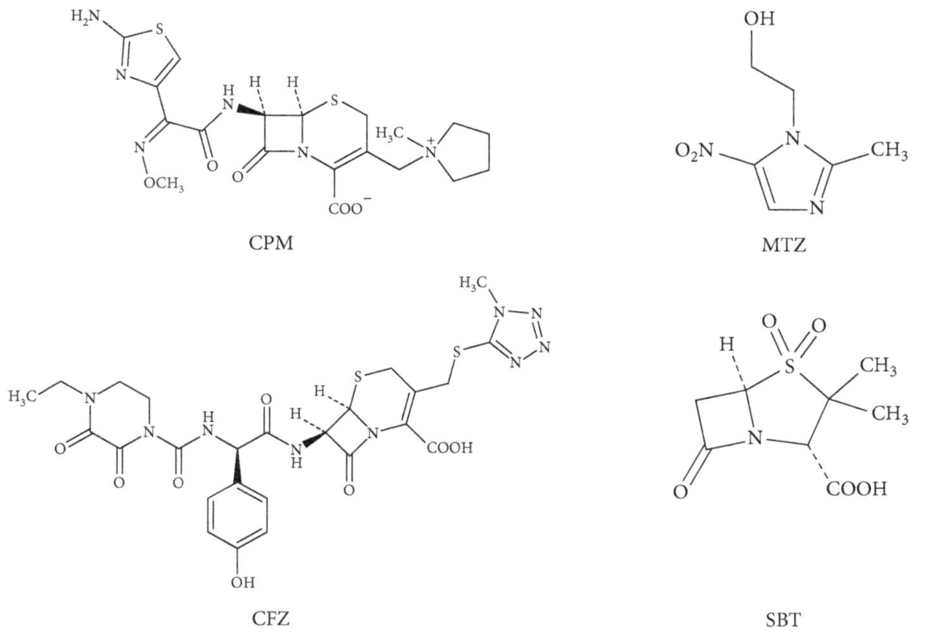

CPM MTZ

CFZ SBT

Figure 1: The chemical structures of CPM, MTZ, CFZ, and SBT.

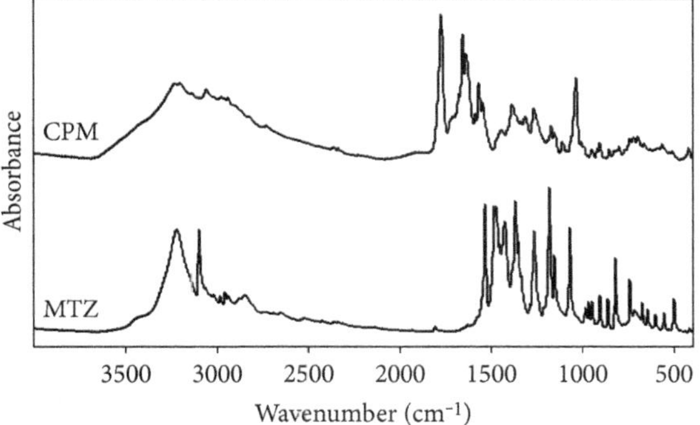

Figure 2: The FTIR spectra of CPM and MTZ in the region of 4000–400 cm^{-1}.

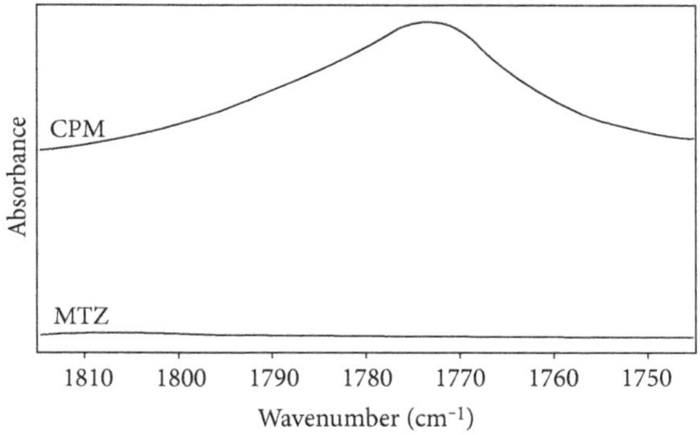

Figure 4: The FTIR spectra of CPM and MTZ in the region of 1810–1750 cm^{-1}.

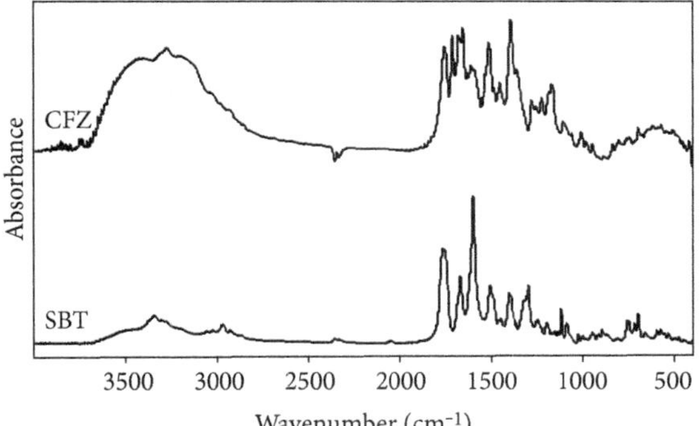

Figure 3: The FTIR spectra of CFZ and SBT in the region of 4000–400 cm^{-1}.

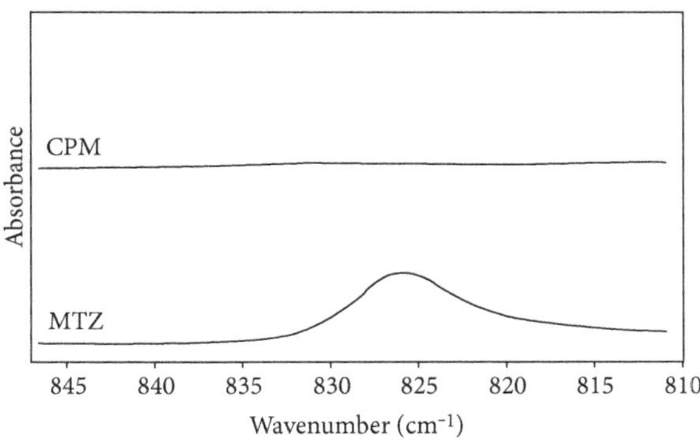

Figure 5: The FTIR spectra of CPM and MTZ in the region of 850–810 cm^{-1}.

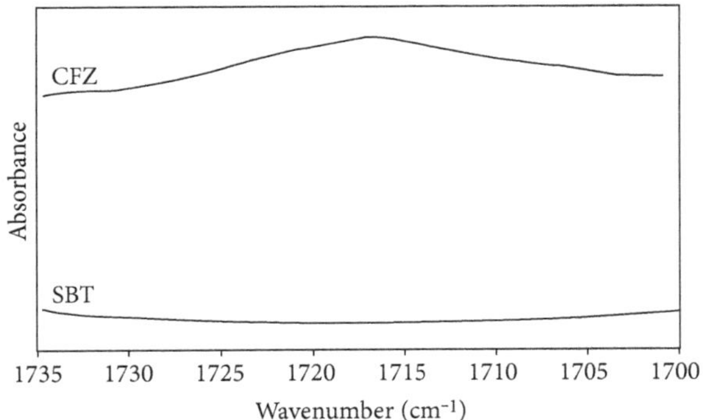

FIGURE 6: The FTIR spectra of CFZ and SBT in the region of 1735–1700 cm^{-1}.

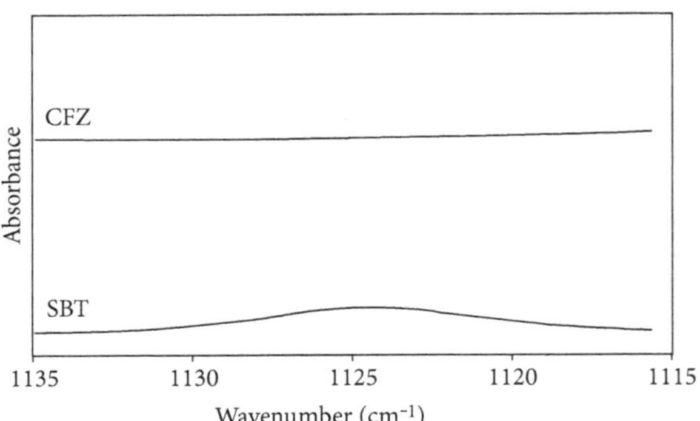

FIGURE 7: The FTIR spectra of CFZ and SBT in the region of 1135–1115 cm^{-1}.

TABLE 1: The distinctive FTIR wavenumbers (cm^{-1}) of CPM and MTZ.

CPM	MTZ	Proposed assignment
3234 *s*	—	υ (NH_2)
—	3230 *sbr*	υ $(OH)_{alcoholic}$
3197 *s*	—	υ (NH)
3056 *ms*	3097 *s*	υ $(CH)_{aromatic}$
2938 *ms*	2950 *ms*	υ $(CH)_{aliphatic}$
1773 *s*	—	υ $(C{=}O)_{lactam}$
1680 *ms*	—	υ $(C{=}O)_{carboxylic}$
1657 *ms*	—	υ $(C{=}O)_{amide}$
—	1535 *s*	υ (NO_2)
—	826 *s*	Ring torsion

m, *s*, and *br* stand for medium, strong, and broad, respectively. υ stands for stretching.

TABLE 2: The distinctive FTIR wavenumbers (cm^{-1}) of CFZ and SBT.

CFZ	SBT	Proposed assignment
3423 *sbr*	—	υ $(OH)_{phenolic}$
3297 *s*	—	υ (NH)
3090 *ms*	3082 *ms*	υ $(CH)_{aromatic}$
2950 *ms*	2964 *ms*	υ $(CH)_{aliphatic}$
1773 *s*	1767 *s*	υ $(C{=}O)_{lactam}$
1717 *s*	1674 *ms*	υ $(C{=}O)_{carboxylic}$
1669 *ms*	—	υ $(C{=}O)_{amide}$
—	1030 *ms*	υ (O=S=O)
—	1124 *s*	Ring torsion

m, *s*, and *br* stand for medium, strong, and broad, respectively. υ stands for stretching.

The aim of the present work is to closely investigate the combinations of CPM-MTZ and CFZ-SBT in the solid state using FTIR spectroscopy as a simple and rapid technique for the first time in comparison with other relevant reported spectrophotometric methods.

2. Experimental

2.1. Chemicals. Cefepime hydrochloride was obtained from Bristol-Myers Squibb Co., Cairo, Egypt. Metronidazole was obtained from Egyptian Int. Pharmaceutical Industries Co., E.I.P.I.CO., 10th of Ramadan City, Egypt; cefoperazone was obtained from Pfizer, El-Thawra St., Almaza, Heliopolis, Cairo, Egypt; sulbactam sodium was obtained from AK Scientific Co.; and potassium bromide was purchased from El-Nasr Pharmaceutical Chemical Co., Abo-Zaabal, Egypt. Solvents and other chemicals were of analytical grade and used as received. All chemicals were stored at room temperature in desiccators over phosphorous pentoxide to avoid any deleterious effects from humidity.

2.2. Pharmaceuticals. Pharmaceutical dosage forms containing the studied drugs were purchased from the local market. Maxipime® vials (Bristol-Myers Squibb Co., Cairo, Egypt) were labeled to contain 1000 mg of cefepime per vial. Flagyl® tablets (Sanofi-Aventis, Cairo, Egypt) were labeled to contain 500 mg of metronidazole per tablet. Sulperazone vials (Pfizer, Cairo, Egypt) were labeled to contain 1000 mg of cefoperazone and 500 mg of sulbactam per vial.

2.3. Disc Preparation and Recording of FTIR Spectra. Mixtures of drugs and KBr (1 : 200) were grinded and mixed well in a glass mortar. The obtained mixtures were diluted to 1000 mg with KBr, then grinded again and pressed under 15000 lbs by a hydraulic pressure system in the die press for 3 min to obtain sample discs. FTIR spectra were collected in the diffuse transmittance mode with potassium bromide as a diluent. The spectra were recorded in the range of 4000–400 cm^{-1} at 4 cm^{-1} spectral resolution with the accumulation of 512 spectral scans. Triplicate spectra were averaged to obtain one spectrum for each sample.

2.4. Binary Mixtures of the Studied Drugs. CPM and MTZ were physically mixed with potassium bromide in various ratios. The calibration curves were constructed by plotting the average peak areas of the characteristic υ (C=O) band at 1773 cm^{-1} for CPM and ring torsion band at 826 cm^{-1} for MTZ and the characteristic υ (C=O) band at 1715 cm^{-1} for CFZ and ring torsion band at 1124 cm^{-1} for SBT as a function of the weight percentage (% w/w) in the range of 5~95.

TABLE 3: Quantitative parameters for the assay of the studied drugs by FTIR spectroscopy in pure forms.

Parameter[a] (n)	CPM	MTZ	CFZ	SBT
Linear range	2.5–18	1.04–10	1.06–10	1.5–12
Intercept (a) ± RMSD	−0.1745 ± 0.0578	−0.2016 ± 0.0150	−0.1003 ± 0.0302	−0.0917 ± 0.0102
Slope (b) ± RMSD	0.2381 ± 0.0047	0.1439 ± 0.0023	0.2812 ± 0.0046	0.0740 ± 0.0013
Correlation coefficient (r)	0.9994	0.9996	0.9996	0.9994
Determination coefficient (r^2)	0.9988	0.9993	0.9992	0.9987
Limit of detection (LOD)[b]	0.80 (μg/mg)	0.35	0.35	0.46
Limit of quantitation (LOQ)[b]	2.40 (μg/mg)	1.04	1.06	1.37

[a]n = three determinations.
[b]The concentration by μg/mg.

TABLE 4: Assay of the studied drugs in binary mixtures by FTIR spectroscopy.

Parameter[a] (n)	CPM	MTZ	CFZ	SBT
Linear range	5–95	5–95	5–95	5–95
Intercept (a) ± RMSD	−0.20175 ± 0.07369	−0.20280 ± 0.01962	0.15207 ± 0.00305	−0.04375 ± 0.01673
Slope (b) ± RMSD	11.958 ± 0.29276	7.1918 ± 0.14795	0.08987 ± 0.00213	2.6896 ± 0.07249
Correlation coefficient (r)	0.9991	0.9994	0.9988	0.9982
Determination coefficient (r^2)	0.9982	0.9987	0.9977	0.9964
Limit of detection (LOD)[b]	0.02	0.009	0.1	0.02
Limit of quantitation (LOQ)[b]	0.06	0.03	0.3	0.06

[a]n = three determinations.
[b]The concentration by % w/w.

TABLE 5: Recovery of standard drugs added to their dosage forms by the proposed FTIR method.

Drug	Dosage form	Declared amount (mg)	Added amount (mg)	Recovery (% ± RMSD)[a]
CPM	Maxipime vials	500	500	99.4 ± 0.77
MTZ	Flagyl infusion	500	500	98.8 ± 0.80
CFZ	Peractam vials	1000	1000	99.2 ± 1.09
SBT	Peractam vials	500	500	99.5 ± 0.86

[a]Values are the mean of three determinations.

The samples were analyzed in triplicates to determine the linearity of the constructed calibration curve.

2.5. Apparatus

2.5.1. FTIR Spectroscopy. FTIR spectra were collected in triplicates using a Nicolet 6700 FTIR Advanced Gold Spectrometer with OMNIC 8 software (Thermo Electron Scientific Instruments Corp., Madison, WI, USA) and Jasco 6000 FTIR (Hachioji, Tokyo, Japan).

All the FTIR spectra were exported to the Galactic SPC format and manipulated using GRAMS AI software (Galactic Industries, Salem, NH, USA, version 7.01)

2.5.2. Spectrophotometry. The absorbance of the studied drugs was measured using UV-1601 PC (Shimadzu, Kyoto, Japan) and Lambda-3 B (Perkin-Elmer Corporation, Norwalk, USA) ultraviolet-visible spectrophotometers with matched 1 cm quartz cells.

3. Results and Discussion

The FTIR spectra of CPM, MTZ, CFZ, and SBT were recorded in the range of 4000–400 cm^{-1} using the transmittance mode of operation. The FTIR spectra of these drugs are shown in Figures 2–7. These spectra have shown noticeable differences which are closely explored in the following subsections.

3.1. FTIR Spectroscopic Investigations of the Studied Drugs. The key FTIR spectral features of CPM are υ (NH_2) band at 3234 cm^{-1}, υ (NH) band at 3197 cm^{-1}, υ $(CH)_{aromatic}$ band at 3056 cm^{-1}, υ $(CH)_{aliphatic}$ band at 2938 cm^{-1}, υ $(C{=}O)_{lactam}$ band at 1773 cm^{-1}, υ $(C{=}O)_{carboxylic}$ band at 1680 cm^{-1}, and υ $(C{=}O)_{amide}$ band at 1657 cm^{-1}. MTZ, in turn, is characterized by υ $(OH)_{alcoholic}$ band at 3230 cm^{-1}, υ $(CH)_{aromatic}$ band at 3097 cm^{-1}, υ $(CH)_{aliphatic}$ band at 2950 cm^{-1}, and υ (NO_2) band at 1535 cm^{-1} and the ring torsion band at 826 cm^{-1}. The distinctive FTIR wave numbers of the combinations of CPM and CFZ are listed in Table 1.

Table 6: The precision of the proposed FTIR method.

Drug	Concentration (μg/mg)	Absorbance Sample number 1	2	3	4	5	Mean	RMSD[a]	CV (RMSD)[b] (%)
CPM	8	0.345	0.346	0.340	0.335	0.332	0.340	0.0061	1.79
MTZ	5	0.646	0.649	0.655	0.640	0.634	0.645	0.0081	1.26
CFZ	5	0.579	0.570	0.583	0.567	0.586	0.577	0.0082	1.42
SBT	8	0.365	0.373	0.360	0.354	0.357	0.362	0.0074	2.06

[a]RMSD: root mean square deviation.
[b]CV (RMSD): coefficient of variation (root mean square deviation).

Table 7: The ruggedness of the proposed FTIR method.

Drug	Recovery (% ± RMSD)[a] Instrument: Nicolet 6700 FTIR	Jasco 6000 FTIR	Interday variation: 1 day	2 days
CPM	99.4 ± 0.77	99.5 ± 0.83	99.4 ± 0.77	99.7 ± 0.67
MTZ	98.8 ± 0.80	99.2 ± 0.66	98.8 ± 0.80	99.1 ± 0.80
CFZ	99.2 ± 1.09	99.6 ± 1.05	99.2 ± 1.05	99.5 ± 1.15
SBT	99.3 ± 0.95	99.6 ± 1.05	99.3 ± 0.96	99.6 ± 0.85

[a]Values are the mean of three determinations ± RMSD.

Table 8: The analysis of investigated drugs in their dosage form using the proposed FTIR and reported methods.

Product	Recovery (% ± RMSD)[a] Proposed method	Reported methods [3, 4]	F-value[b]	t-value[b]
Maxipime vial	99.60 ± 0.29	99.82 ± 0.34	3.04	1.47
Flagyl tablet	98.90 ± 1.24	99.69 ± 1.86	2.42	1.02
Peractam vial	99.40 ± 0.35	99.5 ± 0.8	2.57	1.40

[a]Values are the mean of three determinations ± RMSD.
[b]Theoretical values for t and F at 95% confidence limit ($n = 5$) were 2.78 and 6.39, respectively.

The key FTIR spectral features of CFZ are υ $(OH)_{phenolic}$ band at 3423 cm^{-1}, υ (NH) band at 3297 cm^{-1}, υ $(CH)_{aromatic}$ band at 3090 cm^{-1}, υ $(CH)_{aliphatic}$ band at 2950 cm^{-1}, υ $(C{=}O)_{lactam}$ band at 1773 cm^{-1}, υ $(C{=}O)_{carboxylic}$ band at 1717 cm^{-1}, and υ $(C{=}O)_{amide}$ band at 1669 cm^{-1}. SBT, in turn, is characterized by υ $(CH)_{aromatic}$ band at 3082 cm^{-1}, υ $(CH)_{aliphatic}$ band at 2964 cm^{-1}, υ $(C{=}O)_{lactam}$ band at 1767 cm^{-1}, and υ (O=S=O) band at 1030 cm^{-1} and the ring torsion band at 1124 cm^{-1}. The distinctive FTIR wave numbers of the combinations of CFZ and SBT are listed in Table 2.

3.2. Quantitative Determination and Validation. The FTIR spectroscopy has been utilized for the quantitative determination of the studied combinations. The υ (C=O) band at 1773 cm^{-1} for CPM and ring torsion band at 826 cm^{-1} for MTZ (Figures 4 and 5) and the υ (C=O) band at 1715 cm^{-1} for CFZ and ring torsion band at 1124 cm^{-1} for SBT (Figures 6 and 7) were picked up for their quantitative determination because they are well resolved and free from interferences. The peak areas of the bands of interest were integrated using GRAMS AI package. The developed procedures were validated according to USP 2009 validation guidelines [1] and the International Conference on Harmonization (ICH) guidelines [2].

3.2.1. Linearity and Range. Under the optimal reaction conditions, a series of concentrations of the cited drugs was processed into sample discs and the FTIR spectra were recorded. Calibration curves were constructed by plotting peak areas of the selected FTIR absorption bands as a function of the corresponding concentrations in % w/w. The obtained linear concentration ranges were 1.0–18 and 1.0–12 μg/mg for CPM-MTZ and CFZ-SBT, respectively. The correlation coefficients were in the range from 0.9994 to 0.9996 for the studied drugs in pure forms and from 0.9982 to 0.9987 and from 0.9964 to 0.9977 for CPM-MTZ and CFZ-SBT binary mixtures, respectively.

3.2.2. Limits of Detection and Quantitation. The LOD and LOQ values were determined from the linear calibration range for the studied drugs either alone or in combinations. The calculated LODs and LOQs were in the range of 0.35–0.80 μg/mg and 1.04–2.40 μg/mg for the studied drugs in their pure forms while, in their binary mixtures, they were in the range of 0.009–0.1% w/w and 0.03–0.3% w/w, respectively. The results are presented in Tables 3 and 4.

3.2.3. Accuracy and Precision. The accuracy of the proposed method was assessed by the standard addition method. The recovery values of the added concentrations were 99.4 ± 0.77, 98.8 ± 0.80, 99.2 ± 1.09, and 99.5 ± 0.86 for CPM, MTZ, CFZ, and SBT, respectively, in their pure forms (Table 5) which would indicate the accuracy of the proposed method.

The precision of the method was determined by conducting replicate analysis of five samples of each investigated drug. The coefficient of variation of root mean square deviation (CV (RMSD)) was lower than 2.5%. Accordingly, the proposed method is sufficiently reproducible (Table 6).

3.2.4. Ruggedness. Ruggedness was also evaluated by applying the proposed method to the assay of the investigated drugs using the same procedure but using two different instruments of two different laboratories with different elapsed times. The results were found to be reproducible (Table 7).

3.3. Application of the Analysis of the Pharmaceutical Dosage Forms. The proposed method was applied to the determination of CPM, MTZ, CFZ, and SBT in their commercial dosage forms in the Egyptian market. The results are presented in Table 8. The mean recovery percentages were found to be $99.60 \pm 0.29\%$, $98.90 \pm 1.24\%$, and $99.40 \pm 0.35\%$ for cefepime (Maxipime vial), metronidazole (Flagyl tablets), and cefoperazone-sulbactam (Peractam vial), respectively. The results were compared with those obtained by the reported methods [3, 4] (Table 8) at 95% confidence level. No significant difference was found between the calculated and theoretical values of the t and f tests which indicate good level of precision and accuracy of the proposed method.

4. Conclusions

The key FTIR spectral features of each of the investigated drugs were reliably determined. The FTIR spectroscopy has been utilized for the first time to quantify the studied drugs and their binary mixtures in the solid state. The results were reliably compared with other relevant and previously published spectrophotometric methods. This vibrational spectroscopic technique appears to be a good alternative to other well-established analytical techniques especially in the absence of suitable methods for the determination of the active ingredients that are present in complex matrices as in the pharmaceutical formulations.

References

[1] *United State Pharmacopoeia 36 and National Formulary 29*, Convention, Rockville, MD, 2012.

[2] S. K. Branch, "Guidelines from the International Conference on Harmonisation (ICH)," *Journal of Pharmaceutical and Biomedical Analysis*, vol. 38, no. 5, pp. 798–805, 2005.

[3] R. K. Nanda, D. A. Navathar, A. A. Kulkarni, and S. S. Patil, "Simultaneous spectrophotometric estimation of cefepime and tazobactam in pharmaceutical dosage form," *International Journal of Chemical Technology Research*, vol. 4, pp. 152–156, 2012.

[4] M. R. El-Ghobashy and N. F. Abo-Talib, "Spectrophotometric methods for the simultaneous determination of binary mixture of metronidazole and diloxanide furoate without prior separation," *Journal of Advanced Research*, vol. 1, no. 4, pp. 323–329, 2010.

[5] F. C. Maddox and J. T. Stewart, "HPLC determination of an aqueous cefepime and metronidazole mixture," *Journal of liquid chromatography & related technologies*, vol. 22, no. 18, pp. 2807–2813, 1999.

[6] Pharmacopoeia B., "British Pharmacopoeia Commission London; the Department of Health," *Social Services and Public Safety*, vol. 1, pp. 719–720, 2013.

[7] V. D. Hoang, N. Thi Tho, V. Thi Tho, and M. T. Nguyen, "UV spectrophotometric simultaneous determination of cefoperazone and sulbactam in pharmaceutical formulations by derivative, Fourier and wavelet transforms," *Spectrochim Acta Part A*, vol. 121C, pp. 704–714, 2014.

[8] Chapter, G., 1225, "Validation of compendial methods," in *United States Pharmacopeia 30, National Formulary 25*, The United States Pharmacopeial Convention, Rockville, Md., USA, 2007.

[9] M. K. Ahmed, J. K. Daun, and R. Przybylski, "FT-IR based methodology for quantitation of total tocopherols, tocotrienols and plastochromanol-8 in vegetable oils," *Journal of Food Composition and Analysis*, vol. 18, no. 5, pp. 359–364, 2005.

[10] N. Al-Zoubi, J. E. Koundourellis, and S. Malamataris, "FT-IR and Raman spectroscopic methods for identification and quantitation of orthorhombic and monoclinic paracetamol in powder mixes," *Journal of Pharmaceutical and Biomedical Analysis*, vol. 29, no. 3, pp. 459–467, 2002.

[11] D. E. Bugay, A. W. Newman, and W. P. Findlay, "Quantitation of cefepime 2HCl dihydrate in cefepime 2HCl monohydrate by diffuse reflectance IR and powder X-ray diffraction techniques," *Journal of Pharmaceutical and Biomedical Analysis*, vol. 15, no. 1, pp. 49–61, 1996.

[12] S. Matkovic, G. M. Valle, and L. E. Briand, "Quantitative analysis of ibuprofen in pharmaceutical formulations through FTIR spectroscopy," *Latin American Applied Research*, vol. 35, no. 3, pp. 189–195, 2005.

[13] A. Peepliwal, S. D. Vyawahare, and C. G. Bonde, "A quantitative analysis of Zidovudine containing formulation by FT-IR and UV spectroscopy," *Analytical Methods*, vol. 2, no. 11, pp. 1756–1763, 2010.

[14] F. B. Reig, J. G. Adelantado, V. P. MartnezMartínez, M. M. Moreno, and M. D. Carbó, "FT-IR quantitative analysis of solvent mixtures by the constant ratio method," *Journal of molecular structure*, vol. 480, pp. 529–534, 1999.

[15] Y. Roggo, P. Chalus, L. Maurer, C. Lema-Martinez, A. Edmond, and N. Jent, "A review of near infrared spectroscopy and chemometrics in pharmaceutical technologies," *Journal of pharmaceutical and biomedical analysis*, vol. 44, no. 3, pp. 683–700, 2007.

[16] H. R. Ali, G. A. Saleh, S. A. Hussein, and A. I. Hassan, "In-depth qualitative and quantitative FTIR spectroscopic study of glipizide and gliclazide," *Analytical Chemistry: An Indian Journal*, vol. 14, no. 4, pp. 127–134, 2014.

Permissions

All chapters in this book were first published in JS, by Hindawi Publishing Corporation; hereby published with permission under the Creative Commons Attribution License or equivalent. Every chapter published in this book has been scrutinized by our experts. Their significance has been extensively debated. The topics covered herein carry significant findings which will fuel the growth of the discipline. They may even be implemented as practical applications or may be referred to as a beginning point for another development.

The contributors of this book come from diverse backgrounds, making this book a truly international effort. This book will bring forth new frontiers with its revolutionizing research information and detailed analysis of the nascent developments around the world.

We would like to thank all the contributing authors for lending their expertise to make the book truly unique. They have played a crucial role in the development of this book. Without their invaluable contributions this book wouldn't have been possible. They have made vital efforts to compile up to date information on the varied aspects of this subject to make this book a valuable addition to the collection of many professionals and students.

This book was conceptualized with the vision of imparting up-to-date information and advanced data in this field. To ensure the same, a matchless editorial board was set up. Every individual on the board went through rigorous rounds of assessment to prove their worth. After which they invested a large part of their time researching and compiling the most relevant data for our readers.

The editorial board has been involved in producing this book since its inception. They have spent rigorous hours researching and exploring the diverse topics which have resulted in the successful publishing of this book. They have passed on their knowledge of decades through this book. To expedite this challenging task, the publisher supported the team at every step. A small team of assistant editors was also appointed to further simplify the editing procedure and attain best results for the readers.

Apart from the editorial board, the designing team has also invested a significant amount of their time in understanding the subject and creating the most relevant covers. They scrutinized every image to scout for the most suitable representation of the subject and create an appropriate cover for the book.

The publishing team has been an ardent support to the editorial, designing and production team. Their endless efforts to recruit the best for this project, has resulted in the accomplishment of this book. They are a veteran in the field of academics and their pool of knowledge is as vast as their experience in printing. Their expertise and guidance has proved useful at every step. Their uncompromising quality standards have made this book an exceptional effort. Their encouragement from time to time has been an inspiration for everyone.

The publisher and the editorial board hope that this book will prove to be a valuable piece of knowledge for researchers, students, practitioners and scholars across the globe.

List of Contributors

Luis Zamora-Peredo, Leandro García-González and Julián Hernández Torres
Centro de Investigación en Micro y Nanotecnología, Universidad Veracruzana, Calzada Adolfo Ruiz Cortines 455, Fracc. Costa Verde, 94292 Boca del Río, VER, Mexico

Irving E. Cortes-Mestizo and Víctor H. Méndez-García
Universidad Autónoma de San Luis Potosí, Center for the Innovation and Application of Science and Technology, Sierra Leona 550, Lomas 4a Secc., 78210 San Luis Potosí, SLP, Mexico

Máximo López-López
Centro de Investigación y Estudios Avanzados del IPN, Apartado Postal 14-740, 07360 Ciudad de México, Mexico

Tong Wu and Chao Tan
Key Lab of Process Analysis and Control of Sichuan Universities, Yibin University, Yibin, Sichuan 644000, China

Zan Lin
Key Lab of Process Analysis and Control of Sichuan Universities, Yibin University, Yibin, Sichuan 644000, China
The First Affiliated Hospital, Chongqing Medical University, Chongqing 400016, China

Hui Chen
Yibin University Hospital, Yibin, Sichuan 644000, China

Huo Zhang
School of Mechano-Electronic Engineering, Xidian University, Xi'an, Shanxi 710126, China

Binyi Qin
School of Mechano-Electronic Engineering, Xidian University, Xi'an, Shanxi 710126, China
School of Electronics and Communication Engineering, Yulin Normal University, Yulin, Guangxi 537000, China

Zhi Li
School of Mechano-Electronic Engineering, Xidian University, Xi'an, Shanxi 710126, China
Guangxi Key Laboratory of Automatic Detecting Technology and Instruments, School of Electronic Engineering and Automation, Guilin University of Electronic Technology, Guilin, Guangxi 541004, China

Zhihui Luo and Yun Li
Guangxi Key Laboratory for Agricultural Resources Chemistry and Efficient Utilization (Cultivation Base), Colleges and Universities Key Laboratory for Efficient Use of Agricultural Resources in the Southeast of Guangxi, College of Chemistry and Food Science, Yulin Normal University, Yulin, Guangxi 537000, China

Robert Miotk, Bartosz Hrycak and Mariusz Jasiński
The Institute of Fluid Flow Machinery, Polish Academy of Sciences, Fiszera 14, 80-231 Gdansk, Poland

Jerzy Mizeraczyk
Department of Marine Electronics, Gdynia Maritime University, Morska 81-87, 81-225 Gdynia, Poland

Qiu Li and Gang Jin
Tianjin Key Laboratory of High Speed Cutting and Precision Machining, School of Mechanical Engineering, Tianjin University of Technology and Education, Tianjin 300222, China

Wu Liu
Tianjin Key Laboratory of High Speed Cutting and Precision Machining, School of Mechanical Engineering, Tianjin University of Technology and Education, Tianjin 300222, China
School of Mechanical and Automotive Engineering, Zhejiang University of Water Resources and Electric Power, Hangzhou 310018, China

Wei Qiu
Tianjin Key Laboratory of Modern Engineering Mechanics, School of Mechanical Engineering, Tianjin University, Tianjin 300072, China

Tomislav Vrbanec and Matej Smrkolj
Krka d.d., Novo Mesto, Šmarješka c. 6, SI-8501 Novo Mesto, Slovenia

Primož Šket, Franci Merzel and Jože Grdadolnik
National Institute of Chemistry, Hajdrihova 19, SI-1000 Ljubljana, Slovenia

Zenovia Moldovan, Dana Elena Popa, Iulia Gabriela David, Mihaela Buleandra and Irinel Adriana Badea
Department of Analytical Chemistry, Faculty of Chemistry, University of Bucharest, 4-12 Regina Elisabeta Av., District 3, 030018 Bucharest, Romania

Yijie Peng, Muhua Liu, Jinhui Zhao, Haichao Yuan, Yao Li, Jinjiang Tao and Hongqing Guo
Optics-Electrics Application of Biomaterials Lab, College of Engineering, Jiangxi Agricultural University, Nanchang, Jiangxi 330045, China

Dana Maria Muntean and Ioan Tomut
Department of Pharmaceutical Technology and Biopharmacy, University of Medicine and Pharmacy Iuliu Hatieganu Cluj-Napoca, 41 V. Babes, 400012 Cluj-Napoca, Romania

Cristian Alecu
S.C. Laropharm S.R.L., 145A Alexandriei, 70000 Bragadiru, Romania

Ji Zhang and Yuanzhong Wang
Institute of Medicinal Plants, Yunnan Academy of Agricultural Sciences, Kunming 650200, China
Yunnan Technical Center for Quality of Chinese Materia Medica, Kunming 650200, China

Yan Li
Institute of Medicinal Plants, Yunnan Academy of Agricultural Sciences, Kunming 650200, China
Yunnan Technical Center for Quality of Chinese Materia Medica, Kunming 650200, China
College of Traditional Chinese Medicine, Yunnan University of Traditional Chinese Medicine, Kunming 650500, China

Tao Li
College of Resources and Environment, Yuxi Normal University, Yuxi 653100, China

Tianwei Yang and Honggao Liu
College of Agronomy and Biotechnology, Yunnan Agricultural University, Kunming 650201, China

Violeta Lazic and Massimiliano Ciaffi
Department of FSN-TECFIS-DIM, ENEA, Via E. Fermi 45, 00044 Frascati, Italy

Luisa Sciortino, Simonpietro Agnello, Marco Cannas, Franco Mario Gelardi and Gianpiero Buscarino
Dipartimento di Fisica e Chimica, Università di Palermo, 90123 Palermo, Italy

Michela Todaro
Dipartimento di Fisica e Chimica, Università di Palermo, 90123 Palermo, Italy
Dipartimento di Fisica e Astronomia, Universitàdi Catania, 95123 Catania, Italy

Antonino Alessi
Dipartimento di Fisica e Chimica, Università di Palermo, 90123 Palermo, Italy
Laboratoire H. Curien, UMR CNRS 5516, Universitéde Lyon, 42000 Saint-Etienne, France

Abdolhamed Shahedi and Esmaeil Eslami
Department of Physics, Iran University of Science and Technology, Narmak, Tehran 16846 13114, Iran

Mohammad Reza Nourani
Tissue Engineering Division, Biotechnology Research Center, Baqiyatallah University of Medical Sciences, Tehran, Iran

Reem I. Al-Wabli and Nadia G. Haress
Department of Pharmaceutical Chemistry, College of Pharmacy, King Saud University, Saudi Arabia

Mohamed I. Attia
Department of Pharmaceutical Chemistry, College of Pharmacy, King Saud University, Riyadh 11451, Saudi Arabia
Medicinal and Pharmaceutical Chemistry Department, Pharmaceutical and Drug Industries Research Division, National Research Centre (ID: 60014618), El Bohooth Street, Dokki, Giza 12622, Egypt

Devarasu Manimaran, Liji John and Isaac Hubert Joe
Centre for Molecular and Biophysics Research, Department of Physics, Mar Ivanios College, Thiruvananthapuram, Kerala 695015, India

Yao Lu
Key Laboratory of Coal Processing and Efficient Utilization, Ministry of Education, China University of Mining & Technology, Xuzhou 221116, China
Advanced Analysis & Computation Center, China University of Mining & Technology, Xuzhou 221116, China
School of Chemical Engineering and Technology, China University of Mining & Technology, Xuzhou 221116, China

Xian-Yong Wei and Xing Fan
Key Laboratory of Coal Processing and Efficient Utilization, Ministry of Education, China University of Mining & Technology, Xuzhou 221116, China
School of Chemical Engineering and Technology, China University of Mining & Technology, Xuzhou 221116, China

Hong-Qin Hu and Feng-Jin Xie
School of Chemical Engineering and Technology, China University of Mining & Technology, Xuzhou 221116, China

Yong-Chao Lu
School of Basic Education Sciences, Xuzhou Medical University, Xuzhou 221004, China

Xiaoli Li and Chengwei Li
School of Electrical Engineering and Automation, Harbin Institute of Technology, Harbin 150001, China

Sebastian Karpf
Department of Electrical Engineering, University of California, Los Angeles, Los Angeles, CA, USA

Matthias Eibl and Robert Huber
Institut für Biomedizinische Optik, Universität zu Lübeck, Peter-Monnik-Weg 4, 23562 Lübeck, Germany

Wolfgang Wieser and Thomas Klein
Optores GmbH, Gollierstr. 70, 80339 Munich, Germany

Tianming Yang, Rong Zhou, Du Jiang, Haiyan Fu, Rui Su, Yangxi Liu and Hanbo Su
School of Pharmaceutical Sciences, South-Central University for Nationalities, Wuhan 430074, China

Paweł Frącz, Ireneusz Urbaniec, Tomasz Turba and Sławomir Krzewiński
Faculty of Electrical Engineering, Automatic Control and Computer Science, Opole University of Technology, Prószkowska 76, 45-758 Opole, Poland

A. Tálský, O. Štec, M. Pazderka and V. Kudrle
Department of Physical Electronics, Masaryk University, Kotlarska 2, 61137 Brno, Czech Republic

Yaqiong Zhao, Yilin Gu, Feng Qin, Xiaolong Li, Zhanhong Ma, Pei Cheng, Yang Pan and Haiguang Wang
College of Plant Protection, China Agricultural University, Beijing 100193, China

Longlian Zhao and Junhui Li
College of Information and Electrical Engineering, China Agricultural University, Beijing 100083, China

Qizhou Wu, Yong Jin, Zhaoba Wang and Zhaoqian Xiao
National Key Lab for Electronic Measurement Technology, North University of China, Taiyuan, Shanxi 030051, China

Hassan Refat H. Ali
Department of Pharmaceutical Analytical Chemistry, Faculty of Pharmacy, Assiut University, Assiut 71526, Egypt

Ramadan Ali and Hany A. Batakoushy
Department of Analytical Chemistry, Faculty of Pharmacy, Al-Azhar University, Assiut 71524, Egypt

Sayed M. Derayea
Department of Analytical Chemistry, Faculty of Pharmacy, Minia University, Minia, Egypt

Index

www.ingramcontent.com/pod-product-compliance
Lightning Source LLC
Chambersburg PA
CBHW080644200326
41458CB00013B/4733
* 9 7 8 1 6 8 2 8 5 6 2 3 9 *